Reform in Undergraduate Science Teaching for the 21st Century

A volume in
Research in Science Education
Series Editors: Dennis W. Sunal, *University of Alabama*
Emmett L. Wright, *Kansas State University*

Reform in Undergraduate Science Teaching for the 21st Century

Edited by

Dennis W. Sunal
University of Alabama

Emmett L. Wright
Kansas State University

Jeanelle Bland Day
Eastern Connecticut State University

INFORMATION AGE
PUBLISHING

80 Mason Street • Greenwich, Connecticut 06830 • www.infoagepub.com

Library of Congress Cataloging-in-Publication Data

Reform in undergraduate science teaching for the 21st century / edited
by Dennis W. Sunal, Emmett L. Wright, Jeanelle Bland.
    p. cm. – (Research in science education)
  Includes bibliographical references.
  ISBN 1-930608-85-3 – ISBN 1-930608-84-5 (pbk.)
  1. Science–Study and teaching (Higher)–United States. I. Sunal,
Dennis W. II. Wright, Emmett. III. Bland, Jeanelle. IV. Series.
  Q183.3.A1R44 2004
  507'.1'173–dc22

                                                    2004006396

CONTENTS

part II

Perspectives on Reform in Undergraduate Science   *241*

part III

PREFACE TO THE SERIES

Science education as a professional field has been changing rapidly over the past two decades. Scholars, practitioners, and students find it difficult to keep abreast of knowledge of research, leadership, policy, curricula, teaching, and learning relevant to their special needs. The journals and textbooks that are available report a broad spectrum of diverse science education research, making the search for valid materials on a specific area time-consuming and tedious.

Professionals need to be able to access a comprehensive, timely, and valid source of knowledge about the emerging body of research, theory, and policy in their fields. This body of knowledge would inform researchers about emerging trends in research, research procedures, and technological assistance in key areas of science education. It would inform policy makers in need of information about specific areas in which they make key decisions. It would also help practitioners and students become aware of current research knowledge, policy, and best practice in their fields.

For these reasons, the goal of the book series, *Research in Science Education*, is to provide a comprehensive view of current and emerging knowledge, research strategies, and policy in specific professional fields of science education. This series presents currently unavailable, or difficult to gather, materials from a variety of viewpoints and sources in a usable and organized format.

Each volume in the series presents a juried, scholarly, and accessible review of research, theory, and/or policy in a specific field of science education, K–16. Topics covered in each volume are determined by present issues and trends, as well as generative themes related to current research and theory. Published volumes will include empirical studies, policy analysis, literature reviews, and positing of theoretical and conceptual bases.

PREFACE

Higher education has reached a critical junction in the role it plays in modern society. Stakeholders have raised an increasing wave of criticism derived from political, evidentiary, and accountability concerns implicating faculty and departments in the need for educational change. Reforms have been successful. There are examples of significant innovations in the colleges and universities in the United States. This volume, *Reform in Undergraduate Science Teaching for the 21st Century*, explores issues and research in the planning, development, implementation, and sustaining of reform in science teaching in higher education.

The editors' work and experience in teaching undergraduate science began in the late 1960s and between us incorporates a cumulative knowledge of more than 75 years. We have each been active in teaching, consulting, and reform in higher education during these years. Our concerns during this time have been the slow pace of change in the traditions and beliefs about teaching and learning among college and university faculty of the last half of the 20th century. We were curious about the fact that in every science and engineering department we worked with, there were always one, two, or more faculty members who taught differently and tried out and embraced changes in their classrooms that involved highly interactive teaching and curriculum arrangements. These faculty were always associated with student learning outcomes and student interest in science that were uncharacteristically higher than the department as a whole.

We have attempted to investigate and disseminate these differences in our research and professional writings in the past. This search, and those of many others, crystallized our thinking and understanding of what it takes to

Reform in Undergraduate Science Teaching for the 21st Century, pages xi–xii
Copyright © 2004 by Information Age Publishing

make incremental and continuous change in creating positive learning environments for diverse adults in undergraduate science courses. We have found that reform for science faculty and departments is a long-term endeavor toward publicly expressed goals that best works in a learning community with two or more creative individuals. This volume represents the cumulative experience and research of a national sample of such outstanding faculty, all attempting to explore and share their knowledge of teaching undergraduate science meaningfully and interactively to prepare future scientists, teachers of science, engineers, medical personnel, and scientifically literate college graduates for the remainder of the 21st century.

ACKNOWLEDGMENTS

Undertaking a project like this can be overwhelming. The completion of this volume was made much easier because of our broad network of professional contacts. These contacts have developed over the years through our membership in professional organizations and activities with numerous individuals in higher education institutions throughout the world. Therefore, we wish to acknowledge these organizations and individuals for providing us with a forum to meet, interact, disseminate, and form professional collaboration communities involving individuals with an interest in improving teaching and learning in undergraduate science. We especially thank the National Aeronautics and Space Administration (NASA) and our colleagues in the NOVA program of professional development for university faculty, the national NOVA network of university teams involved in reform, National Association for Research in Science Teaching, American Association for the Advancement of Science, Society of College Science Teachers, and the Association of Educators of Teachers of Science.

We are indebted to the authors who agreed to join us in this endeavor and submitted chapters for review. They added a heavy load to their full schedules by the additional professional task of completing a chapter and making the numerous changes suggested by the reviewers. Our grateful thanks goes to each of the authors who contributed their outstanding professional experiences and expertise.

Special recognition is given to our graduate students at the University of Alabama who in addition to completing their own research and course work undertook tasks that allowed this volume to come to completion. Timothy Cook provided, tirelessly, the enormously important task of for-

Reform in Undergraduate Science Teaching for the 21st Century, pages xiii–xiv
Copyright © 2004 by Information Age Publishing

matting and editing a consistent writing style among each of the chapters, including citations and references. Puneet Gill, Susan Thomas, Misty Nix, Lea Accolagoun, and Robert Mayben provided support on different aspects of the volume and especially those difficult citations and references that seemed to be always missing.

Finally, very special thanks must be given to George Johnson of Information Age Publishers who had the foresight in recognizing the need for this volume in the series on *Research in Science Education*, faith in our ability to complete the work, and who always provided encouragement to complete this volume.

—Dennis W. Sunal
Emmett L. Wright
Jeanelle Bland Day
August 2003

CHAPTER 1

IMPROVING UNDERGRADUATE SCIENCE TEACHING THROUGH EDUCATIONAL RESEARCH

Emmett L. Wright, Dennis W. Sunal, and Jeanelle Bland Day

ABSTRACT

The study of educational reform attempts to determine the interaction between motives, change processes attempted, and the institutional context and culture in relation to the instructor and student outcomes in their science courses. Higher education faculty in science have much to gain by increasing their understanding of educational research on the processes that they spend a great deal of their professional lives perfecting: teaching. To develop expertise in teaching undergraduate science, faculty will require additional knowledge and professional skills of basic issues involving components of successful educational reform in science teaching needed for the 21st century, understanding various perspectives on reform in undergraduate science, and awareness of what works and of models of innovation in undergraduate science.

Reform in Undergraduate Science Teaching for the 21st Century, pages 1–11
Copyright © 2004 by Information Age Publishing

INTRODUCTION

For the United States to maintain its preeminence in the 21st century it is critical that the educational system place itself in the most favorable position to leverage a long-term sustained reform of the science education profession. This will require a fundamental investment in ways we prepare and support quality higher education science faculty to meet the needs of a growing and more diverse college student population in a rapidly changing world (King, 1992). Faculties throughout the country support the need to reinvent their science courses, but most do not have the understandings or strategies to accomplish such change. Increasing numbers of faculty advocate the importance of more quality teaching (Herr, 1988; Ory, 2000, pp. 14–15). Eighty-three percent of faculty in a national survey have indicated that teaching "is one of the most appealing aspects of faculty life, as well as its core undertaking" (Golde & Dore, 2001). A critical concern is the appropriateness of entry-level undergraduate science courses that reflect the outcomes emphasized in the current literature and national reports of many organizations (e.g., American Association for the Advancement of Science [AAAS], 1990, 1993; Boyer Commission, 1998; National Research Council [NRC], 1996a, 1996b, 1997; National Science Foundation [NSF], 1996; Siebert & McIntosh, 2001; Sunal et al., 2001).

Traditional approaches in teaching undergraduate introductory science courses do not work effectively with many of today's students, particularly for the elementary and secondary science majors (Green, 1989; Seymour, 1992a, 1992b, 1995; Tobias, 1992). Helping students understand scientific ideas is a complex task (Hewitt & Seymour, 1991; National Research Council, 2001). For instructors to do it well takes significant pedagogical knowledge and skills in addition to content knowledge and understandings. However, most science faculty have little, if any, professional training in teaching (National Research Council, 2003, pp. 14–15). Effective teaching involves (a) purposeful, research-informed, development of innovative lessons with effective teaching strategies that actively involve students in learning; (b) use of effective strategies of student assessment and feedback; and (c) self-evaluation of the effectiveness of teaching using action research (National Research Council, 2001; Prather, 1995; Sunal & Burry, 1992). Continuous and intensive teacher and student feedback are part of the process. The priorities of effective teaching differ from traditional beliefs about what college instructors do in classrooms.

A major goal of science education reform in the United States has focused on science teaching and learning prior to enrollment in higher education. What has been missing from the reform process is broadening the effort to include the university level. Recent higher education reform movement goals have emphasized increasing the quality of science education, improving participation of minority students and women, increasing the science literacy of all higher education students, and increasing the ability of students to function in an increasingly technological society (Floden, Gallagher, Wong, & Roseman, 1995). The quality of teaching in introductory courses is the single most important factor responsible for university science major dropouts (Tobias, 1990). Because introductory science courses form the basis of science knowledge and science teaching models for the vast majority of college graduates, including undergraduate teacher education majors, it is critical that this level of instruction receive the full attention of the science community, both inside and outside of higher education.

SIGNIFICANCE AND NEED FOR CHANGE

Higher education faces a significant challenge for the 21st century (Goodchild, 1991). As scientific research creates new knowledge and as educational research identifies more effective methods for teaching science to college students, faculty are under pressure to create increasingly effective science teaching at the university level (Magner, 1992; Reinarz, 1991). Recent research in the areas of inquiry teaching, conceptual development, and preconceptions has led to more innovative strategies for college classroom instruction and to new approaches for creating pedagogical change in science courses (Driver, 1986; Driver, Squires, Rushworth, Wood-Robinson, 1994; Duschl, 1990; Floden et al., 1995; NRC, 1997, 2000, 2003; NSF, 1996; Siebert & McIntosh, 2001; Sigel, 1985).

Change is especially needed if the reform movement now underway in pre-university science education is to succeed. K–12 public schools in the U.S. are undergoing reform. In most state curricula guidelines for school science there is a significant departure from the past. These state documents reflect the national reform movement underway as described in *National Science Education Standards* (NRC, 1996a), *Benchmarks for Science Literacy, Project 2061* (AAAS, 1993), and *No Child Left Behind* (U.S. Department of Education, 2001). New guidelines for teacher education program accreditation are reflected in criteria for review by the National Association for Colleges of Teacher Education (NCATE) and the criteria for individual teacher accreditation by the National Science Teachers Association (NSTA). The new science standards are in sharp contrast to the instructional methods and content focus of traditional university science courses (Sunal, Sunal, Jones, & Shratter, 1996). The current science courses in higher education do not provide education majors with basics needed to perform successfully under the new standards as effective teachers after graduation. Nor do they meet the needs of incoming freshmen who have recently graduated from the reformed schools. Can we prepare primary and secondary teachers to teach science effectively under traditional university teaching methods and curriculum now employed in our university science programs?

Even for faculty who attempt to improve the effectiveness of their science courses, the process is slow and the results limited (Barinaga 1991; Bradford, Rubba, & Harkness, 1995; Fedock, Zambo, & Cobern, 1996). Still, many introductory science courses are responsible for driving off many students either from a science major or from taking science courses. Attention must be paid to the common features of these courses that turn off students if progress is going to be made in the retention rate of students.

Reforms can be successful (Fullan, 1993). There are examples of significant innovations in American colleges and universities (Sundberg & Monaca, 1994; Sunal, Freeman, Whitaker, Hedgepeth, & Ewing, 1994; Weimer & Lenze, 1994). Several recent and current reforms have impacted hundreds of universities and hundreds of thousands of students. The National Science Foundation's goal leading to reforms for higher education includes the idea that "all students have access to supportive, excellent undergraduate education in SMT (science, mathematics, and technology) and all students learn these subjects by direct experience with the methods and processes of inquiry" (NSF, 1996, p. 1). The National Research Council's call for reforms focused on the idea that "the ultimate goal of undergraduate education should be for individual faculty and departments to improve the academic growth of students" (NRC, 2003, p. 14). The National Aeronautics and Space Administration program, NASA Opportunities for Visionary Academics (NOVA), leading to national reforms in undergraduate teaching,

> was developed in response to the need to facilitate change in science teaching in higher education by providing assistance to faculty on a national basis. The program's emphasis is on constructing, connecting, and collaborating, using the best of what has been learned through research on faculty professional development ... to enhance science, mathematics, and technology literacy of all undergraduate students ... and focuses on involvement of education majors in entry-level undergraduate courses and building upon the national science and mathematics education standards in K–12 education. (Sunal et al., 2001, p. 250)

These reforms dealt with questions such as, What are the issues, results, and implications in the planning, development, implementation, and sustaining of reform in undergraduate science teaching in higher education? The processes and impact of these reforms need to be explored. Much can be learned.

BARRIERS TO IMPROVING SCIENCE TEACHING

Change is difficult in higher education because the organization of the institution, its expectations, and its social responsibilities inhibit risk taking, ambiguity, and the inquiry required for change to occur (Cohen, 1988; Cuban, 1990, 1999). How different would it be for a student to sit in a college science course today as compared to similar courses 100 years ago? Other than content covered, what differences in instruction would the student experience?

Higher education literature includes descriptions of barriers to change at the institutional, administrative, and policy level, but little is available on barriers to change existing among university faculty at the course and classroom level (Sunal et al., 2001). The research literature that is available at this level includes some descriptions of changes that need to be made in university science teaching but little is known to date about the motivations, perceptions, and processes of change that exist among university teaching faculty. What are the processes affecting change and what are the barriers to change in teaching for science faculty at higher education institutions?

In her study of college science instruction, Tobias (1990) defines two tiers of entering university students. The first consists of those who are going on to earn science degrees and the second, those who have the initial intention and ability to do so but instead drop out or switch to nonscientific fields. The number of students in the second category is very large (Brown, 1995). Tobias concluded that introductory science courses are responsible for driving off many students in the second tier. She found the negative features of the courses include lack of relevance, relegation of students to passivity, emphasis on competition, and a focus on algorithmic problem solving. Knowledgeable and skillful faculty can reach the second tier.

Science faculty in higher education have the potential to be a valuable resource for the improvement of science education at all levels (Barinaga, 1991). Both scientists and university administrators in a National Science Foundation report, *Shaping the Future* (NSF, 1996), concluded that a significant change in science teaching at the undergraduate level is necessary and is possible. The report recommended that scientists make university science courses more meaningful for majors and non-science majors. They noted a particular need among preservice teachers who, in turn, will be more able to make science meaningful for their students. A number of authors have argued that teacher preparation in science at the university level is largely responsible for inadequate science instruction in K–12 schools (e.g., Gilmer, Barrow, & Tobin, 1993). Focused on science content with little attention to instructional practices, many higher education science courses are taught by lecture in rooms holding too many students. Under those conditions, students have little opportunity, if any, for meaningful interaction. Laboratory sections of many science courses are conducted by instructors other than those teaching the lectures. This leads to little communication between instructors, and can be further exacerbated sometimes with scheduling the lecture and laboratory during different semesters (Gilmer et al., 1993). The authors concluded that this ineffective format of teaching science in higher education partially explains why many college graduates have little knowledge of science and have negative attitudes toward it.

Contrary to the perspective that science teachers at universities can solve problems in science education just by bringing their scientific knowledge expertise, the problems with science education go beyond the context of the students themselves. Teachers at the higher education level are part of the problem (Fedock et al., 1996). Effective state and national reform includes developing better or more meaningful learning in science at all levels.

GOALS FOR RESEARCH ON REFORM
IN UNDERGRADUATE SCIENCE

The purpose of science teaching research in higher education biology, chemistry, earth science, and physics classrooms is to understand more fully college and university student learning needs and faculty's cognitive perceptions, teaching skills, and educational knowledge, as well as university culture (Alexander & Dochy, 1995; Tierney, 1988). This research must focus on not only the status and results of reform but the change process during university educational reform and university teachers' efforts at change in response to national mandates to develop science literacy among diverse students (Corno & Snow, 1986). The National Science Foundation has responded in one way to the urgent agenda for reform through requiring all groups they fund in education to commit to serious assessment of their innovations. Particularly, in the Undergraduate Division and in the Teacher Professional Continuum section of the Elementary, Secondary and Informal Education Division, funded groups are required to conduct pedagogical research as related to the project (NSF, 2003).

RECOMMENDATIONS AND IMPLICATIONS
FOR FUTURE RESEARCH FOR REFORM
IN UNDERGRADUATE SCIENCE: SUMMARY
OF QUESTIONS ADDRESSING TENSIONS FOR REFORM

Focus questions to be asked guiding research on undergraduate science teaching are important to creating more effective change. These are critical issues if we are to make real changes in science teaching at the higher education level.

First, *what are we doing now in our classes and for our students to create meaningful learning?* The status of what naturally occurs should be compiled and further documented. What is the institutional context and culture in which the course exists (Neumann, 1995; Peterson & Spencer, 1990)? What are the university teachers' views about curriculum and pedagogy in the classroom? What are their cognitive understandings, perceptions, beliefs, and actions in regard to the teacher and students' role in learning and the origins of knowledge in the higher education science classroom (Peterman, 1993)? How do teachers' instructional actions in classrooms and the structure of their courses reflect their expressed beliefs? What are student beliefs and perceptions about science? What student learning outcomes result from the present beliefs about science teaching?

Second, the question, *What occurs during the reform process?* should be explored. This involves not only faculty and student actions but their beliefs, expectations, and perceptions in science classrooms. A comparison is made with pre-change levels. Information is gathered to attempt to identify barriers and determine approaches to overcoming these barriers to change. How do student beliefs, perceptions, expectations, and meaningful understanding of science change during the reform process? What are the potential effects of reforms on students in general and specifically on students with diverse characteristics (Long, Sadker, & Sadker, 1986; Maple & Stage, 1991; White, 1992)? What is the effect of faculty reflection on pedagogy and curriculum during action research carried out by university faculty involved in staff development efforts (Cross & Steadman, 1996; Hopkins, 1993)? What is the effect of varying reform and change conditions and experiences on perceptions of faculty who attempt to implement innovative curricula and pedagogy? What role must administrators play in the change process? How do individual faculty members' efforts overcome barriers to change during the process of planning, implementing, and sustaining change? What are the perceived, planned for, and real barriers to change?

Third, the question, *What outcomes follow the reform process?* should be investigated in regards to the experiences occurring during the change process. What are the elements of a predictive model of how to assist individual faculty change? What are the elements for a model of group change, creating a learning community in a department or college? Long-term outcomes in instructors' actions in addition to their perceptions, beliefs, and actions in science courses are compared to the change and pre-change levels. Student beliefs, perceptions, expectations, and meaningful learning of science are compared against long-term impact.

SUMMARY

Reforms in teaching science in higher education have been occurring at an increasing rate. Traditional pedagogical approaches of instructors in undergraduate introductory science courses do not work effectively with most of today's students. Helping students understand science ideas is a complex task. Doing it well takes a significant amount of pedagogical, curricular, practitioner research knowledge and skills. Since science faculty have little, if any, professional training in teaching or in implementing reforms in teaching, faculty professional development is needed. Effective teaching involves the purposeful, research-informed, development of innovative lessons actively involving students in learning. Continuous and intensive teacher and student feedback are part of the process. The priorities of effective science teaching differ from traditional beliefs about what college instructors do in classrooms. Reform can best be accomplished in a learning community involving diverse personnel and respect equally for research skills and knowledge in science and science teaching.

REFERENCES

Alexander, P., & Dochy, F. (1995). Conceptions of knowledge and beliefs: A comparison across varying cultural and educational communities. *American Educational Research Journal, 32*(2), 413–442.

American Association for the Advancement of Science. (1990). *The liberal art of science: Agenda for action.* Washington, DC: Author.

American Association for the Advancement of Science. (1993). *Benchmarks for scientific literacy, Project 2061.* New York: Oxford University Press.

Barinaga, M. (1991). Scientists educate the science educators. *Science, 252,* 1061–1062.

Boyer Commission on Educating Undergraduates in the Research University. (1998). *Reinventing undergraduate education: A blueprint for America's research universities.* Menlo Park, CA: Carnegie Foundation for the Advancement of Teaching.

Bradford, C.S., Rubba, P.A., & Harkness, W.L. (1995). Views about science-technology-society interactions held by college students in general education physics and STS courses. *Science Education, 79*(4), 355–373.

Brown, K. (1995). *Who takes college science courses?* Chicago: Chicago Academy of Sciences.

Cohen, D. (1988). *Teaching practice: Plus a change* (Issue paper No. 88-3). East Lansing, MI: National Center for Research on Teacher Education.

Corno, L., & Snow, R. (1986). Adapting teaching to individual differences among learners. In M. Wittrock (Ed.), *Handbook of research on teaching*. New York: Macmillan.

Cross, P., & Steadman, M. H. (1996). *Classroom research: Implementing the scholarship of teaching*. San Francisco: Jossey-Bass.

Cuban, L. (1990). Reforming again, again, and again. *Educational Researcher, 19*(1), 3–13.

Cuban, L. (1999). *How scholars trumped teachers: Change without reform in university curriculum, teaching, and research, 1890–1990*. New York: Teachers College Press.

Driver, R. (1986). *The pupil as scientist.* Philadelphia: Open University Press.

Driver, R., Squires, A., Rushworth, P., Wood-Robinson, V. (1994). *Making sense of secondary science*. New York: Routledge.

Duschl, R.A. (1990). *Restructuring science education*. New York: Teachers College Press.

Fedock, P., Zambo, R., & Cobern, W. (1996). The professional development of college science professors as science teacher educators. *Science Education, 80*(1), 5–19.

Floden, R., Gallagher, J., Wong, D., & Roseman, J. E. (1995). *Project 2061 blueprint for teacher education: A symposium presentation*. East Lansing: Michigan State University.

Fullan, M. (1993). *Change forces: Probing the depths of educational reform*. New York: Falmer Press.

Gilmer, P., Barrow, D., & Tobin, K. (1993, April). *Overcoming barriers to the reform of science content courses*. Paper presented at the annual meeting of the National Association for Research in Science Teaching, Atlanta, GA.

Golde, C.M., & Dore, T.M. (2001). At cross purposes: What the experiences of today's doctoral students reveal about doctoral education. Philadelphia: Pew Charitable Trusts.

Goodchild, L. (1991). What is the condition of American research universities? *American Educational Research Journal, 28*(1), 3–18.

Green, D. (1989). A profile of undergraduates in the sciences. *The American Scientist, 78,* 475–480.

Herr, K. (1988). Exploring excellence in teaching: It can be done! *Journal of Staff, Program, and Organization Development, 6*(1), 17–20.

Hewitt, N., & Seymour, E. (1991, April). *Factors contributing to high attrition rates among science, mathematics, and engineering undergraduate majors*. Washington, DC: Alfred P. Sloan Foundation.

Hopkins, D. (1993). *A teacher's guide to classroom research*. Philadelphia: Open University Press.

King, A. (1992). From sage on the stage to guide on the side. *College Teaching, 41*(1), 30–35.

Long, J.E., Sadker, D., & Sadker, M. (1986, April). *The effects of teacher sex equity and effectiveness training on classroom interaction at the university level*. Paper presented at the annual meeting of the American Educational Research Association, San Francisco.

Magner, D.K. (1992, October 21). A booming reform movement for introductory science courses. *The Chronicle of Higher Education*, pp. 17–18.

Maple, S., & Stage, F. (1991). Influences on the choice of math/science major by gender and ethnicity. *American Educational Research Journal, 28*(1), 37–62.

National Research Council. (1996a). *National Science Education Standards.* Washington, DC: National Academy Press.

National Research Council. (1996b). *From analysis to action.* Washington, DC: National Academy Press.

National Research Council. (1997). *Science teaching reconsidered: A handbook.* Washington, DC: National Academy Press.

National Research Council. (2000). *How people learn: Brain, mind experience, and school* (Extended ed.). Washington, DC: National Academy Press.

National Research Council. (2003). *Evaluating and improving undergraduate teaching in science, technology, engineering, and technology.* Washington, DC: National Academy Press.

National Science Foundation. (1996). *Shaping the future: New expectations for undergraduate education in science, mathematics, engineering, and technology* (NSF 96-139). Arlington, VA: Author.

National Science Foundation. (2003). *National Science Foundation: Where discovery begins.* Retrieved August 18, 2003, from http://www.nsf.gov/start.htm

Neumann, A. (1995). Context, cognition, and culture: A case analysis of collegiate leadership and cultural change. *American Educational Research Journal, 32*(2), 251–282.

Ory, J.C. (2000). Teaching evaluation: Past, present, and future. In K.E. Ryan (Ed.), *Evaluating teaching in higher education: A vision for the future: New directions in teaching and learning* (pp. 13–18). San Francisco: Jossey-Bass.

Peterman, F. (1993). Staff development and the process of changing: A teacher's emerging constructivist beliefs about learning and teaching. In K. Tobin (Ed.), *The practice of constructivism in science education* (pp. 227–245). Hillsdale, NJ: Erlbaum.

Peterson, M., & Spencer, M. (1990). Understanding academic culture and climate. In W. Tierney (Ed.), *New directions for institutional research: Vol. 68. Assessing academic climates and cultures* (pp. 3–18). San Francisco: Jossey-Bass.

Prather, J. (1995). An instructor self-assessment program for improvement of undergraduate science instruction: Providing teachers the opportunity for self-evaluation. *Journal of College Science Teaching, 24,* 410–418.

Reinarz, A. (1991). Gatekeeping: Teaching introductory science. *College Teaching 39*(3), 94–96.

Seymour, E. (1992a). The "problem iceberg" in science, mathematics, and engineering education: Student explanations for high attrition rates. *Journal of College Science Teaching, 21,* 230.

Seymour, E. (1992b). Undergraduate problems in teaching and advising in S.M.E. majors—Explaining gender differences in attrition rates. *Journal of College Science Teaching, 21,* 284.

Seymour, E. (1995). Revisiting the "problem iceberg": Science, mathematics, and engineering students still chilled out: Examining the causes of student attrition in science-based fields on a variety of campuses. *Journal of College Science Teaching, 24*, 392–400.

Siebert, E., & McIntosh, W. (2001). *College pathways to the science education standards.* Arlington, VA: NSTA Press.

Sigel, I.E. (1985). A conceptual analysis of beliefs. In I.E. Sigel (Ed.), *Parental belief systems: The psychological consequences for children* (pp. 345–371). Hillsdale, NJ: Erlbaum.

Sunal D.W., & Burry, J. (1992). *Context-related characteristics of expert science teaching.* Paper presented at the annual meeting of the National Association for Research in Science Teaching, Boston, MA.

Sunal, D., Bland, J., Sunal, C., Whitaker, K., Freeman, M., Edwards, L., et al. (2001). Teaching science in higher education: Faculty professional development and barriers to change. *School Science and Mathematics, 101*, 246–257.

Sunal, D., Freeman, M., Whitaker, K., Hedgepeth, D., & Ewing, L. (1994, October). *Developing scientific literacy through aerospace education.* Paper presented at the annual meeting of the American Institute of Aeronautics and Astronautics, Reno, NV.

Sunal, D., Sunal, C., Jones, L., & Shratter, D. (1996, April). *The effects integrating university science curricula with the National Science Education Standards.* Paper presented at the annual meeting of the National Science Teachers Association, St. Louis, MO.

Sundberg, M.D., & Monaca, G.J. (1994). Creating effective investigative laboratories for undergraduates. *Bioscience, 44*, 698–704.

Tierney, W. (1988). Organizational culture in higher education. *Journal of Higher Education, 59*(1), 2–21.

Tobias, S. (1990). *They're not dumb, they're different: Stalking the second tier.* Tucson, AZ: Research Corporation.

Tobias, S. (1992). *Revitalizing undergraduate science.* Tucson, AZ: Research Corporation.

U.S. Department of Education. (2001, August 21). *No child left behind.* Retrieved August 8, 2003, from http://www.ed.gov/offices/OESE/esea/nclb/titlepage.html

Weimer, M., & Lenze, L. F. (1994). Instructional interventions: A review of the literature on efforts to improve instruction. In K. Feldman & M.B. Paulsen (Eds.), *Teaching and learning in the college classroom* (pp. 653–682). Needham Heights, MA: Ginn Press.

White, P. (1992). *Women and minorities in science and engineering: An update.* Washington, DC: National Science Foundation.

part I

LESSONS FROM RESEARCH ON REFORM IN UNDERGRADUATE SCIENCE

The first section of this volume focuses upon the variety of issues and elements involved in numerous reform efforts in teaching undergraduate science in colleges and universities in the United States. What are the historical and theoretical foundations of current science reform efforts in higher education? What lessons can be gained from a detailed analysis of the research literature on teaching undergraduate science? These questions were addressed by a national sample of researcher-instructors involved in teaching undergraduates while at the same time carrying out significant reforms to improve undergraduate science for all students.

Each chapter focuses on a key issue or element important in current reforms. They collectively provide an historical and theoretical foundation in research, case studies, and examples of different critical components involved in reform. The first chapters begin with the history and overview of factors leading to successful practices for change in undergraduate science in the last part of the 20th century. The remaining chapters examine issues and challenges affecting all science departments and faculty members for the 21st century. The later chapters in this section focus upon critical issues that must be successfully addressed in all undergraduate science courses for the 21st century. These issues involve students, learning, pedagogy, interdisciplinary collaboration and planning, assessment, diversity, use of technology, professional development, and systemic approaches for reform.

In Chapter 2, Bonnie McCormick highlights the importance of reform in science teaching and the STEM pipeline. Higher education has always been a stakeholder in national science teaching reform efforts, yet the science disciplines continually suffer from poor retention rates and a high level of student dissatisfaction with science teaching in undergraduate courses. McCormick weighs in on the importance of reforms in science teaching at the undergraduate level and maintains that only then can we begin to make progress toward preparing future scientists, teachers of science, and scientifically literate college graduates.

Emmett Wright and Dennis Sunal further explore, in Chapter 3, the issues that college and university planning groups and instructors must confront before innovations can be implemented and sustained. The issues affecting change vary according to the size of the institution, the mission of the institution, past experiences of faculty, stage in the reform process, and as well as many contextual factors. These identified factors can be sorted into nine major categories: management, coordination, leadership, faculty, students, curriculum, instruction, budget and resources, and accreditation and certification.

William Harwood, in Chapter 4, focuses on issues of cross-disciplinary collaboration in conducting reforms in undergraduate science. Harwood examines the factors that influence the formation and success of collaborative partnerships, concentrating on collaboration between science and science education faculty members.

Kathleen Fisher, in Chapter 5, explores the relevance of prior knowledge in college science instruction. The persistent failure of understanding seems to derive from strong underlying assumptions that students bring to the college classroom. The chapter examines prior knowledge both as a barrier to understanding and as a foundation on which to build new knowledge. It illustrates the importance of purposeful, in-depth interactions between teacher and students to achieve successful and deep understanding.

In Chapter 6, Dennis Sunal considers reforms in science teaching that are fundamentally different from traditional pedagogy, the transmission of intact ideas. Teaching science so that it is meaningful to students involves helping them reconstruct their prior knowledge in the context of the new science concepts being introduced. The learning cycle is described as a general model of pedagogy in undergraduate science courses that facilitates the meaningful understanding of science concepts.

John Christopher and Ronald Atwood examine, in Chapter 7, the interdisciplinary efforts of a collaborative planning group and the long-term collaboration of a physicist and the science educator. The cultural differences between science and education departments and the different knowledge bases of collaborating faculty from these two fields are viewed as a major strength, when the persons involved have equal status in the longer term. In the short term, the greater expertise each of the collaborators brings to a particular problem makes that person's views valuable.

Lawrence Scharmann, Mark James, and Ann Stalheim-Smith, in Chapter 8, examine the impact of reform in science teaching on student assessment practices. Change in undergraduate science is found to lead to a disconnection with more traditional forms of assessment. Efforts to revise forms of classroom assessment, both formative and summative, must necessarily accompany efforts to reform science instruction.

Lynn Jones Eaton, Chapter 9, examines the need for more diversity in science. Although it is widely held that both females and people of color can do and be whatever they so choose, an overwhelming number choose to not enter the professional fields in science. Jones provides examples of reforms needed to counter that part of the culture of the university faculty that undermines progress toward diversity in both ethnic and gender areas of undergraduate science.

Michael Odell, Scott Badger, Teresa Kennedy, Tim Ewers, and Mitchell Klett, Chapter 10, expand upon ways to reform undergraduate science through the use of information technology. They provide examples of the use of information technology to enhance the impact of reforms on students. Included are technology innovations to assist in inquiry, student teamwork, and real-world problem solving and tools to deepen the classroom learning experience.

The Internet can be both a barrier and a benefit to meaningful learning in undergraduate science courses taught online. Cheryl Sundberg, Chapter 11, examines reforms in undergraduate science through online learning environments in the areas of student support from instructors, diversity of course offerings, course learning experiences that can be accessed at any time, assessments used to evaluate level of reflectivity in discourse, and best practices for the development of online courses.

In Chapter 12, Dianne Raubenheimer investigates ideas about practitioner (action) research held by college and university faculty involved in redesigning science courses to reflect reform-based concerns. Raubenheimer finds that action research produces a wealth of information about successful innovation in undergraduate courses. The outcomes of conducting action research are positive not only in student learning and but also in faculty development in teaching.

Part I concludes with a case study that describes the process and impact of a systemic approach to reform in undergraduate science. In Chapter 13, Dennis W. Sunal, Christy MacKinnon, Dianne Raubenheimer, and Francis Gardner examine a reform program that has changed undergraduate science teaching on a national level, impacting hundreds of courses and thousands of students each semester. The reform program used the best of what had been learned through research on faculty professional development and in teaching and learning to enhance science literacy of undergraduate students in entry-level science courses.

CHAPTER 2

SCIENCE EDUCATION REFORM AND HIGHER EDUCATION

A Historical Perspective

Bonnie McCormick

ABSTRACT

Since the launching of Sputnik in 1957, there has been a continuing debate on the quality of education in the United States. Waves of reform have focused on changing the preparation of students in science, mathematics, engineering, and technology (STEM). These efforts have included numerous reports, position statements, and curriculum development projects by professional organizations and government agencies. The debate on reforming K–12 science and improving teacher preparation for this level has been polarized by competing purposes—educating students to enter the STEM pipeline and scientific literacy for all citizens. Implementing reform has been influenced by competing emphases around the content of the disciplines versus the pedagogy of science teaching. Because of its role in the preparation of scientists, teachers of science, and citizens capable of functioning in an increasingly scientific and technological society, higher education has

Reform in Undergraduate Science Teaching for the 21st Century, pages 17–31
Copyright © 2004 by Information Age Publishing

always been a stakeholder in science teaching reform efforts. Yet science disciplines suffer from poor retention rates and from a high level of student dissatisfaction with science teaching in undergraduate programs. National organizations have called for reform in science teaching to extend to the undergraduate classroom in order to meet the goals of reform. By ensuring that science is taught at the undergraduate level in a manner that is consistent with the reform movements, we can begin to make progress toward preparing future scientists, teachers of science, and scientifically literate college graduates for the 21st century.

SCIENCE EDUCATION REFORM AND HIGHER EDUCATION: A HISTORICAL PERSPECTIVE

Forty-seven years ago, I stood on the High Plains of Texas and watched Sputnik race across the sky. This small dot of light represented both great scientific achievement and great fear that our nation was in danger of losing prominence in the world. This was a time when school children already feared things launched into the air. On a regular basis, we participated in duck-and-cover drills that even third graders knew would not save us from annihilation. What we were unaware of was that our educational system would be called into question. This single event sparked the waves of curricular changes in science education that continue to this day. Since the launching of Sputnik, "Americans have become obsessed with science education reform" (Tobias, 1992, p. 14).

Calls for reform are not new in the history of American education. Three major goals that have focused practices in school science are "understanding scientific knowledge, understanding and using scientific process, and promoting personal-social development" (Bybee & DeBoer, 1993). These goals have shaped the debates on how science education reform should proceed depending on whether the aim of science education has been to prepare scientists, to prepare teachers of science, or to produce a scientifically literate society. The shifting emphases over time on knowledge, process, habits of the mind, and real-world application are driven by these aims.

As a result of the continuing debate, over the past half-century, waves of reform have proposed many new models for the preparation of students in science, technology, engineering, and mathematics (STEM). These efforts have included numerous reports, position statements, and curriculum development projects by professional organizations and government agencies. The debate on reforming K–12 science and improving teacher preparation has focused on the competing purposes of educating students to enter the STEM pipeline or promoting science literacy for all citizens. Improving K–12 science and preparing teachers has been influenced by competing emphases on the content of the disciplines and the pedagogy of science teaching. The key question that drives the debate is, What is worth knowing in science and how do you transfer this knowledge in meaningful ways to precollege and college students?

SCIENCE EDUCATION FROM 1900 TO 1950

During the first half of the 20th century, university departments of science played a role in determining what science should be taught to those students that entered colleges and universities. Debates centered on the types of science courses that should be taught in high schools and the number of science courses required for matriculation to college (DeBoer, 1991). Structure of the disciplines and how knowledge should be taught were the focus of these efforts. The current thinking of university education faculty largely influenced science teaching methods. However, in both cases the debates were about what should be taught in the schools, not in higher education.

During the second decade of the 20th century, John Dewey's educational philosophy referred to as the progressive movement (Shamos, 1995) was a major influence on the teaching of science. Dewey's philosophical contributions to the science education debate were evident in the positions he took on the need for (a) appropriate inquiry science teaching methods, (b) student-centered education, and (c) students addressing relevant social problems in the curriculum (Bybee, 1993). Over the succeeding decades, Dewey's ideas competed and lost to proponents of science content knowledge obtainment as the primary goal of K–12 science. This dominating influence on K–12 science curriculum lasted until the end of World War II and into the late 1950s, and probably would have endured longer if world events had not interceded.

THE POSTWAR CRISIS

When the Soviet Union launched Sputnik in 1957, the educational system of the United States was called into question at a time when our national pride was damaged and our nation's security was considered at risk. Education was blamed for the failure of the United States to win the space race. This one event focused attention on needed reform in science education by modernizing the science curriculum to reflect the structure of disciplines and to increase the number of students choosing science as a career (Bybee & DeBoer, 1993; Duschl, 1990). The reform efforts that began in the 1960s returned the focus of science education to knowledge and the structure of the disciplines from the viewpoint of academic scientists (Shamos, 1995).

Because scientific and technological achievements were viewed as critical to our nation's security, funding from the federal government began to influence reform in science education. In 1958, the National Defense Education Act was passed authorizing the U.S. Department of Education as a separate agency. Congress tripled the education budget of the National Science Foundation (NSF) and authorized the agency to expand its education initiatives to support science, math, and engineering at all levels of the education system. The education initiatives funded curriculum reform projects for the next two decades that focused on the structure of the disciplines and the process of science (DeBoer, 1991).

As a result of the NSF efforts, academic scientists worked with schoolteachers to reform curriculum at the secondary and elementary levels. Education faculty played a secondary role in these efforts. NSF funding helped sustain the efforts of high school curriculum projects such as the Physical Science Study Committee (PSSC), Biological Sciences Curriculum Study (BSCS), and Chemical Bond Approach (CBA). NSF also funded teacher training institutes to strengthen the content knowledge of high school teachers. These curriculum projects were viewed as providing a more accurate view of the nature of science by providing up-to-date information, by focusing on in-depth coverage of significant concepts, and by containing discovery-type investigations (DeBoer, 1991). These curriculum projects ignored the role of science in everyday life.

During the 1960s several elementary science curriculum projects received support from NSF. These projects were widely known as the "alphabet soup" projects and included Elementary Science Study (ESS), Science—A Process Approach (SAPA), and Science Curriculum Improvement Study (SCIS) (Shamos, 1995). Two features that all of these projects had in common were that students had direct experience with materials and objects to develop their knowledge of science concepts and that the curriculum was "teacher proof." In other words, the written materials and activities rather than the teacher would be the most important factor in the learning experience (Bybee, 1993).

The academic scientists dominated both the secondary and elementary curriculum projects. Teachers and administrators tested the curriculum and provided feedback to the scientists (Duschl, 1990). The focus was the structure of the disciplines and the way that scientists thought and created knowledge. The process of inquiry was the means of acquiring that knowledge (Bybee & DeBoer, 1993). Although these projects were widely used and were designed to actively engage children in the learning process, they did not persist. One of the reasons for their failure was that the academic scientists developed the curriculum without equal input from teachers who would implement the programs and with little involvement of science educators (Duschl, 1990).

Winning the "space race" by being the first (and only) country to put a man on the moon 11 years after the launch of Sputnik placed America at the apex of achievement in science and technology, and our national pride was restored. The crisis was over. Since World War II, more than two billion dollars had been spent to revise science curriculum in elementary and high schools to reflect the structure of the science disciplines from the point of view of the academic scientist (Shamos, 1995). Eventually funding levels for reform were reduced and support for science education reform came to an abrupt halt in 1981 when funds to NSF's Education Directorate were cut by the Reagan Administration, whose budget was driven by tax cuts and by a significant increase in defense spending. In 1982, NSF abolished the Directorate of Education (NSF, 2003).

THE NEW CRISIS: A NATION AT RISK

There was, however, a new crisis looming. The National Commission on Excellence in Education (NCEE) published the report, *A Nation at Risk* (NCEE, 1983), which described the threat to the nation caused by an overall decline in the quality of our educational system. The lack of qualified teachers, declining test scores, low standards in the public schools, poor performance of American students when compared to students in other countries, and a decline in functional literacy of American adults were among the indicators of a decline in the quality of our educational system. The report stated,

> If an unfriendly power had attempted to impose on America the mediocre educational performance that exists today, we might well have viewed it as an act of war. We have squandered the gains in achievement made in the wake of the Sputnik challenge. Moreover, we have dismantled essential support systems that helped make those gains possible. We have in effect, been committing an act of unilateral educational disarmament. (NCEE, 1983, p. 5)

The report warned that America could lose its place as leaders in an increasingly technological and global economy. It was clear to the commission that America was not only losing its place in competitive world markets but that we were unprepared to succeed in the information age that was just beginning to take shape. It was noted that although the average citizen in 1983 was better educated than their parents' generation, the average graduate of our schools and colleges was not as well educated as graduates in the previous generation. The commission found this to be alarming. One of the reasons that the United States was a leading world power was because each generation had attained greater economic security and more education than their parents. The fear was that for the first time in American history, this generation would not even equal the educational achievement of their parents.

The NCEE made recommendations on how stakeholders in the system could implement lasting reform over a period of years. These recommendations included that "schools, colleges, and universities adopt more rigorous standards, and higher expectations for academic performance and conduct and that 4-year colleges and universities raise their requirements for admission" (NCEE, 1983, p. 18). All high school students should be required to take three years of science and mathematics. Science content should include not just the concepts and laws of the discipline, but also the methods, applications, and implications of science and technology. The clear message was that government agencies, professional societies, and the entire educational system take responsibility for ensuring that Americans were prepared to meet the challenges of an economy that was becoming increasingly dependent on a workforce that was literate in fields of science and technology. We had entered the era of "science for all" and the beginning of systemic reform that would challenge all parts of the system to collaborate in improving science education at all levels.

Most of the efforts of science education reform prior to the 1980s were directed toward improving the quality of school science in the elementary and high schools. This is likely due to the fact that the funding for higher education was earmarked for curriculum development and for institutes to train teachers to implement the curriculum, and to the fact that college science faculties viewed the problem to be with the preparation of students entering college science programs. What many science professionals failed to recognize was that they were responsible for the content courses that inadequately prepared science teachers to be confident and effective teachers of science content and that they were responsible for the lack of scientific literacy in a country where increasing numbers of students were matriculating to colleges and universities (NSF, 1996).

Since the publication of *A Nation at Risk*, many national organizations have called for reform in science teaching to extend to the undergraduate classroom in order to meet the goals of reform. Higher education has always been a stakeholder in these reform efforts through the preparation of scientists, of teachers of science, and of citizens prepared to function in an increasingly technological society. By ensuring that science is taught at the undergraduate level in a manner that is consistent with contemporary reform movements, we can begin to make progress toward preparing scientists, teachers of science, and scientifically literate citizens.

In the mid-1980s, the National Science Board (NSB) of NSF, commissioned a panel to address the problems of courses in undergraduate science, mathematics, and engineering. NSF had traditionally supported graduate students and precollege programs. The commission's 1986 report, *Undergraduate Science, Mathematics, and Engineering Education,* which is also know as the "Neal Report," called for the Foundation to "bring its programming in the undergraduate area into balance with its activities in the precollege and graduate areas as quickly as possible" (NSB, 1986). As a result of the recommendations in the Neal Report, NSF established a separate directorate for undergraduate education (DUE) and funded proposals that have improved undergraduate education by promoting undergraduate research experiences, developing multidisciplinary curricula, and supporting "access to active learning experiences" (NSF, 1996, p.17).

CONTEMPORARY SYSTEMIC REFORM

In the late 1980s, the American Association for the Advancement of Science (AAAS) established Project 2061 (AAAS, 1990) which had the vision of science literacy for all Americans by the time of the return of Halley's comet in the year 2061. By setting the deadline 75 years in the future, AAAS recognized that reforming the educational system to achieve scientific literacy for all Americans would require a sustained effort over a long period of time. (By contrast, at the education meeting at the National Governor's Conference that same year, President George H. W. Bush set the goal of American students being first in the world in science and mathematics by the year 2000.) By including "all" in its vision, AAAS recognized that reform would have to be systemic and would involve all levels of the educational system.

The first publication of Project 2061, *Science for All Americans (SFA)* (AAAS, 1990), argued for science literacy for all members of society not just because it was crucial to the nation's economic viability, but also because a knowledge of science and technology were crucial to solving global problems such as unchecked population growth, environmental degradation, pollution, and disease. Although the emphasis of this publication is on the nature of science, mathematics, technology, and the basic knowledge required for literacy, the document also addresses the effective use of principles of science teaching and learning that are research-based. The recommendations include being mindful of students' prior knowledge and providing students the opportunity to practice and apply what is learned. "Teaching of science should be consistent with the nature of inquiry" and,

1. Start with questions about nature.
2. Engage students actively.
3. Concentrate on the collection and use of evidence.

4. Provide historical perspectives.

5. Insist on clear expression.

6. Use a team approach.

7. Do not separate knowing from finding out.

8. De-emphasize the memorization of technical vocabulary. (AAAS, 1990, pp. 200–203)

These principles of teaching are echoed in all the reform documents that have been produced since the publication of *SFA*.

SFA also addressed the role that colleges and universities must assume if the goals of science literacy were to be met. Project 2061 recommended that,

> presidents of all colleges and universities establish scientific literacy as an institution-wide priority . . . to ensure that all graduates (from whom, after all tomorrow's teachers will be drawn) leave with an understanding of science, mathematics and technology that surpasses what this report recommends for all high school graduates. (AAAS, 1990, p. 226)

The report also called for college science and mathematics departments to use these principles in designing courses that would provide excellence in preparation for preservice teachers.

The publication of *SFA* set in motion more committees and reports aimed at defining the problems and remedies of STEM education. Many of these were focused on the content and methods of teaching in K–12 science. Project 2061 produced *Benchmarks for Science Literacy* in 1993. With its publication of the *National Science Education Standards* (*NSES*) in 1996, the National Research Council (NRC) also became involved in the drive to develop guidelines for achieving science literacy. The authors of *NSES* stated that science literacy requires the ability to

> use appropriate scientific processes and principles in making personal decisions, engage intelligently in public discourse and debate about matters of scientific and technological concerns, and increase economic productivity through the use of the knowledge, understanding, and skills of the scientifically literate person in their careers. (NRC, 1996)

The *NSES* addressed not only content standards for different grade levels, but also the role of colleges and universities in preparing teachers in science content as well as science teaching methods. As a result of *NSES*, all states have established or are in the process of establishing science standards for K–12 classrooms. Certification requirements are changing in many states and the trend is that prospective teachers receive their degree in the content area. It is critical that teachers experience instruction during their undergraduate years that is consistent with inquiry and student-centered learning practices (NRC, 2001).

EXTENDING REFORM TO UNDERGRADUATE EDUCATION

At the same time as benchmarks and standards were being developed for school science, the NSF and professional societies were examining the problems with undergraduate education in STEM. Recommendations for reform in undergraduate education that came from these groups, and from undergraduate students, focused on the need to change course content and instructional methods employed by undergraduate faculty (Seymour, 1995). For reform to succeed it must be systemic and extend to all levels of education.

Changing the way science is taught in colleges and universities is more difficult than changing school science where decisions are coordinated at the state level (Shamos, 1995). Dissemination of requirements for systemic reform in higher education flows from government agencies such as NSF and the National Academy of Sciences (NAS), and from prestigious professional groups such as AAAS, to individual faculty members. Faculty become aware of systemic reforms through membership in professional societies, participation in professional development, and funding opportunities to improve curriculum and teaching at the undergraduate level. Because of this, implementation of reform in higher education has come from individual faculty members or small groups of faculty who have embraced the need for change in undergraduate teaching and learning.

The Society of College Science Teachers (SCST) examined the problems of the introductory science curricula and formulated recommendations to improve these courses. Their position paper on introductory college science courses stated that the introductory course is where majors are recruited and trained and where preservice teachers learn to "love and appreciate" or hate science (Halyard, 1993). In addition, the SCST specified that these courses should "contribute to the scientific literacy of all college students and provide a conceptual base for subsequent courses" (Halyard, 1993, p. 31). SCST recommended inquiry-based laboratory experiences, exemplary teaching practices that are research-based, and a format that promotes critical thinking, problem solving, and collaborative work on meaningful tasks. Students should be able to "make scientifically based decisions and solve problems drawing on concepts and experiences from relevant areas" (Halyard, 1993, p. 32).

Students are stakeholders in reform of undergraduate education, and their input supports what reform documents indict as the problems with undergraduate education. Research that focuses on the needs of the undergraduate science student shows that learners concur with these recommendations for improving undergraduate education. Elaine Seymour and Nancy Hewitt (1994) conducted a 3-year ethnographic study that included input from more than 450 students on 13 campuses. The subjects of the study were students who were majoring in the natural sciences and engineering, and students who had switched majors from science and engineering. These students were selected based on a minimum math S.A.T. score of 650. More than half of the subjects had switched their major from science and engineering to another field. Their findings show that the greatest concern among both those students that persisted in science and engineering majors and those that left was poor teaching by science, math, and engineering faculty (Seymour, 1995). Students criticized the fact that classes in the sciences were mostly one-way lectures and that there was a lack of applications and discussion of implications. Classes were tedious when the main focus was on memorization of material. Reading or copying material straight from the textbook was identified by the subjects as the worst teaching practice in college science and engineering classrooms (Seymour & Hewitt, 1994).

The NSF report, *Shaping the Future: New Expectations for Undergraduate Education in Science, Mathematics, Engineering and Technology* (NSF, 1996), examined the concerns of all undergraduate students enrolled in undergraduate STEM courses. Focus groups of students confirmed the overall findings of Seymour and Hewitt (1994). Students in these focus groups identified introductory courses as a major barrier, and laboratory exercises were criticized for being mechanical and unconnected to the concepts of the lecture portion of courses.

The NSF review stated that despite the consensus that America's system of graduate education and basic research in science, mathematics, and engineering is the best in the world, its K–12 education system lags behind. The report further stated "America has produced a significant share of the world's great scientists while its population is virtually illiterate in science" (NSF, 1996, p. iii). The authors of the report stated that the imperative for NSF in shaping the future of undergraduate education is that "all students have access to supportive, excellent undergraduate education in science, mathematics, engineering, and technology, and all students learn these subjects by direct experience with the methods and processes of inquiry" (NSF, 1996, p. ii).

Because too many students leave STEM courses that they find dull and exclusive, too many teachers are unprepared to teach these subjects, and too many graduates enter the workforce unprepared to solve real-world problems, to work cooperatively, or to continue learning. In a world where the role of technology is increasing and where information is the "common currency," the opinion of the report was that without an educated citizenry, our society will be at "great risk" and our people will be "denied the opportunity for a fulfilling life" (NSF, 1996, p. ii). The report recommended that NSF increase funding to support curriculum development in undergraduate courses, professional development of teaching faculty, preparation of K–12 teachers, involvement of undergraduates in research programs, and research on learning in undergraduate classrooms.

Blueprints for Reform (AAAS, 1998) addressed the need for reform to extend to undergraduate education by changing the undergraduate curriculum for all students including those who are in teacher education programs. The document recommends that the paradigm of teaching and learning that focuses on the material rather than on the process of learning be changed. Among the recommendations of the report are:

1. Concentrating on the central ideas of each discipline, even at the expense of coverage.
2. Providing all students with an understanding of the interrelationships of human knowledge, including links among fields of science.
3. Providing a more student-centered learning environment, supporting a wider variety of learning styles, and using more varied organizational strategies and teaching materials. (AAAS, 1998, ch. 4)

The justification for this focus was the methods by which scientists themselves work within their discipline by asking questions, devising and performing experiments, and extending their knowledge through interactions with others. Students of science should have the same opportunity to learn by activity, reflection, and practice.

SUSTAINING REFORM IN UNDERGRADUATE CLASSROOMS

It is clear that professional societies such as AAAS, SCST, and NSF are all committed to systemic change that provides quality science instruction for all students including those in colleges and universities. Numerous reports have been written about what should be done. Those have been followed by reports on how systemic reform can be extended to higher education. These reports agree that science literacy should be the primary goal of instruction and that the method of instruction should be based on what research supports as "best practice" in science instruction at all levels of the educational system.

The consensus is that reform in undergraduate courses is critical to systemic reform. These courses are the last place that most of our nations leaders are "asked to think broadly about SME&T in any formal way" (NRC, 1999, p. 13). In addition, all K–12 teachers are trained in these undergraduate content courses. Because the next generation of college faculty will come from majors in science, mathematics, and engineering, they are also affected by the teaching methods modeled by faculty in undergraduate courses. Because many teachers will use the models of instruction that they encounter in those undergraduate experiences, teaching methods at the undergraduate level must change if reform is to be sustained. Currently most faculty have little or no formal training in teaching and are unaware of the literature that can inform and improve teaching (Boyer Commission, n.d.).

Reform of teaching in the K–12 classrooms has been the focus of the reform debate and these changes are mandated through state education systems. The public and state agencies hold these education systems accountable for their performance. The challenge is different in higher education for several reasons that are related to the culture of institutions of higher education. Whereas the performance of students on high stakes tests is critical to K–12 systems, in higher education the reward system for tenure and promotion is focused on success in research and scholarship. Systemic reform will not extend to college and university classrooms unless reward systems change so that scholarly activities that focus on improving teaching and learning are supported and rewarded in the same manner as other types of research (NRC, 2002).

The National Academy of Sciences (NAS), Committee on Undergraduate Science Education (CUSE), in their report *Evaluating and Improving Undergraduate Teaching in Science, Technology, Engineering, and Mathematics* (NRC, 2002), state that

1. Effective teaching in STEM should be available to all students.
2. Design and evaluation of curricula should be the responsibility of faculty.
3. Scholarship that focuses on teaching and learning should be given the same support and rewards as other types of scholarship.
4. Faculty who teach undergraduates should be supported and mentored throughout their careers.

Many agencies and professional organizations have provided funding to change undergraduate science teaching. However, too often, reform efforts are dependent on one faculty member, a "lone ranger" (NSF, 1996, p. 50). In part, this is because professional development programs aimed at improving undergraduate teaching are not available or required at most institutions (Boyer Commission, n.d.; NRC, 2002). In addition to innovation, there needs to be more implementation of methods that are shown by research to be successful in undergraduate science courses. It is critical that faculty have access to supportive faculty development programs at both the institutional and national level and that successful curricula are disseminated at a national level. For this to happen, funding of innovations in STEM education, research of effective practice, dissemination of results, and faculty development programs must continue and increase.

Successful implementation of reforms in undergraduate education will require instructional design that is aligned from the learning objectives to the assessment strategies. The process of alignment requires that needs be diagnosed, objectives and content be specified, content be organized, learning experiences be selected, and appropriate evaluations of the learning experience be determined (Posner, 1988). The first step, diagnosing needs, has largely been completed by AAAS, NSF, and NRC, and other national organizations. Determining objectives and content should be the responsibility of academic departments and individual faculty who should set "clear learning goals for individual courses and for the department's curriculum in general" (NRC, 1999). Implementation of the curriculum is the responsibility of faculty and faculty teams as they work to design the learning experience. Appropriate evaluation of both student learning and teaching effectiveness are crucial to refining the curriculum. The evaluation process can also provide the opportunity for a scholarly approach to the reform process.

CONCLUSION

Twenty years after the publication of *A Nation at Risk* (NCEE 1983), many still believe that we are still at risk—we are still a nation "divided into a technologically knowledgeable elite and a disadvantaged majority" (NRC, 1999, p. 1). We were not first in the world in math and science by the year 2000. An assessment of science achievement conducted by the National Assessment of Educational Progress (NAEP) found that more than 70% of 4th, 8th, and 12th graders failed to reach the proficient level in science on a test designed to assess whether the science standards proposed by national organizations were being met (Vogel, 1997). We are currently facing budgetary problems similar to those faced in 1981. Although the reasons are different, tax revenue has decreased and military spending has increased. State budgets are in crisis and funding cuts are a reality.

We have, however, stayed on a steady course toward systemic reform for the past 20 years. The direction and guiding principles of systemic reform have remained the same. Colleges and universities are moving from the lone ranger model to a model of sustaining and extending change. Faculty teams are participating in professional development through initiatives sponsored by agencies such as NSF and NASA. Professors of science are successfully researching teaching and learning in college science classrooms. Organizations and programs such as NSTA, AAAS, Project Kaleidoscope, and NOVA are disseminating successful strategies and providing professional development for extending reform to higher education. Perhaps by the time Halley's comet returns we will no longer be in a state of crisis and systemic reform will be a reality 104 years after the launch of Sputnik.

REFERENCES

American Association for the Advancement of Science. (1990). *Science for all Americans.* New York: Oxford University Press.

American Association for the Advancement of Science. (1993). *Benchmarks for science literacy.* New York: Oxford University Press.

American Association for the Advancement of Science. (1998). *Blueprints for reform.* New York: Oxford University Press.

Boyer Commission on Educating Undergraduates. (n.d.). *Reinventing undergraduate education: A blueprint for America's research universities.* Retrieved July 24, 2003, from http://huxley.phys.cwru.edu/pcuel/boyer.html

Bybee, R. (1993). *Reforming science education.* New York: Teachers College Press.

Bybee, R., & DeBoer, G. (1993). Research on goals for the science curriculum. In D. Gabel (Ed.), *Handbook of research on science teaching and learning* (pp. 357–387). New York: Macmillan.

DeBoer, G. (1991). *A history of ideas in science education: Implications for practice.* New York: Teachers College Press.

Duschl, R. (1990). *Restructuring science education: The importance of theories and their development.* New York: Teachers College Press.

Halyard, R.A. (1993). Introductory science courses: The SCST position statement. *Journal of College Science Teaching, 23,* 29–31.

National Commission on Excellence in Education. (1983, April). *A nation at risk.* Retrieved July 24, 2003, from http://www.ed.gov/pubs/NatAtRisk/risk.html

National Research Council. (1996). *National Science Education Standards.* Washington, DC: National Academy Press.

National Research Council. (1999). *Transforming undergraduate education in science, mathematics, and technology.* Washington, DC: National Academy Press.

National Research Council. (2001). *Educating teachers of science, mathematics, and technology.* Washington, DC: National Academy Press.

National Research Council. (2002). *Evaluating and improving undergraduate teaching in science, technology, engineering, and mathematics.* Washington, DC: National Academy Press.

National Science Board. (1986). *Undergraduate science, mathematics, and engineering education; role for the National Science Foundation and recommendations for action by other sectors to strengthen collegiate education and pursue excellence in the next genera-*

tion of U.S. leadership in science and technology (Publication No. NSB 86-100). Washington, DC: Author.

National Science Foundation. (1996). *Shaping the future: New expectations for undergraduate education in science, mathematics, engineering and technology* (Publication No. NSF 96-139). Washington, DC: Author.

National Science Foundation. (2003, March 28). *Celebrating 50 years: History.* Retrieved July 24, 2003, from http://www.nsf.gov/od/lpa/nsf50/history.htm

Posner, G.J. (1988). Models of curriculum planning. In L. Beyer & M. Apple (Eds), *The curriculum: Problems, politics, and possibilities* (pp. 56–67). Albany: State University of New York Press.

Seymour, E. (1995). Revisiting the "problem iceberg": Science, mathematics and engineering students still chilled out. *Journal of College Science Teaching, 25,* 392–400.

Seymour, E., & Hewitt, N. (1994). *Talking about leaving: Factors contributing to high attrition rates among science, mathematics, and engineering majors.* Boulder, CO: Bureau of Sociological Research.

Shamos, M.H. (1995). The myth of scientific literacy. New Brunswick, NJ: Rutgers University Press.

Tobias, S. (1992). *Revitalizing undergraduate science: Why some things work and most don't.* Tucson, AZ: Research Corporation.

Vogel, G. (1997). Students don't measure up to standards. *Science, 278,* 794.

CHAPTER 3

REFORM IN UNDERGRADUATE SCIENCE CLASSROOMS

Emmett L. Wright and Dennis W. Sunal

ABSTRACT

There is a major emphasis currently underway in higher education about rethinking undergraduate instruction, particularly the introductory courses that meet the academic needs of elementary and secondary education majors. This effort is being driven by new accreditation policies and by the national standards, professional organizations, and national and state governmental agencies. On the surface, changing undergraduate science courses is perceived as deceptively easy, but in reality, there are many issues that college and university planning groups and instructors must confront before innovations can be implemented and sustained. These issues affecting change vary according to the size of the institution, the mission of the institution, past experiences of faculty, and other factors. Theses identified factors, impeding or supporting variables, of innovation in undergraduate science, addressed in the chapter, can be sorted into nine major categories: management, coordination, leadership, faculty, students, curriculum, instruction, budget and resources, and accreditation and certification.

Reform in Undergraduate Science Teaching for the 21st Century, pages 33–51

INTRODUCTION

Higher education faces significant challenges. As scientific research creates new knowledge and as education research identifies more effective methods for teaching college science, faculties are under pressure to rethink their teaching and become better science instructors (Magner, 1992). Recent research in the areas of inquiry teaching, conceptual development, and preconceptions has led to more innovative strategies for college classroom instruction and to new approaches for creating instructional change in science courses (Driver, 1986; Sigel, 1985).

Although college-level science faculties are attempting to improve the effectiveness of their courses, the process is slow (Barinaga, 1991; Fedock, Zambo, & Cobern, 1996). Introductory science courses are the cause for many students dropping out of the science major or from taking additional science courses. Tobias (1990) reported that common features of courses turning off students include lack of relevance, passive student roles, emphasis on competition, and focus on algorithmic problem solving.

In most states' K–12 science curriculum guidelines there is a significant departure from the past reliance on content goals, reflecting the national reform movement described in the *National Science Education Standards* (National Research Council, 1995) and in *Benchmarks for Science Literacy: Project 2061* (American Association for the Advancement of Science, 1993). These guidelines are integrated into higher education criteria for institutional review by the National Association for Colleges of Teacher Education and in criteria for individual teacher accreditation by the National Science Teachers Association. New science standards contrast with the instructional methods and content focus of traditional higher education science courses (Gilmer, Barrow, & Tobin, 1993).

BARRIERS TO CHANGE

Change is slow to occur in higher education because institutional organization, expectations, and roles inhibit risk taking, ambiguity, and the inquiry required for change to occur (Cohen, 1988; Cuban, 1999; Fullan, 1999). How different is it for a student to sit in a college science lecture today as compared to courses a century ago? What differences in instruction methods would the student experience? Researchers have suggested causes for a similar lack of change in elementary and secondary education instruction over the past century. They include the following: (a) The culture at large creates strong forces inhibiting change; (b) ongoing staff development, follow-up, and monitoring are lacking; (c) the organizational context and structure of the institution shape instructors' practice; (d) the perceived realities of the classroom influence a teacher to institute ineffective incremental changes rather than the major ones needed; and (e) instructors' beliefs and expectations about teaching and learning limit change (Cuban, 1990). Higher education literature includes descriptions of barriers to change at the institutional, administrative, and policy level but little is available on barriers to change existing among university faculty at the course and classroom level.

PERCEIVED BARRIERS TO HIGHER EDUCATION INNOVATION IN SCIENCE INSTRUCTION

Among faculty from around the country, little difference was found concerning barriers to change when one focuses on science courses and programs. The authors and others have conducted 23 national workshop sessions on the topic of barriers to collaboration and innovation with 240 university and college teams over a period of 8 years (1995 to 2003). The workshop activities serve as a component of the National Aeronautics and Space Administration program, NASA Opportunities for Visionary Academics (NOVA), 3-day intensive workshops that explore models for improving undergraduate science instruction. Part of each workshop session was devoted to brainstorming perceived barriers that have to be overcome before changes in science instruction can occur. The major purpose for brainstorming common barrier issues was to sensitize participants to the realization that higher education institutions have both consistent and persistent problems related to change. This divergent thinking process motivated participant teams to develop and share collaborative strategies to overcome barriers that restrict implementing and sustaining change in their home institutions. These higher education faculties (and administrators), from the sciences and education as well as other disciplines (e.g., mathematics and engineering), have consistently raised similar issues over the 8-year period. The issues varied in intensity from consistent problems that are solvable with good planning to critical problems that have no obvious solutions. The perceived barriers also varied according to the size of the institution, the mission of the institution, and the part of the change process the participants were involved in. The identified impediments to innovation in undergraduate science can be sorted into nine major barrier categories: management, coordination, leadership, faculty, student, curriculum, instruction, budget and resource, and accreditation and certification. The nine categories are not exclusive, but work well in prompting discussion and for understanding the extent to which barriers need to be addressed before innovation can be developed, implemented, and sustained in institutions of higher education. One can be better equipped to facilitate change if one first examines the perceived barriers from various perspectives and organizational schemes.

Management Barrier

The management barrier is concerned with issues related to classroom usage and scheduling, administrative turnover, course adoption policies, institutional support for change including resources and time allocated for innovation and change, and shared leadership. Innovative science courses typically need to move away from large lecture halls with separate small laboratories to flexible rooms that accommodate group discussions and collaborative activities. The lack of classrooms and scheduling techniques that are computer-based and flexible to accommodate the space arrangements of innovative courses can become a major problem at many institutions.

In addition to facilities, technology is often a cornerstone of innovative instruction in the science classroom and the lack of access to it can hinder course development and the delivery of the innovation. Immediate access of students to the Internet and the capacity to have real-time electronic exchange of information with each other and the instructors (e.g., data, reports, lesson plans, charts, etc.) as well as state of the art digital probes for data gathering, are all essential in inquiry-driven, constructivist science classrooms. In addition, technologies, such as class personalized response systems, which provide immediate feedback to the instructors in large lectures, is essential for monitoring student progress during real-time class interactions and is a powerful tool in gathering data for formative and summative assessment of course effectiveness.

Changing administration (new department heads, deans, etc.) can be a serious problem if there is no institutional memory to support innovation. Unfortunately, promises made today by one set of administrators are forgotten tomorrow by the new administrators. Many months of work can be nullified overnight.

Until the innovation is approved as a required course in various programs of study, it is vulnerable to unpredictable institutional whims. Faculty governance procedures almost always translate into lengthy adoption time lines (and bureaucratic procedures and paperwork) for new and modified courses and programs. At some institutions, especially large universities, the process can be as long as 2 to 3 years before a course proposal successfully moves through the multilayered approval process. This time lag can be a serious problem to insuring an adequate population of registered students for sustaining the innovation.

Turf wars between the various departments and colleges involved with the innovation can be a serious issue in terms of who gets credit for the student hours, who staffs the course, who pays the bills for the course, and who provides the space. Lack of agreement around these concerns will distract from a rapid approval process and could ultimately inhibit all current and future collaboration. Course and curriculum approval committees, at all levels, will drag their feet if there is discord in the proposing bodies. There needs to be full cooperation among all the units involved. With little tradition in higher education for the faculty from various disciplines to work across department and college boundaries, this can become one of the most serious issues manifesting itself over and over and eventually stopping all progress.

Related to turf issues is the general lack of appreciation of everyone involved concerning the time, resources and creative energy needed for innovation and change. When it is perceived that individuals are being unduly overworked and distracted from other tasks deemed more relevant, serious questions will be raised at faculty and department levels about the worth of the activity. Individual commitments to what is perceived as a non-essential activity are called into question and the lack of fairness between individuals and departments is raised as an issue. Thus, fairness of the assignment of resources and personnel involves effective human resource management (i.e., differential staffing, and assignments as related to reappointment, promotion, tenure, and merit).

Facilitating change occurs best when there is shared leadership. Decisions cannot be made only top down. There must be various avenues for the empowerment of bottom-up decision making. Interdisciplinary faculty teams need to know that their creative work will be addressed fully in the spirit of collaboration across all levels. They need to understand that their superiors are on the same page and will help the faculty overcome resistance to changing the program. The resistance can come from many quarters depending on what vested interests are being questioned—from students, administration at other levels, alumni, and boards of trustees, who, for various reasons, like their program just as it exists, or do not want to risk resources and personnel on untested new programs and courses of study.

It is critical to insure that the planning teams have some members who are highly respected by the decision makers who ultimately approve or disapprove new programs and courses. Young, untenured faculty may have the ideas, energy, and enthusiasm to support innovation but, as a group, they are very vulnerable to those who resist change—whether they are students, faculty, or administrators. The ideal persons to fill the leadership roles are senior faculty who have been very productive in their fields as researchers and teachers, but in the final decade of their careers want, altruistically, to give something back to their society. The key is convincing these individuals that they can play a major role in improving science instruction at the K–16 setting through an evolving science program at their institution. Once this occurs, coupled with the respect and support they can engender from colleagues, administrators, alumni, legislators, and community leaders, the planning team can move full steam ahead knowing that its ideas will have a strong advocate at the highest levels.

Coordination Barrier

The coordination barrier is concerned with issues related to shared workload and decision making of all the vested parties. This can be exacerbated by historical fact that communication rarely occurs between faculty and administrators in education and the sciences. Coordination begins with organizing team planning and writing meetings. Who will organize and sustain the meetings? Who will take responsibility for researching and recording the work of the group? Is the team organized as an informal team or does it acquire status as a formally appointed ad hoc committee? Who decides membership and the terms of membership? Is membership flexible with individuals joining or leaving the team as dictated by time and interest? Workable solutions to these questions will be required to establish long-term working ties for broader collaboration between colleges (and departments) in the university.

The team or committee, if it is going to be relevant, must establish working ties for broader collaboration with college of education science educators who, with expertise in pedagogy and research, can provide missing professional knowledge and skills to the science departments. This is a frequently overlooked criterion for success. In addition, decision makers in the local K–12 school systems can provide an elementary and secondary school perspective that is very important in crafting credible team recommendations for addressing higher education science curriculum issues that are applicable for preparing preservice teachers and enhancing in-service teachers.

Leadership Barrier

The leadership barrier is the process of overcoming a void of leadership—finding the individual or individuals who can provide the skills to promote, implement, and sustain change. A renaissance person is needed to bridge the gaps between disciplines and programs, both within the planning group and within the administration. At the planning team level, if you cannot find a natural leader to do it all, then there are some people who are better at promoting the development of a new vision, while there are other individuals who have highly developed skills to provide leadership for implementing and institutionalizing the change. Getting the right mix of leadership, whether it is one, two, or more people, is essential.

If it does not exist in some form already, it is important to develop a planning structure and organization for change within the institution (to facilitate collaborative efforts). There must be in place a formally recognized process to establish new direction for change within institutions of higher education—new courses and programs that involve interdepartmental and across-colleges cooperation. This is where administrative leadership is critical. There must be an identifiable administrator who is highly respected and willing to counter all aspects of institutional resistance to change. The person must aggressively and faithfully represent the planning team's proposal with peers, central administration, students, the public schools, and alumni. This leader, supported by the planning team as a whole, must also have a vision for establishing the proposed new program as a long-term endeavor that is an essential component of the "big picture"—fulfilling the institution's mission in meeting societal needs, and convincing the decision makers.

Faculty Barrier

The faculty barrier evolves around a series of issues that discourage faculty members from participating in the planning and implementation process. One has to insure that the time and creative effort devoted to new courses and the institution recognizes programs as positive contributions. Young, untenured faculty are particularly vulnerable, but senior faculty can also be affected in terms of teaching load and salary merit. The traditional guidelines and criteria for preparing faculty for tenure and promotion must be flexible enough to accommodate and actually reward the work of planning team members. Merit, promotion, tenure reward systems, and teaching load adjustments must be in place that recognize the work of the planning team before innovation can occur, and more important, for it to be sustained. For example, the publications and conference presentations of innovative pedagogy and curriculum developments need to be accepted by the disciplinary departments as legitimate forms of scholarship. Another example is that co-teaching of science courses must be recognized as legitimate components of teaching loads of education faculty. These factors and many other issues must be addressed. The policies must become explicitly clear to all team members, colleagues, administrators, and so forth, so no doubt exists about the importance of the work and the ultimate reward system.

In some instances a lack of professional respect between administrators and faculty in the college departments of education and the sciences can present a major roadblock to considering change. These problems are far beyond the scope of the team planning process, but if they exist, then the most highly respected member or members of the team (those senior faculty) may choose to discuss the roadblocks with decision makers at higher levels and ask for support. This is high-stakes politics and must be approached most carefully with a well-articulated vision for the future of the program. The vulnerable faculty members must be protected at all costs.

For sure, every effort must be made to inform the traditional, long-term faculty members who do not want change (see no need to change) that the planning process has a high priority for the institution. The same is true for those faculty and administrators who have a different vision. If colleagues disagree, then they can be part of a constructive dialogue sharing their perspectives in an open forum of debate and discussion based on evidence from action research conducted in courses—a long and healthy component of the governance structure of colleges and universities. However, most often resistance comes from a lack of understanding or a fear of change itself. It could be as simple as fear and resistance to technology, working with small interactive groups, or more authentic assessment strategies. Most faculty tend to teach how they were taught, so convincing the recalcitrant faculty member to consider the need for new models of instruction, particularly when no one is currently fulfilling that role, can be a very difficult and long-term process. Change in faculty will not occur unless there is dissatisfaction with existing conceptions of science teaching (Sunal et al., 2001). For this to occur research in courses and student outcomes must become more common. Creating cognitive conflict with existing faculty conceptions of teaching is an important role of successful professional development. Also, there are times when faculty invite the teachers from the K–12 schools, who served on the planning team, to co-teach in innovative courses and programs, rather than inviting other college faculty to become involved. This can be seen as a put-down and cause friction. But ultimately, all parties must understand that fighting the effort behind the scenes in derogatory ways will be considered in a very negative light.

Time for planning and innovation for all faculty (equity in time for planning and innovation) must be agreed on and supported by administrators and colleagues. Workload for those involved in innovation generally far exceeds the time envisioned at the beginning of the activity, so the planning process must be flexible enough to accommodate the instructional and other professional responsibilities of team members. From one semester or even week to the next some faculty will have more or less time to devote to the process. Recognizing the dynamics of the process and developing good will and trust between members will go a long way to moving ahead the planning agenda.

In some institutions, particularly two-year and small private colleges, the lack of full-time, tenured (or tenure-earning) faculty can prove to be a major obstacle to planning for change. Adjuncts, who are minimally paid, in most cases, have little time or interest in planning new curriculum. They typically want to be told what to teach and how to teach, put in their 3 to 4 hours and go home.

Student Barrier

The student barrier includes issues such as fear of science, fear of mathematics, fear of technology, fear of change, inadequate assessment of the background of incoming students, lack of faculty mentorship or leadership of students, lack of student ability or academic preparation for science, mathematics, and engineering, lack of incentives to participate in interdisciplinary innovative programs, and the lack of understanding of student needs (on the part of administrators and faculty). It is important to have the students involved in the planning process from the beginning. Interested students should be recruited for team membership from both the disciplinary and education departments.

Curriculum Barrier

The curriculum barrier issues are concerned with designing an innovation that is congruent with internal and external criteria that must be accommodated by the planning process. Those participating in planning need to recognize the cyclic nature of reform—that planning is a process, not an end in itself. The team must decide what needs to be done first. Next, they have to decide what will last, what is not an ephemeral trend. Members of the team have to take the time early in the process to read extensively the scholarly literature, to share openly with each other and hold brainstorming sessions, to visit other institutions and programs that have successfully redesigned their curricula, and to bring in knowledgeable professionals to share ideas with the team and the broader university audience.

Ultimately, the team will have to decide what is important and doable (based on research and practice). Thus, it is best to move methodically into the process equipped with many viewpoints on how best to plan the innovation. These decisions include whether or not the innovation fits within the constraints of the institution's established general education requirements and degree requirements. Of particular importance is whether or not the innovation lengthens the program of study (adding hours to meet new requirements) or replaces existing coursework. The planning team must seriously consider how the new curriculum proposal impacts on other courses and programs directly or indirectly, and how it impacts long-term faculty expectations for instruction. If the proposal is to replace existing courses or impact other courses (e.g., redefined use of financial resources, space, equipment, personnel [faculty and GTAs]), then colleagues may feel threatened and an extremely difficult turf issue will definitely surface.

The planning team must recognize the power and role of curriculum committees at the department, college, and university levels in supporting or rejecting the innovation. These committees can either serve as preservationist or innovators. A lot depends on how informed they are during the planning process and how generally supportive for the innovation they sense faculty and administrators to be. Thus, it behooves the planning committee to mount a public relations initiative with all the important decision makers at all levels of the institution, preparing them for serious and thoughtful consideration of the request to institutionalize the innovation. In instance after instance, this has been demonstrated to become the unsolvable stumbling block to institutionalizing the proposed changes.

Another issue that the planning team must consider is whether or not there are defined goals for lower-division and upper-division courses in the sciences. If not, how will they go about generating and validating changes with faculty and administrators? If they do exist, then are they congruent with national standards and the vision of the planning team? Is there congruency between the national standards, legislative mandates, and governing board directives, especially where science courses are used for teacher certification? Do the goals call for an appropriate balance between content and methods (or how to combine content and innovative methodology [pedagogy])? Because the innovations incorporated in course changes will be new information for most faculty, serious consideration needs to be focused on staff development opportunities for faculty and administrators both internal and external to the institution. All these curriculum questions and more need to be resolved before proceeding with implementing the new course of study.

The most critical component of successful of curriculum development is field-testing of the innovation and developing evidence that the course of study does what it is purported to do. Who conducts these assessments? What skills are needed for effective assessment and evaluation of innovative curriculum and program outcomes? Possessing good data is very important to convincing individuals outside of the in-group about the value of the innovation. This is particularly important data to provide curriculum committees and administrators.

Instructional Barrier

The instructional barrier issues are directly related to the curriculum issues in the sense of their relationship to teaching and evaluating the curriculum. For example, the lack of pedagogical background for faculty in the disciplines will require considerable professional development, self-study, and practice with science educators to get up to speed. In particular, adapting to inquiry pedagogy, problem-solving small-group instruction (vs. the large lecture/recitation mode of instruction) can become a serious issue for the disciplinary faculty. With de-emphasis on chalk and talk before large groups, another issue could be convincing faculty of the importance of infusing technology into constructivist-based instruction.

Discipline faculty members could enhance their knowledge and pedagogical skills by attending targeted professional development workshops offered, for example, by the NOVA Project (NOVA, n.d.) and Project Kaleidoscope (2003). The NOVA Project, NASA Opportunities for Visionary Academics, is a national network for enhancing science, mathematics, and technology literacy for preservice teachers in the 21st century. The lead institutions of higher education of the NOVA Consortium have focused on developing and sharing innovative undergraduate science, engineering, and mathematics course frameworks. Project Kaleidoscope is an informal national alliance working to build strong learning environments for undergraduate students in mathematics, engineering, and various fields of science.

On the other hand, lack of content background for faculty in education who will be part of the instructional team is another issue that the planning team may have to face. The disciplinary faculty will need to find ways to share their knowledge and expertise in a positive manner that does not alienate the education faculty. In general, the lack of instructors qualified to teach innovative inquiry science courses that are constructivist-based, may prove to become a significant stumbling block to success (particularly in small isolated colleges). The interdisciplinary nature of courses that best meet the needs of preservice teachers can be promoted through sharing expertise between faculty in education and the sciences—the collaborative synergy providing a strong foundation for sustained change.

Budget and Resource Barrier

Budget and resource barrier issues must be considered and resolved before planning gets very far along. Examples of problems that must be faced include the expense associated with innovation and change, the availability of technology and instrumentation, availability of appropriate space (i.e., classrooms and laboratory) for small-group instruction, declining and static budgets that do not permit change, time incentives and resources provided for staff development, lack of technical support, developing competitive proposals for funding, reallocation of resources (e.g., faculty lines, operating funds, equipment funds) for the long term, as well as questions of who controls decisions about the budget and who gets the credit (FTEs).

Accreditation and Certification Barrier

The accreditation and certification barrier issues are important for courses serving as electives or required courses for preservice and in-service teachers. The issues concern fulfilling state board of education accreditation and certification standards, fulfilling local board of education requirements, and meeting NCATE accreditation standards.

RESULTS OF A RECENT STUDY OF BARRIERS TO REFORMS IN UNDERGRADUATE SCIENCE

A recent study of perceived barriers to change involved science, mathematics, and technology faculty at 30 higher education institutions taking part in a national professional development project (Sunal & Hodges, 1997). The results identified barriers similar to those found by Cuban (1990). The most commonly ranked barriers to course level change accounted for 60% of those identified as very important by faculty. They were resources, time, and turf conflicts. Faculty perceived themselves as having little control over these areas. Barriers ranked much less frequently, in the bottom 40%, were (a) students—weak backgrounds, fear of new kinds of instruction, fear of science; (b) personal resistance to change; (c) key personnel—unqualified, uninterested, dead wood, little collaboration; (d) committees—college, department, or curriculum committee approval; (e) lack of training; (f) leadership issues—difficulty in convincing decision makers; (g) changing institution—dwindling resources, credit hours already in the program; (h) tenure and promotion issues; and (i) unavailable curriculum materials. Only in these less frequently ranked areas involving students, personal resistance to change, lack of training, and curriculum materials are barriers under the direct control of the faculty member. Overall, change in higher education was perceived as occurring outside the person and beyond immediate control. Thus for most faculty surveyed, change was thought of as too difficult to accomplish. Although systemic change is a goal in a major institutional reform effort, the place to begin to make immediate and measurable change at the course level is with barriers where personal control is possible and with contextual barriers where neutralization is practical.

In a 2003 follow-up study, institutions who had participated in the professional development, and who had since developed new or modified innovative courses, were reassessed on perceived barriers to innovation during their participation in the annual Leadership Development Conference conducted by the staff of the national professional development project. Twenty-eight institutions of higher education (16 public institutions, 12 private institutions) responded to the post-assessment. Forty individuals, split evenly between education and discipline faculty members, rated the identified issues, ranking them from a 1 (*no problem*) to 5 (*a critical problem that has no obvious solution*). This time the teams had spent up to 5 to 6 years in developing, implementing, and sustaining the innovations. The follow-up study highlights some interesting conclusions. Barriers, perceived and actual, to innovative teaching and course design in undergraduate science change over time. However, barriers do not disappear. Different barriers become critical at each stage of the innovation process. Also, the identified barriers differed between the science discipline and education faculties and between private and public institutions, in several instances.

Even though there were isolated instances reported that management issues were insurmountable, across institutions and faculty, the embedded issues were successfully marginalized by 95% of the institutions, whether public or public institutions. Only in a few instances, such as change in a key administrator, staffing decisions, and turf wars, were there sufficient problems that arose to delay (ranking of 4) or prevent (ranking of 5) the implementation of proposed new courses, without serious modification. Because these data represent only those institutions that had successfully implemented a new course, one would expect that ultimately the management barriers would have been successfully overcome.

Coordination issues remain a problem in public institutions only, particularly among the education faculty. More than 50% of the education faculty still found establishing working ties for broader collaboration between the education faculty and the discipline faculty to be a major issue that was very difficult to overcome. This issue was strongly reflected in the process of organizing team planning and writing meetings.

Leadership issues follow a similar pattern as with coordination issues. Only the public institutions reported a significant problem with leadership. Seventy percent of the public institution faculty members indicated that their institution was resistant to change, ranging from being a consistent problem that is solvable (ranking of 3) to a critical problem (ranking of 5) that has no obvious solution. Of these individuals, 56% ranked the issue a 4 or 5. A similar pattern appears around not having leaders with vision who can promote and implement change—a renaissance person to bridge the gaps between disciplines and programs within the institution. Forty percent rated this issue a 4 or 5. Surprisingly, it was also found that 70% indicated the lack of established processes to set direction for inter-disciplinary cooperation or new collaborative program offerings was a major obstacle that had to be overcome. Forty-three percent ranked this issue a 4 or 5.

A number of faculty issues proved to be continuous problems faced in developing new, innovative courses. Forty percent of the faculty in public institutions thought that their effort was not recognized by the tenure/pro-motion guidelines. This was noted as more of an issue by the disciplinary faculty with 42% ranking the issue a 4 or 5. The highest ranking for educa-tion faculty was a 3 (45%). In private institutions, tenure/promotion was not an issue with only two individuals ranking the issue a 3. All others indi-cated a 1 or 2 ranking. Merit rewards as a form of recognition for the mem-bers of the planning team is ranked consistently a problem across the board. Seventy-one per cent ranked this issue 3 or higher, with about one half indicating a 4 or 5.

A lack of time for planning the new course was seen as a problem by most faculty and institutions, with 71% ranking the issue a 3 or higher (58% a 4 or 5 ranking). The only exception was the disciplinary faculty in private institutions. (It is interesting to note that their collaborators in edu-cation for the most part ranked the issue as a major problem.)

Workload was a major problem identified by 88% of the faculty, 72% with a ranking of 4 or 5. Only the science faculty in private institutions did not see it as a major problem. (Their education collaborators were totally in disagreement with rankings across the board of 4 or 5.)

Other issues that had to be dealt with were faculty fear and resistance to technology (53%), overcoming the trend to teach how one was taught (44%), and long-term faculty who do not want change—see no need for it (68%). Using too many adjuncts instead of full-time tenured or tenure-earning faculty to teach the introductory courses was an issue for 53% of the respondents and spread evenly across types of institutions. An underly-ing issue that was only reflected in public institutions was the lack of profes-sional respect of and between administrators and faculty in education and the sciences, particularly from the education faculty (55%) who ranked respect as a major or severe problem.

Student issues that persisted throughout the development and imple-mentation phases include fear of science (47% ranked 4 or higher), fear of mathematics (50% ranked 4 or higher, with 53% of the responses a 5), fear of technology (75% ranked 3 or higher), lack of student ability or aca-demic preparation (32% ranked 4 or higher), lack of incentive to participate in interdisciplinary science courses (38% ranked 4 or higher), fear of change (65% ranked 3 or higher), and lack of faculty mentorship to develop student leaders (82% ranked 3 or higher with 38% 4 or higher). Part of the problem was the lack of understanding about student needs on the part of administrators and faculty (60% ranked 3 or higher), with particular concern reflected by the education faculty where several ranks of 4 appeared. This probably relates to inadequate assessment of incoming students being ranked 3 or higher by 50% of the respondents.

There are several curriculum issues ranked high by the respondents. Issues include fitting the new course within the constraints of established general education requirements and degree requirements (47% ranked 3 or higher) and impact on the total number of hours in the program of study (75% ranked 3 or higher, with 67% of these individuals indicating a ranking of 4 or 5). The education faculty in both the public and private institutions and the discipline faculty in public institutions were the most sensitive to these issues. In both cases, the private science faculty did not note an impact. More than 60% (rank of 3 or higher) indicated a general lack of opportunities for staff development that promote the development of innovative, constructivist-based new courses. This appears to be more of a problem with discipline and education faculty in public institutions and with education faculty in private institutions. Except for the discipline faculty in private institutions, the respondents indicated that a big issue was the preservationist view held by curriculum committees at all level.

Another interesting finding was the lack of defined goals that support innovation for lower division courses in public institutions (50% of the respondents rank the issue 3 or higher, with several 4's and 5's). This was not identified as an issue in the private institutions. A similar pattern was reflected around the issue of how to develop a balance of content and pedagogy that supports innovation. None of the private institution faculty saw this as an issue. In public institutions, 67% of the respondents indicated this to be an issue between the discipline and education faculties (rank of 3 or higher). On the other hand, the education faculty in private institutions noted a lack of congruency between national standards and internal mandates that drive curriculum development (57% ranked the issue 4 or higher). Public institution faculty never ranked the issue higher than a 3. In a similar vein, accreditation and certification issues for teacher education (such as state board of education requirements, NCATE accreditation and local board of education requirements) were not seen as barriers to change, but in fact, offering support for meeting these expectations.

Another issue was the preservationist role played by curriculum committees at all levels. Eighty percent (rankings 3 or higher, with 42% of the individuals with rankings of 4 or 5) saw this as a problem in institutionalizing the new innovative course. A consistent problem that was identified by faculty, except for the private institution discipline respondents, was the lack of skills needed for effective assessment of the constructivist-based curriculum and program outcomes, with 40% ranking the issue 4 or higher.

Curriculum is interpreted through the instruction offered students. The most critical of the instructional issues that surfaced during the planning and implementation phases was the lack of pedagogical background by the faculty in the science disciplines. Sixty-five percent of the respondents, across all faculties and institutions, ranked the problem 3 or higher (with 30% of the rankings at the 4 or 5 level). On the other hand, particularly the discipline faculty members did not perceive the lack of content background for faculty in education as a problem. Seven of the education faculty (37%) did rank the issue a 3 but nothing in the 4 or 5 range. The discipline faculty rankings were typically a 1 or 2.

Another problem that arose was adapting instruction to hands-on, problem-solving small-group instruction (versus the large lecture/recitation mode of instruction). This was a particular issue with the education respondents, with 63% ranking the issue 3 or higher (45% a ranking of 4 or 5).

Budget and resource issues dominate the landscape of change in both private and public institutions of higher education. It is expensive to develop and implement constructivist-based, innovative courses, both in terms of faculty time and the need for space and resources to support the effort (44% ranked the issue a 4 or 5). In a time of declining or, at best, static higher education budgets, it is very difficult to support real change. across the board, 56 percent of the respondents, indicated declining budgets to be a crucial problem. More than one half ranked this issue a 4 or 5. This is reflected in the unavailability of technology and instrumentation that support the innovations (47% ranking the issue 3 or higher), the lack of technical support for the preparation of quality, competitive proposals for funding to sustain the change (65% ranking the issue 3 or higher, with 70% of these respondents indicating 4s and 5s), availability of appropriate space for small-group instruction and long-term laboratory projects (59% ranked this a 4 or 5), and the lack of time incentives and resources provided for staff development (44% ranked this 3 or higher).

Various turf conflicts are a source for a number of serious issues. Who controls the budget decisions for support of innovation can be a problem (44% ranking this issue a 3 or higher). The reallocation of internal resources for the long term, including faculty lines, operating funds, equipment funds, and training funds, is perceived as a major stumbling block by the planning groups, with 47% ranking this category a 4 or 5. Ultimately, who gets the credit for the innovation (FTEs) determines whether or not the new course is institutionalized (over one half see this as a major problem with rankings of 4 or 5).

SUMMARY AND IMPLICATIONS FOR CREATING CHANGE IN UNDERGRADUATE SCIENCE

Several important outcomes are evident in conducting reform in undergraduate science. Major barriers are perceived at all institutions. Some critical barriers appear to have a more local character. Major barriers are commonly perceived as related to leadership, faculty, student, curriculum, instruction, and budget and resources concerns. Many of these issues never go away. Even through several years of work on the new curriculum, the issues persist and make the work of the planning group very difficult in many instances. In spite of the major hurdles, the individual planning teams continue with the work, ultimately designing, implementing, and institutionalizing the innovations.

Less critical barriers are related to management, coordination, and accreditation issues. Change in faculty will not occur unless there is dissatisfaction with existing conceptions of science teaching. Innovative pedagogical ideas for course change must be made clear and plausible through a variety of collaborative experiences if faculties are to attempt their use. The use of a team approach involving several types of professional development approaches in a long-term program is effective in creating innovative change in course design and teaching. Effective change is enhanced when administrators value the innovative ideas and skills faculty bring to their courses. Leadership must be more flexible and personal to adjust to individual faculty. Faculty members cannot operate as isolated individuals and be effective teachers in higher education. Change likely will not take place unless faculty work with their colleagues to negotiate and create common understandings related to reform. Leaders cannot make change alone.

While most higher education institutions attempt to implement innovative change in coursework, the success is limited and elements are unconnected. Their effectiveness can be improved. It is time to move to more effective teaching in higher education. A collaborative team approach, effective faculty development, and growth in extending the idea of research into the dissemination and teaching aspect of the science discipline all form the foundation of reform in undergraduate science courses.

IMPLICATIONS AND FUTURE RESEARCH FOR REFORM IN UNDERGRADUATE SCIENCE

In beginning the process of change—developing an innovative new course in science for undergraduates, particularly teacher education candidates—what one can do to maximize the chances for success depends on where one goes for help. Broad-based approaches to build on include the following:

1. Developing an interdisciplinary *collaborative team* that is constructed from a group of faculty, students, administrators, and public school personnel that join together to develop a vision and mission statement for the change. This team may change over time but each member is selected for the unique insights that he or she can provide for the planning process and implementation of the new course.

2. Identifying an *institutional constituency* (administrator, faculty, school systems, feeder colleges, etc.) that will support and foster the work of the collaborative team.

3. Linking up with various *groups and networks* that are involved in change, such as the NOVA Network of institutions, by examining the exemplary curricular materials they produced and attending organized meetings where one can gain intellectual support for ideas for change.

4. Making as many *site visitations* that can possibly be scheduled for the purpose of seeing firsthand innovative courses being taught under a variety of conditions including institutions similar to one's own setting.

5. Designing and conducting *action research* projects where one learns to assess the effectiveness of various aspects of one's curriculum in meeting stated goals and objectives, particularly one's learning outcomes for students.

6. Relying heavily on national, state, and institutional standards to set the direction for an innovative course, and as an authoritative source of evidence to confront and dilute criticisms that will arise from within and outside of the institution about the relevance of the course for the target audience.

7. Reading widely through conducting a thorough *literature and Internet search* of relevant ideas that support new course innovations, and discussing these ideas thoroughly within the collaborate team, and the broader institutional audience.

AUTHOR NOTE

This work was in part supported by NASA Opportunities for Visionary Academics (NOVA), a program funded by the National Aeronautics and Space Administration, although the views expressed here are the authors' only.

REFERENCES

American Association for the Advancement of Science. (1993). *Benchmarks for scientific literacy: Project 2061.* New York: Oxford University Press.

Barinaga, M. (1991). Scientists educate the science educators. *Science, 252,* 1061–1062.

Cohen, D. (1988). *Teaching practice: Plus a change* (Issue Paper No. 88-3). East Lansing: Michigan State University, National Center for Research on Teacher Education.

Cuban, L. (1990). Reforming again, again, and again. *Educational Researcher, 19*(1), 3–13.

Cuban, L. (1999). *How scholars trumped teachers: Change without reform in university curriculum, teaching, and research, 1890–1990.* New York: Teachers College Press.

Driver, R. (1986). *The pupil as scientist.* Philadelphia: Open University Press.

Fedock, P., Zambo, R., & Cobern, W. (1996). The professional development of college science professors as science teacher educators. *Science Education, 80*(1), 5–19.

Fullan, M. (1999). *Change forces: The sequel.* Philadelphia: Falmer Press.

Gilmer, P., Barrow, D., & Tobin, K. (1993, April). *Overcoming barriers to the reform of science content courses.* Paper presented at the annual meeting of the National Association for Research in Science Teaching, Atlanta, GA.

Magner, D.K. (1992, October 21). A booming reform movement for introductory science courses. *The Chronicle of Higher Education,* pp. 17–18.

National Research Council. (1995). *National Science Education Standards.* Washington, DC: National Academy Press.

NOVA. (n.d.) *NASA Opportunities for Visionary Academics.* Retrieved August 18, 2003, from http://nova.ed.uidaho.edu

Project Kaleidoscope. (2003). Retrieved August 18, 2003, from http://www.pkal.org

Sigel, I.E. (1985). A conceptual analysis of beliefs. In I.E. Sigel (Ed.), *Parental belief systems: The psychological consequences for children* (pp. 345–371). Hillsdale, NJ: Erlbaum.

Sunal, D., & Hodges, J. (1997, January). *Summary of national reports of innovative changes in college science teaching.* Presentation at annual national conference of the NOVA Leadership Forum, College Park, MD.

Sunal, D., Bland, J., Sunal, C., Whitaker, K., Freeman, M., Edwards, L., et al. (2001). Teaching science in higher education: Faculty professional development and barriers to change. *School Science and Mathematics, 101,* 246–257.

Tobias, S. (1990). *They're not dumb, they're different: Stalking the second tier.* Tucson, AZ: Research Corporation.

CHAPTER 4

SCIENCE EDUCATION REFORM

Factors Affecting Science and Science Education Faculty Collaborations

William S. Harwood

ABSTRACT

Collaborations between science and science education faculty members have been tried for many years. Some fail and some are successful, but too often collaborations have only local effect (do not transfer beyond the collaborative group) or fail to continue beyond the initial collaborative partnership. Lately, there are new forces that encourage cross-disciplinary collaboration on behalf of improving the preparation of future science teachers. This chapter presents a series of factors that influence the formation and success of collaborative partnerships between science and science education faculty. These factors are classified under two areas: internal factors and external factors that can encourage or discourage successful ongoing collaboration.

Reform in Undergraduate Science Teaching for the 21st Century, pages 53–68
Copyright © 2004 by Information Age Publishing

INTRODUCTION

Calls for the reform of teacher preparation are not new (American Council on Education, 1999; Bestor, 1985; Conant, 1963; Haycock & Brown, 1993; Kramer, 1991; Lieberman, 1995; Rickover, 1960; Tafel & Eberhart, 1999). Historically, criticism of teacher preparation has been directed at schools of education. A new element in the discussion of teacher preparation reform, however, is the perspective that the whole university is responsible for the preparation of teachers (Fallon, 1999; Riley, 1999; U.S. Department of Education, 2000). One result of this recent discussion is to view discipline area departments from the arts and sciences as sharing responsibility for teacher preparation rather than looking solely to schools of education, particularly as regards teacher content knowledge (McDiarmond, 1992).

If we accept the premise that responsibility for preparing future teachers is shared across the whole university, then we must develop strong collaborations among the faculty from arts and sciences and the school of education. A number of such collaborations have been developed (Beckmann et al., 2001; Carbone, 2000; Cole, Ryan, Serve, & Tomlin, 2001; Davis et al., 2003; Duggan-Haas, Smith, & Miller, 1999; Haruta & Stevenson, 1999; Harwood, 2003) and have experienced varying levels of success. The goals of such collaborations may include any of the following areas of focus: preparation of elementary teachers in science, preparation of secondary science teachers, professional development of in-service teachers, and preparation of science majors and college science teachers.

This chapter explores some of the key factors that influence the success of such cross-disciplinary collaborations. These factors can be divided into internal factors and external factors, much like Fullan's inside and outside influences on education reform (Fullan, 2000). Internal factors represent issues and circumstances that exist within the campus setting and community. External factors are those that arise from institutions and organizations off-campus. Depending on context, several of the factors discussed below can have either a positive or negative influence on the success of collaboration (see Table 4.1).

A BRIEF DISCUSSION OF CROSS-DISCIPLINARY COLLABORATION

As indicated above, there is a growing sense that university and college communities as a whole are responsible for teacher preparation. There have been a variety of calls for action that encourage cross-disciplinary collaboration as the means to achieve the goal of improved science teaching and learning (Boyer Commission, 1998; Edmundson, 1991; National

Table 4.1.   Internal and External Factors Affecting Science/Science Education Collaboration

Internal Factors	External Factors
Administrative support	Institutional accountability
Stakeholders present	Administrative support
Awareness of need for reform	National and state standards
Shared goals	External funding
Awareness of science education research base	
Faculty development programs	
Individual relationships	

Research Council [NRC], 2001). Research is just beginning to be published that demonstrates the challenges and rewards from collaborative efforts at reform. Seymour (2001), for example, provides a discussion of several theoretical models of science teaching reform. She focuses attention on the academic department and its culture as a key element required for improving college-level teaching practice. In addition, several external groups and organizations are identified that can have profound influence on the process of science teaching reform. In contrast to focusing on departments and groups that influence change, Gess-Newsome and colleagues have provided a model of reform that focuses on the individual faculty member (Gess-Newsome, Southerland, Johnston, & Woodbury, in press; Woodbury & Gess-Newsome, 2002). These researchers focus on the need for change to proceed at a gradual pace that allows individuals to reflect upon their teaching and to engage in a process of conceptual growth regarding teaching. This view of a graduate collaborative process for transforming undergraduate science teaching is echoed by others (Taylor, Gilmer, & Tobin, 2002).

Documenting the success of collaborations between science and science education with solid data and analysis continues to be a challenge (Williams, 2002). Wyckoff (2001) reports early and positive results from a collaborative program (Arizona Collaborative for Excellence in the Preparation of Teachers [ACEPT]) at Arizona State University. Reports from other programs that began in the late 1990s suggest that the collaboration itself is valued by the faculty participants (Beckmann et al., 2001; Carbone, 2000; Cole et al., 2001; Duggan-Haas et al., 1999; Haruta & Stevenson, 1999). These reports identify positive outcomes from the collaboration with regard to grant writing and obtaining grants that foster change as well as courses that are undergoing collaborative reform. It remains to be seen what outcomes for student learning come from these efforts.

Based on the literature and my own experience, I present here 10 factors that can affect the success of cross-disciplinary reform efforts. In this regard, I am focusing on the college setting and will not be addressing the K–12 school context directly. The factors discussed below have been divided into internal and external factors (Table 4.1). Note that *Administrative support* is listed in Table 4.1 as both an external and internal factor. I double listed this one factor because of the ambiguous situation for some administrators, especially department chairs. In collaborations between science and science education, department chairs may have multiple roles that place them on both the inside and the outside of the collaborative community. I discuss this factor only once, however. Also, each factor raises issues that can have either a positive or negative effect on the success of collaboration.

EXTERNAL FACTORS

Institutional Accountability

One motivating factor for institutions to encourage cross-disciplinary collaboration is a change in the landscape of criticism for teacher education programs (Fallon, 1999). If one feels that teachers are underprepared in a discipline area, it is no longer believed to be the fault of the school of education solely. Rather, the institution as a whole is viewed as failing that teacher (McDiarmond, 1992). The discipline area department is as much a part of such problems as is the school of education. This external accountability provides a pressure for deans and other administrators to encourage discipline area faculty and their counterparts in the school of education to work together. A concern, however, is that senior administrators may not have a genuine interest in implementing the results of faculty collaborative effort. This leads to reports and recommendations gathering dust on a shelf. If collaboration is not perceived by the faculty to be valued by senior administration, it lessens the likelihood that faculty members will choose to participate in a new collaborative effort.

Administrative Support

Support from key administrators is a requirement for successful implementation of collaboration. This can occur through support for reforms of courses based on issues such as retention of majors (Haruta & Stevenson, 1999), science for all (Beckmann et al., 2001; Duggan-Haas et al., 1999; Seymour, 2001), or improvement of the preparation of future college science teachers (Williams, 2002). Support can come from university or college presidents, but deans and department chairs are perhaps more important in this regard (Seymour, 2001). These are the administrators who are able to gather resources, human and financial, that can support initial collaborative efforts. They also set the "tone" for their units in ways that can encourage reform-minded faculty to invest themselves in collaborative efforts of reform.

At Indiana University, administrative support has been essential for faculty collaboration through a program called the 21st Century Teachers Project (Harwood, 2003). This project focuses principally on the arts and sciences courses for preservice teachers (both elementary and secondary preservice teachers). The project was initiated by Indiana University President Myles Brand and is a key element in the strategic plans of both the College of Arts & Sciences (COAS) and the School of Education. The deans of COAS and Education actively encourage participation and internal funds were made available for intensive summer work on course revision.

Early on, faculty raised concerns regarding this program. They worried that their team's plans might not be enacted for reasons that are structural, such as our budgeting system (resource-centered management). The deans have strongly and publicly indicated that they view their task as administrators to find ways for overcoming such obstacles in order to support faculty-approved goals. This is a powerfully motivating message to faculty who do not want to see their work become a report that gathers dust on a bookshelf. This stance also indicated to the faculty that the project is not a top-down directive. Such theories for change, when used alone, are ineffective (Seymour, 2001). It also demonstrated to the faculty that they are "in the driver's seat" on this collaborative project, giving it the feel of a grassroots effort that empowers faculty. Together, a faculty-led effort with administrative support presents a powerful combination more able to effect change than either process alone (Fullan, 2000; Seymour, 2001).

STANDARDS

One of the external factors that encourages collaboration and informs collaborative efforts is the development and adoption of national and state standards in science. The National Science Education Standards (NRC, 1996) indicate that teachers of science must plan an inquiry-based science program for their students (p. 30, Teaching Standard A). At the same time, the standards indicate, "What students learn is greatly influenced by how they are taught" (p. 28). The implication, then, is that the science courses that preservice teachers take must use methods of instruction that are inquiry-based. Making such a change at large institutions requires input and investment from both arts and sciences faculty and faculty in education. Based on these insights from the National Standards, many institutions developed partnerships involving K–12 schools, education programs, and science departments to improve the preparation of elementary teachers in science (Shroyer, Wright, & Ramey-Gassert, 1996; Williams, 2002).

Many institutions have also worked on reforming courses that science majors take (see Wyckoff, 2001, for one example). Individual faculty may make substantial and successful efforts to improve the student learning that takes place in their course. Documenting these improvements is not always accomplished or may not have the rigor that educational researchers prefer to see. Consequently, good ideas are not heard or adopted beyond the institution. Larger programs, such as ACEPT (Wyckoff, 2001) and others (Davis et al., 2003; Seymour, 2001; Williams, 2002) do gather data that can demonstrate the impact of their reforms. These data help to demonstrate in what ways the members of the collaboration value partnerships with education researchers.

EXTERNAL FUNDING

In the world of the research university there is, perhaps, no more important influence than external funding. An external grant provides the resources necessary to accomplish a project. Receipt of a grant also provides an important validation of the worthiness of an idea and, by extension, the worthiness of the participants. Landers, Weaver, and Tompkins (1990) identified external funding for collaborative partnerships as one way to address real or perceived inequities in the promotion and tenure process.

Funding agencies can have a profound influence on building collaborations focused on college science teaching and learning. The efforts of funding agencies such as the National Science Foundation are very helpful. There has, for example, been a long-standing policy of providing science researchers with supplemental funds if they extend their work into an educational venue, or if they bring teachers or undergraduate students into their laboratories. Large collaborative projects such as the Centers for Excellence in Teacher Preparation (CETP) are now beginning to show results (Davis et al., 2003; Wyckoff, 2001) as are other collaborative projects that contain elements aimed at college science teaching (Seymour, 2001; Williams, 2002). There are also new initiatives that focus on collaborative partnerships (Davis et al., 2003; Mervis, 2002). Additional good news is the developing interest of other funding agencies (Howard Hughes Medical Institute, for example) in promoting collaboration between science and science education.

One challenge, however, is in how collaboration is viewed by the funding agency, grant reviewers, and the grant writing team. One scientist, the principal investigator for a funded project, once told me that the educational component was important because that's "how we get the money." Clearly, that component of the project was not as highly valued to him as other aspects of the project. But the story raises the concern regarding science education research and development being considered an "add-on" rather than a valued essential component of the project.

INTERNAL FACTORS

Stakeholders Present

In approaching the K–16 education reform many states have developed a variety of collaborative partnerships (Tafel & Eberhart, 1999). A concern with these large statewide collaborative efforts is to assure that the voices of a wide range of "stakeholders" are heard. In focusing more tightly on reform in science education, it is also important that we support a diversity of perspectives at the table (Fullan, 1999). In the 21st Century Teachers Project described above, three voices are at the table: arts and sciences faculty, education faculty, and K–12 faculty (Harwood, 2003). The result is lively discussion that provokes team members and produces innovative ideas (Stacey, 1996).

Sometimes, however, collaborations are missing an important voice at the table. An example of one such incomplete partnership comes from a project funded by a private foundation. The project focuses on an initiative between an urban school system and the science departments at the local major university. The science education faculty were specifically excluded from participation in that project apparently because the research scholarship and expertise of the education faculty was not valued by the scientists at the foundation or the university (see *Awareness of Science Education Research*, below). Buck (1999) has pointed out that confusion over the roles of science and science education faculty can be a challenge for collaboration.

In my view, the lack of involvement of science educators unnecessarily hampers the project and lessens its likelihood for long-term sustainable success. The model for professional development chosen by the urban project described above is contrary to the expectations expressed in the National Science Education Standards (NRC, 1996, Professional Development Standards D). The standards call for a program that is coherent and integrated in ways that should include all three groups: science faculty, education faculty, and K–12 faculty. A similar criticism can be made of professional development programs that emanate from schools of education but that do not bring in faculty from the sciences.

Awareness of the Need for Reform

One of the first challenges for science/science education collaboration is for individual faculty members to realize that there is something that needs to be done regarding the way in which science is presented to students (Gess-Newsome et al., in press; Seymour, 2001; Taylor et al., 2002). This is especially true when looking at the science courses students take at the college level. Craig Nelson (1997) suggests that teaching is like a form of "love." First, different concerns emerge at different levels of mastery and maturity. Second, it is a taboo subject where our academic culture pretends that there is nothing to be known that can make a major difference in teaching. This silence makes engagement on the subject of student learning and faculty teaching in science difficult.

Thus, in choosing faculty members to work on collaborative efforts for reform, it is important to find those people who agree that change is needed. A willingness to seek reform is clearly associated with one's beliefs about teaching and learning (Ballone & Czerniak, 2001; Gess-Newsome et al., in press; Slate, Jones, & Charlesworth, 1990). They need not agree on direction or degree of change, but should be unwilling to maintain the status quo. In this regard, administrative support can come into play as a positive or negative influence. Deans and department chairs that create an atmosphere that fosters discussion of issues provide much needed support for the process of collaborative change.

Shared Goals

Gore (1987) suggests that collaboration between education and the arts and sciences is not one that is grounded in shared meaningful goals and is, therefore, not worth pursuing. More recently, Fullan (1999) states that shared goals are not required for successful collaboration. A like-minded consensus does not provide access to different perspectives and ideas. At the same time, however, confusion regarding the collaboration's purpose that is left unresolved can have a negative impact on the success of the collaboration (Buck, 1999). Fullan (1999) suggests that the anxiety produced by several perspectives, if it occurs in a supportive and trust-building environment, can be a key driving force toward innovative solutions to problems identified by the collaborative group.

Still, the idea that participants must have a strong realization of shared goals persists and can be a challenge for efforts at the college level. This was alluded to in the factor, Stakeholders Present, where some science faculty perceive, with some justification, that faculty in the school of education do not have an interest in teaching and learning issues in the college science classroom (Redish, 1999). Redish (2001) comments, "The education schools have other fish to fry" (p. 1). This general statement may be true for some institutions and, where true, I would encourage those schools of education to consider part of their mission to include the college level. The teaching and learning that preservice teachers experience at the college level has a profound impact on their practice in their future precollege classroom. Most secondary science teachers were science majors and many end up teaching a "watered down" version of freshman science as their course (Shumba & Glass, 1994). That is, they teach the way they were taught. Preservice elementary teachers report that part of their antagonism toward science comes from their experience with secondary science courses (Reiff, 2003). College science teaching, then, is part of a feedback cycle affecting the whole K–16 system.

The goods news, based on my experience, is that many science education faculty members are interested in working with college science faculty on college science issues. Certainly the literature is full of projects undertaken by science education faculty that explore issues of learning science with college students. These are, after all, the courses that all secondary teachers are taking to prepare them in their content area. Indeed, science faculties also teach the science courses taken by preservice elementary teachers (Gess-Newsome et al., in press; Shroyer et al., 1996; Woodbury & Gess-Newsome, 2002). Moreover, a number of educational researchers have specific interests that focus solely on issues of teaching and learning at the college level.

Awareness of Science Education Research

In our 21st Century Teachers Project (Harwood, 2003) we have found that one result of having the three faculties at the table is a change in the nature of the conversation. Education faculty have reported to me that in previous formal exchanges with arts and sciences faculty members, they feel the need to explain the K–12 setting and issues. In other words, they serve as apologists for K–12 teachers. With the K–12 faculty members at the table, their perspective is delivered without translation or interpretation. This allows the education faculty members to bring in their scholarship and understanding of the research literature to aid in discussion of issues. It is a surprise, however, for some of the arts and sciences faculty members to discover that there *is* a solid research base in education (Nelson, 2000; Wyckoff, 2001). Indeed, some science faculty members may believe that there is little or no body of knowledge (King, 1987) and that they will need to develop it on their own (Redish, 1999, 2001). This attitude appears connected with the popular view expressed by King (1987, p. 6) that, "education is everyman's area of expertise." This perspective arises because everyone has had experience as a student in a school setting. Individuals therefore believe their experience makes them knowledgeable about issues of teaching and learning. This belief may also extend to a view that little is required to learn how to conduct educational research—a possible holdover from the sloppy work of some early education researchers (King, 1987).

Faculty Development Programs

Like many universities, Indiana University provides a variety of faculty development programs to which faculty members may apply or be nominated to participate. Some of these are fellowships and are competitive, thereby providing recipients with a level of distinction. A common goal of all these programs is to provide faculty from across a campus the opportunity to reflect upon and develop specific courses and improve their teaching practice.

Such programs often seed later opportunities for collaboration. Faculty members in these sorts of programs are typically in a cohort and have the rare opportunity to discuss teaching issues and methodology. They share concerns and ideas with each other as well as the stories of their classroom experiences. The bonds made between the faculty participants from disparate departments are remembered and valued when other opportunities to collaborate on teaching and learning issues come up.

Many universities have offices dedicated to improving faculty members' teaching effectiveness. (For example, see *Center for Teaching Excellence*, 2002. James Greenberg, an extraordinary scholar and teacher of teachers, directs this program at the University of Maryland.) Faculty can seek out these professional staff and/or faculty colleagues for assistance. Occasionally, faculty members with a poor track record of teaching are sent to get help from their department chair.[1]

Another supportive element is the opportunity to examine and discuss one's teaching in a scholarly way. At Indiana University, as elsewhere, this comes through a program of Scholarship of Teaching and Learning (*SOTL*, n.d.). SOTL is a national movement and supported by organizations such as the Carnegie Foundation. At Indiana University there are seminars scheduled throughout the year where faculty can report on their SOTL projects. Moreover, there are workshops and one-on-one support for faculty who wish to develop a study of their teaching practice and its impact on student learning. One challenge in these efforts is the uneven quality of the studies. Education researchers view many of these SOTL studies as poorly designed or superficial. This is to be expected for faculty in arts and sciences who are unused to the techniques and instruments of educational research and, as discussed above, may be unaware of the research literature in education. If SOTL is understood as a learning process for arts and science faculty members, then perhaps the education faculty can feel more comfortable with this beginning. If handled as a process of conceptual change, involvement in SOTL projects may help dispel faculty myths regarding the ease with which anyone can do quality educational research.

Individual Relationships

Even when all necessary stakeholders have been invited to participate, it is crucial that there is a sense of mutual respect and trust for a collaboration to succeed (Buck, 1999; Fullan, 1999). Individuals can be identified who have a shared sense of a need for change and who are willing to put their time and energy into making changes. An important aspect of this is a willingness to make changes in their own beliefs and practices. Difficulty in implementing change can occur because individuals have not developed new beliefs regarding science teaching and learning (Gess-Newsome et al., in press). Here again is an area where administrators, who are often in a position to identify participants for collaboration, can provide helpful support.

The individuals engaged in the collaborative must also work toward demonstrating that each collaborator is a valued member of the group. Individuals need to be valued for their ability to bring ideas and information that will move the collaborative toward its goals. Basing a value on individuals according to where one works or whom one teaches can be harmful to the successful establishment of a collaborative community (Davis et al., 2003). This is not to suggest that collaborations should not have an organizational structure, but rather that the structure should be one that explicitly values its members.

CONCLUSION

Among the challenges that remain is the need to develop solid research results regarding the products and impact of collaborative reform on students (Williams, 2002). Most studies understandably have focused on describing a collaborative and discussing some measures of its success, including the success of the process. Several such studies are cited in this paper and the work of those researchers has helped me identify the factors discussed here. One wonders, however, which factor among the several discussed in this paper is most essential.

I have suggested in this paper that a successful collaborative must have administrative support but be led by a genuine team of stakeholder-participants. This assertion needs to be examined more deeply. Successful, unsuccessful, and partially successful collaborations can all provide pertinent information regarding the relative importance—or weight—each factor has on the shape of the whole collaboration.

For example, trust and respect among participants is an important component of any collaboration that is embedded in the factor, Individual Relationships. Davis et al. (2003) describe the primary mode of failure for a partially successful collaborative as resulting from participants' lack of respect for each other and lack of acceptance of members of different stakeholder groups as equals. In this study the participants have shared goals for their collaborative along with a focus on standards, receipt of external funding, and administrative support. This suggests that individual relationships are a key factor without which a collaborative is likely to fail to fully meet its goals. Further research may reveal how some other factors can be key components of either failure or success for a collaborative of this type. Understanding both circumstances, success and failure, in collaborative reform efforts may help unlock the mystery regarding why innovations developed at one institution are not readily adapted (transferred) to other institutions (Fullan, 1999; Seymour, 2001).

Many science faculty members care deeply about the issue of science education reform at all levels. Their concern is becoming focused through a variety of individual and partnership efforts (Duggan-Haas et al., 1999; Gess-Newsome et al., in press; Seymour, 2001; Williams, 2002; Wyckoff, 2001). In these partnerships faculty often are delighted to discover colleagues in science education with knowledge to share and who can work effectively to move ideas forward. This increased interest in collaboration on issues of science teaching and learning bodes well, I feel, for the development of future scientists.

AUTHOR NOTE

During the past 15 years, I have taught at a private liberal arts college, a regional public college, and two public research universities. I have spent 12 of the past 15 years at Research I institutions. Both universities are well regarded for the quality of their faculty in science and education and both of these research universities have developed exciting collaborative efforts that address science learning at all levels, K–16. Each institution, however, has a distinct culture, history, and set of challenges and opportunities that inform the nature and long-term success or failure of their collaborative efforts.

I have functioned as a faculty member and as a campus-level administrator in both research institutions. These different positions helped me to identify and look across some common elements that affect the ability of faculty to successfully engage in collaborative partnerships regarding science teaching and learning. I am involved in several collaborations involving science and science faculty members (see Harwood, 2002) and currently administer a collaborative effort directed at improving teacher preparation in arts and sciences courses (Harwood, 2003).

Finally, it may be important to a reader to know that I am one of those faculty members that has crossed over the bridge between science and education research. My PhD thesis work was in an area of basic chemical research investigating multiple bonds between metal atoms. It was with my postdoctoral work that I began to become involved in science education. My involvement in science education research increased when I took a position in the chemistry and biochemistry department at an eastern research university. There, I worked on issues of science education with state committees and university system committees regarding core learning goals in science for high school students. Although my academic appointment remained fully in the Department of Chemistry & Biochemistry, two graduate students in science education completed their doctorates under my direction. I came to Indiana University as an administrator; however, when moving into a full-time faculty position, it made the most sense for my academic home to be in the Science Education Division of the Department of Curriculum & Instruction. I also hold an adjunct appointment in the Department of Chemistry and have taught one course per year in that department. Thus, I have grown, as have research faculty members from many science disciplines, to a point where I look for connections among the disciplines and try to focus on issues that affect teaching and learning in science more generally.

NOTE

1.   Really! I am aware of cases involving full professors. No one enjoys being a poor teacher or being viewed as such by most of one's students.

REFERENCES

American Council on Education. (1999). *To touch the future: Transforming the way teachers are taught.* Washington, DC: Author.

Ballone, L. M., & Czerniak, C. M. (2001, December). Teachers' beliefs about accommodating students' learning styles in science classes. *Electronic Journal of Science Education, 6*(2), Article 3. Retrieved August 10, 2003, from http://unr.edu/homepage/crowther/ejse/balloneetal.pdf

Beckmann, S., Davion, V., Desmet, C., Harrison, S., Hudson-Ross, S., Mewborn, D. S., et al. (2001, March). *Expanding the "Great Conversation" to include arts and sciences faculty.* Paper presented at the annual meeting of the American Association of Colleges for Teacher Education, Dallas, TX. (ERIC Document Reproduction Service No. ED452152)

Bestor, A. (1985). *Educational wastelands: The retreat from learning in our public schools* (2nd ed.). Urbana: University of Illinois Press.

Boyer Commission on Educating Undergraduates in the Research University. (1998). *Reinventing undergraduate education: A blueprint for America's research universities.* Stony Brook: SUNY Stony Brook. Retrieved August 10, 2003, from http://naples.cc.sunysb.edu/Pres/boyer.nsf/

Buck, G. A. (1999). *Collaboration between science teacher educators and science faculty from arts & sciences: A phenomenological study.* Paper presented at the annual meeting of the National Association of Research in Science Teaching, Boston.

Carbone, R. E. (2000, February). *Collaborations between the College of Arts and Sciences and the College of Education at Clarion University of Pennsylvania.* Paper presented at the annual meeting of the American Association of Colleges for Teacher Education, Chicago. (ERIC Document Reproduction Service No. ED440070)

Center for Teaching Excellence. (2002). Retrieved August 10, 2003, from the University of Maryland Web site: http://www.cte.umd.edu

Cole, D. J., Ryan, C. W., Serve, P., & Tomlin, J. A. (2001, June). *Collaborative structures between the Colleges of Education and Human Services and Science and Mathematics.* Paper presented at the combined Standards-Based Teacher Education Programs Conference of the American Association of Colleges for Teacher Education and the Council for Basic Education, Washington, DC. (ERIC Document Reproduction Service No. ED455199)

Conant, J. (1963). *The education of American teachers.* New York: McGraw-Hill.

Davis, K. S., Feldman, A., Irwin, C., Pedevillano, E. D., Capobianco, B., Weiss, T., et al. (2003). Wearing the letter jacket: Legitimate participation in a collaborative science, mathematics, engineering, and technology education reform project. *School Science and Mathematics, 103*(3), 121–133.

Duggan-Haas, D., Smith, E., & Miller, J. (1999). *A brief history of the collaborative vision for science and mathematics education at Michigan State University.* East Lansing: Michigan State University. (ERIC Document Reproduction Service No. ED443655)

Edmundson, P. J. (1991). *What college and university leaders can do to help change teacher education.* Washington, DC: AACTE Publications.

Fallon, D. (1999, September 15). *Our grand opportunity: Remarks on teacher education for college and university chief executives.* Paper presented to the President's Summit on Teacher Quality, University of Maryland. Retrieved August 10, 2003, from http://www.ed.gov/inits/teachers/conferences/fallon.html

Fullan, M. (1999). *Change forces: The sequel.* Philadelphia: Falmer Press.

Fullan, M. (2000). The three stories of education reform. *Phi Delta Kappan, 81,* 581–584.

Gess-Newsome, J., Southerland, S. A., Johnston, A., & Woodbury, S. (in press). Offering a model of reform: Tracing the interaction of factors that impact scientists' practice of reform-based teaching. *American Educational Research Journal.*

Gore, J. (1987). Liberal and professional education: Keep them separate. *Journal of Teacher Education, 38*(1), 2–5.

Haruta, M. E., & Stevenson, C. B. (1999). *Integrating student-centered teaching methods into the first year SMET curriculum: The University of Hartford model for institution-wide reform. Summative evaluation.* Chaplin, CT: Curriculum Research and Evaluation. (ERIC Document Reproduction Service No. ED440977)

Harwood, W. S. (2002, June 4). *William S. Harwood.* Retrieved August 10, 2003, from http://php.indiana.edu/~wharwood

Harwood, W. S. (2003). *The 21st century teachers project: A K–16 collaboration for teacher preparation.* Manuscript in preparation.

Haycock, K., & Brown, N. (1993). *Higher education and the schools: A call to action and strategy for change.* (ERIC Document Reproduction Service No. ED369356)

King, J. A. (1987). The uneasy relationship between teacher education and the liberal arts and sciences. *Journal of Teacher Education, 38*(1), 6–10.

Kramer, R. (1991). *Ed school follies.* New York: Free Press.

Landers, M. F., Weaver, R., & Tompkins, F. M. (1990). Interdisciplinary collaboration in higher education: A matter of attitude. *Action in Teacher Education, 12*(2), 25–30.

Lieberman, A. (1995). Practices that support teacher development: Transforming conceptions of professional learning. *Phi Delta Kappan, 76,* 591–596.

McDiarmond, G. W. (1992). *The arts and sciences as preparation for teaching* (Issue Paper No. 92-3). Retrieved August 10, 2003, from the National Center for Research on Teacher Education Web site: http://ncrtl.msu.edu/http/ipapers/html/pdf/ip923.pdf

Mervis, J. (2002, January 11). U.S. programs ask faculty to help improve schools. *Science, 295,* 265.

National Research Council. (1996). *National Science Education Standards.* Washington, DC: National Academy Press.

National Research Council. (2001). *Educating teachers of science, mathematics, and technology.* Washington, DC: National Academy Press.

Nelson, C. (1997). Tools for tampering with teaching's taboos. In W. E. Campell & K. A. Smith (Eds.), *New paradigms for college teaching* (pp. 51–77). Edina, MN: Interaction.

Nelson, C. (2000). Must faculty teach in ways that make them easily dispensable? *The National Teaching & Learning Forum, 9*(6), 4–5.

Redish, E. F. (1999). Millikan Lecture 1998: Building a science of teaching physics. *American Journal of Physics, 67,* 562–573.

Redish, E. F. (2001, February). *Building a science of teaching.* Paper presented at the Indiana University SOTL Symposium, Bloomington. Retrieved August 10, 2003, from http://www.physics.umd.edu/perg/talks/redish/SOTL.pdf

Reiff, R. (2003). *If inquiry is so great, why isn't everyone teaching it?* Manuscript submitted for publication.

Rickover, H. G. (1960). *Education and freedom.* New York: Dutton.

Riley, R. (1999). *Remarks as prepared for delivery by U.S. Secretary of Education Richard W. Riley: President's Conference on Teacher Quality.* Retrieved August 10, 2003, from http://www.ed.gov/inits/teachers/conferences/rwraddress.html

Scholarship of Teaching and Learning. (n.d.) Retrieved August 10, 2003, from Indiana University, Bloomington Web site: http://www.indiana.edu/~sotl

Seymour, E. (2001). Tracking the processes of change in U.S. undergraduate education in science, mathematics, engineering, and technology. *Science Education, 86,* 79–105.

Shroyer, M. G., Wright, E. L., & Ramey-Gassert, L. (1996). An innovative model for collaborative reform in elementary school science teaching. *Journal of Science Teacher Education, 7*(3), 151–168.

Shumba, O., & Glass, L. W. (1994). Perceptions of coordinators of college freshman chemistry regarding selected goals and outcomes of high school chemistry. *Journal of Research in Science Teaching, 31*(4), 381–392.

Slate, J. R., Jones, C. H., & Charlesworth, J. R. (1990). Relationship of conceptions of intelligence to preferred teaching behaviors. *Action in Teacher Education, 12*(1), 25–29.

Stacey, R. (1996). *Complexity and creativity in organizations.* San Francisco: Berrett-Koehler.

Tafel, J., & Eberhart, N. (1999). *Statewide school-college (K–16) partnerships to improve student performance. Strategies that support successful student transitions from secondary to postsecondary education.* Denver, CO: State Higher Education Executive Officers. (ERIC Document Reproduction Service No. ED434611)

Taylor, P. C., Gilmer, P. J., & Tobin, K. (Eds.). (2002). *Transforming undergraduate science teaching.* New York: Peter Lang.

U.S. Department of Education. (2000). *Eliminating barriers to improving teaching.* Washington, DC: Author.

Williams, V. L. (2002). *Merging university students into K–12 science education reform.* Arlington, VA: RAND.

Woodbury, S., & Gess-Newsome, J. (2002). Overcoming the paradox of change without difference: A model of change in the arena of fundamental school reform. *Educational Policy, 16,* 763–782.

Wyckoff, S. (2001). Changing the culture of undergraduate science teaching. *Journal of College Science Teaching, 30,* 306–312.

CHAPTER 5

THE IMPORTANCE OF PRIOR KNOWLEDGE IN COLLEGE SCIENCE INSTRUCTION

Kathleen M. Fisher

ABSTRACT

I was waiting for a bus on a street corner in London when I struck up a conversation with the man standing next to me. I said, "I don't go anywhere without my Macintosh." He said enthusiastically, "Neither do I." Our conversation continued for possibly several minutes before we realized that I was talking about my *Macintosh computer* (hanging on my right shoulder) while he was talking about his *Macintosh rain gear* (draped over his left arm). We had a good laugh. This illustrates the nature of prior knowledge and the way it can interfere with communication. Misunderstandings can often be quickly clarified in ordinary conversations. But when they occur in one-way information delivery (as in lectures or books), they can persist for weeks or semesters or quite often indefinitely. Sometimes the misunderstandings are direct as in this case, arising from words that have multiple meanings, where each individual associates a different meaning with the word. Sometimes the influence of prior knowledge is quite indirect. For example, the majority of students who study photosynthesis fail to understand that carbon derived from carbon dioxide in the air is used by plants to construct themselves, first

Reform in Undergraduate Science Teaching for the 21st Century, pages 69–83
Copyright © 2004 by Information Age Publishing

by incorporating the carbon into sugars and then incorporating the sugars into cellulose. The persistent failure of understanding seems to derive from strong underlying assumptions that air (including carbon dioxide) has no weight, and therefore cannot be used to create a massive tree. This chapter examines prior knowledge both as a barrier to understanding and as a foundation on which to build new knowledge. It illustrates the incredible importance of interaction between teachers and students to achieve successful and deep understanding.

INTRODUCTION

Hestenes, Wells, and Swackhammer (1992) used students' ideas about physical phenomena to develop a test called the Force Concept Inventory. This apparently simple test has been used in hundreds of college physics classes across the nation. When instructors first review the test, they tend to think that it is much too easy and their students will ace it. In most cases, however, instructors are shocked by their students' poor performance. The test draws upon research into students' prior knowledge, using common naive ideas as distracters for correct responses. The test reveals that even students who get A's and B's in physics classes often do not understand the most basic physics concepts. The authors conclude,

> Every student begins physics with a well-established system of commonsense beliefs about how the world works derived from years of personal experience ... these beliefs play a dominant role in introductory physics. Instruction that does not take them into account is almost *totally ineffective*, at least for the majority of students. (p. 141, italics added)

Researchers have discovered that in college science classes, the prior knowledge of learners determines to a large extent what each individual can learn from a particular situation. It is not productive simply to try and pour facts into their brains. Each student must assimilate and make sense of new ideas by connecting them to what they already know. Researchers have recognized that learning science is an effortful process that is promoted by engagement with phenomena and dialogue with peers. Eliciting and finding ways to challenge prior knowledge can be critical for successful learning. Researchers also realize that knowledge construction is hard work, especially when the knowledge is counterintuitive and defies everyday experience.

These observations are consistent with the prevailing learning theory known as constructivism (Ausubel, 1963, 1968; Ausubel, Novak, & Hanesian, 1978; McComas, 1997; Osborne & Wittrock, 1983; von Glaserfeld, 1987; Wandersee, Mintzes, & Novak, 1994; Wittrock, 1974a, 1974b). People construct meaning from their experiences in everyday life. Their knowledge is stored in long-term memory in part in the form of semantic networks. Research with the Force Concept Inventory and other assessment tools has shown that student learning increases significantly when students are given the opportunity to construct meaning about science in their science classes. Further, an emerging theory of the mind explains why prior knowledge can be so difficult to alter and provides a theoretical explanation for the importance of eliciting prior knowledge. These things are described in more detail below, but first, a bit of history.

The scientific community initially recognized the importance of prior knowledge in the late 1970s and early 1980s, especially with respect to its interference in learning. Various misconceptions were identified in the thinking of many students, and these conceptions proved difficult to change (Clement, 1982; McCloskey, 1983). These misconceptions were characterized as ideas that (a) tend to be shared by a significant proportion of a population, (b) produce consistent error patterns, and (c) are remarkably resistant to being "taught away" (Clement, 1982; Fisher & Lipson, 1986; McCloskey, 1983). While many other types of errors (such as slips of the tongue, miscalculations, or confusion between similar terms) can be readily corrected, misconceptions are different. They persist. Further, good students can often memorize correct answers and regurgitate them on multiple-choice tests, earning A's and B's, but if they have not understood why those responses are correct, the information will quickly fade away. A new field of study was born in 1983 when the first international seminar on misconceptions in science and mathematics was held at Cornell (Helm & Novak, 1983). Thousands of studies have since been done to elucidate naive preconceptions in fields of science (Fisher & Moody, 2000; Novak, 1987, 1993; Wandersee et al., 1994; Wandersee & Fisher, 2000).

In every domain of science, commonsense beliefs are sometimes at odds with scientific theories. For this reason, effective instruction requires technical knowledge, not only of the subject being taught, but also about how students think and learn. No matter how often a piece of information is given or how cleverly it may be presented in a lecture, it is not learned, in some cases, by a significant fraction of the students. This resistance to instruction was perplexing to early researchers. Creative sleuthing was necessary to uncover the problem. It turned out that students have many underlying assumptions about how the world works. Sometimes those assumptions are wrong. Often they are implicit rather than explicit. In many cases students are not consciously aware of their assumptions and therefore are unaware of conflicts they have with what the instructor is telling them. They simply discount or discard the new information, or distort it to fit.

Consider this example: In the Harvard/Smithsonian *Minds of Our Own* video series (Schnepps, 1997), an interviewer shows fourth graders first a seed and then a dry log from a tree. "Where does the weight of the tree come from?" she asks. The fourth graders say, "the sun, the soil, rain, nutrients." The interviewer presents the same question to students graduating from MIT and Harvard and *gets the same answers from most of them*. Every student has had high school biology. Some have studied biology in college. Some have even majored in biology. Chances are they have all memorized the formula for photosynthesis in which carbon dioxide goes in and sugar comes out. But their answers are similar to those provided by fourth graders. These college graduates from prestigious universities do not understand that in the process of photosynthesis, plants take carbon dioxide from the air through their leaves, then break the CO_2 apart and use the carbon to produce sugar molecules. The sugar molecules produced by photosynthesis support all living things. Biology professors who have asked this question in their classrooms, even after spending a week or two on photosynthesis, generally find that a large proportion of their students are unable to answer the question correctly. The students memorize details about photosynthesis including the formula, but *they miss the main idea*.

The video goes on to show an interview with a middle school student. It is gradually discovered that the student believes that air has no weight. How can you take a substance from this weightless, invisible air and create something as massive and heavy as a tree? It does not make sense. Since it makes no sense, students dismiss that part of the formula. At the same time, students attribute the weight of the tree to the soil because soil is known to have weight, even though there is no soil in the formula for photosynthesis. The same idea was explored by Jean Baptiste van Helmont about 400 years ago (National Science Teachers Association, 1998). Naive intuitive ideas observed in students often have historical roots. Notice that the misconception is not directly connected to photosynthesis. It focuses on the nature of air. These distant misconceptions are the most difficult to discover and to change. Further, a deep-seated misconception about the nature of air interferes with the learning of many things such as the conservation of matter (physics), change of state (physics, chemistry), nutrient cycles (biology, geology, oceanography), and so on.

Alternative conceptions such as this one impact students abilities to learn every subject. Hestenes and Wells (1992) developed a multiple-choice Mechanics Diagnostic Test to assess student understanding of Newtonian mechanics. The test was further refined to create the Force Concept Inventory (Hestenes et al., 1992). The test was revised in 1995 (Halloun, Hake, Mosca, & Hestenes, 1995). It provides a forced choice between Newtonian concepts and commonsense alternatives. These tests have been given in hundreds of college and high school physics classes involving thousands of students. They have provided substantial evidence for the relative ineffectiveness of lecture instruction in introductory physics and for the relative value obtained with well-designed, hands-on, student-centered lessons. For example in one study by Hestenes and Halloun (1995), traditional college students earned an average score of 34 in calculus-based physics while students in a modeling section had an average score of 60. Hestenes and Halloun describe modeling or working with models as a major process for constructing and employing physics knowledge. Hake (1998) has summarized outcomes in 62 physics courses enrolling 6,542 students at three levels: (a) high school, (b) two- and four-year college, and (c) PhD-granting university. Of these, there are 14 traditional courses ($n = 2,084$) and 48 interactive engagement courses ($n = 4,458$). A consistent analysis over diverse student populations is obtained by using normalized gain scores.[1] The 14 traditional courses had an average gain of 0.23 +/– 0.04 (standard deviation). In contrast, the 48 interactive classes were almost two standard deviations higher with an average gain of 0.48 +/– 0.14.

Diagnostic tests are useful for assessing what students know as they enter and leave a course. Ideally such tests would be used in every college course to provide a measure of relative success, both within course (from year to year) and between similar courses. These tests are designed on the basis of many years of research on students' naive conceptions. The tests use those naive conceptions as distracters (incorrect responses). Working with a series of students over a period of seven years, for example, I developed a Conceptual Inventory of Natural Selection (CINS), a 20-item test that assesses understanding of 10 concepts intrinsic to natural selection. We reviewed the literature and then gave open-ended, short-essay tests to students. We also interviewed majors and nonmajors. From this response data combined with data from actual studies of evolution, we constructed a series of multiple-choice tests. These tests represented several iterations of testing and revision. The test was finally polished and evaluated extensively in collaboration with my graduate student, Dianne Anderson (Anderson, Fisher, & Norman, 2002). The CINS test is most appropriate for use as a pretest and posttest in nonmajors' college courses (or in high school biology) or as a pretest in introductory biology for majors.

Many other diagnostic tests in science have been developed and can be downloaded at the FLAG Web site (n.d.). Physics, chemistry, and mathematics are especially well represented in the test bank. Our CINS test will soon appear on the Web site and we look forward to several other tests that have been developed in biology becoming available as well. Extensive use of diagnostic tests has confirmed the value of interactive teaching methods.

As noted above, researchers have discovered that the learning process is facilitated when students can observe and interact with phenomena, ask questions, engage in dialogue, and move at a comfortable pace (Duit,

Goldberg, & Niedderer, 1992). Interacting with the object or phenomenon being studied can support thinking, generate conversation, and provide a basis for developing a shared language among learners. Such experiences convey a far richer view of the topic of study than can be captured in words alone. Images, smells, sequences, interactions, events, and feelings can significantly elaborate knowledge about scientific phenomena and therefore increase the likelihood of both understanding and retrieval from memory.

Students also learn a great deal by making predictions. Predicting what will happen requires students to construct a mental model for a phenomenon that draws upon each student's prior knowledge. A professor presents a situation to students, asks them to discuss their predictions in small groups, and then asks each student to write down his/her individual prediction. This can be done in large lectures as well as small classes. Once the model is tested by observing the phenomenon and outcome, students are asked to discuss their interpretations in small groups and to explain individually and in writing why their prediction was or was not correct. This process can help initiate conceptual change in students when they find that their mental models do not accurately predict what happens. When this occurs in small classes where students are conducting the experiment themselves, their first conclusion is that they did the experiment wrong. Only after they learn that all groups got the same result are they willing to consider the possibility that their mental model may be erroneous. Even a large-class demonstration may need to be repeated for students to be persuaded that the outcome is "real." They say, "Seeing is believing." But researchers have discovered that in many cases "believing is seeing," you see what you expect to see. These interactive sessions provide great teaching and learning moments. Many researchers have moved beyond mental models to facilitate knowledge construction and engage students in working with physical models, computer-based models, or paper-and-pencil models, as in the modeling physics approach cited above.

What do we mean by the term "knowledge construction"? All individuals construct knowledge about science and other academic subjects in their conscious working memory and store that knowledge in long-term memory. The prevailing model for the way in which denotative knowledge is stored in memory is the semantic network (Quillian, 1967, 1968, 1969). Quillian's model grew out of 50 years of research on free word association (Deese, 1965). One way to visualize the model is with the SemNet (Fisher, 2000; Fisher et al., 1990) or Semantica (2002) software . SemNet was designed in the early 80s by my research group as a tool to promote meaningful learning of biology among students in a five-unit biology course at a major university. The software was patterned directly after Quillian's model and was designed for ease of use by busy students. Semantica is the modern reincarnation of the SemNet software (Semantic Research, 2003), being cross-platform and Java-based. Most semantic networking programs aim to make computers more intelligent (Brachman & Levesque, 1985). The most recent practical application of semantic networks aims to make the World Wide Web behave more intelligently (Berners-Lee, Hendler, & Lassila, 2001). In contrast, Semantica is designed to help people think more clearly. Semantica knowledge structures are complex, multidimensional webs of ideas (see Figure 5.1).

Figure 5.1.   An overview of four Semantica knowledge structures. From left to right they describe the human body (411 instances), a family tree (476 instances), a series of events (Attack on America, 739 instances), and biology (3,632 instances).

A semantic network or knowledge structure is created with three primitives: concepts, relations, and instances. A concept may be defined as any idea or thought and is typically represented as a noun or noun phrase (e.g., acceleration, dog, evolution). A relation describes the link between two concepts and is typically represented by a *bidirectional* verb or verb phrase (e.g., has a part/is a part of). When joining two concepts together with a relation, a bidirectional instance is created (see Figure 5.2). A fourth primitive, a knowledge object (including images, texts, and URLs) can be attached to any element.

The computer-based Semantica network serves as a mirror of the mind: As knowledge is assembled in the computer, it either reflects existing knowledge or supports the creation of new knowledge in the mind. Since Semantica allows individuals to off-load their thinking to the computer, they then have increased capacity to reflect on, talk about, think about, point to, and revise their thinking (metacognition). As McAleese (1998) says, Semantica provides an arena in which individuals and groups can manipulate their thoughts. Modern human-oriented knowledge representation has been evolving since the 1880s, when Charles Peirce developed semiotics and existential graphs (Peirce, 1960).

Semantica is a useful tool for capturing a learner's prior science knowledge. The graphic frame view serves as the primary working interface, showing one central concept with all of its links to related concepts (see Figure 5.3). In one study, 19 college graduates enrolled in a teaching credential program were asked, as a homework assignment in a course on

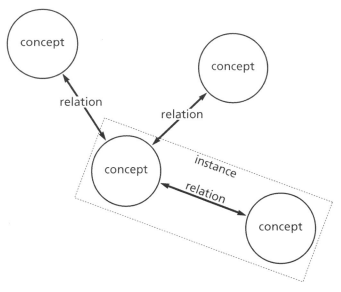

Figure 5.2.   Three primitive elements in a Semantica knowledge structure are *concept*, *relation*, and *instance*. Two concepts joined by a bi-directional relation constitute an instance. A fourth element, knowledge objects, can be attached to any of the primitives.

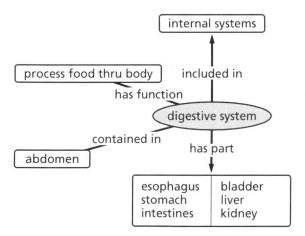

Figure 5.3.   Display of one aspect of one student's prior knowledge in the graphic frame view of the SemNet software, showing the belief held by 4 of 19 graduate students studied, that the kidneys and bladder are part of the digestive system. This is one of 186 graphic frames in a network containing 186 concepts, 6 relations, and 213 instances. This was the only student to include the liver in the digestive system; it is not included by experts but does play a role in removing ingested toxins from the blood. (Relation directions have been moved from original for ease of viewing).

technology, to create a knowledge structure about the human body (Fisher, 2003). They had four days to complete the assignment. Their knowledge structures ranged in size from 39 to 374 concepts, with an average of 119 concepts (Semantica keeps count of all elements in a knowledge structure). Students varied significantly in their knowledge about the human body. Differences in their prior knowledge were most striking in (a) the size of their knowledge structure and (b) the relative proportions of concepts used to describe external versus internal features of the body. The most common error was the inclusion of kidneys or bladder in the digestive system by 21% of the students (Figure 5.3). Interviewed students imagined that ingested fluids are diverted via a shunt from the intestines to the kidneys or bladder, without ever entering the bloodstream. I think of this "shunt" hypothesis as a naive conception but not a misconception, since I believe it is relatively easy to teach away.

For students to effectively express their prior knowledge in Semantica, three conditions must be met. First, students must be generating knowledge from their heads, without reference to texts or other instructional materials. Second, they must feel free to express their thoughts, knowing they will not be graded with respect to scientific correctness. Third, they must be sufficiently familiar with the software so as to be able to express their knowledge effectively. Under these conditions, students are willing to include personal as well as objective knowledge—in this case referring to surgeries, back pain, split ends, and so forth.

Can knowledge construction with Semantica promote college students' conceptual learning about science? Absolutely. For example, Gorodetsky and Fisher (1996) studied two classes of a capstone biology course in which the SemNet section used the SemNet software as a study tool and the reference (control) section used traditional study methods. The sections were taught by two different experienced teachers. Students were prospective teachers in an upper division university biology course. Students in both sections were given a survey about learning and Schmeck's Inventory of Learning Processes (Schmeck & Ribich, 1978; Schmeck, Ribich, & Ramanaiah, 1977) at the beginning and end of the semester.

As part of their final exams, students in the two sections were asked to describe the digestive system in a short essay. Both sections had studied the digestive system with similar (but not identical) hands-on activities, and SemNet students had summarized their knowledge in SemNet. In their essays, SemNet students wrote about twice as many sentences (10.3 sentences vs. 5.3 on average) and used almost twice as many words as reference students (79 words vs. 49 on average). The richer descriptions of the digestive system generated by SemNet students suggests that either their knowledge is more extensive or they were better able to retrieve it or both. SemNet students also used shorter sentences (relative to reference students) to describe their knowledge—likely a consequence of their knowledge organization. Not surprisingly, when the digestive system essays were evaluated in terms of *quality* of the descriptions, the SemNet section outperformed the reference section by more than a factor of two. As a control, students were asked to write short essays on two other topics that were studied in both sections but not represented in SemNet in either section. Their average performances on these two essays were indistinguishable.

In response to a question on their learning survey about how to study, SemNet students used terms like "integrate," "relate," or "make associations" twice as frequently at the end of the semester (34 times) as at the beginning (15 times), whereas there was no change for reference students (15 before and 13 afterwards). In addition, SemNet students made significant gains on the deep processing scale of Schmeck's Inventory of Learning Processes, whereas reference students did not make gains on any of the four scales.

Students who used SemNet reported that they acquired two newfound abilities, first to *recognize* that some ideas are more important than others are, and second to be able to *identify* the important ideas. They said that these newly acquired abilities had a significant effect on their note-taking strategies in other classes, suggesting that knowledge representation experiences may be quite valuable if introduced earlier in students' careers. Another comment students often made is that SemNet makes biology concepts more concrete. Perhaps this can be explained by Wilensky (1991) who suggested that concreteness is a function of a person's relationship to an object rather than a property of the object itself. That is, the concreteness perceived by an individual varies with the extent of their relatedness to the object, the richness of their representations, and the number of their interactions and connections with the object.

This study confirms Anderson's (1990) theoretical claim that students who engage in explicit knowledge construction should be better able to retrieve and apply their knowledge than students who do not. It is also consistent with the instructional strategies recommended by the National Research Council (2000, p. 207), to provide scaffolding to enhance learning, and to give students and teachers many opportunities for diagnosis, feedback, reflection, and revision. As instructors see what students are thinking, they can facilitate learning much more effectively.

My classroom research revealed confusion about relations by second-language learners. ESL students often use relation rays such as "has a part" and "is a part of" either backwards or randomly. These basic relations seem trivial to native speakers, but language learners including both first (Gentner, 1978, 1982) and second-language learners, including children and adults (Rosenthal, 1996) regularly learn nouns before verbs. Given that the whole/part relation is one of three relations used 50% of the time in describing biology knowledge (Faletti & Fisher, 1996; Fisher, 1988), getting this right can make a huge difference in students' abilities to learn. Once the problem is identified and corrected, students are able to master the relations quickly. Diagnosis is the key.

Is student use of Semantica valuable? David Jonassen, who is one of the world's leading authorities on cognitive tools, describes it in the following way:

> Perhaps the greatest benefit of Semantica is that most students find building concept maps both illuminative and enjoyable. I have collected numerous testimonials from students regarding the clarity of their understanding following the production of concept maps with Semantica.
>
> We are soon engaged in a project at the University of Missouri integrating Semantica into large, undergraduate lecture classes. Innovation in those contexts is very problematic. Our belief is that constructing concept maps with Semantica is probably the single most effective innovation that we could implement.
>
> The only potential drawback of using Semantica is that it causes students to think hard, especially if they build coherent concept maps. Lamentably, some students are not inured to thinking hard. I believe that it is time they should. (D.H. Jonassen, personal communication, April 14, 2003)

Cognitive research may provide a simple explanation for the amassed observations about misconceptions. Psychologists have long been aware that the brain has vast capabilities for storing, processing, and retrieving information, yet conscious working memory is surprisingly small (can perceive five to nine "ideas") and also is constrained (can only focus on one train of thought at a time, although can switch quickly). These observations are incorporated into global workspace theory (Baars, 1997). According to this theory, conscious working memory is both the "bottleneck" and the "activist." It is the one place in the brain that provides conscious access to all the subconscious modules. Global workspace theory suggests that we are unable to modify knowledge structures in the subconscious modules unless we transfer that knowledge into conscious working memory where we can reconsider it. This would explain why lectures are so ineffective. Students listen, take notes, and assimilate some new ideas. But they have no stimulus or time to bring up their prior knowledge and examine it, much less restructure it. Thus while some new ideas get assimilated and stored, they are "less competitive" than the well-established, familiar ideas which are also still there in the brain. The new ideas typically decay with disuse.

Some students, of course, are spontaneous meaning makers. They are rather successful at doing these things on their own. In my classes of prospective elementary students (primarily senior undergraduates), there were usually about six such students out of 30 (20%). In other majors the proportion may be higher, ranging up to 30 to 35%, I would guess. The majority of students has been conditioned by our systems and by their natural inclinations to cram, memorize, and then forget.

Global workspace theory makes sense of the practices that have been found to be effective in addressing alternative conceptions. Most strategies involve eliciting prior knowledge. This brings prior knowledge into conscious memory so it can be examined and potentially modified. Interacting with carefully selected phenomena (chosen specifically to challenge common alternative conceptions) can prompt the reflections that can produce conceptual change (Posner, Strike, Hewson, & Gerzog, 1982). It allows identification of anchoring conceptions in prior knowledge (Clement, 1983). In such settings, researchers and teachers employ strategies to promote conversations among students about their perceptions, interpretations, and conclusions. The dialogue allows give and take, question and response, clarification and negotiation, all of which can lead to discovery of new insights, adoption of common language, and identification and remediation of formerly unrecognized preconceptions. This mode of teaching is less effective than lecturing when effectiveness is measured by rate of information transfer. But it is more effective than lecturing when effectiveness is measured by the appearance of desirable long-term changes in students' thinking.

The accumulated research points to a variety of research questions such as the following: Is it possible in the United States to create educational systems that prompt students to think deeply from a young age? Can we refocus at all levels on depth of understanding rather than breadth as the ultimate test of rigor? Can we begin to test the hypothesis that global workspace theory does account for alternative conceptions, and if so, can it provide further insights regarding how to address them? Can teachers recognize that *teaching students without knowing what they are thinking is like driving a car with their hands tied behind their back and a blindfold over their eyes?* They have no idea what they are doing or where they are going, but they can do it at top speed!

NOTE

1.   A normalized gain score is defined as the ratio of the average actual gain (% post—% pre) to the maximum possible average gain (100—% pre).

REFERENCES

Anderson, D.L., Fisher, K.M., & Norman, G.J. (2002). Development and evaluation of the Conceptual Inventory of Natural Selection. *Journal of Research in Science Teaching, 39,* 952–978.

Anderson, J.R. (1990). *Cognitive psychology and its implications.* New York: W.H. Freeman.

Ausubel, D.P. (1963). *The psychology of meaningful verbal learning.* New York: Grune & Stratton.

Ausubel, D.P. (1968). *Educational psychology: A cognitive view.* San Francisco: Holt, Rinehart, & Winston.

Ausubel, D.P., Novak, J.D., & Hanesian, H. (1978). *Educational psychology: A cognitive view* (2nd ed.). New York: Holt, Rinehart, & Winston.

Baars, B.J. (1997). In the theatre of consciousness: Global workspace theory, a rigorous scientific theory of consciousness. *Journal of Consciousness Studies, 4,* 292–309.

Berners-Lee, T., Hendler, J., & Lassila, O. (2001, May). The semantic web: A new form of Web content that is meaningful to computers will unleash a revolution of new possibilities. *Scientific American.* Retrieved August 20, 2003, from http://www.sciam.com/article.cfm?colID=1&articleID=00048144-10D2-1C70-84A9809EC588EF21

Brachman, R.J., & Levesque, H.J. (Eds.). (1985). *Readings in knowledge representation.* Los Altos, CA: Morgan Kaufmann.

Clement, J. (1982). Students' preconceptions in introductory mechanics. *American Journal of Physics, 50*(1), 66–71.

Clement, J. (1983). A conceptual model discussed by Galileo and used intuitively by physics students. In D. Gentner & A.L. Stevens (Eds.), *Mental models* (pp. 325–339). Hillsdale, NJ: Erlbaum.

Deese, J. (1965). *The structure of associations in language and thought.* Baltimore, MD: Johns Hopkins University Press.

Duit, R., Goldberg, F., & Niedderer, H. (Eds.). (1992). *Research in physics learning: Theoretical issues and empirical studies.* Kiel, Germany: University of Kiel, Institute for Science Education.

Faletti, J., & Fisher, K.M. (1996). The information in relations in biology. In K.M. Fisher & M. Kibby (Eds.), *Knowledge acquisition, organization, and use in biology* (pp. 182–205). Berlin: Springer-Verlag.

Fisher, K.M. (1988, April). *Relations used in student-generated knowledge representations.* Paper presented at the annual meeting of the American Educational Research Association, New Orleans, LA.

Fisher, K.M. (2000). SemNet® semantic networking. In K.M. Fisher, J.H. Wandersee, & D. Moody (Eds.), *Mapping biology knowledge* (pp. 143–166). Dordrecht, The Netherlands: Kluwer.

Fisher, K.M. (2003). *Expert/novice knowledge about the human body as seen with knowledge visualization.* Manuscript in preparation.

Fisher, K.M., & Lipson, J.I. (1986). Twenty questions about student errors. *Journal of Research in Science Teaching, 23,* 783–803.

Fisher, K.M., & Moody, D. (2000). Student misconceptions in biology. In K.M. Fisher, J.H. Wandersee, & D. Moody (Eds.), *Mapping biology knowledge* (pp. 55–76). Dordrecht, The Netherlands: Kluwer.

Fisher, K.M., Faletti. J., Patterson, H.A., Thornton, R., Lipson, J., & Spring, C. (1990). Computer-based concept mapping: SemNet software—A tool for describing knowledge networks. *Journal of College Science Teaching, 19,* 347–352.

Gentner, D. (1978). On relational meaning: The acquisition of verb meaning. *Child Development, 49,* 988–998.

Gentner, D. (1982). Why nouns are learned before verbs: Linguistic relativity versus natural partitioning. In S. Kuczaj (Ed.), *Language development: Language, cognition, and culture* (pp. 301–334). Hillsdale, NJ: Erlbaum.

Gorodetsky, M., & Fisher, K.M. (1996). Generating connections and learning in biology. In K.M. Fisher & M.R. Kibby (Eds.), *Knowledge acquisition, organization, and use in biology* (pp. 135–154). New York: Springer-Verlag.

Hake, R.R. (1998). Interactive-engagement vs. traditional methods: A six-thousand-student survey of mechanics test data for introductory physics courses. *American Journal of Physics,* 64–74.

Halloun, H., Hake, R., Mosca, E., & Hestenes, D. (1995). Force Concept Inventory (Rev.). Retrieved August 20, 2003, from http://modeling.asu.edu/R&E/Research.html by e-mailing for permission to Dukerich@asu.edu

Helm, H., & Novak, J.D. (Eds.). (1983, June). *Proceedings of the International Seminar: Misconceptions in Science and Mathematics.* Ithaca, NY: Cornell University, Department of Education.

Hestenes, D., & Halloun, I. (1995, April). *Modeling instruction in physics.* Presented at the annual meeting of the National Association for Research in Science Teaching, San Francisco.

Hestenes, D., & Wells, M. (1992). A mechanics baseline test. *The Physics Teacher, 30,* 159–163.

Hestenes, D., Wells, M., & Swackhammer, G. (1992). Force concept inventory. *The Physics Teacher, 30,* 141–158.

McAleese, R. (1998). The knowledge arena as an extension to the concept map: Reflection in action. *Interactive Learning Environments, 6,* 1–22.

McCloskey, M. (1983). Naive theories of motion. In D. Gentner & A.L. Stevens (Eds.), *Mental models* (pp. 299–324). Hillsdale, NJ: Erlbaum.

McComas, W.F. (1997). 15 myths of science: Lessons of misconceptions and misunderstandings from a science editor. *Skeptic, 5*(2), 88–95.

National Institute for Science Education. (2004). Field-Tested Learning Assessment Guide. Retrieved February 20, 2004, from http://www.flaguide.org.

National Research Council. (2000). *How people learn: Brain, mind, experience, and school* (Expanded ed.). Washington, DC: National Academy Press.

National Science Teachers Association. (1998). *Van Helmont's experiment: It only took him five years!* Retrieved August 20, 2003, from http://www.nsta.org/Energy/find/primer/primer2_3.html.

Novak, J.D. (Ed.). (1987). *Proceedings of the Second International Seminar: Misconceptions and Educational Strategies in Science and Mathematics: Vol. II.* Ithaca, NY: Cornell University, Department of Education.

Novak, J.D. (Ed.). (1993). *Proceedings of the Third International Seminar: Misconceptions and Misconceptions and Educational Strategies in Science and Mathematics: Vol. III.* Ithaca, NY: Cornell University, Department of Education.

Osborne, R.J., & Wittrock, M.C. (1983). Learning science: A generative process. *Science Education, 67,* 489–508.

Peirce, C.S. (1960). *Collected papers of Charles Sanders Peirce.* Cambridge, MA: Harvard University Press.

Posner, G.J., Strike, K.A., Hewson, P.W., & Gerzog, W.A. (1982). Accommodation of a scientific conception: Toward a theory of conceptual change, *Science Education, 66,* 211–227.

Quillian, M.R. (1967). Word concepts: A theory and simulation of some basic semantic capabilities. *Behavioral Sciences, 12,* 410–430.

Quillian, M.R. (1969). The teachable language comprehender. *Communications of the Association for Computing Machinery, 12,* 459–475.

Rosenthal, J.W. (1996). *Teaching science to language minority students.* Bristol, PA: Multilingual Matters.

Schmeck, R. R., & Ribich, F. D. (1978). Construct validation of the Inventory of Learning Processes. *Applied Psychological Measurement, 2,* 551–562.

Schmeck, R. R., Ribich, F., & Ramanaiah, N. (1977). Development of a self report inventory for assessing individual differences in learning processes. *Applied Psychological Measurement, 1,* 413–431.

Schnepps, M. (Producer). (1997). *Minds of our own: Lessons from thin air* [Video]. Cambridge, MA: Harvard University, Science Media Group.

Semantic Research. (2003). Retrieved July 25, 2003, from http://www.semanticresearch.com

Semantica [Computer software]. (2002). San Diego, CA: Semantic Research.

von Glaserfeld, E. (1987). Learning as a constructive activity. In C. Janvier (Ed.), *Problems of representation in the teaching and learning of mathematics* (pp. 215–227). Hillsdale, NJ: Erlbaum.

Wandersee, J. H., & Fisher, K. M. (2000). Knowing biology. In K. M. Fisher, J. H. Wandersee, & D. Moody (Eds.), *Mapping biology knowledge* (pp. 39–54). Dordrecht, The Netherlands: Kluwer.

Wandersee, J. H., Mintzes, J. J., & Novak, J. D. (1994). Research on alternative conceptions in science. In D. Gabel (Ed.), *Handbook of research in science teaching and learning* (pp. 177–210). New York: Simon & Schuster Macmillan.

Wilensky, U. (1991). Abstract mediations on the concrete. In I. Harel & S. Papert (Eds.) *Constructionism: Research reports and essays, 1985–1990* (pp. 193–203). Norwood, NJ: Ablex.

Wittrock, M.C. (1974a). A generative model of mathematics learning. *Journal for Research in Mathematics Education, 5,* 181–196.

Wittrock, M.C. (1974b). Learning as a generative process. *Educational Psychologist, 11,* 87–95.

CHAPTER 6

INNOVATIVE PEDAGOGY FOR MEANINGFUL LEARNING IN UNDERGRADUATE SCIENCE

Dennis W. Sunal

ABSTRACT

College students bring to any science course their own experiences, ideas, and skills related to the discipline. Teaching science so that it has meaning to students involves helping them reconstruct their prior knowledge in the context of the new science concepts being introduced. The learning cycle is described as a general model of pedagogy that facilitates the meaningful understanding of science concepts by assisting students in this conceptual reconstruction. It is fundamentally different from traditional pedagogy, the transmission of intact ideas that attempts to add new knowledge to a student's blank mind. Planning course lessons with specific examples using the learning cycle of guided inquiry is provided to illustrate meaningful learning in undergraduate science courses.

Reform in Undergraduate Science Teaching for the 21st Century, pages 85–122

INTRODUCTION

As a result of major reforms in science teaching taking place today, many college instructors have asked the questions, How can we get students engaged in learning science? and, What pedagogical strategy is most effective in helping students to engage in science? To address the concern of student engagement in science, theory and research has to consider the nature of knowledge and conditions that promote meaningful understanding of science knowledge. Knowledge is stored in the minds and bodies of beings (Johnson, 1987). Learning is the construction of knowledge by individuals as sensory data are given meaning in terms of prior knowledge. Everyone's "reality" is a personal construction that what is held to be truth will be based on what "works" for the individual. It is an interpretive process, involving constructions by individuals through social collaboration (Tobin, Briscoe, & Holman, 1990, p. 411). We know only what our mind constructs. These statements outline a constructivist philosophy based on the ideas that people make their own "world view" and therefore their own knowledge, that new learning depends on and is mediated by prior knowledge, that social interaction is important in creating knowledge, and that meaningful learning occurs through solving real-world problems (Applefield, Huber, & Moallem, 2001; Green & Gredler, 2002; Vermette & Foote, 2001). Although there are differences in the interpretations of constructivist theory, there is agreement on these four statements.

Seen from this constructivist viewpoint, beginning with John Dewey (1916), learning and teaching are only loosely connected in a way similar to the relationship between advertising and buying an expensive product, perhaps a new automobile. One can have buying (meaningful learning) with and without advertising (teaching). Much of the time and effort devoted to advertising (teaching) occurs without buying (meaningful learning). Just as the advertising executive strives to increase advertising's effectiveness to produce more buying, the college instructor strives to increase the effectiveness of his or her teaching to produce more meaningful learning.

Meaning can only be created by students in their own minds (Saunders, 1992). Students cannot be passive during learning. College instructors cannot transmit knowledge to students, as in the traditional transmission model of pouring in knowledge using a funnel. Instructors, however, can facilitate meaningful learning by planning and using experiences that engage students in working with ideas in their own minds and with their own actions and interactions (Yager, 1991). Meaningful learning in science courses must be an active construction process.

Meaningful learning can be facilitated by a pedagogy that involves and depends on (a) the *background knowledge* the learner brings to a situation,

(b) whether the learner's *attention is focused* on the ideas being presented, and (c) the *mental and physical actions* of the learner as he or she works with objects, ideas, and events in testing prior or new trial versions of a science idea to be learned. See Chapter 4, "Lessons from Research: The Importance of Prior Knowledge" by Kathleen Fisher for additional background.

This constructivist pedagogy is based on the understanding that students bring both conceptions and misconceptions to class about the material being studied because of their unique prior experiences. These existing student understandings and the associated beliefs need to be identified and addressed before meaningful learning can occur. This approach must be student centered.

Applying constructivist pedagogy in the classroom involves activities such as providing lab activities before discussing the results students are expected to find (exploration and discovery), removing lab data tables so that students generate or organize their own data, changing exams to require more application and transfer of ideas learned, asking students to design an experiment to answer a question or problem, and placing students in a situation requiring group debate, discussion, research, and sharing.

STUDENTS LEARNING IN UNDERGRADUATE SCIENCE

One important component of scientific literacy in higher education is changing students from novice "seeing-as" observers when they arrive in class to informed "seeing-that" observers. What "seeing" meant to scientists in the 1800s and 1900s, seeing-as observations, is different from what seeing means to scientists in the 21st century, seeing-that observations. A fundamental change has occurred in scientific observation and, thus, there are changes in what counts as observational evidence in the formulation of science generalizations and scientific theories (Duschl, 1990).

The role of human senses in data collection has become de-emphasized in favor of instruments and methods based on fundamental theories in science. This de-emphasis leads to the result that it is increasingly more difficult in undergraduate science to explain what and how we know.

Seeing-as observations are carried out with little prior knowledge. It involves a literal description of patterns in nature that is possible to even the uninitiated in science. Seeing-as is observing the irregular pattern in the placement of stars in the sky, the similarities and differences between structures used to build airplane wings, and the angle of reflection of light off surfaces. Data are represented by things that can be sensed or patterns developed directly from observations. These data are used to develop generalizations such as constellations, lift in the flight of aircraft, and the refraction and velocity of light.

Seeing-that observations can only be interpreted with extensive prior knowledge. Not everyone can see this way. It takes an extensive set of experiences. For example, this type of observing occurs when "seeing" the disease-prone nature of crops in satellite data, observing the movement of crustal plates of the earth, seeing the magnetosphere of the earth, or detecting the presence of a black hole in a galaxy. In seeing-that, data evolve from theoretical considerations that are abstract and not sensed directly. These data lead to the development of theory such as bosons and hadrons making up all matter as viewed with particle accelerators, plate tectonics as derived from magnetic anomalies and complex earthquake patterns, and biological evolution as derived from statistical probabilities in the diversity of life on Earth.

In the college classroom much of what we do should involve moving students from novice seeing-as observers or naive seeing-that observers to informed seeing-that observers (Duschl, 1990).

This is what today's new pedagogy using conceptual reconstruction is attempting to do. Learners, students or scientists, revise and replace their knowledge as they acquire experience and understanding. A student's prior knowledge must be linked to every phase of the conceptual restructuring process. We must teach students to think critically about new ideas presented and to creatively apply and evaluate the knowledge of science. To know something in science is to construct and reconstruct meaningful relationships among concepts.

The type of science presented in many textbooks and traditional lectures consist of many final form statements (Duschl, 1990). These final form statements consist of rules and algorithms defining concepts and generalizations about the world. Species is a class of individuals having common attributes and having a common name. The angle of incidence equals the angle of reflection. Both are final form statements. There is little added to describe the original situations, or cases where these rules are found in the world and less on their application contexts and limitations. Final form statements that fill traditional textbooks and classrooms do not convey meaning. This type of science teaches the conclusions of science as fact. Final form science is traditionally taught and learned through memorization. Science knowledge is taught without the procedures for understanding the meaning of the content. The important goal of effective modern science teaching is not only to understand the new ideas meaningfully, which are developing at an ever quickening pace, but also to understand the changes to our meanings of the investigative process and the characterizations of knowledge—what we mean by observation, the nature of evidence, the methods of science, and what we know (our theoretical viewpoint). We must assist students in learning *how* to learn if they are to learn and apply meaningful knowledge of 21st-century science.

Vygotsky (1962) described the sources of knowledge as that constructed from everyday interaction with the environment and through formal instruction. Everyday knowledge is intuitive and connected to language and culture, often referred to as naive knowledge forming the person's view of the world. Formal instruction, or classroom knowledge, is an "expert's" interpretation of the world, another person's reality. Prior knowledge is the mixing of naive knowledge and classroom knowledge. For learning to occur, classroom experiences must be perceived by the student, who must mentally reconstruct it in relation to already existing prior knowledge. Only when this new knowledge representation is modified to fit in with what is already known in the student's mind is the classroom learning experience completed. Students, individually, have now constructed meaning. If the classroom experience has been planned to take both naive knowledge and classroom knowledge into account, then scientific knowledge has been given meaning and students will be able to apply this new knowledge usefully in settings different from where it was learned.

Meaningful learning is learning with understanding which involves creating knowledge of the environment that makes it possible to have an awareness, appreciation, or ability to make decisions and successfully participate in everyday life (National Research Council [NRC], 1996). This kind of learning is very different from rote memorization. Learning through students' active mental and physical involvement is appropriate for more complex forms of learning beyond rote memory and recall learning. This type of learning involves the following:

1. Constructing concepts based on experiencing, identifying, selecting, and classifying facts (e.g., understanding that various species in an area can have several habitats and thus rely on differing food sources for energy).

2. Constructing generalizations by linking concepts (e.g., representing groups of species in energy flow systems forming food webs, food chains, and ecosystems).

3. Constructing theories of how things work (e.g., the flow of energy in systems of living organisms can predict the evolution of life systems and how systems change).

4. Developing inquiry thinking skills (e.g., classifying living organisms based on previous knowledge of internal functions).

5. Developing attitudes and dispositions about the world (e.g., willing to suspend judgment about an event or problem situation until sufficient evidence is available to form a reasonable conclusion). (See Table 6.1 and Figure 6.1.)

Table 6.1.   Types of Science Knowledge

Methodological	Conceptual
Searching for patterns and relationships among generalizations—"seeing-that"; developing scientific attitudes and dispositions	Theory (and schemata)
Searching for patterns and relationships among concepts and developing inquiry skills	Generalizations
Categorizing and defining	Concepts
Observations—"seeing-as"	Facts

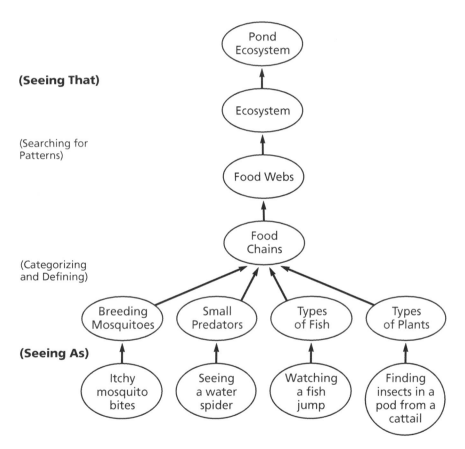

Figure 6.1.   Examples of the knowledge representations for the knowledge structure "pond." The arrows suggest some of the relationships found in this knowledge structure. Actually, many more could be drawn in. Every part of the knowledge structure is related to every other part in some way.

RECONSTRUCTION OF PRIOR KNOWLEDGE AND ELEMENTS OF AN EFFECTIVE PEDAGOGICAL STRATEGY: USING THE LEARNING CYCLE IN COLLEGE SCIENCE COURSES

Meaningful learning of new science ideas requires willingness and an effort of the student to change his or her own ideas about the world. Teaching involves helping students reconcile their own prior knowledge with the new science idea and therefore must focus on changing prior knowledge rather than only on introducing new science knowledge. The starting point of teaching is the ideas students bring with them. The ideas students bring to the classroom are termed "alternative conceptions" of reality. They are naive, poorly related to the real world, or are misconceptions. They bring these alternative conceptions to biology, chemistry, and physics classrooms. See Tables 6.2 and 6.3 for sample student conceptions found among students.

College instructors must go far beyond traditional teaching strategies to create meaningful learning in students. When presented with science ideas in lectures, science textbooks, or Internet readings, students have to modify and reconstruct their prior knowledge in order to accommodate the new ideas. Providing information about an idea through lecture or reading, followed by review, gives priority to mental recall actions and produces knowledge that is not useful beyond the first quiz. Students must be involved in learning experiences that foster restructuring of prior knowledge and making connections to the new science ideas (Stofflett, 1998). The underlying college science pedagogy (National Science Foundation, 1996) that supports this constructivist approach to teaching and learning is an inquiry science-teaching model.

Four general conditions need to exist to help students to discard or modify prior knowledge and accept a new idea based on real-world fact: (a) A student must be dissatisfied with an existing idea or misconception (confrontation), (b) a new idea must be comprehensible to the student (conceptual clarity), (c) a new idea must appear as plausible as the student's own idea (plausibility), and (d) a new idea has to be more useful than the previously held belief for solving problems or making predictions (validity). These conditions can be expanded to describe the mental restructuring required to create meaningful learning. They involve motivating students to (a) recall prior knowledge, (b) become aware of their prior knowledge, (c) make comparisons (confrontations) of prior knowledge with the new idea (often producing discrepancies) and analyze the problems that develop, (d) make connections of the new idea to students' prior knowledge and consider tentative solutions to the problems of discrepancies, (e) become aware of discrepancies with prior knowledge, (f) seek opportunities for students to test the new idea and evaluate the effectiveness of the tentative solutions to the discrepancies, (g) reconstruct prior knowledge and use new approaches when trials of the first tentative solutions are not successful, (h) construct their own new knowledge representation of the idea, (i) apply the new idea successfully, and (j) apply the new knowledge (transfer) in ways that are different from the situation in which it was learned (NRC, 2000; Posner, Strike, Hewson, & Gertzog, 1982; West & Pines, 1985).

Table 6.2.   A Sample of Alternative Conceptions That Students Bring to the Classroom (modified from Driver, Squires, Rushworth, Wood-Robinson, 1994).

Physical Science—Sample misconceptions

1. Force and motion misconceptions
   a. Force and pressure and energy are the same thing.
   b. Constant motion requires a constant force.
   c. If an object is not moving, there are no forces on it.
2. Gaseous states Misconceptions
   a. Air molecules are motionless.
   b. Air molecules expand.
   c. Compressed air is considered heaped up or shriveled.
   d. Volume and mass are the same thing.
   e. Gases can't be heated.
   f. Gases exert force in only one direction.
   g. The atmosphere exerts a pressure only on a surface.
   h. Nature abhors a vacuum.
   i. Air sucks things up.
   j. Pressure is density.
3. Matter in the gaseous phase misconceptions
   a. There is no space between gas particles.
   b. Air particles/molecules do not have intrinsic motion.
   c. Partial vacuum does not exist, must be vacuum or normal air.

Mathematics—Sample misconceptions

Ratio and probability misconceptions
   a. Ratio in a physical event is determined by addition or subtraction, (e.g., 4 is to 6 as 6 is to 8).
   b. All events must show change for a variable to have an effect.

Biological Science—Sample misconceptions

1. Osmosis misconceptions
   a. Do not understand the term concentration.
   b. Hydrostatic potential is not understood.
   c. Insist on an observable movement of solutions.
   d. Membranes cannot be semipermeable.
   e. Solvents always move, but solutes never move.
2. Amino acids misconceptions
   a. Amino acids, not enzymes, are products of translation.
   b. Amino acids come from the cytoplasm.

Table 6.3.   Aspects of Students' Prior Knowledge About Heat to Watch for (modified from Driver, 1986a)

Students may hold the alternative view that

1. "Heat " and "temperature" can be used synonymously: Some students may think that these words have the same meaning or that more heat means higher temperature.

2. The sensation of coldness is due to transfer of cold towards the body: The sensation of hotness is also due to heat transfer towards the body.

3. Heat and cold are seen as opposite, fluid materials: Some students use the word "heat" as a noun and write about it as though it flows into and out of objects. This may reflect little more than customary use of language. On the other hand it may reflect an underlying conception of heat as "stuff." This conflicts with the accepted view of *heat as energy in the process of transfer* and may make it more difficult for students to differentiate at a later stage between heat and the internal energy possessed by matter. Many students refer to cold and heat as though they are opposite substances.

4. Some substances are "naturally" colder than others: Some students do not understand the idea that substances in thermal contact are at the same temperature. They suggest that substances have a "natural' temperature (e.g., metal is naturally colder than plastic).

5. There are no degrees of conductivity: Students appear to think that either an object conducts heat (like metal) or it does not (like wood).

6. Metals have a greater capacity for heat than other materials: Because metals feel cold students think that metals draw heat towards them, or naturally hold heat more effectively.

7. Particles of matter have macroscopic attributes: Students may think that as heat is transferred to a substance, constituent particles expand, or melt.

8. Substances have "natural" melting and boiling points: The fact that a substance based on some observable property has reached the"natural" temperature at which melting occurs may be a sufficient explanation for some students (e.g., wax melts at a low temperature because it is soft).

9. Change of state occurs over a range of temperatures: Despite identifying a melting point correctly from a graph many students suggest that melting would occur over a temperature range. Everyday experience may be confusing here, because rarely are substances that are observed melting or boiling maintained at uniform temperatures.

Several inquiry pedagogy models have been devised that center on conceptual reconstruction. They all fit under the general name of "learning cycle" (Karplus, 1979). The original model was first proposed by Atkin and Karplus (1962) and later used in the innovative science program, Science Curriculum Improvement Study (SCIS). These models are similar in that they center on a strategy for reconstruction of knowledge that involves *experience, interpretation,* and *elaboration* in a specific sequence of learning experiences. The learning cycle sequence has been used effectively with college students in science courses (Allard & Barman, 1994; Zollman, 1990, 1997; and in this volume, chapters by McCormick & MacKinnon, ch. 22; Waggoner, Schaffner, Keller, & McArthur, ch. 23; Gabel, ch. 24; Zollman, ch. 25; Bland, ch. 28; and Karr & Sunal, ch. 30). The learning cycle assists college instructors in planning and using pedagogy to help students create meaningful learning and apply knowledge gained in the college classroom to new areas or to new situations. The learning cycle strategy involves the instructor in helping student (a) become more aware of their own reasoning, (b) search for new patterns in phenomena and new ideas, (c) develop and explore ideas in safe and supportive learning environments, (d) cooperatively and interactively experience course activities, (e) recognize shortcomings in their prior knowledge as a result of being encouraged to test it against other ideas, (f) apply procedures found in solving problems, and (g) apply new ideas in other areas and in new settings (Sunal & Sunal, 2003). Instruction must strengthen these elements of teaching in all students and discourage unquestioning acceptance of poorly understood concepts, theories, and thinking skills.

LEARNING CYCLE MODELS

Using the organizational sequence of the learning cycle can assist in the planning of a specific class lesson or set of lessons designed to modify students' prior knowledge. Particular sequences of classroom activities are required for conceptual change. However, what we know about prior knowledge suggests that these sequences cannot be planned until we have found out just what it is that individual students already know, their prior knowledge.

A pedagogical framework has been proposed by a number of researchers in the past for planning of science class lessons designed to change student conceptions. The common theme in this framework centers on a learning and teaching strategy and sequence that involves *experience, interpretation*, and *elaboration*. The learning and teaching models that fit under this framework have the general name of the "learning cycle" (Karplus, 1979). The models vary in that those developed earlier focus less on students confronting their own prior knowledge. In the later models students are overtly confronted and assisted in their struggle to create deeper meaning (Lawson, 1995; Osborne & Freyberg, 1985). See Table 6.4 for a summary of learning cycles models. Originally developed by Robert Karplus and Herbert Thier (1967), the learning cycle is an inquiry-based pedagogy that engages students in constructing meaningful science knowledge.

The earliest learning cycle models focused originally on the cognitive development of the student. One such model was proposed by Karplus (1977) and was influenced by Piagetian theories of development (Piaget, 1969, 1971). Karplus proposed that science learning is a process of self-regulation in which the student forms new reasoning patterns. All meaningful learning results from reflection on student actions and interactions with phenomena and with the ideas of others. Karplus developed a three-phase learning cycle sequence:

1. The first phase is one of exploration in which pupils learn through their own actions and reactions with minimal guidance, while the teacher anticipates few specific learning outcomes. The students are expected to raise questions that they cannot answer with their present ideas or reasoning patterns.
2. In the second phase of the Karplus model, the concept is introduced and explained. Here the teacher is more active, and learning is achieved by explanation.
3. Finally, in the application phase, the concept is applied to new situations and its range of applicability is extended, and transfer is encouraged in new contexts. Learning is developed by repetition and practice so that new ideas and ways of thinking have time to stabilize.

Renner (1982) suggested that the most common practice of teachers, both at the college and precollege levels, is to pass on to their pupils the content as understood by the teacher, the transmission model. The sequence of this approach to learning is that the science is provided to the student as information; the student verifies it through observation of a demonstration or lab observation; and finally, the information is applied in some way that involves practicing and recalling. The transmission model follows three steps:

1. The first stage, informing or telling, is usually completed through the teacher's introduction to a science activity.
2. The second stage is a science activity. This is not an experiment as in an investigation, however, but a verification of what both pupils and the teacher already know.
3. The final stage is an application of terms or definitions that usually involves answering questions and solving quantitative problems from a textbook for a quiz or exam.

Renner describes the learning sequences as that of a guided tour where the guide, the teacher, points out all the sights to be observed and the learner is discouraged from taking any detour that, in the guide's view, is not productive (Renner, 1982).

However, if we must construct our own meaning we have about an idea, Renner proposes a different teaching sequence as more effective in creating understanding:

1. The first stage of his teaching model involves students gaining experience with the new idea. Students are provided with relevant experiences in order to help create meaning of the idea for themselves.

2. In the second stage, the learner is provided an explanation and specific terminology in relation to the idea being investigated. The instructor uses this to assist the learner to interpret what has been found.

3. In the third stage, the new ideas are integrated with existing knowledge in order to expand both that knowledge and the newly acquired idea. Additional elaboration experiences are an essential part of the sequence. These experiences would parallel an investigation in that the outcomes would not be known.

Renner's view is that traditional science teaching is simply a training process that involves telling, confirming, and practicing (Osborne & Freyberg, 1985). The critical limitation is the lack of important activities that involve originating experiences, interpretation, and elaboration.

For Driver (1986b) an idea or preconception will not be rejected until there is something more appropriate and reliable to replace it. Students can be given experiences which conflict with their expectations, but these experiences do not of themselves help the pupils to reconstruct an alternative view of the system. Driver proposes the following sequence of instruction:

1. In the first stage, evidence or data should be presented so that students have an opportunity to explore and discover a new idea that challenges their old way of thinking.

2. In the second stage, the instructor must provide explanations of the new ideas because many times students are left with an incomplete construction or none at all after the first stage.

3. In the third stage, students are encouraged to gain a deeper understanding of the ideas by applying them in a range of activities. This will take the greatest amount of time; otherwise, students will simply memorize the new ideas.

Nussbaum and Novick (1981, 1982) have also suggested a three-stage sequence. Their sequence focuses on what happens as students change their conceptions during a lesson. The sequence is designed around the idea that science concept learning involves cognitive restructuring (accommodation) of prior knowledge surrounding the new science idea.

1. In the first stage, the student's prior knowledge is made public to the student and his or her peers. Nussbaum and Novick relate this stage to Ausubel's view that preconceptions are tenacious and resistant to extinction (Ausubel, 1968), and that such preconceptions often interfere with the teacher's learning outcomes. Thus, Nussbaum and Novick propose that the first step in facilitating cognitive reconstruc-

tion should be to ensure that every student is aware of his/her own preconceptions. To them, this is most easily achieved if some event can be devised which requires learners to make explicit their existing ideas in order to interpret it. Students are encouraged to describe their own views verbally and pictorially, and the teacher assists them to state these ideas clearly, in order to recognize what they can and cannot explain. Students are encouraged to confront and argue the various views represented by all of their fellow learners, in order to better understand the features of each view.

2. In the second stage, following learner dissatisfaction with their existing ideas in the first stage, the teacher provides additional experiences that will lead to conceptual conflict. The conflict must be great enough to motivate students to recognize that their existing ideas, prior knowledge, require modification.

3. In the third stage, cognitive reconstruction will develop from students' search for a solution to their conflicting ideas. In the Nussbaum and Novick sequence, making prior knowledge public, creating conceptual conflict, and encouraging cognitive reconstruction create concept learning.

Erickson (1979) proposed a similar sequence:

1. The first stage of his model is exposure to a set of activities, experiential maneuvers, which allow the students to become familiar with a wide range of phenomena. They might make public their intuitive ideas or beliefs. In this stage, the activities should be experienced in sufficient depth to allow the students to clarify their ideas and to make predictions.

2. The second stage involves anomaly maneuvers, including the creation of discrepant events that lead to unexpected outcomes. An element of uncertainty is introduced; the learner needs to restructure his/her views.

3. The third stage is a set of restructuring maneuvers that are designed to help students in accommodating the unexpected outcomes. Restructuring, in this strategy, might be created by, for example, group discussions and teacher intervention.

Barnes (1976) argues that students must be responsible for the formulation of their own knowledge. They must take the lead, it cannot be done by others. To reduce the teacher's control over learning, Barnes proposes that students should work in small groups. He proposed the following sequence:

1. Focusing stage: The teacher, through student actions, introduces alternative knowledge.
2. Exploratory stage: This involves much reflection on and discussion of the activities, including experimentation.
3. Reorganizing stage: The teacher refocuses attention on the alternative and student prior knowledge and provides guidelines to the groups on discussing the alternative knowledge, how they will report, and how long they have to prepare for it.
4. Public stage: The groups present their findings to one another, and this leads to further discussion.

A sequence proposed by Rowell and Dawson (1983) focuses on the confrontation between students' science and scientists' science. They proposed the following sequence:

1. During the first stage the teacher establishes, through questioning, the ideas that children bring to the problem situation. Awareness of these ideas is of value to both the teacher and the student.
2. During the second stage, the ideas of the students are accepted by the teacher as possible solutions.
3. During the third stage students are asked to retain their ideas, and the teacher states that he or she is going to put forward another possibility that may help the students. Together they will evaluate this new possibility.
4. During the fourth stage the "new" idea is taught by linking it to a basic idea already held.
5. During the last stage when the new idea is available for students, the old ideas are recalled for comparison with each other and with the real world.

Rowell and Dawson propose that students are less threatened by this approach since both "old" and "new" ideas are the pupils' own in the sense that all are the groups' knowledge. Old theories are rarely defeated by contrary evidence, but only by better theories. Rowell and Dawson argue that students with several ideas available to them are in the best possible situation to accept the scientific one when it is tested against the others.

The Generative Learning Model of teaching (GLM sequence proposed by Osbourne and colleagues, Osborne & Freyberg, 1985; Osborne & Wittrock, 1983) has four steps:

1. In the preliminary stage, before beginning any formalized instruction, teachers assess students' ideas and conceptual explanations.

2. In the focus stage, the teacher provides experiences related to the particular concept that motivates the students to explore their level of conceptual understanding.
3. Next, the teacher helps students exchange points of view and challenges students to compare and contrast their ideas and support their viewpoints with evidence (the challenge stage).
4. In the application stage, students use their newly refined conceptual understandings in familiar contexts.

The Riverina-Murray Institute of Higher Education (Boylan, 1988) presents a five-stage model of learning and teaching that students must pass through as they develop a new level of conceptual understanding. The stages are

1. The teacher identifies the student's naive ideas about a selected concept.
2. Based on that information, the teacher selects events, situations, and activities for the learner to explore.
3. The exploratory phase provides a practical base upon which the learner begins to develop a new understanding. In Stage 3 the student is encouraged to make the concept explicit and also is introduced to new language and symbols.
4. In Stage 4 the student organizes the new idea and establishes links with relevant prior knowledge; a new mental scheme emerges.
5. In Stage 5 the student practices and applies the new idea in novel situations to consolidate the newly developed understanding.

Hewson and Hewson (1988, p. 607) after reviewing studies on science learning, summarize "key points in instructional strategies that help students overcome their naive, inappropriate conceptions." Teachers must (a) diagnose students' thoughts on the topic at hand, (b) provide an opportunity for students to clarify their own thoughts, (c) directly contrast students' views and the desired view through teacher presentation or class discussion, (d) immediately provide an opportunity for students to use the desired view to explain a phenomenon, and (e) provide an immediate opportunity for students to apply their newly acquired understanding in novel situations.

Anton Lawson (1988), Michael Abraham (1989), and colleagues (Lawson, Abraham, & Renner, 1989; Renner, 1982) have described a three-step learning cycle based on the original model first proposed by Atkin and Karplus (1962):

1. The first stage uses a laboratory experiment, activity, or demonstration to expose students to the concept to be developed. Abraham calls this the exploration or gathering data phase.
2. Next, the students and/or teacher derive the concept from the data, usually a classroom discussion (the conceptual invention phase).
3. The final phase, expansion, gives the student the opportunity to explore the usefulness and application of the developing concept.

Lawson (1988) and others prefer to call the second phase "term or concept introduction" because they recognize that, while teachers can give students new terminology, ultimately the student must actively invent or generate the concept. Lawson has subsequently proposed three kinds of learning cycles: descriptive, empirical-deductive, and hypothetical deductive (Lawson et al., 1989). The sequence of learning-teaching events is essentially the same in each.

Driver and Oldham (1986) describe a constructivist teaching sequence used in the Children's Learning-in-Science Project and propose that it be viewed as a flexible outline because the demands of different conceptual areas and the time available for learning and teaching will vary. In the orientation phase, students are motivated to learn the topic. In the elicitation phase, students make their ideas explicit through discussions, creation of posters, or writing. In the restructuring phase, teacher and students clarify and exchange views through discussion, promote conceptual conflict through demonstrations, exchange ideas, and evaluate alternative ideas. In the application phase, students use their new ideas in familiar and novel settings. The review phase allows students to reflect on how their ideas have changed. The model incorporates several aspects of technological problem solving, decision making, notable evaluation of alternative ideas, and reflection at the end of the learning sequence.

Extensive research using the learning cycle model has been conducted over the past 30 years. Renner and associates carried out a series of studies on the Science Curriculum Study Program that demonstrated the viability of the learning cycle in learning science for younger students (Renner et al., 1976). Other studies in classrooms of various science disciplines concluded that the form and sequence of the learning cycle created higher achievement than other varieties of learning strategies or organizations for learning science (Abraham & Renner, 1986; Renner, Abraham, & Birnie, 1985, 1988; Renner & Marek, 1990). Smith and Lott (1983) and Abraham (1989) provide an analysis of the use of this particular learning cycle within college classrooms. The structure and organization of the learning cycle are effective in enhancing student achievement.

Table 6.4. Changing Student Ideas: A Variety of Learning Cycle Frameworks

Phase	Renner	Karplus	Driver	Nussbaum & Novick
1	Experiences	Exploration	Discovery	Exposing Alternatives
2	Interpretation	Explanation	Presentation	Creating conceptual conflict
3	Exploration	Application	Application	Encouraging cognitive accommodation

Phase	Erickson	Barnes	Rowell & Dawson	Osbourne & Wittrock GLM Model
1	Experiential maneuvers	Focusing	Establish initial ideas	Assess student ideas
2	Anomaly maneuvers	Exploration	Introduce new ideas	Exchange points of view
3	Restructuring maneuvers	Reorganizing	Comparison of ideas	Use ideas
4		Public		

Phase	Riverina & Murry	Hewson & Hewson	Lawson & Abraham	Driver & Oldham
1	Identify naive ideas Select events	Diagnose	Exploration	Orientation and motivation
2	Exploratory activities	Opportunity to clarify and contrast	Conceptual invention	Elicitation of ideas
3	Organize ideas and establish links	Practice new idea		Restructuring ideas through exchange
4	Practice and apply new idea	Apply idea	Expansion	Application and review

PLANNING SCIENCE USING THE LEARNING CYCLE

The general model of the learning cycle (Karplus, 1979) involves students in a sequence of activities beginning with experiencing (exploration) of an idea or skill, leading to a more guided explanation or interpretation (invention) of the idea or skill, and culminating in elaboration of the idea or skill through application and transfer to new settings (see Figure 6.2). This sequence represents a single lesson on one concept lasting one to several instructional periods depending on the complexity of the science idea (Sunal & Sunal, 1991a, 1991b). Because of what occurs in each phase, the three parts of the learning cycle are called exploration (experience), invention (interpretation), and expansion (elaboration). A college instructor has a large number of choices in deciding how to provide learning activities for

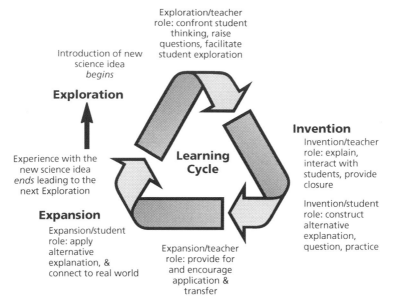

Exploration/teacher
role: confront student
thinking, raise
questions, facilitate
student exploration

Introduction of new
science idea
begins

Exploration

Invention

Invention/teacher
role: explain,
interact with
students, provide
closure

**Learning
Cycle**

Experience with the
new science idea
ends leading to the
next Exploration

Invention/student
role: construct
alternative
explanation,
question, practice

Expansion

Expansion/student
role: apply
alternative
explanation, &
connect to real world

Expansion/teacher
role: provide for
and encourage
application &
transfer

Figure 6.2.   The learning cycle.

each phase of the learning cycle. Methods for helping students restructure their prior knowledge and link it with a new idea include broadening the range of application of a prior conception, differentiating a conception from another, building experiential bridges to a new conception, relating and updating prior knowledge related to new conceptual problems, inventing or importing a different model or analogy to help with the new idea, progressively shaping a prior conception to more closely simulate the new idea, and providing experience to help students construct a new idea. The selection of learning activities to use in each phase of the learning cycle (e.g., lecture, inquiry lab investigation, discussion in a small group, Internet simulation, reading of textbook, etc.) should be determined by (a) the type of idea(s) or skill(s) to be taught, (b) the cognitive level and specific learning needs of the student, (c) the part of the learning cycle the college instructor is involved with, (d) the form and content of students' prior knowledge, (e) the number and kind of learning activities needed to create conceptual restructuring, and (f) the type of knowledge representation required for the idea to be understood (Sunal & Sunal, 2003).

Beginning the Learning Cycle: The Exploration Phase

Exploration provides students with the opportunity to use their prior knowledge to explore a range of phenomena for themselves and experience a confrontation to their own way of thinking (Driver, 1986b). The goal is to produce some disequilibrium with experienced events so that prior knowledge is brought to the forefront and is available to the student for restructuring. The instructor is not concerned with right or wrong but rather with facilitating the gathering of data and observations of the students. The experience itself may produce restructuring in a few students, based on their own initiatives.

When a college instructor plans the exploration part of a lesson, several decisions are made. See Tables 6.5 and 6.6 for typical decisions and learning activities useful for planning the exploration phase. An objective is selected or developed that is relevant to the scientific ideas that are key to the lesson and to the students' past experiences. Sometimes it is best to begin by thinking of a desired activity to be performed by students. With this activity in mind, it may be easier to develop a focus for the lesson objective. Most experienced instructors plan their lesson focus this way (Clark & Peterson, 1987). For example, the lesson may be planned around a textbook chapter that discusses the topic of soil types. The instructor may be thinking about an activity in which students are asked to dig a small hole in an undisturbed piece of forest ground. Imagining the activity or trying it out will lead the instructor to focus on the properties of the soil, layers of soil, and how the soil gets to be the way it is now seen. The activity will easily lead to the thinking skills needing development along with the science concepts. The skills would include observation and classification of various soil properties, communication and inferences relating to patterns found in the information, and prediction and hypothesis leading to generalizations made about soils.

Next, the instructor decides how the initial part of the lesson can best be used to prepare students' minds for accomplishing the lesson's objective. In making this decision, the instructor thinks about five elements necessary in planning an effective sequence of learning. What initial activities will

1. *Focus* students' attention on experiences related to the new idea or skill to be taught?

2. *Confront existing knowledge* of students? Start with a "key" question that involves them in a physical or mental activity that focuses their attention.

3. Encourage students to *recall and relate previous knowledge* to new knowledge?

4. *Bring out and make public* what the students now know, their prior knowledge?

5. Provide an opportunity for students to *try out prior knowledge?*

Table 6.5.   Steps in Planning a Class Lesson Using the Learning Cycle for a New Idea or Thought Process (Thinking Skill/Algorithm)

The learning cycle sequence is not a blueprint for teaching, but a set of decision points that all college instructors must address in the planning process if they are to adequately help students learn important ideas. Recognizing these decision points assists instructors in deciding to act in keeping with what is known about how learning takes place.

Exploration

- Attempt to *confront existing knowledge* of students. Start with a "key" question that involves them in a physical/mental activity which focuses their attention.
- *Focus students' attention* on experiences related to the new idea or skill to be taught.
- Encourage students to *recall and relate previous knowledge* to new knowledge.
- *Bring out and make public* what the students now know, their prior knowledge.
- Provide an opportunity for students to *try out their prior knowledge in the new setting.*

Invention

- Ask students to *reflect* on and *discuss* the results of Exploration activity to provide connections to focus the idea of the lesson.
- Provide a *clear explanation using multimedia* and interacting with students where possible describing aspects, analogies, contexts, and uses.
- Provide *clear examples or model* the new skill.
- Provide *student practice* using the new knowledge.
- Provide a concise brief *closure.*

Expansion

- Provide *additional student practice* activities if needed. Use personally relevant examples, not abstract, repetitive practice.
- Provide *student application activities* in new relevant contexts. Multiple activities should *transfer* the new knowledge to increasingly real world situations and involve *more relevant* to students personal and professional needs.
- Provide a *summary* that highlights and focuses attention on the experiences where the new knowledge was learned.

Focusing attention and confronting existing knowledge.   As a classroom lesson begins, students' minds are in different places. They may be thinking about what they are going to do tonight, eat for lunch, or about a test later that day. When students' attention is focused on the intended lesson objective, greater learning will be likely to occur. The first few minutes of any lesson are important. Sensory information is held in the student's memory for less than one second before the mind decides what to hold and what to forget. Most sensory information is not perceived by the student and is forgotten. If the instructor does not help a student focus on the lesson's key ideas and skills the student may not obtain information relevant to the meaningful

Table 6.6.   Sample Characteristics of Each Phase of a Learning Cycle

Exploration Phase

Purposes:

To provide background experience and learning through students' own actions and reactions and

To introduce aspects and values of a new idea—concept, variable, generalization, or thinking skill (enhances assimilation).

Exploration allows students to confront and make evident their own thinking/representation of the idea or skill to be learned.

Characteristics include

1. Encourages learning through students' own inquiry and focuses interest.
2. Involves minimal guidance or expectation on the instructor's part.
3. Often provides an experience that confronts students' old way of thinking.
4. Begins with a well-planned "key" question from the instructor.
5. Involves students working in cooperative learning groups.
6. Encourages observation of natural world.
7. Raises questions for the students.
8. Provides for student action with hands-on materials, collecting and organizing data.
9. Encourages students' mental actions in selecting resources discussion and debate.
10. Encourages trying out prior ideas, suspending judgment, predicting, hypothesizing and testing.
11. Provides students with adequate time to relate prior knowledge with new idea.
12. Allows students to know the purpose and objective of the lesson.
13. Allows instructor to know present student understanding in the lesson objective area.

Invention Phase

Purpose:

To explain an alternative (new) idea or situation leading students to mentally construct new patterns of reasoning (encourages accommodation).

Invention builds on the Exploration by guiding the students through a more direct teaching format, to experience and develop the concept or skill more fully or to a higher order.

Characteristics include

1. Continues development of the new idea or skill (reasoning pattern) in students through instructor directed reflection and discussion of Exploration experience.
2. Involves communication of information and ideas offering alternative ideas (solutions) for the confrontation.
3. Allows learning from "explanation" which includes an interesting variety in teaching actions, multimedia, and interactions with students describing aspects, ranges, contexts, and uses of the new idea or skill.

Table 6.6.   Sample Characteristics of Each Phase of a Learning Cycle

4. Introduces idea or skill in a structured manner through additional student experience using a variety of demonstrations, analogies, audiovisual materials, sense modalities, textbook readings, or other medium.

5. Encourages students to develop as much of the new reasoning pattern as possible through providing one or more complete cycles of explanation, giving clear examples, modeling, and checking for understanding.

6. Offers students time to question, try out and practice the new alternative explanation.

7. Ends with a concise closure describing the main idea or skill introduced.

Expansion Phase

Purpose:

To apply and transfer the new reasoning pattern (idea or skill) to other example(s), situations and contexts extending the range of applicability to help stabilize (make permanent) the new knowledge.

Expansion activities allow student's to practice, apply, and transfer the idea or skill just explained in the Invention.

Characteristics include

1. Provides for learning by additional practice where students use labels, definitions, explanations, and skills in new, but similar situations. It is important to use personally relevant examples, not abstract, repetitive practice.

2. Provides additional time and experiences for students to ask questions, observe, record, use explanations, make decisions, and design experiments to apply the new idea or skill in new, but similar situations.

3. Encourages transfer of the new knowledge to various real-world contexts and other times different from where the new idea or skill was explained.

4. Relates student activities to personally and professionally relevant settings, thus, helping complete abstraction from classroom and textbook concrete examples.

5. Ends with a lesson summary which highlights and focuses attention on the experiences where the new knowledge was learned.

Note: If a phase is eliminated or all students are expected to demonstrate specific accomplishments after each one, then the overall effectiveness of the learning cycle will be compromised.

perception and later construction of the ideas into knowledge. This is necessary to help students connect the lesson's activities to their prior knowledge. The student's attention will always focus on something. The instructor is responsible for focusing it on the key idea of the lesson.

Students' attention can be focused in many ways. Students performing or demonstrating discrepant events is effective at this point in the lesson. For example, to focus students' attention on temperature, the instructor can ask students to touch different objects near them and describe how warm they feel. The instructor might also use verbal statements of procedure, thinking aloud, a demonstration, or a problem situation to focus students' attention. Such activities can be omitted if students are already focused on the intended objectives and are ready to continue learning.

Bring out and make public what the students now know.   While it is possible to carry out a lesson without knowledge of a student's prior knowledge, meaningful science learning will usually not result. Students who have not systematically experienced soil properties, for example, will not meaningfully understand concepts such as soil type and soil layering. In previous experiences students will have had to observe several examples of different soils and then classify them by common properties. These properties may have included size, color, shape, stickiness, and roughness to the touch. This activity may be related on class day to layers of soil the students discover while handling soil types or digging a hole, or to a simulated example of soil layering that students construct.

While students are actively participating in these experiences, an effective instructor will observe their interaction with the materials, the content of their discussion, and the meaning they begin to make of their explorations. This information will help the instructor decide how relevant the new ideas are to the students in view of their existing knowledge and the extent of instructor guidance needed in the next part of the lesson.

Relate previous learning to new learning and try out prior knowledge.   In order to help students remember what they have learned or transfer it to other things they know, it is important to use open-ended questions and examples of materials which call attention to their past experiences to try to help students recall them from their long-term memory. The instructor should help students retrieve as many related ideas or skills from long-term memory as possible. The retrieval of relevant information provides a knowledge structure into which new material learned can be placed. The retrieval also makes the student's prior knowledge public. The student is made personally aware of her prior knowledge, in a nonthreatening way, as are her peers and instructor. This is important. The student must check the ability of her prior knowledge to predict real-world events. This will allow the student to determine the adequacy of the view or suggest a possible better one to replace it. Unless students confront their misconceptions about the real world, they will keep them. They are likely to use the new knowledge on the next test but will forget it afterward since they are unable to transfer it to any new setting.

Recall from long-term memory alone is not a sufficient beginning for a lesson. Asking students, for example, to recall what happens when they have dropped different objects from a height is not sufficient to relate previous learning about gravity to new learning about this idea. However, encouraging students to explore gravity with a set of objects in free fall and on inclines of varying angles, encouraging them to make predictions about their speed, weights, and time of falling motion, and to record the results of tests, could result is an effective first activity introducing the concept of gravity. A discussion of their experiences after students have worked with the objects will tie the activity more completely into their prior knowledge. Often one activity at the beginning of a lesson can accomplish all three purposes: focus, bring out, and relate old learning to new learning.

Continuing the Learning Cycle: The Invention Phase

Here the instructor should introduce a competing "scientific" conception to the student's prior knowledge. The invention should help students organize their information from the exploration phase. When planning the invention part of the lesson instructors make decisions on the following questions: (a) How can the *exploration experiences be developed to focus* on the basic idea or skill to be taught? (b) How is the idea or skill *best explained?* (c) How should the idea or skill be modeled or demonstrated? (d) What strategies or techniques should be used to make sure all students understand it? (e) What *student practice* is needed using the new knowledge? and (f) What would be a concise, brief *closure?* This part of the lesson is more instructor guided. The instructor provides students with clear explanations and examples, then completes this phase of the lesson with a closure in which the idea or skill being taught is defined and clearly stated. See Tables 6.5 and 6.6 for typical decisions and learning activities useful for planning the invention phase.

Providing an explanation. Explanations may be provided in a variety of ways, including discussion of findings resulting from the exploration activities, lecture, multimedia presentations, computer simulation, viewing a videotape, explaining sections of a textbook, and focused student activities. Since the short-term memory has a limited capacity, instructors must make certain that only important information is provided. When visuals and graphic organizers such as pictures, graphs, and demonstrations accompany verbal explanations, more information can be stored efficiently. All aspects, conditions, and contexts of a concept or experience with a generalization should be provided to students in order to concretely demonstrate its basic structure. An example of a concept might be "cold-blooded animals." A generalization might be "the lower the temperature of the surroundings, the slower the pace of the bodily functions of a cold-blooded animal." Note-taking guides often help make the organizational structure of the lesson more concrete when provided at the beginning of the invention phase. This is especially helpful when much of the explanation is done using lecture, textbook, or video presentation.

Providing examples and practice.   Students need to see and practice clear examples of what the new ideas or skills represent so they may easily compare this new idea with their prior knowledge. One or more examples demonstrating the idea or skill should normally be presented at this point in the lesson. Sometimes this consists of demonstrating knowledge or skill through analogies, using working models or mathematical modeling. It also could involve taking the students through a step-by-step process. The more ways in which an idea or skill can be modeled for students, the more meaningful it will be to them (Clarke, 1990). The instructor should help the students distinguish between the varieties of alternative conceptions possible.

An example of the concept of "cold-blooded animals" should involve students in experiences in which these animals take on the temperature of their surroundings, a range of cold-blooded animal types is observed, and the range of temperatures where these animals can function effectively is noted. To learn the generalization, "the lower the temperature of the surroundings, the slower the pace of the bodily functions of a cold-blooded animal," the students should be asked to make predictions and test out hypotheses about the breathing rate of an animal as the temperature is reduced. This could be done through simulated data or by placing a small goldfish in a jar of water, adding an ice cube, counting the gill slit movements for short intervals of 10 seconds, and measuring the temperature of the water with a thermometer. Communicating and discussing the results of predictions should be done in small groups and in whole-class meetings.

Closure.   It is important to plan activities to make certain that students have a clear description of the science idea or skill they have been working with. An instructor should accomplish this at the end of the invention by using one of the following strategies. Stating the idea in a clear form to the students, asking students to state the main idea of the lesson orally or in writing the idea, or asking the students to demonstrate the skill may accomplish a concise description.

Completing the Learning Cycle: The Expansion Phase

The expansion phase is perhaps the most important, but most overlooked, part of the lesson in traditional teaching. The goal is to help students finish restructuring old beliefs, prior knowledge. After the invention or explanation part of the lesson, it is important to help students apply and transfer the new idea to new situations. This additional practice will help students retrieve it from their memory when they need to do so. This learning phase will require some time. The instructor must decide how to provide the practice necessary for accomplishing transfer into long-term memory. See Tables 6.5 and 6.6 for typical decisions and learning activities useful for planning the expansion phase. The instructor should stress the importance of thinking and talking about the significance of the experiences. The instructor should always act as the mediator between the student's prior knowledge and the scientific view of the new idea. Types of practice include manipulative activities, paper-and-pencil problems, question-and-answer discussions, field trips, games, computer simulations, and a return to the confrontation met in the exploration phase.

Additional practice and application activities. At first, the instructor should guide student practice of an idea or skill. This enables them to receive feedback that tells them when they are accurate and successful. Without such guidance, students might practice errors, creating misconceptions that will require a great amount of effort to unlearn. Concrete or computer simulation activities should be used for practice. For example, students can be shown an example of a parallel electrical circuit. Then, they can be asked to find out which bulbs will light using this circuit. Or, the instructor could demonstrate a problem circuit and ask, "What is wrong with this circuit?" These activities let the instructor know how accurately the students understand the lesson objective—the idea or skill taught. An important part of this practice includes asking students to explain their answers and to describe their evidence for making the explanation, whether correct or incorrect.

At this point in the lesson, the instructor must decide whether the students have sufficient experience with the new idea to transfer it to a new context. Observation of student performance allows the instructor to decide which students are ready to move on to an application activity and which students need more practice or even reteaching.

Transfer activities. In order for an idea or skill to be remembered and used automatically from long-term memory, sufficient application and transfer is needed, spaced out over time and in different contexts. After students perform the new skill or use the new idea in the classroom context, they are ready to transfer the new ideas to different situations and times. Often, this step is omitted in traditional college courses because students have given some evidence of recall learning of the new idea earlier in the lesson. This amount of evidence is usually thought to be sufficient. However, students need extended experience in using the new idea in a new context over a period of time before an idea or skill can be stabilized in the long-term memory (Perkins & Salomon, 1991). An example of practicing a new idea in another situation occurs when an instructor follows up the circuit activities described above by having students investigate the inside of simple, electrically wired devices and circuit diagrams to determine what types of circuits are used or to draw the possible electrical wiring for a room in a house.

Giving a summary.   Following the expansion activities a brief summary of the lesson should be given. The summary should include the sequence of the important ideas and events experienced in the lesson. Students can be asked to summarize or the instructor can give the summary.

PLANNING WITH AND USING THE LEARNING CYCLE: PREREQUISITES AND LIMITATIONS

The learning cycle approach is best used as part of a pedagogical strategy that also stresses learning for all students, effectively uses cooperative learning groups, provides continuous assessment and feedback, and evaluates and rewards development of thinking skills such as critical thinking, creativity, development of self-worth, self-reliance, and respect for the opinions of others. In order for conceptual restructuring to occur and for meaningful learning to result from a learning cycle strategy, there are several prerequisites that need to be fulfilled:

1. Change in an idea or skill should not be too great. Students should be challenged but not overwhelmed.

2. Lesson content should be related to the background experiences of the students. This will help create a mental structure to which the new idea or skill can be tied. This is important if the new idea or skill is to be easily retrievable from long-term memory.

3. Many concrete examples should be used in real situations during a learning cycle. Technology should be very helpful here.

4. Students should be provided with opportunities to work through practice situations using real or simulated actions. Learning should usually take place in cooperative learning groups.

5. Students should be given time to reflect, make mistakes, and form the new ideas or skills.

6. Teaching using the learning cycle requires less coverage of content but increased understanding of basic concepts. The alternative is to teach using traditional procedures that foster memorization and a greater coverage of content in much less depth.

7. Using the learning cycle requires that the instructor select only those concepts that are basic to understanding the principle or theory. If time is a problem, it is possible to teach the other less important concepts in a more traditional style with some increased understanding.

8. Addressing all phases of the learning cycle is required in the order listed. Deleting a phase will create significantly less meaningful learning.

Other prerequisites may exist in various situations but these eight are always essential.

While most students benefit from use of the learning cycle, gifted students require less dependence on use of the learning cycle and more open general inquiry activities for conceptual reconstruction to take place. The learning cycle may be planned to teach different types of objectives. Teaching a thinking skill, concept, generalization, or a theory will require the learning cycle to include different learning activities. For instance, generalizations require an investigative set of activities starting with a hypothesis and testing in the exploration and leading to verification of the hypothesis in different contexts during the expansion phase.

The following learning cycle lesson on light and reflection from mirrors provides a detailed example for a freshman undergraduate physical science course. Other examples of the use of learning cycle inquiry strategies can be found in Chapters 22, "A Model for Reform in Teaching in the Biological Sciences: Changing the Culture of an Introductory Biology Course"; 23, "A Model for Reform in Teaching in the Biological Sciences: Infrastructure for Inquiry in an Introductory Biology Laboratory"; 24, A Model for Reform in Teaching Chemistry: With a Focus on Prospective Elementary Teachers; 25, "A Model for Reform in Teaching Physics: Large-Enrollment Physics Classes"; 28, "A Model for Reform in Teaching in Engineering and Technology: Creating Links Among Disciplines for Increased Scientific Literacy"; and 30, "A Model for Reform in Teaching in Engineering and Technology: Artificial Intelligence Systems in Science" of this volume.

LEARNING CYCLE EXAMPLE:
REFLECTION OF LIGHT AND MIRRORS

(American Association for the Advancement of Science, 1993; Arms & Camp, 1982; Kowalski, 1993; Parker, 1994).

Student Alternative Conceptions (Misconceptions) Addressed by the Lesson

1. Sight "comes from our eyes" with or without the need for a light source.
2. There is no pattern for how light reflects off an object.
3. We see simply because light is reflected off the surface of an object.
4. These alternative conceptions of students result from other, possibly more basic conceptions (e.g., You can see in the dark. Reflected light

is not important in seeing images.), and lead to more complex alternative conceptions and misconceptions, relating to real and virtual images and the functions and characteristics of lenses.

Lesson Goal:

Students will investigate light reflection in mirrors and determine how light is transferred to the eye.

Prerequisites:

Students are familiar with the concepts of *plane mirror*, parts of the human *eye*, and lab safety procedures with mirrors and light sources.

Exploration—Begin Lesson

Objective: Students will make predictions and then test them regarding the size of a mirror and the distance needed to the full length of a person standing in front of it.

Materials: For each group:

Mirrors (one 12–20 cm & one 100 cm)	Worksheet 1 & 2
Graph paper	Meter stick
Rulers (30 cm)	Colored pencils
Protractor	

Procedure:

A. Place the students in lab groups of four and assign roles or ask student groups to decide on roles: materials manager, task coordinator, spokesperson, and a data recorder. All members will be observers.
B. Show the students a mirror.
C. Students as Individuals:

    1. State the key questions: "How large does a mirror have to be in order for you to see your whole body in it?" and "Does the distance you stand from a mirror make a difference?" How?
    2. Have the individual students predict the results, write them down, and illustrate their results in a drawing on Worksheet 1.
    3. Ask the students to discuss their individual predictions in their groups.

D. Students in Groups:

    Ask the groups to perform the following activity, discuss the results in a group, and write down what they find. Students should record their data in a form that will show their results clearly on group Worksheet 2. The data should be recorded in a data table and in a descriptive narrative.

1. Begin by asking the question, "What is the length of a mirror needed to see your whole body?"
2. Allow the groups to begin their investigation of the question in 1.
3. When groups have a result, challenge them with the question, "Does the distance you are from the mirror make a difference in the size needed to see your whole body?"
4. Have the groups continue their investigation of the questions in 1 and 3.

E. Ask each group to discuss the results of Activity D and their individual responses to Activity C above.
F. Have students turn in their individual responses on Worksheets 1 and 2.

Evaluation: Evaluate Worksheets 1 and 2. Each student will have completed predictions for the two key questions. Students' predictions should be evaluated for their prior knowledge. The instructor should monitor student participation in the group by observing if the groups started quickly, stayed on task while working, and whether each person performed their assigned role in the activity.

Invention—Continue Lesson

Objective.   The students will relate their findings in the exploration to a description of the way in which we see light in a mirror.

Materials: for each group:

Graph paper	Protractor
Ruler (30 cm)	Meter stick
Colored pencils	Worksheets 3 & 4
Mirrors (three 5 cm, one 20 cm, & one 100 cm)	Light source with slit

Procedure:

A. Continue the student group structure used in the exploration.
B. Have the spokesperson from each group present their results to the questions above, relating their investigation results to their original predictions.
C. The instructor should discuss the various student results found as a whole class, illustrating the responses on the board. Ask students to provide evidence for their ideas as to why their results may not have met the original predictions made on Worksheet 1.
D. Ask the students to solve the following problem and determine a "rule" which will always make them successful. *Set up two mirrors and a 3×5 card in the front of a light source in such a way that when the light is*

turned on, the beam will bounce off Mirror 1 to Mirror 2 and onto the 3 × 5 card. See Worksheet 3. More difficult problems should be given to groups who solve the first problem quickly. Have groups report results to the class through a spokesperson.

E. Provide a geometric explanation demonstrating the results the students found in D, if not presented in C above, using the concepts of light rays, reflection, plane mirror, light entering the eye, and images formed in a mirror. Answer any questions the students may have with these concepts and their relationships.

F. Again ask the groups the following thought questions: "Does the size of the mirror used make a difference in how much you were able to see?" and "Does the distance you stood from the mirror have an effect on the size of the image?" and "How does this relate to the rule you just discovered?" Ask the groups to investigate and discuss these questions again using the available materials on group Worksheet 4.

G. Have the groups report on the results of the questions investigated through a spokesperson.

H. Provide a geometric explanation of the results of viewing self in a mirror, if not presented clearly by the student groups. Use a handout, video, laserdisc, or software pertaining to the eye and how we see to reinforce the concepts. A combination of the above would be most beneficial.

I. Closure: Provide an explanation of light as reflected from plane mirrors.

Evaluation: Evaluate Worksheets 3, 4, and 5. Each group of students will have completed all procedures for the invention activities. Their worksheets should be evaluated for their completeness in attempting to explain the phenomena using the new information. Additionally, monitor participation in the group by observing if the groups stay together on task, and whether each performed their role in the activity.

Expansion—Complete Lesson
Objective: The students will explain what is happening when observing light from a Reflectance Box.
Materials: For each group:

Reflectance Box
Worksheet 5
Graph paper
Colored pencils

Procedure:

A. Place the students in groups of four, and assign the roles of materials manager, recorder, group leader, and reporter.

B. Describe for the students the materials and instructions needed for student groups to carry out the activity of using the Reflectance Box.

C. State the key questions: "What happens to the light in the mirror as it is turned?" and "Why does this happen?"

D. Have the groups perform the following activity:

    1. Place the mirrored, three-sided pyramid in the Reflectance Box.

    2. Start with a flat side facing the student viewing the pyramid in the box. Be sure that there is enough light entering the box to see a reflection.

    3. Have one of the members of the group rotate the pyramid slowly from the hole in the back of the box. Repeat this so that *all* members of the group have both viewed and rotated the pyramid.

E. Have the groups record their observations on Worksheet 5 and relate them to the questions in C.

F. Have the student groups report, discuss, and show diagrams of light rays of their responses to Worksheet 5 with the class.

G. Ask the students to describe and explain how two of the following devices or activities work: periscope, reflecting telescope, rearview mirror, putting on makeup, dentist's mirror, or security mirror. Have the groups write down their answers to turn in.

H. Summarize the results of the lesson by describing some of the alternative conceptions the students demonstrated for how light is reflected by a mirror and the activities experienced during the lesson.

Evaluation: Have the students work in their groups to come up with an answer to the following questions: "How can this technique be used to determine the distance to the moon?" Their answer should be evaluated for their correctness at explaining the phenomena observed in the expansion using the information at their disposal. Additionally, you should monitor their participation in the group by observing if the groups stay together on task, and each person performed their role in the activity. In addition, monitor groups to see if they review what to do before starting and that they talk about each student's ideas and why it may or may not work, but do not criticize the person.

Sample quiz item: Design, describe, and illustrate a way to observe a person coming down a hall without being seen using only mirrors.

Abbreviated Versions of Student Worksheets

Worksheet 1.   Make these predictions by yourself. Write them down and illustrate your predictions in a drawing below.

1. How large does a mirror have to be in order for you to see your whole body in it?
2. Does the distance you stand from a mirror make a difference? How?

Worksheet 2.   Perform these investigations with your group. Describe the investigation plan and the results and illustrate your investigations in a drawing below.

1. How large does a mirror have to be in order for you to see your whole body in it?
2. Does the distance you stand from a mirror make a difference? How?

Worksheet 3.   Perform these investigations with your group. Describe the investigation plan and the results and illustrate your investigations in a drawing below.

1. Set up two mirrors and a 3 × 5 card in the front of a light source in such a way that when the light is turned on, the beam will bounce off Mirror 1 to Mirror 2 and onto the 3 × 5 card. Describe what you did to solve the problem. Draw your trials and your successful attempt. What is the "rule" for success?
2. Make up a more difficult problem to solve using the same rule.
3. Describe the rule for reflection from mirrors.

Worksheet 4.   Perform these investigations with your group. Describe the investigation plan and the results and illustrate your investigations in a drawing below.

1. Does the size of the mirror used make a difference in how much you were able to see?
2. Does the distance you stood from the mirror have an effect on the size of the image?
3. How does this relate to the mirror rule you just discovered?

Reflectance Box

Perform these investigations with your group. Describe the investigation plan and the results and illustrate your investigations in a drawing below.

1. What happens to the light in the mirror as it is turned?

    (a) Place the mirrored, three-sided pyramid in the Reflectance Box.

(b) Start with a flat side facing the student viewing the pyramid in the box. Be sure that there is enough light entering the box to see a reflection.

(c) Have one of the members of the group rotate the pyramid slowly from the hole in the back of the box. Repeat this so that *all* members of the group have both viewed and rotated the pyramid.

2. Why does this happen?

SUMMARY FOR INNOVATIVE PEDAGOGY FOR MEANINGFUL LEARNING IN COLLEGE SCIENCE COURSES

Before students experience any formal teaching about science concepts in the college science classroom, *they are likely to have formulated intuitive ideas about the concepts that enable them to explain and predict familiar phenomena to their own satisfaction.* This intuitive prior knowledge is reinforced by students' everyday use of language, and is context dependent and very persistent. When students are presented with ideas in a science course, *they make them fit into their prior knowledge,* and the result may be a mix of classroom science and intuitive science.

To create meaningful learning in college science courses, *students have to actively modify and restructure their own ideas.* This requires a willingness and effort on the part of the learner. If the ideas held by the students are to be taken into account, teaching cannot simply be seen as the "telling" or "giving" of knowledge to passively sitting students. Teaching involves helping each student to construct for her or himself the accepted ideas. The starting point of a teaching sequence is the prior knowledge students bring with them. Planning and use of the learning cycle strategy as a general pedagogical model helps students restructure their ideas and fosters meaningful learning. Specific examples of instruction using the learning cycle of guided inquiry are provided to illustrate meaningful learning in undergraduate science courses. The following points summarize the learning cycle. To accomplish meaningful learning,

1. Most students need a learning sequence different from traditional pedagogy used to recall facts of science.

2. The worthwhile objectives (concepts, generalizations, theory, thinking skills, or dispositions) in the science topics taught require a strategy of instruction different from traditional classroom presentation. Identify these topics in advance and plan lessons using the learning cycle.

3. The learning cycle consists of exploration, invention, and expansion (EIE). The student must experience each phase in the order described.

4. A learning cycle must be planned for a specific idea or skill. Do not mix several important concepts or generalizations in a single learning cycle.

5. Demonstrate a questioning and reflecting attitude toward the content you teach. Generate hypotheses, examine alternative explanations and encourage your students to do the same. Ask students, "What do you know about . . .?" "Why do you think . . .?" "How do you explain . . .?" "What evidence do you have?" Reward appropriate responses from students.

Having determined the prior knowledge held by students in a class, *the college instructor then becomes a prescriber and facilitator of the appropriate student learning activities.* The professional ability of the college instructor to make such decisions about the needs of the students, difficulty or level of the content, and the pedagogical sequence and strategy, is of greater value in the teaching and learning process than the textbook or other curriculum materials provided in the traditional college classroom.

AUTHOR NOTE

This work was in part supported by NASA Opportunities for Visionary Academics (NOVA), a program funded by the National Aeronautics and Space Administration, although the views expressed here are the author's only.

REFERENCES

Abraham, M.R. (1989). Research and teaching: Research on instructional strategies. *Journal of College Science Teaching, 18,* 185–187.

Abraham, M., & Renner, J. (1986). The sequence of learning cycle activities in high school chemistry. *Journal of Research in Science Teaching, 23*(2), 21–43.

Allard, D.W., & Barman, C.R. (1994). The learning cycle as an alternative method for college science teaching. *Bioscience, 44*(2), 99–101.

American Association for the Advancement of Science. (1993). *Benchmarks for scientific literacy.* New York: Oxford University Press.

Applefield, J.M., Huber, R., & Moallem, M. (2001). Constructivism in theory and practice: Toward a better understanding. *High School Journal, 84*(2), 35–54.

Arms, K., & Camp, P. (1982). *Biology sense organs.* Philadelphia: Saunders College Publishing.

Atkin, J.M., & Karplus, R. (1962). Discovery or invention? *The Science Teacher, 29*(5), 45–51.

Ausubel, D. (1968). *Educational psychology.* New York: Holt, Rinehart, and Winston.

Barnes, D. (1976). *From communication to curriculum.* Hammondsworth, UK: Penguin Books.

Boylan (1988). Enhancing learning in science. *Research in Science and Technological Education, 6*(2), 205–217.

Clark, C., & Peterson, P. (1987). Teachers' thought processes. In M. Wittrock (Ed.), *Handbook of research on teaching* (3rd ed., pp. 225–296). New York: Macmillan.

Clarke, J.H. (1990). *Patterns of thinking, integrating learning skills in content teaching.* Boston: Allyn & Bacon.

Dewey, J. (1916). *Democracy and education: An introduction to the philosophy of education.* New York: Free Press.

Driver, R. (1986a). *Children's learning in science project.* Leeds, England: University of Leeds, Centre for Science and Mathematics Study.

Driver, R. (1986b). *The pupil as scientist.* Philadelphia: Open University Press.

Driver, R., & Oldham, V. (1986). A constructivist approach to curriculum development in science. *Studies in Science Education, 13,* 105–122.

Driver, R., Squires, A., Rushworth, P., & Wood-Robinson, V. (1994). *Making sense of secondary science.* New York: Routledge.

Duschl, R.A. (1990). *Restructuring science education* (pp. 30–80). New York: Teachers College Press.

Erickson, G.L. (1979). Children's conceptions of heat and temperature. *Science Education, 63,* 221–230.

Green, S.K., & Gredler, M.E. (2002). A review and analysis of constructivism for school based practice. *Social Psychology Review, 31*(1), 53–71.

Hewson, P.W., & Hewson, M.G. (1988). An appropriate conception of teaching science. *Science Education, 72,* 597–614.

Johnson, M. (1987). *The body in the mind: The bodily basis of meaning, imagination, and reason.* Chicago: University of Chicago Press.

Karplus, R. (1977). *Science teaching and the development of reasoning.* Berkeley: University of California Press.

Karplus, R. (1979) Teaching for the development of reasoning. In A.E. Lawson (Ed.), *1980 AETS yearbook: The psychology of teaching for thinking and creativity* (pp. 174–215). Columbus, OH: ERIC/SMEAC.

Karplus, R., & Thier, H. D. (1967). *A new look at elementary school science: Science curriculum improvement study.* Chicago: Rand McNally.

Kowalski, F. (1993). Reflectance demonstration. *The Physics Teacher, 31,* 153.

Lawson, A. (1988). Student reasoning, concept acquisition, and a theory of science instruction. *Journal of College Science Teaching, 17,* 314–316.

Lawson, A. (1995). *Science teaching and the development of reasoning.* Belmont, CA: Wadsworth.

Lawson, A.E., Abraham, M.R., & Renner, J.W. (1989). *A theory of instruction: Using the learning cycle to teach concepts and thinking skills* (Monograph No. 1). Atlanta, GA: National Association for Research in Science Teaching.

National Research Council. (1996). *National Science Education Standards.* Washington, DC: National Academy Press.

National Research Council. (2000). *Inquiry and the National Science Education Standards: A guide for teaching and learning*. Washington, DC: National Academy Press.

National Science Foundation. (1996). *Shaping the future: New expectations for undergraduate education in science, mathematics, engineering, and technology* (NSF 96-139). Arlington, VA: Author.

Nussbaum, J., & Novick, J. (1981). An assessment of children's concepts to invent a model: A case study. *School Science Review, 62*, 771–778.

Nussbaum, J., & Novick, S. (1982). Alternative frameworks, conceptual conflict and accommodation: Toward a principled teaching strategy. *Instructional Science, 11*, 183–200.

Osborne, R.J., & Freyberg, P. (1985). *Learning in science*. Auckland, New Zealand: Heinemann.

Osbourne, R.J., & Wittrock, M.C. (1983). Learning science: A generative process. *Science Education, 67*, 489–508.

Parker, S.P. (Ed.). (1994). *McGraw-Hill concise encyclopedia of science & technology* (3rd ed.). New York: McGraw-Hill.

Perkins, D., & Salomon, G. (1991). Teaching for transfer. In D. Costa (Ed.), *Developing minds: A resource book for teaching thinking* (pp. 215–223). Alexandria, VA: Association for Supervision and Curriculum Development.

Piaget, J. (1969). Foreword. In H.G. Furth, *Piaget and knowledge: Theoretical foundations*. Englewood Cliffs, NJ: Prentice-Hall.

Piaget, J. (1971). *Biology and knowledge* (B. Walsh, Trans.). Chicago: University of Chicago Press.

Posner, G., Strike, K., Hewson, P., & Gertzog, W. (1982). Accommodation of a scientific conception: Toward a theory of conceptual change. *Science Education, 66*, 211–227.

Renner, J. (1982). The power of purpose. *Science Education, 66*, 709–716.

Renner, J.W., & Marek, E.A. (1990). An educational theory base for science teaching. *Journal of Research in Science Teaching, 27*, 241–246.

Renner, J.W., Abraham, M.R., & Birnie, H.H. (1985). The importance of form of student acquisition of data in physics learning cycles. *Journal of Research in Science Teaching, 22*, 303–326.

Renner, J.W., Abraham, M.R., & Birnie, H.H. (1988). The necessity of each phase in the learning cycle in teaching high school physics. *Journal of Research in Science Teaching, 25*, 39–58.

Renner, J.W., Stafford, D.G., Lawson, A.E., McKinnon, J.W., Friot, E., & Kellog, D.H. (1976). Research teaching and learning with the Piaget model. Norman: University of Oklahoma Press.

Rowell, J.A., & Dawson, C.J. (1983). Laboratory counter examples and the growth of understanding in science. *European Journal of Science Education, 5*, 203–215.

Saunders, W. (1992). The constructivist perspective: Implications for teaching strategies for science. *School Science and Mathematics, 92*(3), 136–141.

Smith, E.L., & Lott, G.W. (1983). Teaching for conceptual change; some ways of going wrong. In H. Helm & J.D. Novak (Eds.), *Proceedings of the International Seminar on Misconceptions in Science and Mathematics* (pp. 57–66). Ithaca, NY: Cornell University, Department of Education.

Stofflett, R.T. (1998, December). Putting constructivist teaching into practice in undergraduate introductory science. *Electronic Journal of Science Education, 3*(2), Article 3. Retrieved December 31, 2001 from http://unr.edu/homepage/jcannon/ejse/stofflett.html

Sunal, D.W., & Sunal, C.S. (1991a). Backyard aesthetics. *Science Scope, 15*(1), 25–29.

Sunal, D.W., & Sunal C. (1991b). Tree growth rings: What they tell us. *Science Activities, 28*(2), 19–26.

Sunal, D.W., & Sunal, C.S. (2003). *Teaching elementary and middle school science.* Columbus, OH: Merrill Prentice Hall.

Tobin, K., Briscoe, C., & Holman, J. (1990). Overcoming constraints to effective elementary science teaching. *Science Education, 74,* 409–420.

Vermette, P., & Foote, C. (2001). Constructivist philosophy and cooperative learning practice toward integration and reconciliation in secondary classrooms. *American Secondary Education, 30,* 26–37.

Vygotsky, L. (1962). *Thought and language* (E. Hanfmann & G. Vaker, Trans.). Cambridge, MA: MIT Press.

West, L., & Pines, L. (Eds.). (1985). *Cognitive structure and conceptual change.* New York: Academic Press.

Yager, R. (1991). The constructivist learning model: Towards real reform in science education. *Science Teacher, 58*(6), 52–57.

Zollman, D. (1990). Learning cycles in a large enrollment class. *The Physics Teacher, 28,* 20–25.

Zollman, D. (1997). Learning cycle physics. In E.F. Redish & J.S. Rigden (Eds.), *The changing role of physics departments in modern universities* (pp. 1137–1149). College Park, MD: American Institute of Physics.

CHAPTER 7

INTERDISCIPLINARY CURRICULUM PLANNING IN A COLLEGE COURSE

John E. Christopher and Ronald K. Atwood

ABSTRACT

This chapter describes the interdisciplinary collaborative planning and subsequent implementation of a standards-based physical science course for preservice elementary teachers. Perspectives of a science teacher educator, elementary practitioners, and university faculty from geology, chemistry, and physics informed the work. Varied and sustained efforts to obtain formative and summative evaluation data are described. Results across several semesters were consistently positive. Conclusions and implications are suggested.

INTRODUCTION

In far too many colleges and universities responsibility for the preparation of preservice elementary teachers to teach science rests almost solely with a college or department of education. Students in these programs do complete science coursework in science departments. However, the courses

Reform in Undergraduate Science Teaching for the 21st Century, pages 123–135

taken usually are surveys for non-science majors. By design they provide a superficial coverage of many science topics and allegedly serve a general education function for students pursuing other academic majors. In this situation the focus of a conscientious instructor is likely to be on how to efficiently present information through lecture rather than on how to make the content functional for particular groups of users. Certainly an instructor for this type of course could not be expected to feel a sense of ownership in the institution's elementary education program. The previous characterization, valid for the science preparation of preservice elementary education majors at the University of Kentucky (UK) prior to the collaborative efforts described in this chapter, is at odds with the needs of elementary teachers and with recommendations of prestigious national organizations (National Research Council [NRC], 2000; National Science Foundation [NSF], 1996).

A science educator at UK had been concerned for many years about the weak science background of many preservice elementary teachers enrolled in an elementary science methods and materials course. Students frequently began the course with a lack of scientific understanding of science concepts they might reasonably be expected to teach. Content deficiencies were evident across the life, earth, and physical sciences, but seemed to be greatest for topics generally included in elementary physics. Further, students in the course seemed unaccustomed to applying concepts, communicating applications, supporting assertions with data, and utilizing other inquiry skills which now are associated with national science education standards (NRC, 1996; American Association for the Advancement of Science [AAAS], 1990). Considering the nature of their general studies science coursework, how could it be otherwise? Although the senior-level science methods and materials course used inquiry strategies to address important life, earth, and physical science concepts, it was clear that inquiry and science content needs could not be addressed adequately in this three-credit course. Conversations with other science educators over the years indicate the problem is pervasive.

Voicing concern about these deficiencies to colleagues in the science departments generally produced guarded expressions that seemed to reflect a limited understanding of the problem and even less commitment to addressing it. The substance of statements made in defense of the status quo typically included a lack of funds for smaller classes, faculty shortages, and the difficulty in knowing which concepts to emphasize in depth and which to omit.

However, an exception to the pattern occurred following focused conversations about the problem with a physicist who had demonstrated a strong interest in innovative teaching over several years. As examples, he had utilized small-group active-learning exercises in large lecture classes, including activities similar to the "Concept Test" (Mazur, 1997) and "Context Rich Problems" (Heller & Hollabaugh, 1992). In his introductory physics classes he had field-tested early versions of *Tutorials in Introductory Physics* (McDermott & Schaffer, 1998). During study visits that focused on innovative curricula and teaching he had developed a better understanding of some of the most innovative instructional development efforts in physics education, including *Physics by Inquiry* (McDermott, 1996), "Workshop Physical Science" (Jackson & Laws, 1997), and tutorials. He was intrigued with the conceptual change work being done in physics education (Hestenes, Wells, & Swackhammer, 1992; McDermott, 1991). This interest provided common ground with science educator, whose teaching experience and research supported the conclusion that traditional instruction is ineffective in promoting conceptual understanding (Atwood & Atwood, 1995, 1996). Common interests and complementary understandings provided a basis for more frequent discussions of deficiencies in the conceptual understanding of preservice elementary teachers, and eventually for action for needed changes. In fact, strongly shared common interests and complementary knowledge bases of the principal collaborators, a physicist and a science educator, are judged to be essential to the success of the larger interdisciplinary collaboration.

In order to tap the expertise of other collaborators and acquire a mandate from a university administrator in an influential position, the dean of the College of Arts and Sciences was presented with a written statement about the weak conceptual understanding in science of preservice elementary teachers. Further, the dean was requested to appoint a committee, including elementary teachers and scientists from other disciplines, to address the problem. When several months passed, and no action was taken, the physicist took the initiative and appointed an ad hoc committee to address the problem in the physical and earth sciences. Biological science was not included in order to make the task more manageable and because, at the time, interest in the problem among faculty in biological science was not at the level deemed necessary for successful collaboration. The committee included, in addition to the physicist and science educator, two elementary teachers with strong interests in science, and a chemist and a geologist who had shown an interest over time in the content preparation of teachers.

DIRECTION FROM THE COMMITTEE

The committee devoted several sessions to discussing the problem and possible solutions. AAAS publications, *Science for All Americans* (1990) and *Benchmarks for Science Literacy* (1994), and a preliminary draft of the

National Science Education Standards (NRC, 1996) were particularly influential in the committee's deliberations. These documents advocate addressing a modest number of important concepts through constructivist, inquiry strategies that simultaneously develop scientific skills and conceptual understanding. The planning of the committee also was informed by other literature (Bendall, Goldberg, & Galili, 1993; Bisard, Aron, Francek, & Nelson, 1994; Driver, Asoko, Leach, Mortimer, & Scott, 1994; Driver & Oldham, 1986; McDermott, 1991).

After considerable deliberation the committee decided the desired instruction would be provided through a course to be offered by the physics department and a course to be offered by the geology department. Together, the new courses would prepare elementary teachers to address the K–4 physical and earth science standards from the *National Science Education Standards* (NRC, 1996). These standards, judged to be similar to the AAAS standards, were being adopted as state standards in Kentucky. The most important considerations in the placement of a course in the physics department and a course in the geology department, rather than in the chemistry department, were the relatively modest amount of chemistry content in the K–4 standards, and the availability of interested faculty in geology and physics to lead the course planning and teach the courses.

Committee members agreed that addressing the national science education standards meant utilizing an inquiry approach to teach standards-based science concepts. The group worried that science courses designed especially for preservice elementary teachers might be assumed by reviewers to be "watered down" as the new courses subsequently moved through the university's approval process. The collaborators were in complete agreement that the courses must be substantive and rigorous while being perceived as useful to elementary practitioners. It was thought if links to the standards were clearly made and many of the materials and activities used in the courses for preservice teachers could be easily modified to teach science to children, then preservice teachers likely would view the course as useful.

The vision for the new courses was decidedly interdisciplinary. The modest chemistry content in the K–4 standards was integrated into the course offered by the geology department. Some space science, astronomy by sight, was integrated into the course offered by the physics department. The former seemed logical because of the heavy emphasis on properties and some attention to chemical tests in geology. A study of astronomy by sight was judged to be facilitated by an understanding of light phenomena, which made it logical to strategically place the topic in the physics course. Although education is not a discipline in the sense that chemistry, geology, and physics are disciplines, it is a broad field of study with a knowledge base which greatly influenced the collaborative planning. Specifically, it was agreed that inquiry-oriented instructional strategies consistent with a constructivist approach would be utilized (Brooks & Brooks, 1993; Driver, 1986; Driver & Oldham, 1986). Direct experiences and interpretive, sense-making discussions were to be central strategies of the courses. Further, commonly held scientific misconceptions were to be confronted using these strategies (Hewson, 1981; Hewson & Hewson, 1983).

With this framework in mind planning of the two courses proceeded. When completed, the respective science departments and the College of Arts and Science approved the new courses. This approval involved administrative as well as faculty support, which is judged to be critical to the long-term survival of the courses. The courses also were approved enthusiastically by the College of Education as requirements in the elementary teacher education program. Gaining approval of the courses as program requirements was viewed as essential to institutionalizing the courses. For the first time it was clear that the physics and geology departments were partners with the College of Education in the preparation of elementary teachers. The remainder of the chapter focuses on the new course offered by the physics department.

DESIGN, IMPLEMENTATION, AND EVALUATION
OF THE PHYSICS COURSE

Several recent initiatives have produced very promising instructional materials and strategies for teaching introductory college physics. Included are *Physics by Inquiry* (McDermott, 1996), *Powerful Ideas in Physical Science* (American Association of Physics Teachers, 1996), "Workshop Physical Science" (Jackson & Laws, 1997), and "How Can Computer Technology Be Used to Promote Learning and Communication Among Physics Teachers?" (Goldberg, 1997). For the new physics course at UK it was decided to make extensive use of promising research-based, commercially available instructional materials. In addition to benefitting preservice elementary teachers who would enroll in this physics course, the use of widely available instructional materials was expected to facilitate the replication of the course by others attempting to address similar needs. Considering the expectations of the collaborative planning committee, commercially available materials, the topics to be addressed, and the characteristics and needs of the population to be served, the committee selected *Physics by Inquiry* as the primary source of instructional ideas. Potential users should know that guidelines for use of *Physics by Inquiry* are incorporated in the two-volume set of books. Further, the University of Washington Physics Education Group, which developed the materials, sponsors workshops on recommended use.

The five standards-based content topics addressed in the three-credit physics course include some basics on the motion of objects; magnetism; electrical circuits; the behavior of light, which includes work with mirrors; and astronomy by sight. The *National Science Education Standards* (NRC, 1996) document, from which these topics were selected, does not specify a complete curriculum. In fact it does not suggest a scope and sequence for a desirable study of any topic included in the standards. For example, the physical science content standards for K–4 include the directive that children should develop an understanding of magnetism. However, the only elaboration on magnetism in the standards document is that "magnets attract and repel each other and certain kinds of other materials" (NRC, 1996, p. 127). The principal collaborators and the planning committee had agreed that teacher's breadth and depth of understanding of any content topic should exceed what they are expected to teach. Therefore, in addition to making a judgment about what teachers might reasonably try to teach about magnetism to K–4-level children, a judgment also was made on how those expectations should be exceeded in the preparation of teachers.

These judgments resulted in a list of expectations to guide instruction for magnetism and the other four topics. A small grant from the Partnership for Reform Initiatives in Science and Mathematics (PRISM) supported the preparation and distribution of a document connecting the lists to state and national standards (Kovash, Christopher, & Atwood, 1998). The grant also supported the development of the concepts test described later in this chapter. This grant was the only extramural support sought for the development and evaluation of the physics course.

Most students who enroll in the physics course are in the 2nd or 3rd year of a 4-year undergraduate program. During the early implementation years approximately 90% were female; approximately 90% were White, and 6% were African American. The average composite ACT score for students enrolled in the elementary education program during this period was 23.

During 5 of the required 6 hours of instruction per week in the experimental course, students work in groups of three, collaborating in hands-on investigations and interpretive discussions of predicted and actual results. Structured discussions, called checks, occur at critical points between a group of students and an instructor, in order to determine whether group members can show evidence of having the desired conceptual understanding. Preparation for a check, which occurs 4 to 5 times a week, requires each group member to prepare a written response to a prompt. Interpretive discussions occur frequently within the group as the written response is being prepared. If preparation for the check is done well, key concepts will be applied and the applications will be communicated verbally and in writing. Extensive probing by the instructor occurs during a check and involves all group members. The oral portion of the check requires a defense of the written response, and frequently is extended by the instructor to make connections with previously studied concepts. Daily home assignment tasks are related to instructional activities and the checks, and also require application of one or more major concepts. Additional related application tasks are included on examinations, which are given at the conclusion of the study of each of the five topics. All homework assignments and examination solutions must include clear and complete descriptions and explanations. Written work is rigorously evaluated and promptly returned to the students.

A check, homework assignment, and examination task provide successive opportunities to apply the concepts introduced through instructional activities. A major goal of these multiple applications in similar, but somewhat novel contexts, is conceptual understanding of standards-based concepts. Increasing students' ability to utilize inquiry skills, such as supporting assertions with data, is another major goal.

The checks, homework assignments, and periodic examinations also provide timely formative evaluation data. If the initial instructional activities are not effective for a small group, it will be apparent on the check. Deficiencies are addressed by having students repeat or reinterpret instructional activities; group members then expand their written response and redo the oral component of the check. Mastery by a group on a check is the expectation, and mastery must be demonstrated before the group can move to the next set of instructional activities. The homework provides an opportunity for individuals to apply the concepts in a different context than was used in a check. Fortunately, *Physics by Inquiry* (McDermott, 1996) includes far more potential homework and examination tasks than one possibly could use in a course, and they range in difficulty. So, if several groups struggle with the check for a particular concept, the assigned homework task will be less difficult than if most groups performed well on the check. Subsequently the choice of examination tasks for individuals also is influenced by student performance on the checks and homework assignments. This system of successively obtaining and utilizing formative evaluation data, first from small groups and then from individuals, is viewed as a key factor in the success of the course.

Over several semesters many small modifications in activities and checks from *Physics by Inquiry* were made based on results of the checks and homework assignments. More substantial changes in the course also were made. For example, locally modified materials from "Workshop Physical Science" (Jackson & Laws, 1997) recently have been utilized for the study of forces and motion. This change was based on the need to connect forces with motion very explicitly in response to the standards and student performance. The kinematics section initially used from *Physics by Inquiry* did not do that.

The principal collaborators were impressed with the Force Concept Inventory (Hestenes et al., 1992), a multiple-choice test which assesses conceptual understanding of Newtonian mechanics. Popular misconceptions are embedded in the distracter options of the test items, making them very attractive to persons who lack deep scientific conceptual understanding of the underlying concepts. Results of giving the test to thousands of college physics students (Hake, 1998) indicate that traditional introductory physics instruction has little impact on conceptual understanding of Newtonian mechanics.

As the new physics course was being planned, the principal collaborators decided to develop a multiple-choice test to provide summative evaluation data on conceptual understanding. Using the Force Concept Inventory as a model, popular misconceptions for each of the major topics were identified from the literature and embedded in the distracter options of the test items. Three interested graduate students in physics and one in education joined the principal collaborators in contributing to the development of a 32-item test called Survey of Selected Concepts in Physical Science (SSCPS). Previous work provided important ideas for test items (Goldberg & McDermott, 1986; McDermott, 1996; Osborne & Freyberg, 1985; Thornton & Sokoloff, 1998).

Administering the SSCPS from the second semester that the course was offered, through the fifth semester, provided very encouraging results as shown in Table 7.1. Note that, using pre and post matched data, the mean percentage of items answered correctly on pretests ranged from 41.9% to 48.1%, and the mean percentage of correct responses on posttests ranged from 79.4% to 82.5%. Thus, students consistently showed evidence of the desired conceptual understanding on about 80% of the items on the posttest. The normalized gains (Hake, 1998), shown in Table 7.1, indicate from 62% (spring 1998) to 67% (fall 1997) of the gains possible, based on SSCPS scores, were achieved. Based on posttest data, Kuder-Richardson reliability coefficients ranged from .70 to .75 over the four semesters. As is often the case for experimental courses, an appropriate control group was not available for conducting an experimental study. Thus, we cannot rule out the possibility that some of the gains may be attributable to factors other than the physics instruction. However, no program component other than the physics course provided instruction on the five topics during the semesters the course was completed.

If resources were available to do so, it would have been very advantageous to obtain individual interview data to assess conceptual understanding for each major component of the course. Interviews can provide rich descriptions of conceptual understanding (Atwood & Atwood, 1996), and

Table 7.1. Summary of Results for Several Semesters for Survey of Selected Concepts in Physical Science (SSCPS)

Student group	Mean 32 items	SEM[a] 32 items	SE[b] 32 items	K-R[c]	Mean %	NG[d] %
pretest fall 1997 (46 students)	15.4	2.5	0.59	0.61	48.1%	
posttest fall 1997	26.4	1.9	0.56	0.75	82.5%	0.67
pretest spring 1998 (46 students)	14.8	2.5	0.49	0.44	46.3%	
posttest spring 1998	25.4	2.0	0.55	0.70	79.4%	0.62
pretest spring 1999 (47 students)	13.4	2.5	0.64	0.66	41.9%	
posttest spring 1999	25.6	2.0	0.55	0.72	80.0%	0.66
pretest fall 1999 (54 students)	14.9	2.5	0.47	0.48	46.6%	
posttest fall 1999	26.0	1.9	0.51	0.73	81.3%	0.65

Notes:
[a]Standard error of measurement.
[b]Standard error.
[c]Kuder-Richardson Reliability Index.
[d]Normalized gain = (Posttest score % – pretest score %)/(100% – pretest score %).

would have helped to cross-validate evaluative data from the SSCPS. Thus far, individual interviews have been completed and analyzed only for a major component of astronomy by sight, the cause of moon phases (Trundle, Atwood, & Christopher, 2002). That study documented conceptual understanding in ways that have both formative and summative value. As an example of data used for formative purposes, pre-instruction interviews surprisingly revealed that only 38.6% of the sample understood the moon orbits the earth. In response, reinforcement of this idea was added to the course at the point where students developed models to explain two months of their own moon observations. As a result, during post-instruction interviews, 95.2% of the sample provided evidence of understanding the moon orbits the earth.

Perhaps the most important summative finding in the moon phase study is that the instruction on the cause of moon phases was highly effective. Before instruction only 9.5% of 42 students showed evidence of holding a scientific conception of the cause of moon phases. Three weeks after instruction 95.2% showed evidence of holding a scientific conception. The research report (Trundle, Atwood, & Christopher, 2002) provides a detailed description of the moon-related instruction that produced these results. Previous studies of preservice elementary teachers, who also used individual interview procedures to evaluate the impact of instruction on the cause of moon phases, have reported far less instructional success (Callison & Wright, 1993; Targan, 1988). Strong evidence of student achievement as a result of the interdisciplinary collaboration is viewed as a major factor in sustaining the work over a period of years.

Until the 2002–2003 academic year the physicist, who was a principal collaborator in planning the course, served as the principal instructor for all sections, and he trained the graduate assistants and peer facilitators (undergraduate students who successfully completed the course) who assisted. During the 2002–2003 academic year two other physicists served as principal instructors for sections of the course. Joint planning sessions involving all instructors and assistants have helped maintain a high level of consistency across course sections. This consistency by multiple instructors is very important to the long-term institutionalization of the course as evaluated.

CONCLUSIONS AND IMPLICATIONS

The interdisciplinary collaborative efforts of the larger planning group and the longer term, continuing collaboration of the physicist and the science educator is judged to be highly productive. The plans, evaluation, and institutionalization of an effective standards-based physics course for preservice elementary teachers provide support for that assertion. This conclusion supports the view that effective change efforts begin with a shared goal to be accomplished (Sunal et al., 2001). The varied data used to evaluate the course are judged to be functional in informing course modifications and providing important summative evidence that the course is effective. The strategies employed in collecting and using formative and summative data are recommended to others with similar course evaluation needs. Settling for a typical university course evaluation—student opinions on a Likert scale instrument and testimonials—never seemed like a meaningful option for the collaborators, and seems totally inadequate if conceptual understanding is a major instructional goal. Inclusion of the course as a requirement in the elementary teacher education program is considered to be a major program improvement and essential to the institutionalization of the course. The commitment of the physics department to support the labor-intensive course over time represents an important recognition of the department's obligation to the education of teachers. Similar commitments are needed from other departments within this institution and many other teacher preparation institutions.

It should again be noted that *Physics by Inquiry* (McDermott, 1996) is research based and reflects several years of research and development.

Making extensive use of these materials was surely a major factor in the success of the physics course. Interdisciplinary collaborative efforts outside of physics apparently have not had the benefit of comparable instructional materials, and collaborators have had to do more local development (Krockover, Shepardson, Adams, Eichinger, & Nakhleh, 2002). Better mechanisms are needed for sharing local R & D work in order to reduce the resource requirement and increase the quality of future work.

The cultural differences between science and education departments (Carr, 2002) and the different knowledge bases of collaborating faculty from these two fields are viewed as a major strength, when the persons involved have equal status in the longer term. In the short term, the greater expertise one of the collaborators brings to a particular problem makes that person's views more valuable.

Sustained interdisciplinary collaborative effort provides numerous opportunities for professional growth. In this case both the physicist and the science educator learned a great deal about strategies and materials for teaching and assessing physics and astronomy concepts. They have made several joint presentations on their work at regional and national meetings, and they continue to do conceptual change research collaboratively.

This report supports the growing body of literature that suggests the education of teachers can greatly benefit by becoming more of an interdisciplinary institutional concern. In fact, it is difficult to envision how the desired level of teacher competence in science can be attained without the extensive collaboration of faculty from education, practicing teachers, and the content specialists who teach the science courses. Substantial improvements in introductory science courses, especially in physics, have been documented in several colleges and universities across the nation. However, far too few science course offerings for teachers, including introductory physics, have been reformed to the extent needed (NSF, 1996). Higher education can and must do better (Seymour, 2001).

REFERENCES

American Association for the Advancement of Science. (1990). *Science for all Americans.* Oxford, England: Oxford University Press.

American Association for the Advancement of Science. (1994). *Benchmarks for science literacy.* Oxford, England: Oxford University Press.

American Association of Physics Teachers. (1996). *Powerful ideas in physical science.* Washington, DC: Author.

Atwood, V.A., & Atwood, R.K. (1995). Preservice elementary teachers' conceptions of what causes day and night. *School Science and Mathematics, 95*, 290–294.

Atwood, R.K., & Atwood, V.A. (1996). Preservice elementary teachers' conceptions of the cause of seasons. *Journal of Research in Science Teaching, 33*, 553–563.

Bendall, S., Goldberg, F., & Galili, I. (1993). Prospective elementary teachers' prior knowledge about light. *Journal of Research in Science Teaching, 30,* 1169–1187.

Bisard, W.J., Aron, R.H., Francek, M.A., & Nelson, B.D. (1994). Assessing selected physical science and earth science misconceptions of middle school through university preservice teachers. *Journal of College Science Teaching, 24,* 38–42.

Brooks, J., & Brooks, M. (1993). *The case for constructivist classrooms.* Alexandria, VA: Association for Supervision and Curriculum Development.

Callison, P.L., & Wright, E.L. (1993, April). *The effect of teaching strategies using models on preservice elementary teachers' conceptions about earth-sun-moon relationships.* Paper presented at the meeting of the National Association for Research in Science Teaching, Atlanta, GA.

Carr, K. (2002). Building bridges and crossing borders: Using service learning to overcome cultural barriers to collaboration between science and education departments. *School Science and Mathematics, 102,* 285–298.

Driver, R. (1986). *The pupil as scientist.* Philadelphia: Open University Press.

Driver, R., Asoko, H., Leach, J., Mortimer, E., & Scott, P. (1994). Constructing scientific knowledge in the classroom. *Educational Researcher, 24,* 5–12.

Driver, R., & Oldham, V. (1986). A constructivist approach to curriculum development in science. *Studies in Science Education, 13,* 105–122.

Goldberg, F.M. (1997). How can computer technology be used to promote learning and communication among physics teachers? In E.F. Redish & J.S. Rigden (Eds.), *The changing role of physics departments in modern universities: Proceedings of ICUPE* (pp. 375–392). Woodbury, NY: American Institute of Physics.

Goldberg, F.M., & McDermott, L.C. (1986). Student difficulties in understanding image formation by a plane mirror. *Physics Teacher, 24,* 472–480.

Hake, R.R. (1998). Interactive-engagement versus traditional methods: A six thousand-student survey of mechanics test data for introductory physics courses. *American Journal of Physics, 66,* 64–74.

Heller, P., & Hollabaugh, M. (1992). Teaching problem solving through cooperative grouping. Part 2: Designing problems and structuring groups. *American Journal of Physics, 60,* 637–644.

Hestenes, D., Wells, M., & Swackhammer, G. (1992). Force concept inventory. *Physics Teacher, 30,* 141–158.

Hewson, P.W. (1981). A conceptual change approach to learning science. *European Journal of Science Education, 3,* 383–396.

Hewson, M.G., & Hewson. P.W. (1983). Effect of instruction using students' prior knowledge and conceptual change strategies on science learning. *Journal of Research in Science Teaching, 20,* 731–743.

Jackson, D.P., & Laws, P.W. (1997). Workshop physical science: Project-based science education for future teachers, parents, and citizens. In E.F. Redish & J.S. Rigden (Eds.), *The changing role of physics departments in modern universities: Proceedings of ICUPE* (pp. 623–630). Woodbury, NY: American Institute of Physics.

Kovash, S.S., Christopher, J.E., & Atwood, R.K. (1998). *Teaching university physics in a KERA environment.* Lexington: University of Kentucky Department of Physics and Astronomy.

Krockover, G., Shepardson, D., Adams, P., Eichinger, D., & Nakhleh, M. (2002). Reforming and assessing undergraduate science instruction using collaborative action-based research teams. *School Science and Mathematics, 102,* 266–284.

Mazur, E. (1997). *Peer instruction: A user's manual.* Upper Saddle River, NJ: Prentice-Hall.

McDermott, L.C. (1991). Millikan Lecture 1990: What we teach and what is learned: Closing the gap. *American Journal of Physics, 59,* 301–315.

McDermott, L.C. (1996). *Physics by inquiry: An introduction to physics and the physical sciences* (Vols. 1–2). New York: Wiley.

McDermott, L.C., & Schaffer, P.S. (1998). *Tutorials in introductory physics* (Preliminary ed.). Upper Saddle River, NJ: Prentice-Hall.

National Research Council. (1996). *National Science Education Standards.* Washington, DC: National Academy Press.

National Research Council. (2000). *Inquiry and the National Science Education Standards.* Washington, DC: National Academy Press.

National Science Foundation (1996). *Shaping the future: New expectations for undergraduate education in science, mathematics, engineering and technology* (NSF 96-139). Washington, DC: Author.

Osborne, R., & Freyberg, P. (1985). *Learning in science.* London: Heinemann.

Seymour, E. (2001). Tracking the processes of change in U.S. undergraduate education in science, mathematics, engineering, and technology. *Science Education, 86,* 79–105.

Sunal, D.W., Sunal, C.S., Whitaker, K.W., Freeman, L.M., Odell, M., Hodges, J., et al. (2001). Teaching science in higher education: Faculty professional development and barriers to change. *School Science and Mathematics, 101,* 246–257.

Targan, D. (1988). *The assimilation and accommodation of concepts in astronomy.* Unpublished doctoral dissertation, University of Minnesota, Minneapolis.

Thornton, R.K., & Sokoloff, D.R. (1998). Assessing student learning of Newton's laws: The Force and Motion Conceptual Evaluation and the evaluation of active learning laboratory and lecture curricula. *American Journal of Physics, 66,* 338–352.

Trundle, K.C., Atwood, R.K., & Christopher, J.E. (2002). Preservice elementary teachers' conceptions of moon phases before and after instruction. *Journal of Research in Science Teaching, 39,* 633–658.

CHAPTER 8

ASSESSMENT IN COLLEGE SCIENCE COURSES

**Lawrence C. Scharmann, Mark C. James,
and Ann Stalheim-Smith**

ABSTRACT

Those who embrace active forms of learning in their science classrooms quickly discover a disconnection with more traditional forms of assessment. Efforts to revise forms of classroom assessment, both formative and summative, have necessarily accompanied efforts to reform science instruction. The authors present an overview of literature and research in this area and a summary of their own experiences with alternative forms of assessment. Instructional "best practices" including innovative efforts with assessment techniques that promote student-centered forms of teaching and learning are described within the contexts of a science teacher preparatory program and college science discipline.

INTRODUCTION

The essential characteristic of well-designed assessments is that the processes used to collect and interpret data are consistent with the purpose of the assessment. (National Research Council, 1996, p. 78)

Reform in Undergraduate Science Teaching for the 21st Century, pages 137–152
Copyright © 2004 by Information Age Publishing

This quote is taken from the *National Science Education Standards*, a document that promotes efforts to reform science instruction through inquiry teaching, active forms of learning, and a revision in our thinking with respect to forms of assessment. The need to make an explicit statement regarding a match between the type of assessment and its purpose implies that such a match does not occur with the frequency that perhaps it should.

The *National Science Education Standards* and a companion document, *Inquiry and the National Science Education Standards* (National Research Council [NRC], 2000), each provide an outstanding blueprint for educational reform. There is a pressing need for secondary school teachers to become versed in national- and state-level standards and assessment alternatives for the following reasons:

- School accreditation is tightly aligned with achievement of national and state standards at a proficient level.
- Teachers are expected to be familiar with standards and how to appropriately assess them—through the use of alternative, performance, and authentic assessments.
- Federal (and state) funding agencies demand that schools (and hence teachers) be held more accountable for students' learning outcomes.

Efforts to revise undergraduate instruction toward more active forms of learning and accompanying performance assessment have, unfortunately, met with far greater resistance at the collegiate level for a variety of reasons, among them being fear of change, ignorance of alternatives to traditional assessment, and unwillingness to make undergraduate teaching a real priority. Yet those who embrace more active forms of learning quickly discover a disconnection with more traditional forms of assessment—low-level multiple-choice and fill-in-the-blanks examinations do not adequately match higher level, performance learning tasks.

In this chapter the authors present an overview of the relevant research literature and a summary of their experiences with alternative forms of assessment. In the section below, we consider an argument for assessment reform, what to assess, and approaches to alternative assessment in college science courses. The two sections that immediately follow present two cases. The first case concerns action research, which reconstructs the evolution of a secondary science teaching methods course with respect to assessment. The second case describes a professor's efforts to employ instructional "best-practices" including innovative formative assessments in large-lecture college biology courses. We then end this chapter with our recommendations to individuals considering changes in their assessment practices at the collegiate level.

AN ARGUMENT FOR ASSESSMENT REFORM

Imagine a college basketball season that consists of only one game that is played on the last day of the year. Imagine further that this basketball game is not the same game of basketball that is played by professionals. Rather, it is a game composed of a series of concise drills and plays which experts have determined to be valid and reliable predictors of a basketball player's ability to play basketball. Imagine student players who have no way to determine which of the hundreds of drills and plays they have studied will be included in the game. Further, imagine a secret scoring system that is only available to the assessors of the performance. Finally, imagine that the players do not receive their ultimate scores until days or weeks after the game has been played. This dysfunctional variant of college basketball is offered by Wiggins (1998) to illustrate limitations of the traditional approach to assessment in science classes. Students are primarily assessed through high-stakes secure tests that are prepared in secret by their instructor. At best, these tests are psychometrically valid and reliable indicants of a student's relative comprehension of course content. At worst, the tests are obstacles, which subvert the efforts of students to learn science. The primary goal of traditional testing is to give grades to students that reflect the degree to which they have mastered the various skills and concepts in the course. However, even if this goal is realized by using secure, easy-to-score testing schemes, such testing neglects the needs of the student for engaging in useful and interesting work. Furthermore, by compelling students to focus their efforts on being able to recall scientific facts and apply algorithms to solve hypothetical problems out of context, the assessment presents students with a false image of science where learning science content is a goal unto itself rather than a means to an end.

Tobias (1990) echoes Wiggins' grim view of traditional assessment in her landmark study of undergraduate students who switch their college majors from science to nonscientific fields. Contrary to conventional wisdom, Tobias found that most of the attrition from science majors was not due to the inherent difficulties of learning science. Instead, students who were at first interested in studying science were turned off by a course structure that required passivity in the classroom and intense competition for grades. Students became frustrated by a frenetic pace which demanded mindless memorization of problem solving algorithms in place of the development of conceptual understanding.

Dickinson and Flick (1998) continued with Tobias's theme in a study depicting how a traditional assessment system could undermine the pedagogical goals of a well-meaning instructor. In this study of an introductory physics course, the blistering fast pace of content coverage and emphasis on scoring for "the right answers" on tests, laboratories, and homework assignments forced initially enthusiastic students to give up their hopes of learning physics in favor of developing creative procedural strategies that enabled them to obtain passing grades without developing understanding. Mazur (1997) used the Force Concept Inventory (Hestenes, Wells, & Swackhammer, 1992) to show that a significant fraction of students in traditional university physics classes who are able to successfully solve quantitative problems do not possess even a rudimentary grasp of the underlying physical concepts (Crouch & Mazur, 2001).

Perhaps the greatest objection traditional instructors have to an assessment system that rewards the development of conceptual understanding in place of problem-oriented tests is the additional time required. However, a growing research base suggests that assessment systems that limit content coverage in order to focus on deeper understanding result in enhanced student performance on standardized content tests (NRC, 1996). In a study of a large number of high school students enrolled in a science course in which assessment was based on student projects, Schneider, Krajcik, Ronald, and Soloway (2002) found that the focus on understanding allowed students to outperform a closely matched group of students from traditional classes on a national standardized science test. Taylor and Watson (2000) studied the presence or absence of traditional testing in two groups of otherwise similar elementary methods of teaching science classes and found that the presence of traditional testing increased the level of student anxiety and decreased the level of student participation. In contrast, the absence of traditional testing was found to be associated with an increased sense of class relevance. Taylor and Watson also found that including traditional testing did not enhance science understanding as measured by two independently developed tests designed for preservice elementary teachers.

Since the focus of traditional assessment is on gauging "whether students know," rather than probing "what they know," it has been criticized for not providing instructors with critical feedback about the nature of their students' prior knowledge (McDermott, 1991, McClymer & Knoles, 1992; Pride, Vokos, & McDermott, 1997; Tobias, 1990). In traditional assessment, students are taught to reproduce the "correct" answers that are valued by teachers as evidence of learning rather than constructing their own explanation of physical phenomena. Tobias observed that students in traditional classrooms often develop two independent lines of reasoning to describe physical phenomena, one that reflects their own alternate conceptions, and another incompatible line of reasoning that reflects the teacher's seemingly improbable "scientific" view. By probing for quality of understanding, rather than ability to replicate low-level answers to stereotypical questions, nontraditional assessment strategies seek to provide instructors with ongoing feedback, which can be used to match teaching to learning.

WHAT TO ASSESS?

At the heart of assessment reform is the belief that the primary role of assessment should be to support learning rather than to merely audit learning (Shepard, 2000). However, the fundamental question of precisely what ought to be learned remains unresolved.

Aristotle first articulated the distinction between knowledge as a superficial collection of information, and understanding as a deep grasp of causal relationships (Aristotle, 1995). Adapting this idea to a contemporary setting, Perkins (1992) coined the term "fragile knowledge" to refer to the naive, ritualistic, inert, or ephemeral knowledge which students often obtain in traditional settings. Many argue (Bloom, Englehart, Furst, Hill, & Krathwohl, 1956; Gardner, 1981) that in order to assess genuine student understanding, a student must be challenged to apply knowledge to a novel circumstance. Wiggins (1998) suggests that by allowing such a novel circumstance for assessment to emerge from an authentic situation that emulates a real life professional setting, the assessment provides learners with its own rationale for learning. Wiggins develops the position further by suggesting that all curriculum development should begin with an authentically based assessment design, providing just the pedagogical elements that allow students to succeed in the authentic assessment circumstance. From this perspective, learning only becomes meaningful when it is directed toward some authentic goal.

Taking a more theoretical tack, Delandshere (2002) has sought to clarify the philosophical and epistemological assumptions that have been made about teaching and learning in the context of assessment. In so doing, Delandshere rejects the notion of outcome-based assessment reform as incompatible with a nondeterministic epistemology. She goes on to suggest the need for a framework for assessment that is itself based on inquiry and the constructivist view that "what we know" is inseparable from "how we know." Such an interdependent connection between inquiry, assessment, and constructivism is fully recognized by the National Research Council in the development and implementation of the National Science Education Standards (NRC, 1996, 2000).

The types of assessment that are currently being implemented in contemporary college science classes are constantly evolving as university instructors seek to reform their teaching methods to embrace an inquiry-oriented philosophy of learning. While the types of assessment that are being used differ widely, several techniques have emerged to form a basis for further efforts. The power of peer-to-peer interaction in the development of coherent knowledge schemes has been utilized in the Collaborative Learning and Peer Instruction techniques (Berry & Nyman, 2002; Mazur, 1997). Portfolios have been used as an assessment tool that encourages learners to compile and organize artifacts that illustrate an individualized learning path (Childers & Lowry, 1997; Slater, Ryan, & Samson, 1997). Concept maps have been used as a vehicle for students to articulate conceptual understanding (Nash, Liotta, & Bravaca, 2000; Noh & Scharmann, 1997). Finally, scoring rubrics have been used to empower students with the ability to continuously self-assess their own work (Wiggins, 1998).

Having provided a contextual summary of relevant literature, we now turn our attention to two cases. These cases document two different instructors' struggles to implement alternative assessments that reflect the literature and represent instructional best practices.

CASE ONE: EVOLUTION OF AN UNDERGRADUATE SCIENCE TEACHING METHODS COURSE

In an ideal form, a science teaching methods course should model best instructional practices and articulate a curricular design that enhances preservice teachers' opportunities to observe, plan, and perform these same instructional skills. There would seem no better a class in which to model appropriate forms of alternative, performance, and authentic assessment than a teaching methods course. Such an ideal, however, depends on the expertise of the course instructor to organize and deliver such a class. In the case below, narrated from the instructor's point of reference, we delineate critical instructor decisions made, via action research, to enhance a science teaching methods course with respect to assessment.

Case One: Description

A typical science teaching methods class, at least by mid-1980s standards, was text-based, taught by lecture/recitation, and provided little to no opportunities for preservice teachers (PTs) to gain access to real secondary students in natural school-based classrooms until their final semester of undergraduate studies. Expected assessment techniques were fairly traditional—mid-term and final exams supplemented by lesson/unit planning exercises. Alternatives to traditional assessment, where they existed at all, consisted of lesson planning complemented by performance-based "microteaching" opportunities. Microteaching meant planning and delivering a brief 10-to-20-minute lesson in front of peer PTs. This is the science methods class I inherited when

I began my university teaching career; it was, however, neither the methods class in which I had participated as an undergraduate myself nor the one I wanted my future undergraduate PTs to experience.

I had instead, as a preservice teacher myself, two 2-week opportunities to be in a real science classroom setting with real secondary students. In the first 2-week period, which occurred about 1 month into the semester, I was expected to observe and assist the experienced teacher in a role much like that of an aide. My classmates and I came back to campus, possessing insights from our time in the classroom, for an additional 6 weeks before going back to the schools for the second 2-week period. In the latter school-based visit, we were expected to plan and teach two different lessons on two different days. In reflection, although not perfect, my own under-graduate experience was closer to the ideal end of the performance-authentic assessment continuum than were my first attempts to teach my own PTs in a science teaching methods class. I decided that changes in my approach were in order if I were to have my students make a better transition from college student to professional teacher.

I quickly discovered that it was difficult to enact immediate change—the institutional barriers on both the public school and university sides were great, relationships were not in place, administrators were skeptical, and resistance to change the way things had been done was high. I was faced with the recognition that, instead of a rapid overhaul, gradual changes over time needed to occur. Thus, began the evolution of my science teaching methods course.

Jettisoning Traditional Tests (1988–1995). Where I had begun in 1988 with traditional tests and a few lesson planning efforts, my first curricular revisions were focused on the creation of more performance-based assessment alternatives. I gave up the use of a "methods" textbook and the quizzes and exams that accompanied the lecture/recitation mode of delivery. I retained only a final exam for the course, but modified it to better reflect the changes that had taken place in the rest of the course curriculum.

Determined also to make the course a closer approximation of the decision-making processes exemplary teachers use in designing instruction, the "methods" textbook was replaced by shorter professional readings from primary source journals such as *The Science Teacher, American Biology Teacher, Journal of Chemistry Education, The Physics Teacher,* and so forth. An important ramification of this decision was that I found myself interacting with my PTs as if they were already novice teachers rather than students who would soon be teachers.

Assignments embedded in the chapters of the "methods" textbook were also ultimately replaced over a 6-year period. Instead of practice drills, represented by the textbook worksheet pages associated with each discrete chapter theme, I attempted to make my assignments more practical and applicable to actual instructional planning performed by professional teachers. The result was a set of more complex, synthesis-level assignments, shorter in length, yet more difficult for PTs to construct. Each assignment required good decision-making rather than serving merely as comprehension/application of a textbook writer's suggestions. Grading these more complex assignments, however, required creative assessment tools—and rubrics had not come to my attention as yet. I settled on developing a series of scoring keys, which specified the performance criteria by which I would grade. These scoring keys became cover pages to be submitted with each written assignment. Thus, my PTs knew what my expectations were in advance; my cards were more overtly laid on the table.

The final examination was an analysis-level assessment. It required PTs to analyze, based on their experiences, a series of instructional vignettes to determine if they agreed with the instructional approaches taken in each episode and what made the vignette one of quality instructional planning/performance. Alternately, if they disagreed with the approaches taken, PTs were to suggest revisions and to justify their decisions. In retrospect, the final exam associated with this initial wave of revision was well received compared with those tests used in prior semesters. The final exam, nonetheless, became the single course artifact with which I became the most dissatisfied because it was inconsistent with all other assignments given to my PTs. However, with no good alternatives, I retained it for another 7 years.

Integration of Relevant Field Experiences (1995–1997). Although I had been able to make improvements in my methods course, I had not as yet accomplished incorporating time for PTs with real students in actual classrooms during the science teaching methods semester. This all changed when a series of conversations took place to examine simultaneous improvement through shared resources. These conversations wended their way around such issues as charter schools, laboratory schools, and so forth, ultimately landing on the concept of the professional development school (PDS). The PDS concept commits time and resources to the mutual benefit of a public school and university partnership (Darling-Hammond, 1994; Goodlad, 1994; Holmes Group, 1995). The infrastructure of a PDS permitted me to place my PTs into public school science classrooms for significant time periods, while spending my own personal time renewing expertise/working in a high school science classroom. It also allowed me to bring exemplary high school teachers in contact with my PTs to become a more direct part of the science teaching methods curriculum. To gain both access and time for my PTs to be in real classroom situations meant that I needed to revise the assignments PTs performed because their experiences were now more authentic. This also meant that the final exam, in its present form, was a relative waste of my PTs' time. Changes were, once more, accomplished in the major course assignments; however, their articulation with one another and an as-yet undetermined final project left me with the sense that something was still missing.

The Final Project: Attacking Three Birds with One Stone (1995 to present).   I elicited the participation of PTs and in-service teachers to brainstorm concerning the design of the final project. We settled on the construction of a long-range plan that would ultimately match with the first unit each PT would be expected to initially implement during their student teaching semester. The final project was thus designed to represent reflection, synthesis, and application according to the following:

1. Reflection—on each of the major independent assignments required earlier in the semester (e.g., concept mapping, learning cycle "inquiry" lesson planning, etc.);

2. Synthesis—of the "tools" represented and introduced in each of these individual assignments (i.e., using the "tools" interdependently rather than independently); and

3. Application—construction of a unit to have direct application during the student teaching semester yet also to serve as a critical item to be placed in a portfolio and taken to prospective interviewers as a representation of curricular planning quality, teaching philosophy, and creativity.

With the creation of this final project, I realized I was finally challenging my PTs in a manner commensurate with my intentions for the course, ensuring that they had the planning skills to survive both their student teaching and 1-year teaching experiences.

The systematic decisions, undertaken as an action research agenda, had finally netted the intended results. But why had the process taken nearly 10 years? Wiggins and McTighe (1998) in their text *Understanding by Design* describe the most effective curricular designs as "backward." Wiggins and McTighe clearly illustrate the intent of backward design in the following:

> Why do we describe the most effective curricular designs as "backward"? We do so because many teachers *begin* with textbooks, favored lessons, and time-honored activities rather than deriving those tools from targeted goals or standards. We are advocating the reverse: One starts with the end—the desired results (goals or standards)—and then derives the curriculum from the evidence of learning (performances) called for by the standard and the teaching needed to equip students to perform.

> This backward approach to curricular design also departs from another common practice: thinking about assessment as something we do at the end, once teaching is completed. Rather than creating assessments near the conclusion of a unit of study (or relying on the tests provided by textbook publishers, which may not completely or appropriately assess our standards),

backward design calls for us to operationalize our goals or standards in terms of assessment evidence as we *begin* to plan a unit or course. (Wiggins & McTighe, 1998, p. 6)

The notion of "backward" planning is clearly consistent with the results obtained from 10 years of action research conducted by this science methods instructor. The final project in science methods promotes instructional best practices, integrates the *National Science Education Standards* (NRC, 1996) for content, teaching, and assessment, and creates the need to reconstruct each of the major assignments of the course leading up to the final project. PTs now routinely recognize the practical value of the final project on day one of the course—as a result they begin to think and plan accordingly. They also realize that each independent assignment provides them with an opportunity to try out and receive feedback on each of the "tools" that they will ultimately need to incorporate into the final project.

CASE TWO: A LARGE LECTURE CLASS IN BIOLOGICAL SCIENCE

The adoption of more active forms of learning and the use of alternative forms of assessment carry a higher level of risk, according to Bonwell and Eisen (1991), but they also create a more exciting instructional climate. In the previous case, risk taking was manageable because the instructor was comfortable with promoting active learning within a class size that averages 10 students each semester. What happens when the class size increases to 160?

Large lecture classes do not lend themselves easily to active forms of learning and as a consequence, professors, with little background in the use of anything but traditional lecture, often dismiss alternative forms of assessment as something done in colleges of education. Why should a collegiate science professor voluntarily make grading more difficult and time-consuming when traditional assessments are an adequate fit with traditional instruction? The answer lies in the satisfaction professors can gain as a result of being more responsive to the needs of undergraduate students. We relate in the case below, a successful attempt made by a biology professor to engage students in active learning through the use of formative assessment techniques.

Case Two: Description

I have always been interested in teaching and learning processes. Indeed, my passion as a college professor has been in working with undergraduate students—some 160 of them each semester in a course called "Human Body." The relationship one can forge with undergraduates takes time away from other pursuits (e.g., research, service) but it has always been worth my time. I have, for the past 10 years, made formative assessment a focus of my instructional planning. I am not sure when I decided to make the use of formative assessment more systematic. It probably was sparked by my attendance at professional meetings, in which I always made it a priority to attend one or more paper sessions devoted to teaching and learning. I have used many techniques over the years to create an active learning climate for my students. Among the more successful, in increasing order of management difficulty, are the following:

1. The "One-Minute Paper," in which students write brief responses to the questions, What is the most important thing you have learned in class today? (e.g., enzymes are proteins) and What is the question that is uppermost in your mind? (e.g., How do enzymes work?).

2. Simulation/Model Building, in which students serve as active participants to construct a model or to be part of a simulation.

3. "Points to Ponder" (or "Buzz Groups") provides students with a problem-based scenario and asks small groups of students sitting close to one another in a large lecture to come to a consensus on possible answers.

Students are not always receptive to active learning at first, especially in a large lecture class in which they are accustomed to passively taking notes. After a month or more, however, some students express their individual appreciation for having me make them think while other students would simply prefer to take notes. I use active learning strategies to hopefully benefit my students directly; however, many of my students may initially remain unaware of any tangible benefits. My work, as the instructor of the course, is greatly enhanced through the use of active learning strategies as formative assessments of student progress to provide me ongoing feedback upon which to base my instructional decisions.

The "One-Minute Paper," for example, forces me to reflect on student concerns and to always follow up with a response to the most common critical questions. Sometimes, depending on students' needs, demonstrations or analogies are constructed to make critical points using other perspectives and/or appealing to different learning modalities (Fleming & Mills, 1992). A further example of using learning modalities to my advantage is the use of simulation/model building, which represents the *kinesthetic* modality. Using the enzyme question listed above, students are asked to play the part of enzymes acting on a protein chain (represented by balloons connected by string). Although some students may find this latter activity silly, others have commented over the years how much this kind of example is personally helpful in understanding complex functions.

Finally, in the "Points to Ponder" strategy, students first individually consider a problem being posed and then accept or reject each of the multiple-response options. Once completed, groups are formed. Students must now negotiate individual responses and contribute to the determination of a group consensus. Groups are randomly selected to share their consensus response; other groups request further clarification (if necessary). This form of critical thought, based on the work of Robert D. Allen (Vice President for Instruction, Victor Valley College, CA, in his workshop, "Teaching Critical Thinking Skills in Biology"), takes time to develop and students will resist it for a variety of reasons—too hard, afraid to be wrong, and so forth. It is a mistake, however, for any professor interested in getting students to take a more active role in class, to give up on this active learning strategy. Once students accept that I will continue to persist with the format, they begin to take it more seriously and with less anxiety about being wrong in front of their peers. They eventually accept it as a regular part of the format of the class and begin to enjoy it. My mid-semester evaluations reveal that 90% of the students indicate that the "Points to Ponder" strategy is a good learning tool for them.

REFLECTIONS AND RECOMMENDATIONS

We began with a quote concerning well-designed assessments—that they are a conscious match between the purpose of the assessment(s) and the means by which the data is collected and used to inform assessment decisions. We then explored an argument for assessment reform, which delineated some potential strategies for making better assessment decisions. Finally, we presented two cases—one of action research in a small-enrollment, methods of teaching science course and the other in a large-enrollment lecture biology class.[1] We close this chapter with the following recommendations (in ascending order of risk/benefit) concerning assessment design:

1. Revision of an individual course (low risk/modest reward). Consider reading the book *Understanding by Design* (Wiggins & McTighe, 1998). Use the practical suggestions contained in this resource to revise a course syllabus and accompanying assessments to better

reflect instructor intentions concerning students' learning. A version of this document can be found online. Another excellent resource to consider is *Active Learning: Creating Excitement in the Classroom* (Bonwell & Eisen, 1991). Changes in a course can be documented through action research; revisions, like those described in the first case study, are continuously accomplished using a continuous improvement model.

2. Participate in peer collaboration (moderate risk/modest to great reward). Consider working with a colleague within your own discipline to mutually review and revise one another's courses. The process of peer collaboration requires more risk than the first recommendation since it opens up one's work to the scrutiny of another person. It is no different, nonetheless, than asking a colleague to critique a manuscript one might submit for potential publication in a peer review journal. The difference is the kind of scholarship in which we engage. While publications might represent a scholarship of application or discovery, peer collaboration in course design represents the scholarship of teaching (Boyer, 1990). Typical of peer collaboration models, colleagues begin by sharing and discussing course philosophy and providing a written reaction to one another's course syllabi. A typical follow-up action is to make one to several visits to one another's class sessions; the visitor to the colleague instructor usually shares a written reaction. Finally, colleagues share student work samples that represent a high, middle, and low achiever. Each stage of peer collaboration provides colleagues with opportunities to reflect upon assessment practices as they relate to course goals and learning outcomes. The degree of reward obtained depends greatly, nonetheless, on the willingness of an individual to be open to constructive criticism.

3. Engage in collaboration with colleagues in other academic disciplines (high risk/modest to great reward). This process is similar to that of peer collaboration but extends the risk to a justification of course philosophy, classroom visits, and student work samples to individuals outside of one's primary discipline. When successful, insights are gained that are invaluable because colleagues outside one's discipline seldom view the discipline as would a colleague trained within one's discipline. The risk is high because dissimilar colleagues may not easily find common ground, in which case the reward is minimal. The reward is great, however, in cases where common ground is found through mutual trust, respect, open-mindedness, and willingness to consider change(s).[2]

NOTES

1. In this chapter, we introduce several formative and performance/authen-
    tic assessments. These include the "One Minute Paper," "Simulation/
    Model Building," and "Points to Ponder" from the second case study and
    the use of exclusively performance-based assessments leading up to an
    authentic assessment final project from the first case study. Creating appro-
    priate and challenging assessments takes time, effort, and patience. Sci-
    ence professors (and secondary school science teachers) looking for
    additional examples of alternatives to traditional "tests" might consider
    using one of the following strategies:

    1. Collaborative learning and peer instruction (Berry & Nyman, 2002;
        Mazur, 1997).
    2. Concept maps (Nash et al., 2000).
    3. Independent research projects or long-term laboratory experiments (as
        opposed to traditional "cookbook" verification laboratory exercises).
        See Uyeda, Madden, Brigham, Luft, and Washburne (2002) for use of
        an authentic assessment example.
    4. Portfolios (Childers & Lowry, 1997; Slater et al., 1997).

    What each of these strategies has in common is that the assessment of stu-
    dent work requires the use of a scoring key (or "rubric") that specifies the
    performance criteria to be assessed and the level of work expected in order
    to attain specific point values. Bednarski (2003) provides a good example
    for creating a scoring rubric for assessing performance tasks. For a large col-
    lection of discipline specific alternative testing strategies, see Tobias and
    Raphael (1997).

2. In the College of Education at Kansas State University we have worked to
    establish the mutual trust, respect, and so forth, necessary to collaborate
    with public school teachers and colleagues in outside colleges. Using a U.S.
    Department of Education grant, College of Arts and Sciences faculty, Col-
    lege of Education faculty, and public school teachers collaborate to provide
    simultaneous improvement opportunities. The final two years of this 5-year
    grant were focused on alternative, performance, and authentic assessments,
    as described by Bonwell and Eisen (1991) and Wiggins and McTighe
    (1998). At the time of this writing, College of Arts and Sciences colleagues
    recognize how heavily their courses and curricula depend on traditional
    forms of assessment. Small steps have been taken and modest changes have
    been accomplished.

REFERENCES

Aristotle. (1985). Metaphysics, Book I. In J. Barnes (Ed.), *The complete works of Aristotle: The revised Oxford translation* (p. 981). Princeton, NJ: Princeton University Press.
Bednarski, M. (2003). Assessing performance tasks. *The Science Teacher, 70*(4), 34–37.

Berry, J., & Nyman, M.A. (2002). Small-group assessment methods in mathematics. *International Journal of Mathematical Education in Science and Technology, 33,* 641–649.

Bloom, B.J., Englehart, M.D., Furst, M.D., Hill, E.J., & Krathwohl, D.R. (1956). *Taxonomy of educational objectives: The classification of educational goals. Handbook 1: Cognitive domain.* New York: David McKay.

Bonwell, C.C., & Eisen, J.A. (1991). *Active learning: Creating excitement in the classroom.* Washington, DC: George Washington University, Graduate School of Education and Human Development.

Boyer, E. (1990). *Scholarship reconsidered: Priorities of the professoriate.* Princeton, NJ: Princeton University Press.

Childers, P., & Lowry, M. (1997). Engaging students through formative assessment in science. *The Clearing House, 71*(2), 97–102.

Crouch, C.H., & Mazur, E. (2001). Peer instruction: Ten years of experience and results. *American Journal of Physics, 69,* 970–977.

Darling-Hammond, L. (1994). Professional development schools: Schools for developing a profession. New York: Teachers College Press.

Delandshere, G. (2002). Assessment as inquiry. *Teacher College Record, 78,* 1461–1484.

Dickinson, V.L., & Flick, L.B. (1998). Beating the system: Course structure and student strategies in a traditional introductory undergraduate physics course for nonmajors. *School Science and Mathematics, 98,* 238–246.

Fleming, N.D., & Mills, C. (1992). Not another inventory, rather a catalyst for reflection. *To Improve the Academy, 11,* 137–149.

Gardner, H. (1981). *The unschooled mind: How children think and how schools should teach.* New York: Basic Books.

Goodlad, J.I. (1994). *Educational renewal.* San Francisco: Jossey-Bass.

Holmes Group. (1995). *Tomorrow's schools of education.* East Lansing, MI: Author.

Hestenes, D., Wells, M., & Swackhammer, G. (1992). The Force Concept Inventory. *The Physics Teacher, 30,* 141–158.

Mazur, E. (1997). *Peer instruction: A user's manual.* Upper Saddle River, NJ: Prentice-Hall.

McClymer, J.F., & Knoles, L.Z. (1992). Ersatz learning, inauthentic testing. *Excellence in College Teaching, 3,* 33–50.

McDermott, L.C. (1991). Millikan Lecture 1990: What we teach and what is learned—Closing the gap. *American Journal of Physics, 59,* 301–315.

Nash, J.G., Liotta, L.J., & Bravaca, R.J. (2000). Measuring conceptual change in organic chemistry. *Journal of Chemical Education, 77,* 333–337.

National Research Council. (2000). *Inquiry and the National Science Education Standards.* Washington, DC: National Academy Press.

National Research Council. (1996). *National Science Education Standards: Observe, interact, change, learn.* Washington, DC: National Academy Press.

Noh, T., & Scharmann, L.C. (1997). Instructional influence of a molecular-level pictorial presentation of matter on students' conceptions and problem-solving ability. *Journal of Research in Science Teaching, 34,* 199–217.

Perkins, D. (1992). *Smart schools: From training memories to educating minds.* New York: Free Press.

Pride, T.O., Vokos, S., & McDermott, L.C. (1997). The challenge of matching learning assessments to teaching goals: An example from the work-energy and impulse-momentum theorems. *American Journal of Physics, 66,* 147–157.

Schneider, R.M., Krajcik, J.M., Ronald, W., & Soloway, E. (2002). Performance of students in project-based science classrooms on a national measure of science achievement. *Journal of Research in Science Teaching, 39,* 410–422.

Slater, T.F., Ryan, J.M., & Samson, S.L. (1997). Impact and dynamics of portfolio assessment and traditional assessment in a college physics course. *Journal of Research in Science Teaching, 34,* 255–271.

Shepard, L.A. (2000). The role of assessment in a learning culture. *Educational Researcher, 29*(7), 2–14.

Taylor, A.R., & Watson, S.B. (2000). The effect of traditional classroom assessment on science learning and understanding of the processes of science. *Journal of Elementary Science Education, 12*(1), 19–32.

Tobias, S. (1990). *They're not dumb, they're different: Stalking the second tier.* Tucson, AZ: Research Corporation.

Tobias, S., & Raphael, J. (1997). *The hidden curriculum—Faculty-made tests in science: Lower division courses.* New York: Plenum Press.

Uyeda, S., Madden, J., Brigham, L. A., Luft, J.A., & Washburne, J. (2002). Solving authentic science problems. *The Science Teacher, 69*(1), 24–29.

Wiggins, G. (1998). *Educative assessment: Designing assessments to inform and improve student performance.* San Francisco: Wiley.

Wiggins, G., & McTighe, J. (1998). *Understanding by design* (Introduction, chaps. 1 & 2). Retrieved June 6, 2003, from the Association for Supervision and Curriculum Development Web site: http://www.ascd.org/readingroom/books/wiggins98book.html

CHAPTER 9

TEACHING FOR DIVERSITY IN UNDERGRADUATE SCIENCE

R. Lynn Jones Eaton

ABSTRACT

The need for more diversity in science, technology, engineering, and mathematics (STEM) has long been documented. Although it is widely held that both females and people of color can do and be whatever they so choose, an overwhelming number choose to not enter the professional fields of STEM. Professors of undergraduate students in natural sciences are often known as the "gatekeepers" of the professional areas associated with STEM and perceive a duty to "weed out" those whom they deem unworthy of a major in STEM. This chapter focuses on concerns associated with gender and ethnic diversity in STEM and provides suggestions for change.

INTRODUCTION

A widely held belief states that anyone, regardless of personal or sociocultural biography, can do and be whatever he or she so chooses. Despite this, an overwhelming number of both females and people of color choose to

Reform in Undergraduate Science Teaching for the 21st Century, pages 153–166

not enter the professional fields of STEM. The culture associated with academic areas of science, technology, engineering, and mathematics (STEM) consists of university faculty who are socialized to believe they are societal experts in their chosen fields and often wear the mask of omnipotence (Pang, Anderson, & Martuza, 1997). Removing that mask is not always easy because it can threaten an individual's security and institutional power.

GENDER EQUITY IN STEM

Participation of Females in STEM

It is reasonable to expect that girl's precollege experiences in science and mathematics will have consequences for their subsequent experiences as undergraduate or graduate women (Baker, 1990). Much of the precollege gender research finds that boys consistently receive more attention, praise, critical feedback, and support for assertive behavior. The learning experiences of girls are more passive, less demanding and less experiential—even in all-girls' schools (Jones & Wheatley, 1990; Morse & Handley, 1985; Tobin & Garnett, 1987). It has also been documented that different expectations are exhibited for boys and girls by mathematics and science teachers.

The consequences of these processes are discernable by ninth grade, though prior to this, girls and boys are almost identical in mathematics and science achievement. Thereafter, girls and boys diverge, both in the number of science and mathematics classes they take (especially in advanced mathematics and the physical sciences) and in their academic performance in these subjects. (For a review of this literature, see White, 1992.) The effects are also clear in the low ratios of women-to-men among college freshmen indicating an intention to major in science, mathematics, technology, or engineering—five or six men to one woman in engineering, and two or three men to one woman in the sciences, and still, even more discrepant ratios in technology and mathematics (College Board, 1988; Dey, Astin, & Korn, 1991; Tobias, 1993).

Instructor-Student Interaction in STEM

In university classes, women are interrupted more often by professors, asked fewer questions, coached less, and given less time to respond to questions than male students. Professors also make more eye contact with males and ask them more "higher level" questions, while females are asked "lower level" questions (Funk, 1993).

It has been noted that women consistently chose the word "discouragement" to describe their reaction to the experience of weed-out classes in STEM, and especially to faculty's refusal to interact with them as individual learners (Seymour & Hewitt, 1997). Faculty may or may not realize the critical role that they play in the persistence of undergraduate women, both as a source of ongoing support and at times of crisis. Many female undergraduate STEM majors plummet into depression, confusion, and uncertainty, due to their experiences, and seek the counsel of their professors about whether or not they should continue. They are prepared to accept their professor's assessment of their ability and performance, so long as it is conveyed in a manner that suggests he or she cares one way or another about their well-being. And surprisingly, just as it is seen in the precollege years that not only male, but female science teachers as well, may exhibit certain types of discouraging behaviors toward their female students, it is also noted among professors of undergraduate female students (Sadker & Sadker, 1994). Young women who are looking for encouragement to bolster their self-confidence, but who cannot evoke it from faculty tend to feel discouraged even though faculty may have said nothing negative to them. There is no perceived neutral ground: Failure to encourage is taken as discouragement. Depending on teachers for performance evaluation, reassurances about progress, and as a basis for motivation, constitutes a serious handicap for the many women who enter college having learned how to learn in this manner. Although such dependence is learned and can be outgrown, it seldom happens without feelings of anxiety and frustration, which can often lead to leaving as majors of STEM.

Self-Confidence in the Sciences

Females suffer an early loss in confidence in their ability to "do science" or to "do math" (American Association of University Women, 1992). This is sometimes seen as early as middle school years, but is made more vivid in their high school years. Females attribute success in science and math to effort, and failure to lack of ability, which is just the opposite for males (Fennema, 1990). For instance, females may say, "I did well because I tried so hard," or they may say, "I didn't do well because it was just too difficult and I couldn't do it." On the other hand, males may say, "I did well because it was easy for me," or they may say, "I didn't do well, because I just didn't want to put the time in to do well. But, I could have done it had I tried harder." These are very different feelings of confidence in their abilities.

Females tend to accept the role of spectator, but not that of participant and in doing so, hesitate to take risks and learn independence (Funk, 1993). They tend to give more self put-downs when answering questions, for example, "I'm not sure if this is what you want" or "This probably isn't right, but..." Even when their performance is adequate or good, women who have an undeveloped sense of their abilities in mathematics or science have difficulty in knowing that they are "doing okay" without the teacher's reassurance. Teacher-dependent students (whether women or men) draw upon the feeling that the teacher cares about them as a way to motivate themselves. They work hard to please their teacher and use the teacher's praise and encouragement as the basis for their self-esteem. Deprived of positive relationship and exchange with professors, certainty about self in science and mathematics is lost until the relationship is reconstructed with another supportive teacher, or a more independent self-concept is developed (Seymour & Hewitt, 1997).

ATTRITION RATES IN STEM

Who Leaves?

Much of the research on attrition rates of undergraduates in science, mathematics, and engineering has been documented by Seymour and Hewitt (1997). They found that the attrition rate in undergraduate science majors is egregiously high and often expected and accepted among college faculty. Again, the "gatekeepers" and "weed-out" classes take their toll on both male and female science, mathematics, and engineering majors.

As to the gender differences in losses from the sciences, Strenta, Elliott, Matier, Scott, and Adair (1993) reported that the persistence rates of men in STEM majors varied between 61% for highly selective institutions, to 39% for national samples, while the comparative rates for women ranged between 46% and 30%. Astin and Astin (1993) observed that absolute losses were greater among men, but, because the proportional loss of women was greater, their underrepresentation increased during undergraduate STEM education. In the same report, they documented high loss rates among that smaller proportion of STEM entrants who are Hispanic, Black, or Native American. Only one third of Hispanics, one half of Blacks, and one half of Native Americans who enrolled in STEM majors graduated in them. During college, the highest risk of STEM switching occurred in the transition from freshman to sophomore year. From the start of junior year to graduation, the attrition rate dwindled. Very few students transferred into STEM majors after college enrollment, and there was always a net loss.

Why Do They Leave?

Research conducted by Seymour and Hewitt (1997) found that there are six major factors contributing to the attrition rates of students. They are (a) rejection of careers based on STEM majors, and of the lifestyles they are presumed to imply; (b) the choice of non-STEM careers which seem more appealing; (c) doubt as to whether the rewards of an undergraduate STEM degree will adequately compensate for the effort required to complete it; (d) switching as "system-playing"; (e) concerns about the financial problems of completing a STEM major; and (f) concerns about the length of time required to finish a STEM major. These considerations reflect the traditional role of undergraduate education in the exploration of major life issues and the making of major life choices.

Contrary to the assumption that most switching or leaving is caused by personal inadequacy in the face of academic challenge, one strong finding is the high proportion of factors cited as significant in switching decisions which arise either from structural or cultural sources within institutions, or from students' concerns about their career prospects. The most commonly cited concerns are listed below in rank order.

- lack or loss of interest in science,
- belief that a non-STEM major holds more interest, or offers a better education,
- poor teaching by STEM faculty, and feeling overwhelmed by the pace and load of curriculum demands,
- choosing a STEM major for reasons that prove inappropriate,
- inadequate departmental or institutional provisions for advising or counseling about academic, career, or personal concerns,
- inadequate high school preparation, in terms of disciplinary content or depth, conceptual grasp, or study skills,
- financial difficulty in completing STEM majors,
- conceptual difficulties with one or more STEM subjects,
- the unexpected length of STEM majors (i.e., more than 4 years),
- language difficulties with foreign faculty or teaching assistants (TAs).

Degree Trends

Participation of women and minorities in science and engineering (S&E) higher education continues to rise, yet this involvement is not yet equivalent to their representation in the U.S. population of 18–30-year-olds. The number of S&E bachelor's degrees earned by women and under-represented minorities increased between the 1980s and mid-1990s.

Although still underrepresented in S&E in 1995, the number of S&E bachelor's degrees received by women (175,931) was a 36% increase over the number earned a decade earlier (128,871). The number of underrepresented minorities earning S&E doctorates rose 68% from 711 in 1985 to 1,194 in 1995 (Olson, 1999).

Seymour and Hewitt (1997) report a declining enrollment in advanced STEM degrees by American-born students. Hilton and Lee (1988) described the failure of able STEM undergraduates to continue into graduate school as the second greatest source of loss from the pipeline. The 1989 Office of Technology Assessment report blamed stagnation in the academic job market and observed that graduate enrollments have been sustained largely by foreign students who have helped to compensate for the decline in enrollments by U.S. citizens. The 1992 edition of national science indicators (National Science Foundation [NSF], 1993) also reported that between 1971 and 1991 the number of science and mathematics doctorates awarded to non-U.S. citizens rose 135% (170% in engineering), while those awarded to U.S. citizens fell by 10% (19% in engineering). NSF (1993) also reported that in 1988, foreign students accounted for more than 28% of PhDs in science, mathematics, and engineering.

Evidence of declining scientific literacy in the population, and of reduced numbers of STEM graduates available for research, development, or teaching, has generated expressions of concern that America's international competitiveness in the science and technology-dependent sectors of the U.S. economy would be undermined as a consequence of these trends.

The response of the academic and professional community has been differently expressed. A series of commissions, task forces, conferences, and working groups—sponsored by the National Science Foundation, the National Academy of Sciences, Sigma Xi, the National Association of State Universities and Land Grant Colleges, and the American Association for the Advancement of Science, and others—began to collectively brainstorm the causes and consequences of low interest in, and high attrition from, mathematics and science at all educational levels. The most influential of these include the "Neal Report" (National Science Board [NSB], 1986), the Report of the Disciplinary Workshops on Undergraduate Education (NSF, 1988), and the Sigma Xi "Wingspread Conference" of the National Advisory Group (1989). Each represents the collective wisdom and experience of higher education administrators and educators, officers of learned bodies, and representatives of the scientific community, industry and government. The Neal Report pointed to flaws in the undergraduate experience: Lab instruction, at worst, was said to be "uninspired, tedious and dull"; lab facilities and instruments were described as limited and "obsolete"; teaching was inadequate and poorly organized, and reflected little knowledge of modern teaching methods; teaching materials were out of date and curricular content failed to meet students' varied and emergent career needs. The report segmented its account by types of institution, disciplines, and to some degree, by the special difficulties of underrepresented groups. It also noted a decline in the number of STEM graduate students choosing academic careers, and thus a growing shortage of engineering faculty since 1976 and of mathematics faculty since 1981 (Seymour & Hewitt, 1997, p. 5). Surveying the condition of undergraduate STEM education overall, it warned that "all sectors of undergraduate education in mathematics, engineering, and the sciences are inadequately responsive to either its worsening condition, or to the national need for revitalization and improvement" (NSB, 1986, p. 3).

ETHNIC DIVERSITY IN STEM

Instructor Expectations and Interactions

Faculty are the second most important influence on the development of students, with family being the largest influence (Astin, 1993). As in the case of women's interactions with faculty, students of color often have little or no relationship with professors of STEM (Seymour & Hewitt, 1997).

One theory offered to explain poor retention by students of color (and women) is that they have different learning styles from those usually encountered in mainstream science and mathematics education. Straight lecture (the predominant teaching style in STEM higher education) however, does not work particularly well for students of *any* ethnicity. However, students who had attended predominantly White schools (regardless of ethnicity) had been socialized to anticipate it. Furthermore, it was not cultural differences in learning styles, but features in the educational socialization of STEM students that explained their learning difficulties. When Seymour and Hewitt (1997) interviewed students from predominantly minority high schools, the pedagogy of STEM college professors rated poorly compared with that of their high school teachers. As with women's descriptions of good teaching, these students stressed individual attention from teachers. In contexts where few students planned to attend college, teachers used their relationships with students to motivate them to continue in mathematics and science. Many students from such high schools had not learned to use peer study groups because they had become highly reliant upon their relationship with particular teachers. Teachers in these settings also motivated students by rewarding effort, as well as performance, with good grades, and their students carried this expectation into college. Thus, they responded to their first experiences of "objective" grading in STEM classes by defining STEM faculty as unfeeling or discriminatory. While the personalized attention that students receive in high schools may motivate them to graduate, it does not prepare them for the competitive culture of STEM courses in a large university where C and D grades are the norm and many students experience discouragement and lowered self-confidence.

Students from high schools where their own racial or ethnic group was dominant were those most at risk of a learned overdependence on teachers. Having sought in vain for individual attention from STEM college teachers, they sometimes found emotional support through peer groups. However, they had often not learned how to work in groups. Without a supportive relationship with faculty or peers, these students were at special risk of switching majors or of leaving college altogether.

Language and Dialects

In America, discrimination based on language is fairly widespread. The United States has ranged from a grudging acceptance of language diversity to outright hostility (Nieto, 1992). It also occurs in colleges with professors, TAs, and students who are STEM majors (Seymour & Hewitt, 1997). Just as college professors act more negatively toward students who speak English as a second language than toward those who speak it as a first language, STEM college students also cite language barriers of their professors and their TAs as reasons for switching majors (Seymour & Hewitt, 1997). Although most STEM students cite poor teaching by American-born faculty as a primary reason for switching majors, much of the high level of dissatisfaction of undergraduates with faculty pedagogy puts more stress on teaching assistants to compensate for those perceived inadequacies. Rather than allowing communication barriers to influence them to switch majors, many just dropped or avoided classes associated with foreign faculty or TAs and retook the class later. However, the communications problems with foreign faculty and TAs most commonly expressed is the difficulty that American students have in adjusting to different cadences in pronunciation by people who are fluent or native in a nonstandard or non-American English. Regardless of linguistic differences, the willingness of foreign-born professors and TAs to work with undergraduates on an individual basis, is most important.

Culture, Values, and Behavioral Styles

Minority college students often see their culture as a source of shame, due to discriminatory treatment by nonminority students and teachers

(Nieto, 1992). The assumption that culture is the primary determinant of academic achievement implies stereotyping and can be over-simplistic, dangerous, and counterproductive. An internalized negative stereotype can cause those who receive it to have serious doubts about their abilities. Many non-White minority STEM students internalize stereotypes perpetuated by White students and faculty that, for example, deem Black and Hispanic students as academically less capable than their White counterparts or Asian American students. The stereotypical generalizations of Blacks, Hispanics, or Native Americans as lazy, unintelligent, or lacking in ambition are not applied to Asian Americans, who are portrayed as overly ambitious, competitive, clannish, and intelligent in self-interested ways. When internalized, these stereotypes damage the self-concept of students by undermining their confidence to persist, regardless of their actual level of ability or preparation.

Seymour and Hewitt (1997) found that the culture, values, role conflicts, and interaction patterns of minority students in STEM, was important in the determination of whether or not they persisted as STEM majors. Some of the values found to weigh heavily on non-White (Black, Hispanic, Asian Americans, and Native Americans) STEM students are listed below:

- Obligation to serve community (Blacks, Hispanics, Native Americans).
- Obligation to be a role model (Blacks, Hispanics, Native Americans).
- Conflict between student and family roles (Blacks, Hispanics, Native Americans).
- Educational goals defined by parents (Asian Americans).
- Encouraged to be self-assertive (Blacks).
- Encouraged to be self-reliant and autonomous (Blacks, Native Americans).
- Supportive, effective peer group culture (Hispanics, Native Americans).

Helping students to confront and resolve these conflicts requires an understanding that is subculturally specific. Broad programs of "minority" support (including advising or mentoring), that lack understanding of the needs and perspectives of particular student groups will not improve retention. When program directors, counselors, or advisors can understand the origin and nature of the specific issues of different cultural groups, they can be anticipated, planned for, and preempted.

SUGGESTIONS FOR CHANGE

The issues and concerns presented throughout this chapter, though complex and many, are by no means exhaustive. This section seeks to provide suggestions for change in STEM higher education classes to directly improve the retention of females and non-White students. It has become increasingly evident in the past two decades that a focus on retention and not recruitment is necessary for the successful matriculation of female and non-White students in the areas of STEM. The suggestions for change offered below are centered on change to benefit all students of STEM. The approach is taken that a diverse STEM working population should mirror its college-prepared diverse STEM population and that truly "good teaching" is good for *all* students.

Incorporate Pedagogy

STEM faculty have a reward system based on research and development. The organizational context and structure of the institution shapes instructors' practice in ways that lead to ineffective teaching (Sunal et al., 2001). Teaching is not a part of that system, yet it is a very important part of what they do each day. Though faculty sometimes like to begin a program of reform with discussions of curriculum content and structure, this is unlikely to improve retention unless it is part of a parallel discussion of how to secure maximum student comprehension, application and knowledge transfer, and give students meaningful feedback on their academic performance (Seymour & Hewitt, 1997). There are no illusions that faculty will find this easy. Even to begin so fundamental a debate requires a willingness to explore the body of knowledge about how people learn that has largely been developed by faculty who are not members of STEM departments. Incorporating pedagogical professional development into the repertoire of beginning STEM faculty is necessary and should be ongoing in the sense that faculty should be encouraged to seek to improve their teaching. Why not make good use of those dreaded teaching evaluations and suggestions that students provide each semester? STEM departmental resources must be in place to address the specific needs of their students.

Shift From Competitive to Cooperative Learning Model

Shifting from a competitive to a more cooperative learning model is necessary for better retention of all students in STEM. Many females and nonminorities do not enter colleges with the knowledge and skills necessary to participate in peer and group study. This is not due to their ability, but to their lack of experience in cooperative or group study. Assigning or encouraging study groups among all students would foster learning communities and provide the support necessary for the retention and success of all students (Treisman, 1992). This is especially helpful in large science and mathematics classes where students seldom have contact with faculty regarding class questions and coursework. Faculty can provide the impetus for students to form these groups, thereby aiding female and non-White minorities who do not feel confident to approach STEM classmates because they are of a different race or ethnicity. This would lessen the isolation that these students generally feel, possibly rendering this no longer an issue or impediment to their success. Small groups provide a forum in which students ask questions, discuss ideas, make mistakes, learn to listen to others' ideas, offer constructive criticism, and summarize their discoveries in writing.

Observe Classroom Dynamics

What happens in college STEM classrooms (especially large ones) is often not seen by the faculty who are teaching in them. They are so heavily focused on imparting their knowledge (lecture style) to students in the given amount of time that to focus on whether or not students actually understand what they are saying goes largely unnoticed. Faculty must be deliberate in their attempts to change the dynamics of their classrooms. They can encourage class participation of all students by allowing a wait time before choosing someone to answer a question. They can seek outside feedback from a colleague to become more aware of whom they call upon and whose questions receive more response. Again, it is difficult to concentrate on the content, delivery, and patterns of interactions with students when lecturing. Monitoring language and materials is also important. By referring to the contributions of a diversity of professionals in STEM, male and female, of majority and minority ethnic groups, faculty provide a greater feeling of connection and inclusion.

Draw on Experiences and Knowledge of All Students

It must be acknowledged by STEM faculty that the United States is becoming an increasingly diverse society. Given the opportunity, many STEM students may be able to provide alternative views and applications to the theories and designs often required throughout their coursework. Diverse populations of students will certainly bring rich, diverse experiences to the classrooms of STEM professors everywhere. Real-world applications of the noted theories taught in STEM classes are sometimes nonexistent. However, real-world applications are necessary for all students to truly comprehend and assimilate what their professors are teaching.

Provide Role Models

It is no secret that the majority of required STEM textbooks are written by White males. Given this fact, professors must be diligent in their quest to provide role models for their STEM students. This would include the contributions of females and non-White minorities throughout their curriculum. Of course, for many professors of STEM, this would entail a reeducation of sorts on their behalf and would require some time to produce. There are sparse resource materials that provide the contributions of such professionals (females and people of color) to the areas of STEM, but they do exist.

Allowing practicing STEM-area professionals (of all gender and ethnic backgrounds) who are outside of academia to be guest lecturers or provide colloquia would also show a commitment to diversity within and across the profession. And as stressed earlier, to do this in introductory classes would show the applicability of theories being taught.

Last, but not least, to hire a diverse body of faculty members would serve as a sign of the importance and appreciation of the contributions of all qualified professionals. It is important that STEM students see themselves as having a "real" chance at success in STEM very early in their college career. It is possible that all STEM faculty members can serve as role models and mentors for their students, even those of a different gender or ethnicity. The time and willingness to help students are the only requirements.

REFERENCES

American Association of University Women. (1992). *How schools short-change girls: The AAUW Report*. Washington, DC: National Education Association.

Astin, A.W. (1993). *What matters in college? Four critical years revisited*. San Francisco: Jossey-Bass.

Astin, A.W., & Astin, H.S. (1993). *Undergraduate science education: The impact of different college environments on the educational pipeline in the sciences*. Los Angeles: University of California at Los Angeles, Higher Education Research Institute.

Baker, D. (1990, April). *Gender differences in science: Where they start and where they go*. Paper presented at the meeting of the National Association for Research in Science Teaching, Atlanta, GA.

College Board. (1988). *College-bound seniors: 1988.* New York: College Entrance Examination Board.

Dey, E.L., Astin, A.W., & Korn, W.S. (1991). *The American freshman: Twenty-five year trends.* Los Angeles: University of California at Los Angeles, Higher Education Research Institute.

Fennema, E. (1990). Justice, equity, and mathematics education. In E. Fennema & G.C. Leder (Eds.), *Mathematics and gender* (pp. 1–9). New York: Teachers College Press.

Funk, C. (1993, June). *What do women students want? (and need!): Strategies and solutions for gender equity.* Paper presented at the annual American Association of University Women Symposium, Minneapolis, MN.

Hilton, T.L., & Lee, V.E. (1988). Student interest and persistence in science: changes in the educational pipeline in the last decade. *Journal of Higher Education, 59,* 510–526.

Jones, M.G., & Wheatley, J. (1990). Gender differences in teacher-student interactions in science classrooms. *Journal of Research in Science Teaching, 27,* 861–874.

Morse, L.W., & Handley, H.M. (1985). Listening to adolescents: Gender differences in science classrooms. In L.C. Wilkinson & C.B. Marrett (Eds.), *Gender influences in classroom interactions* (pp. 116–127). Madison, WI: Academic Press.

National Advisory Group of Sigma Xi. (1989, January). *An exploration of the nature and quality of undergraduate education in science, mathematics and engineering.* Report to the Scientific Research Society, Wingspread Conference, Racine, WI.

National Science Board, Task Committee on Undergraduate Science and Engineering Education, Homer A. Neal (Chairman). (1986). *Undergraduate science, mathematics and engineering education; Role for the National Science Foundation and recommendations for action by other sectors to strengthen collegiate education and pursue excellence in the next generation of U.S. leadership in science and technology* (Report No. NSB 86-100). Washington, DC: National Science Foundation.

National Science Foundation. (1988). *Changing America: The new face of science and engineering* (Interim and Final Reports, The Task Force on Women, Minorities, and the Handicapped in Science and Technology). Washington, DC: Author.

National Science Foundation. (1993). *Indicators of science and mathematics education, 1992* (Division of Research, Evaluation and Dissemination, Directorate for Education and Human Resources). Washington, DC: Author.

Nieto, S. (1992). *Affirming diversity: The sociopolitical context of multicultural education.* New York: Longman.

Olson, K. (1999). *Despite increases, women and minorities still underrepresented in undergraduate and graduate S & E education* (Report No. NSF-99-320). Washington, DC: National Science Foundation. (ERIC Document Reproduction Service No. ED427958)

Pang, V.O., Anderson, M.G., & Martuza, V. (1997). Removing the mask of academia: Institutions collaborating in the struggle for equity. In J.E. King, E.R. Hollins, & W.C. Hayman (Eds.), *Preparing teachers for cultural diversity* (pp. 43–52). New York: Teachers College Press.

Sadker, M., & Sadker, D. (1994). *Failing at fairness: How our schools cheat girls.* New York: Simon & Schuster.

Seymour, E., & Hewitt, N.M. (1997). *Talking about leaving: Why undergraduates leave the sciences.* Boulder, CO: Westview Press.

Strenta, C., Elliott, R., Matier, M., Scott, J., & Adair, R. (1993). *Choosing and leaving science in highly selective institutions: General factors and the question of gender* (Report to the Alfred P. Sloan Foundation). New York: Alfred P. Sloan Foundation.

Sunal, D.W., Hodges, J., Sunal, C.S., Whitaker, K.W., Freeman, L.M., Edwards, L., et al. (2001). Teaching science in higher education: Faculty professional development and barriers to change. *School Science and Mathematics, 101,* 246–257.

Tobias, S. (1993). *Overcoming math anxiety: Revised and expanded.* New York: Norton.

Tobin, K., & Garnett, P. (1987). *Gender-related differences in science activities. Science Education, 71,* 91–103.

Treisman, U. (1992). Studying students studying calculus: A look at the lives of minority mathematics students in college. *College Mathematics Journal, 23,* 362–372.

White, P.E. (1992). *Women and minorities in science and engineering: An update.* Washington, DC: National Science Foundation.

CHAPTER 10

INTEGRATING INFORMATION TECHNOLOGY INTO UNDERGRADUATE SCIENCE

Michael Odell, Scott Badger, Teresa Kennedy, Timothy Ewers, and Mitchell Klett

ABSTRACT

Information technology (IT) is helping to transform undergraduate science education. Innovations in software, increases in bandwidth, and universal access to the Internet by faculty and students is providing the tools to extend the classroom and facilitate inquiry, collaboration, and real-world problem solving. New pedagogies utilizing IT are impacting student attitudes, achievement, and increasing access and efficiency.

INTRODUCTION

Changes in science education at the precollege level are beginning to have an impact on postsecondary education. National standards and benchmarks in science (American Association for the Advancement of Science,

Reform in Undergraduate Science Teaching for the 21st Century, pages 167–180
Copyright © 2004 by Information Age Publishing
All rights of reproduction in any form reserved.

1993; National Research Council [NRC], 1996), mathematics (National Council of Teachers of Mathematics, 1989, 1991), and technology (International Technology Education Association, 2000) have been developed and translated by state departments of education to create state standards. The National Science Education Standards (NRC, 1996) emphasize inquiry as a means of learning fundamental scientific concepts. Inquiry is "a set of interrelated processes by which scientists and students pose questions about the natural world and investigate phenomena" (NRC, 1996, p. 214). When engaging in inquiry, students develop ideas and construct explanations, test those explanations against current scientific knowledge, and communicate their ideas to others. They consider alternative explanations by examining their assumptions. Students develop an understanding of science by combining scientific knowledge with reasoning and critical thinking skills. Inquiry is an active process, requiring students to be engaged in the science. This is a shift away from the traditional lecture and laboratory approach utilized on most higher education campuses, especially in introductory courses. The K–12 arena has also been active in integrating technology into science instruction. University science courses historically have consisted of lecture, recitation, and laboratory. Students are very passive in this approach. The reform literature indicates that lecture may contribute little to some students' learning. In physics, there is evidence to suggest that lectures do not facilitate student understanding of fundamental physics concepts (McDermott, Shaffer, & Somers, 1994). The logistics caused by large lecture sections can also make it difficult to incorporate inquiry into many undergraduate science courses. Traditionally, college courses have tried to incorporate more active learning in lab sections that complement the lecture.

Likewise, the laboratory component of many undergraduate science courses can also negatively contribute to developing students' conceptual understanding. Experiments and activities in many laboratory courses are often structured in a step-by-step cookbook fashion. In essence, laboratory is not inquiry; it is an exercise in verification. Even in curricula designed around the reform initiatives, laboratory exercises are often self-contained and emphasize verification. Additionally, it is common that laboratory sessions do not align with material presented in lecture sessions (Hilosky, Sutman, & Schmuckler, 1998). Inquiry as presented in the K–12 standards promotes students conducting real scientific research projects extending over several days or weeks. How can science educators transform undergraduate science instruction? One solution may be the introduction of information technology (IT) into the science program.

THE IMPACT OF INFORMATION TECHNOLOGY

Information technology is dramatically transforming the scientific enterprise. Over the last few decades the ability to handle massive amounts of data, collect data continuously in extraordinarily fine detail, perform millions of computations per second, visualize information in three and four dimensions with rotations and translations, and simulate immensely complex phenomena with multivariate, multidimensional models has become commonplace (NRC, 2002). The Internet through e-mail allows scientists to collaborate over long distances. Remote-access technologies allow scientists to control specialized instruments such as telescopes and collect data from distant locations. Over the last decade IT has changed how undergraduate science is learned and taught. Universities have enhanced the IT infrastructure allowing students to access resources and classes anytime and anywhere through computers accessing the Internet. Laptop computers, calculators, and palm technologies are commonplace, and are often required of students at many universities. Student assignments are completed utilizing computers or calculators that can perform almost all of the calculations expected on homework or during exams. Due to reforms in K–12 education, many students already know how to use productivity software (e.g., word processors, spreadsheets) and calculators upon entering college. As a result, college instructors may be pressured to utilize these same tools in science courses. Science instructors are using course management systems (CMS) such as Interactive Virtual Courseware (IVC), WebCT, and Blackboard to post lectures, reading lists, and assignments on the Internet. They also utilize CMS tools to create online discussion groups and to use IT in other ways to deliver courses.

Effective instruction requires that IT be embedded in instruction, not just provided as an additional activity to a standard course or program (NRC, 2002). For example, science professors and instructors are using more than one technological innovation to help students learn. Students access course materials prior to class and complete preassessments, fostering better use of class time by instructors. Earth scientists utilize the Internet to take students on virtual field trips. They also utilize geographic information systems (GIS), global positioning systems (GPS), and remote sensing data to enhance their instruction. Students enrolled in the earth science courses can utilize LandSAT images, GIS, and GPS to conduct local watershed studies and mapping exercises. Data can be collected with handheld technologies and probeware to collect water quality data. The collected data can be downloaded into a database and analyzed using commercially available software to look for patterns and trends. It is becoming evident that IT enhances learning in multiple ways. Some IT tools such as the Internet increase students' ability to access information, while other IT tools such as GIS and GPS involves them in the doing of science instead of just reading or hearing about it, and yet other tools help students relate to what real scientists experience on a regular basis.

The widespread availability of these new technologies and recent scientific findings about how people learn are forcing science faculty members to rethink what and how they teach. Though a small percentage of professors are still reluctant to retool their teaching strategies, most are eager to learn how to integrate these technologies into their curriculum. Federal agencies such as NASA have funded projects to assist science faculty in transforming how undergraduate science is taught. Through NASA's Opportunities for Visionary Academics (NOVA) program, teams of academic faculty and science educators are beginning to use new pedagogical practices. In addition to incorporating the latest findings on how people learn, NOVA Network faculty are utilizing IT to support student learning through visualization, simulation, problem solving, collaboration, and inquiry.

INFORMATION TECHNOLOGY IN SCIENCE INSTRUCTION

University science faculty are redesigning the undergraduate science curricula to tap the broad range of content accessible through IT. They are understandably taking advantage of learning opportunities beyond the classroom. Students now have access to digital libraries, databases of raw information, GIS, GPS, simulations, and other IT tools (NRC, 2002). Instructors are also taking advantage of the visualization capabilities of IT. Chemistry professors, for example, are utilizing workstations that support molecular modeling programs. Molecular modeling is one way IT assists students' understanding of the more difficult and abstract concepts. Also, students often have difficulty understanding the relationship between the macroscopic appearance of a substance, its symbolic representation as a chemical formula, and its molecular shape and structure (Gabel, 1990). In physics, instructors are making use of video-based laboratories (VBL) to analyze and graph motion on video. Using VBL to see both the concrete motion event and the abstract graphical representation, students are better able to make cognitive links between the two and address graphing misconceptions they may have (Beichner & Abbott, 1999).

In the earth sciences, three-dimensional physical models have been used to assist students in visualizing large-scale concepts such as faulting, folding, and plate tectonics (Hewitt, Odell, & Worch, 1995). IT allows for these physical models to be represented on the computer screen so that students can actually see the motion of plates as an enhancement to physical models. In addition, graphic software offers a way to bring static two-dimensional representations of three-dimensional drawings into a dynamic medium that more closely simulates reality.

A study conducted by Williamson and Abraham (1995) found that students who used computer-based molecular modeling tools developed better conceptual understanding of phenomena such as diffusion of intermolecular forces than did students who received traditional instruction. Consider the earth sciences where GIS databases are being used in the classroom to construct detailed, precise computer maps. GIS allows students to overlay different types of data so students can look for patterns and trends.

Other visualization tools are also being utilized in undergraduate science. Spreadsheets and graphing calculators are used for mathematical modeling. Large databases of all types of images are accessible to students and faculty from all over the world. NASA databases such as the Distributed Active Archive Center (DAAC) at NASA Goddard are used at colleges and universities around the country. Other databases contain images taken by the Hubble Space Telescope. These images can be downloaded for use in astronomy courses. An introductory astronomy course at the University of Idaho has developed an online course site that links students to the NASA databases.

Technologies that rely on interactivity with the World Wide Web are rapidly being developed and are supporting pedagogical change. Java applets make it possible for students to perform a wide variety of interactive animations and other tasks online. Applets are considered a technological breakthrough because they function without having to send a user request back to the server, saving downloading time. The Institute for Mathematics, Interactive Technologies and Science at the University of Idaho has created Java applets for introductory physics and astronomy courses. For example, students can plot points on a sky chart to track the path of Mars through the sky to help explain retrograde motion, or mix colored light to better understand the visible spectrum (NOVA, n.d.). The Constructing Physics Understanding Project (CPU, 2000) has developed a number of applets for students to explain how light behaves (Goldberg & Bendall, 1995). Many computer-based visualization tools are available free on the Internet, transforming the ways in which many science faculty teach their courses.

ENHANCING LEARNING THROUGH SIMULATIONS

Simulation programs let students view multiple aspects of complex systems simultaneously or sequentially (NRC, 2002). For example, SimEarth a commercial software game popularized in the 1990s allowed students to manipulate environmental variables such as the level of CO_2 in the earth system.

The output data could then be displayed in graphical form on maps concurrently. More advanced simulations model earth systems such as the water or carbon cycles. This "simulated" earth engages students in the exploration of open-ended problems giving the students opportunities for data analysis, visualization, and the ability to experiment with multiple variables. What happens to the atmosphere if a large asteroid falls into the Atlantic Ocean? Students can investigate this research problem, ask "what-if" questions, pose new problems, and explore fundamental concepts.

The NASA-sponsored GLOBE program is being used on a number of campuses to facilitate extended scientific inquiry. GLOBE allows students to collaborate with each other and with research scientists around the world. Students collect environmental data and enter that data into the GLOBE database via the Internet (Kennedy & Odell, 2000). Students and instructors are able to use visualization software available from GLOBE to create graphical representations of the data to highlight patterns in that data over time. The site also provides simulations with real-world data that allow the learner to incorporate theoretical and empirical data to understand complex phenomenon such as climate.

Fortner, Schar, and Mayer (1986) compared the effectiveness of computer simulations to non-simulations with respect to perception of environmental relationships by undergraduate students enrolled in an introductory conservation resource course ($N = 110$). The treatment group utilized simulations and workbooks that were embedded in the course as individual learning modules. The control group utilized comparable workbook modules and traditional reference materials. Students were assessed on content and presentation techniques based upon content gains on sub-test instruments and an environmental relationship perception survey. Results showed that the experimental group performed significantly higher than the control group on knowledge and application assessment.

Computer simulations can also allow students to perform experiments that are too dangerous or too expensive in a traditional laboratory environment. Simulations even permit the ability for students to slow down or speed up time allowing for complete examination of certain phenomena.

Simulations are an efficient and cost-effective instructional tool. Simulations should not replace authentic laboratory experiences, but should instead enhance them. "Using simulations, students can replicate aspects of historically important models and classical experiments. They also can learn long-term strategies of scientific research using strategic simulations. Finally, simulations are a means of enhancing conceptual integration and conceptual change" (NRC, 2002 p. 43).

CHALLENGES FOR UNIVERSITY FACULTY

It can be easily argued that one of the most important goals in the reform efforts in science education involves an increased focus on problem solving in a real-world context. This means that students are encouraged to tackle real scientific problems. Part of the problem-solving process now involves students' utilization of a wide variety of information technologies such as images, databases, special interest discussion groups which often house experts, and other IT tools used by practicing scientists.

Problems have become increasingly complex and multifaceted while professional scientists have generally become increasingly specialized in their field. Consequently, they have found that collaborative teaming has become a more productive problem-solving strategy compared to working in isolation. This is why science faculty are increasingly encouraging students to foster their ability to work effectively in the context of a team project. It is certainly fortuitous that these needed collaborations are so greatly facilitated by IT. The Internet allows students to communicate in synchronous or asynchronous fashion while recording their threaded conversations. Simple access to the Internet means that useful links, files, images, databases, and software are easily and quickly disseminated between team members.

Like it or not, science instructors in higher education are faced with the fact that the traditional teaching method incorporating lectures and confirmatory laboratory experiences are becoming less effective for a variety of reasons. First and foremost, there has been an experiential and perceptual shift from the students of yesterday to the students of today. Students today bring a different set of experiences and expectations to the classroom. Policy changes in the K–12 system have exposed most students to inquiry-based approaches to learning, and students who have grown accustomed to this pedagogical style tend to be less receptive to lecture-based teaching. In addition, by the time students enter their freshman year of college, they are often very familiar and comfortable with many information technologies such as productivity software and Internet search engines, often more so than the professors from whom they take freshman courses. Science instructors unwilling to adapt to the needs and expectations of their students by embedding IT into their curriculum cannot reasonably expect optimal results.

For those science faculty members who have yet to acquire the needed IT skills or who lack the confidence to incorporate these skills into their teaching, part of the transformation process is simply becoming familiar with the broad range of innovative technologies available. Some of these innovations include using computer-interfaced probes to monitor and measure events, software commonly used to collect and analyze data, modeling and simulation software, and portable data collection devices.

Once faculty are sufficiently familiar with IT, it is important that they go beyond using it as a simple supplement to their traditional pedagogies, and develop projects that are truly technology-based and relevant to the real world. There is now a considerable body of research that can guide our attempts to optimize the teaching of undergraduate science courses (Gabel, 1990) and part of the solution clearly involves embedding IT into the curriculum so that the greatest gains in learning can be achieved. Innovations in IT seem particularly suited to facilitate inquiry-based science recommended by the science education reform documents.

USING IT TO TRANSFORM UNDERGRADUATE SCIENCE AT THE UNIVERSITY OF IDAHO

The integrated science (INTR 103) course at the University of Idaho was created to provide science content to students, especially those in teacher preparation through a grant from the National Science Foundation. The course was designed by faculty from physics, geology, life sciences, chemistry, and education with the goal of creating a course that reflected the vision of the K–12 science standards. The premise for creating the course was that students must experience science inquiry as defined in the reform documents if they were to teach science in this manner upon entering the teaching profession. INTR 103 is not a pedagogy course; the course instructors simply modeled the pedagogy that is described in the reform documents. The course integrated lecture and laboratory sessions to avoid the problems discussed previously in this chapter. The course was organized around the content standards and utilized an Earth systems approach to integrate the physical, earth, and life sciences in an extended field study. The course covers the following topics:

- nature of science and scientific inquiry
- physics and chemistry of atoms and molecules
- electrical circuits
- physics and biology of light
- magnets and motors
- forces shaping the earth
- ecology

In 1997, the course was updated to integrate IT with funding from NASA's Opportunities for Visionary Academics (NOVA) Program. Information technologies were integrated into the course to enhance teaching of abstract concepts, enhance communication with students, facilitate inquiry, and improve assessment and instructor efficiency.

The IVC system was the central organizing tool for the course. Because there are multiple instructors for the course, a course management system allowed the instructors to organize the course from various locations on campus. Class resources, including assignments, readings, and resources, were uploaded into the Library feature of the system. A decision was also made to utilize the system to structure students' out-of-class time. The university has a guideline that students should spend 2 hours outside of class for every contact hour.

USING IT TO IMPROVE STUDENT READINESS

Instructors determined that a number of classroom activities could be done outside of class. These included small-group discussions of readings, certain assessments, and journaling activities. For example, students in the class were preassessed on their inquiry skills. Results of that preassessment showed that the majority of students needed remediation on basic science process and inquiry skills. In past years, the instructors spent the first four to five class sessions (15 hours) remediating these skills. It was hypothesized that online remediation outside of class was as effective as in-class remediation.

The remediation modules were placed online and students were issued materials kits so they could complete the activities at home. Ruchti (2001) randomly assigned students enrolled in the course ($N = 29$) to two groups. Students either completed the modules online (treatment) or in class with traditional instruction (control). The online group completed the modules completely online outside of class and submitted their assignments electronically. The in-class group completed the same modules physically in the classroom with an instructor. Students were assessed using a pre and post multiple-choice content test as an indicator of achievement, as well as the Computer Attitude Questionnaire (CAQ) as an indicator of the students' attitude toward computers. Results showed a significant difference between the pretest and posttest for the whole class on both the achievement and attitude measures but indicated no significant difference in either achievement or attitudes toward computers between online and in-class groups. Achievement was defined as understanding of the content and was measured through a "native" or locally written 39-question multiple-choice exam over content found in the module. The implications of the Ruchti study were that online instruction was an efficient and effective way to remediate students in science inquiry skills and better prepare them for the class.

The IVC course management system was also used to assist students in completing assigned readings prior to attending class. Based on years of experience, many students come to class expecting the instructor to go over course reading materials as part of the lecture. The integrated science course minimizes lecture and requires students to have materials read prior to class. Prior to using IT, instructors relied on quizzes to ensure students had read the material. Many students saw this as punitive and indicated this on end-of-course evaluations. The course instructors developed a new strategy of monitored online discussions over the readings prior to class. Students were organized into small groups of 6 to 8. The instructors posted a list of questions or issues to be discussed and monitored the discussions. This approach served the same purpose as quizzes and did not result in negative feedback on the end-of-course evaluations. Another outcome of this strategy was that instructors communicated and interacted with every student in the course. Since the implementation of IT into the integrated science course, student evaluations have improved from an average of 2.7 to 3.5 on a 0–4 scale. Although we cannot attribute IT as the sole cause for this improvement, anecdotal evidence leads us to believe there has been a positive impact.

INFORMATION TECHNOLOGY IN EXTENDED INQUIRY

One feature of the course is an extended inquiry project that is conducted throughout the entire semester. Students conduct a watershed study of a local stream system. The Paradise Creek Watershed is studied from its source to its end (10 miles). There are six study sites along the stream and students in the course have utilized the same sites for 5 years. Students in teams study the water quality of the stream, the landcover, and soil of their site. Paradise Creek's source is located on a mountain a few miles north of town and flows down the mountain through wheat fields. Once in town it flows through residential neighborhoods and into the heart of the city. As it exits town, it flows past the water treatment facility and finally into the Palouse River. Students are given the opportunity to see the impacts of humans on the environment. Students utilize GPS to find their study sites and GIS to map the watershed.

To facilitate this activity, instructors chose to utilize the GLOBE program. Although GLOBE is a K–12 program, its online database, visualization tools, and science protocols were appropriate for the study. Selecting GLOBE also served a second purpose. Since most of the students enrolled in the course will become teachers, it provided them training in an inquiry-based technology-rich program they could utilize upon entering the profession. Students collected data with traditional science equipment and hand-held computer-

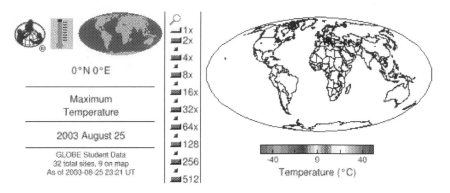

Figure 10.1.   Sample GLOBE visualization.

based probes. Hydrology data collected included transparency, temperature, dissolved oxygen, pH, electrical conductivity, nitrates, and alkalinity.
Landcover included biometry measurements, GPS location, and site photos
for comparison over time as well as determination of Modified UNESCO
Classification (MUC) code. Biology and biometry measurements included
the determination of dominant and co-dominant vegetation types, tree
height and circumference, grass biomass, canopy cover, and ground cover.
Soil moisture, temperature, and soil pH were also collected. These measurements are all entered into the GLOBE database and used by GLOBE scientists. The GLOBE site (GLOBE Program, n.d.) also provides graphing and
visualization tools. Students also use the graphing and visualization tools to
create a scientific poster that is uploaded into the electronic portfolio tool
of the IVC course system (see Figure 10.1). Because the data is in a large
database, they can access data for their study site from prior semesters. To
date, students in the class have taken more than 10,500 measurements
(GLOBE, n.d.). Students can access the other study sites on the stream as
well to provide a total watershed perspective. Utilizing an online program
such as GLOBE actually made it possible to do extended inquiry in the class
by providing the protocols, database, and visualization tools.

ENHANCING LEARNING THROUGH INTERACTIVE
JAVA APPLETS

The course also uses a number of Java applets to illustrate hard-to-under-
stand concepts. In the light and color unit, students utilize a colored light
applet to allow them to mix colored light on the computer. Student understanding of color is based on their experience with pigments, so when presented with mixing colored light they often rely on their knowledge of

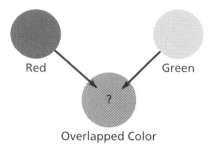

Figure 10.2.   Colored Light Mixer Applet sample activity

mixing pigments. Prior to developing the applet, the instructors for the course used three slide projectors with color filters to demonstrate the mixing of colored light. Because the slide projector method was a demonstration, students were passive learners and typically reverted back to their experience with colored pigments when tested. The applet allowed every student to actively explore colored light. Instructors created an online module that asked students to develop a theory of colored light. Students were asked to mix two colors at first and then three. Students were required to use the spectrum of visible light to predict results and finally develop their own theory of colored light (see Figure 10.2). After mixing a number of light combinations and examining the spectrum, students used the following instructions to compare their ideas to the scientific explanation:

1. Imagine that you observe two patches of colored light on a dark screen: One is red and the other is green (see below.) If the two patches are made to overlap, what color will you see?

2. Test your prediction. Open the color simulator and mix the patches of light.

The addition of the colored light applet has led to the development of a number of other applets for the course. The fact that these are online allows students to refer back and utilize them when studying for exams or preparing homework assignments.

INFORMATION TECHNOLOGIES AND ASSESSMENT

Another way in which IT is enhancing the course is in the area of assessment. In addition to preassessment mentioned earlier, the IVC system is utilized for formative and summative assessments. Instructors utilize online journals to document student ideas and promote reflection. Prior to class students are asked questions related to the topic. Students write down their ideas and instructors scan the ideas prior to class to identify misconceptions that need to be addressed in class. After class students are asked to reread their entry and post their new understanding and any questions they may have. This process promotes self-reflection and provides a safe medium for students to ask questions they may otherwise not ask. The process also allows the instructors to ascertain where to focus instruction and evaluate the effectiveness of instruction for each class period.

CONCLUSION

These are just a few examples of curricular innovations based on information technology affecting undergraduate science instruction. Although not a complete list of innovations, the provided examples illustrate the ways that science instruction is incorporating and using IT to provide students new and authentic science learning experiences. These innovations also are consistent with recommendations made in science reform documents that encourage visualization, simulation, real-world problem solving, and inquiry. These innovations are consistent with what research has shown about how students learn. More studies need to be conducted into the strategies integrating IT into undergraduate science courses and their impact on science learning.

REFERENCES

American Association for the Advancement of Science. (1993). *Benchmarks for science literacy.* New York: Oxford University Press.

Beichner, R.J., & Abbott, D.S. (1999). Video-based labs for introductory physics. *Journal of College Science Teaching, 29*(2), 101–104.

CPU Project. (2000). *Constructing Physics Understanding.* Retrieved August 21, 2003, from http://cpucips.sdsu.edu/web/CPU/default.html

Fortner, R.W., Schar, J.F., & Mayer, V.J. (1986). *Effect of microcomputer simulations on computer awareness and perception on environmental relationships among college students.* Columbus: Ohio State University, Office of Learning Resources. (ERIC Document Reproduction Service No. ED270311)

Gabel, D.L. (1990). Students' understanding of the particle nature of matter and its relationship to problem-solving. In *Empirical research in mathematics and science education* (pp. 92–105). Proceedings of the International Seminar. Dortmund, Germany: University of Dortmund.

GLOBE Program. (n.d.). *The GLOBE Program: An exciting, worldwide, hands-on education and science program.* Retrieved August 21, 2003, from http://www.globe.gov

Goldberg, F., & Bendall, S. (1995). Making the invisible visible: A teaching/learning environment that builds on a new view of the physics learner. *American Journal of Physics, 63,* 978–991.

Hewitt, P.A, Odell, M.R.L., & Worch, E.A. (1995). Models make it better: Three-dimensional laboratory models. *Science Scope, 19*(3), 26–29.

Hilosky, A., Sutman, F., & Schmuckler, J. (1998). Is laboratory-based instruction in beginning college-level chemistry worth the effort and expense? *Journal of Chemistry Education, 75*(1), 100–104.

International Technology Education Association. (2000). *Standards for technology education.* Retrieved July 27, 2001, from http://www.iteawww.org/TAA/TAA.html

Kennedy, T.J., & Odell, M.R.L. (2000). GLOBE: An international technology-based interdisciplinary program. In H.E. Klein (Ed.), *Case method research and application: Creative interactive teaching, case method and other techniques* (Vol. 12, pp. 215–218). Madison, WI: Omni Press.

McDermott, L.C., Shaffer, P., & Somers, M. (1994). Research as a guide for curriculum development: An illustration in the context of the Atwood's machine. *American Journal of Physics, 62,* 46–55.

NASA Opportunities for Visionary Academics. (n.d.). Retrieved August 21, 2003, from http://education.nasa.gov/nova

National Council of Teachers of Mathematics. (1989). *Curriculum and evaluation standards for school mathematics.* Reston, VA: Author.

National Council of Teachers of Mathematics. (1991). *Professional standards for teaching mathematics.* Reston, VA: Author.

National Research Council. (1996). *National Science Education Standards.* Washington, DC: National Academy Press.

National Research Council. (2002). *Enhancing undergraduate learning with information technology.* Washington, DC: National Academy Press.

Ruchti, W. (2001). *A comparison of online and traditional delivery of science process skills on achievement.* Unpublished master's thesis, University of Idaho, Moscow.

Williamson, V.M., & Abraham, M.R. (1995). The effect of computer animation on the particulate mental models of college chemistry students. *Journal of Research in Science Teaching, 32,* 521–534.

CHAPTER 11

TEACHING UNDERGRADUATE SCIENCE ONLINE

C.W. Sundberg

ABSTRACT

The chapter outlines definitions, benefits, and barriers of learning science online. Typical benefits of the use of the Internet in undergraduate science classes include increased student participation in critical, reflective course dialogue and "round the clock" and "round the globe" availability of courses and assistance from Web sites and course instructors. In the research, a model for evaluating levels of reflectivity in discourse was outlined as a method of quality assurance in online courses. Emergent recommendations for developing effective online courses are included.

INTRODUCTION

Budget crises and an increase in the percentage of nontraditional students have resulted in rapid proliferation of distance education courses delivered in some manner via the Internet (Van Dusen, 2000). With the demands of tightened budgets, many administrators laud online learning environments as cost saving. However, with the cost of the technology, cost

Reform in Undergraduate Science Teaching for the 21st Century, pages 181–197
Copyright © 2004 by Information Age Publishing
All rights of reproduction in any form reserved.

of access, time requirements for course development and teaching, and an emerging research base in learning online, caution should be exercised to ensure the quality of online courses. Evaluation of online courses for quality control is currently a point of concern with instructors. Unfortunately, early studies in online learning focused on affective learning outcomes and the development of the hardware structure for online learning environments and not on the extent of cognitive growth in participants. However, more recent studies have focused on student cognitive and other learning outcomes.

"E-learning," or "online learning," is used to describe a continuum of instruction via the Internet: (a) "Web-assisted" posting of course documents (e.g., syllabus and assignments); (b) "hybrid" classes that meet online about half the time; and (c) fully online courses, "virtual classrooms," requiring intensive and extensive interaction between course participants and with curricular materials (Hall, 2001; Stock-McIsaac, 2002). "E-moderators" is the term used to describe online instructors, tutors, or course facilitators (Creanor, 2002) and "online support" is the dynamic processes involving the scaffolding dialogue between students and course facilitators (Thorpe, 2002). The online educational experience may be "synchronous" (basically in real time) or "asynchronous" (accessed at different times).

TEACHING AND LEARNING SCIENCE ONLINE

Benefits of Online Science Learning Activities

A rise in proliferation of fully online courses and degree programs is mirrored in an increase in concerns about quality in undergraduate science teaching and learning. In a review of 873 undergraduate science syllabi (biology, chemistry, earth science, and physics) available online through "Syllabus Finder" (Cohen, 2002), a random selection revealed few fully online courses. Of the syllabi identified as fully online, most involved dialogue on the topic with peers, instructors, and experts or extensive uses of online databases (see Table 11.1).

In the hybrid and fully online science courses reviewed in the literature search, typical online assignments included (a) computer lab simulations (Carnevale, 2003), (b) Internet notes (Borman, 2003; Manteri, 2000), (c) visualization aides and proportional reasoning in chemistry (Gonzalez, Arasasingham, & Wegner, 2003), (d) video and animations of science phenomenon (Chasteen, 2001), (e) online tutoring (Littlejohn, Suckling, Campbell, & McNicol, 2002), and (f) online problem sets (Schaeffer, Bhargava, Nash, Kernes, & Stocker, 2001).

Table 11.1.   Analysis of Science Syllabi Available Through a "Syllabus Finder" Online Search

	Total number of syllabi	Random number of selected syllabi	Syllabi categorized as Web-assisted: posting of online documents and syllabus	Syllabi categorized as Hybrid: Includes course documents, syllabus, online quizzes, online problem sets, approx. half of course is online	Syllabi categorized as Fully Online: Most or all of the course is conducted online
Biology	381	29	9	15	5
Chemistry	141	30	9	20	1
Earth Science	172	10	7	2	1
Physics	242	20	8	12	0

Note. Some items located from the search were not actually syllabi.

Online resources have been used in many ways. Probably the most common use of the Internet in courses is the use of Web resources to post lecture notes and presentations for easy student access. According to Borman (2003) and Manteri (2000) the use of Internet notes and presentations provides efficient delivery of content and improves test scores. Lab simulations, another typical assignment, allow students the ability to change parameters and to conduct experiments that would be otherwise too dangerous to perform (Carnevale, 2003). Simulations are not limited to traditional laboratory experiences with Bunsen burner and beaker; simulations are also useful for case studies. Lundeberg, Bergland, Klyczek, and Hoffman (2003) reported the use of lab simulations of genetic diseases was a good way to incorporate real-life problems in an undergraduate biology course. In a similar manner, simulations can be used to explain complex constructs. According to Chasteen (2001) and Gonzalez et al. (2003), visualization aids, videos, and animations can facilitate understanding of complex scientific phenomenon. While, sending and receiving assignments via the Internet (e-mail, discussion board, or interactive course Web site like WebCT, Interactive Virtual Courseware [IVC], or Blackboard) is used quite frequently, a more interactive assignment was described by Laine (2003). In an undergraduate biology course, teams of students wrote and published laboratory journal articles to a course Web site, a process which mirrors the activity of professional organizations. Laine (2003) noted a positive outcome of the online assignment and ownership of their laboratory journal articles. Finally, a typical assignment involves online tutoring and problem sets (Littlejohn et al., 2002; Schaeffer et al., 2001). Benefits of online tutoring and online problem sets involve needed scaffolding at any time or location. In a mathematics course for engineers, Karr, Weck, Sunal, and Cook (2003) reported students in the online section of the course performed better in the analytical portion of the course because the online students had to study more intensely to understand the material.

Benefits of Online Dialogue

In general, the research base indicated student learning in an online learning environment was at least equivalent with a traditional face-to-face setting (Jewett, 1998). Online courses offer an unique opportunity for extended critical dialogue between peers and instructor (Hoadley, 2000) with a higher percentage of students involved in online dialogue than dialogue in a traditional face-to-face class (Bohrer, Colbert, & Zide, 1998). Along with increased participation, the asynchronous online dialogue may be more thoughtful because the discourse is written, allowing additional time for reflection (Bos, Krajcik, & Patrick, 1995; Hall, 1997). The discourse fosters collegiality and corporative conversation, important for increased student learning (DeSimone, Schmid, & Lou, 2000; Jewett, 1998; Jiang & Ting, 1998; McLellan, 1996; Tetter, 1997; Wegner, Holloway, & Crader, 1997; Wegner, Holloway, & Garton, 1999).

Barriers in Online Learning and Teaching

A review of the research on the use of online learning at colleges and universities revealed many barriers. There is considerable expense in the initial cost of equipment and subsequent upkeep (Van Dusen, 2000). Faculty members require training and release time in order to develop a quality Web presence (McNaught, Kenny, Kennedy, & Lord, 1999). Not only does it require time to develop a quality Web presence, responding to discussion prompts and e-mail messages requires a substantial amount of time for both students and teachers (Mende, 1998; Russett, 1994; Tetter, 1997). As new technology has emerged, often there has been a lack of vision and mission for the new technology (Tiene, 2002). Another problem occurs when new technological developments may not be congruent with cultural values in another society (Hedberg & Brown, 2002; Stock-McIsaac, 2002). In addition, there are issues with inequitable access based on factors such as age, gender, geographical location, poverty, and race (Hedberg & Brown, 2002; Stock-McIsaac, 2002). Students complain of a lack of personal contact with the course instructor (Bohrer et al., 1998; Guy & Lima, 1999), concerns that can be addressed by online office hours and if possible, periodic face-to-face meetings (Riffell & Sibley, 2003). Students and instructors may lack prerequisite skills needed to successfully access, respond, and maneuver in the online learning environment. Other research indicates students are more successful if they have adequate computer and Internet access at home (Cuban, Kirkpatrick, & Peck, 2001; Russett, 1994). In addition to a lack of prerequisite skills and adequate access to the technology and the Internet, students without a high personal degree of self-regulation are often not as successful in an online environment as in a traditional classroom (Hargis, 2000). In terms of performance, most of the research indicates there was no statistical difference in achievement between online groups and traditional face-to-face instruction (Jewett, 1998; Tetter, 1997; Wegner et al., 1997, 1999).

Research in Science Teaching and Learning Online

Not only do inherent barriers in utilization of the technology exist, there is a significant need for research in teaching and learning online (Geisen, Wright, Sunal, & Sundberg, 2002; Sunal, Sunal, Odell, & Sundberg, 2003; Sundberg, Mays, Sunal, Odell, & Ruchti, 2002). Many of the studies reviewed by Sunal et al. (2003) simply were "how-to-do-it" (p. 5) descriptions of techniques used to develop online courses; few of the studies actually used standard research analysis to evaluate the effectiveness of the online courses. In fact, evaluation of online courses has typically involved affective surveys or analysis of technical and navigation problems, not student learning (Geisen et al., 2002). Sunal et al. (2003) and Geisen et al. (2002) reported the evidence for course effectiveness was not clearly delineated, data collection and data analyses were not fully outlined or were missing in entirety, and the bulk of the investigations was not based on research in teaching and learning. How well students learn is the crux of formal education, therefore, the body of research in online learning should be expanded to reflect the central focus of institutions of higher education and indeed, learning at all levels.

The research base for online learning is still emerging. As previously noted, the use of the Internet for lecture notes and presentations provide efficient delivery of the content and aids in preparing students for exams (Borman, 2003; Manteri, 2000). Laboratory simulations, another typical use of Internet resources, allow students to conduct dangerous experiments (Carnevale, 2003) and interact with colleagues in various types of investigations (Lundeberg et al., 2003). Simulations and other visual aides can facilitate knowledge construction according to Chasteen (2001) and

Gonzalez et al. (2003). Interactive journaling is beneficial (Laine, 2003), as are online tutoring and problem sets (Karr et al., 2003; Littlejohn et al., 2002; Schaeffer et al., 2001).

AN INVESTIGATION IN TEACHING AN EARTH SYSTEM SCIENCE COURSE ONLINE

The literature review of research in online teaching and learning revealed a need for a model to evaluate the effectiveness of student learning (Geisen et al., 2002; Sunal et al., 2003). In order to evaluate the effectiveness of online courses, data gathering and analysis should focus on student learning. This emphasis is based in the context of current theory and research in effective science teaching and learning pedagogy for face-to-face teaching and learning (Sundberg et al., 2002). The research reported here involved two case studies of a university earth system science course taught entirely online. The course was a content course for in-service teachers on the interrelationships among the earth's spheres, focusing on change as it occurs in one or more spheres (atmosphere, biosphere, hydrosphere, and lithosphere). The central purpose of the research was to develop a method for evaluating the online course in terms of student achievement and subsequently to implement change in the course content and delivery in later semesters.

CASE STUDY 1

Purpose

The purpose of this investigation focused upon development of a model for evaluating the effectiveness of online courses through the use of critical dialogue. Critical dialogue can be defined as a disequilibrating discourse between peers and mentors and is crucial for learning (Hoadley, 2000; John-Steiner & Meehan, 2000). The outcome of this type of analysis would provide course developers and online instructors with a process for formative and summative evaluation that is designed to facilitate effective change in online courses. The central question addressed was, Is the research model effective in measuring reflective level for each student?

Participants and Setting

The participants were students enrolled in the online earth science course at three major research universities located in the Midwest, Northeast, and Southeast. The participants were teachers involved in teaching science in middle schools grades 5 to 8. Of the 27 teachers in the study, 16 were female, and 11 were male. The average teacher had 13 years of experience (ranging from 1 to 33 years) and taught in a school (small city or rural) with a predominantly White student population of between 200 and 500 students.

Online Course Material Used in the Study

In addition to the online learning environment (interactive discussion board, e-mail, course documents, syllabus, and assessment rubric), the students interacted with other media throughout the course. Print materials, CD-ROMs, online databases, and other identified Web sites were utilized to construct meaning about the impact of a change in one sphere, atmosphere, for example, on another sphere, biosphere, hydrosphere, or lithosphere.

Methodology

The online earth system science course content focused on the interactions between the earth's spheres (atmosphere, biosphere, hydrosphere, and lithosphere) related to concepts such as deforestation, hurricanes, and volcanoes. The assignments were distributed among students in a jigsaw fashion (Slavin, 1990) and each group member researched a different effect, for example, deforestation effects on one of the earth's spheres. The members discussed the dynamic interaction between the spheres as a result of deforestation on the discussion board and upon reaching consensus, posted the findings to the whole class discussion board. To ensure individual accountability, each student e-mailed a summary of the discussion to the instructor and produced a culminating activity, the preparation and implementation of a lesson plan involving sphere analysis in a middle school classroom.

The qualitative study design used naturalistic analysis (Patton, 1990) of artifacts (e-mail dialogue, transcripts of telephone interviews, final portfolios, and field notes from course facilitators). The variety of artifacts was used to establish "investigator triangulation" (Denzin, 1978; Joyce & Showers, 1995; Patton, 1990). "Methodological triangulation" (Denzin, 1978; Patton, 1990) was established by the use of the same data collection methodology over the course of the semester (Joyce & Showers, 1995). To establish content validity and content trustworthiness, the midterm interview questions were developed by a committee of experts (science educators from the three research universities, who had previously conducted research in the area of online learning). The data analysis used psycholinguistic analysis of the artifacts where the dialogue was coded and analyzed for emergent patterns (Patton, 1990) and for reflection (Lamy & Goodfellow, 1999; Surbeck, Han, & Moyer, 1991). Evidence of reflection was indicated when the conversation centered on effective teaching and learning.

Results

One typical problem noted by course facilitators and participants was equipment failure and server access. Julie, course facilitator, in comments regarding the earth systems science course noted, "Many of the students had older computers, were using old versions of browsers, and did not have reliable Internet service. This needs to be assessed ahead of time." Another problem often encountered in Internet courses was a lack of technical skill. Some students had great difficulty in accessing the Internet and navigating the site. Again, Julie commented: "Because of the problems many of the students encountered with the technology, I would recommend a module that familiarizes them with the technology and navigation around the online course."

The participants collaborated with colleagues at the three universities. However, portions of the dialogue indicated a need for personal contact.

> One thing I did already is dealing with the water quality, with some sampling of the water. Danny and I did some water sampling and we talked in class about what the water quality is here compared to other parts in the state, and also incorporated it with some kids from Russia. (Susie, midterm interview)

Also see Table 11.3, social discourse.

Of central importance to the research was the analysis of the data for evidence of reflective thought. Most of the dialogue between the in-service teachers was coded as non-reflective. For example, analysis of the e-mail dialogue revealed the majority of the messages centered on technical concerns (48%) (see Tables 11.2 and 11.3).

Table 11.2.   Basic Categories and Summary of E-mail and Interview Discourse (Adapted from Lamy & Goodfellow, 1999)

	Percent of coded messages
Non-reflective dialogue	
Social	08%
Technical	48%
Course Assignments, Assessment, or Required Course Content	42%
Reflective (dialogue focused on student learning, e.g., lesson plans and student projects)	02%

Table 11.3.   Levels of Reflectivity of E-mail and Interview Discourse in Dialogue from ESS course (adapted from Surbeck et al. [1991])

Little to No Reflectivity

1. Positive Feeling	16%	"I am thrilled that went through. My computer told me I had made an error" (E-mail).
2. Negative Feeling	8%	I was frustrated, and continue to be frustrated with the design, that we are using. (Rachel, midterm interview)
3. Technical query/ comment	46%	"Tonight was a rough one for me and my computer. I tried to install software for my new Palm Pilot and AOL wouldn't work. Had to uninstall everything to get back online." (E-mail)
4. Course content or assignments	6%	Maybe if we'd have put dates, like you did this last time instead of week one and week two, put October and the date, it would have been better. (Kathryn, midterm interview)
5. Assessment— Primarily requests for grades in course	18%	I am getting about 3 e-mails a day asking me what we are supposed to be doing…I am a little confused as to what I am really supposed to be doing and what we will be graded on. (E-mail)

Low Reflectivity

1. Concrete— Dialogue on course content beyond course require- ments	4%	"We had tornado sirens going off last night. Pretty scary…A hurri- cane would be too." (E-mail)
2. Comparative— Compare prior knowledge to course content or practice of teaching	0%	
3. Generalized to theories— Compare course content to learning theory	1%	I used the structure of the online course in my class. The students loved it. I went back in and said: "tell me what you've learned from this - what have you learned?" and they would say, "Well I learned about the particular sphere and how one thing, one event, can really affect me and how I can play on the others." So, 7th grade, this was very applicable to them, but you really have to break it down so that they can see. (Valerie, midterm interview)

More Reflective

1. Personal focus— Compare current situation to per- sonal professional life	1%	The way that I would correct it in my own classroom, as within the cooperative learning groups, each person has responsibility. In a true jigsaw activity, the puzzle does not come together unless every- body has their piece of the puzzle. For instance, when we were in sphere groups, we were supposed to go out and learn about our spheres, come back and share what we had learned, and then cre- ate causal chains. Each person within each of the sphere group should have had a more specific job, in detail. That person's piece of the puzzle would have fit better. If someone did not contribute, then it would not have worked. (Lynn, midterm interview)

continues

Table 11.3.   Levels of Reflectivity of E-mail and Interview Discourse in Dialogue from ESS course (adapted from Surbeck et al. [1991])

More Reflective (continued)		
2. Professional focus—Impact of current situation on the profession	0%	
3. Social/ethical focus—Moral implications of the experience	1%	"…Although you should have the web sites predetermined so they do not go where they should not be, such as, porn sites or student sites with incorrect information."
Social	3%	Stephanie is having computer troubles so we worked together to post one thing tonight. We spent hours at the kitchen table eating approximately one ton of Smarties…☺

Besides technical concerns, many messages coded as non-reflective focused on confusion about course requirements and grading of assignments. For example,

> I am getting about 3 e-mails a day asking me what we are supposed to be doing.… I am a little confused as to what I am really supposed to be doing and what we will be graded on. (E-mail)

The lack of reflectivity in the dialogue was problematic because reflective dialogue can be an indicator of student learning (Lamy & Goodfellow, 1999; Surbeck et al., 1991). Julie, course facilitator (instructor), noted,

> The dialogue on the content indicated low reflectivity. It is unclear whether or not the teachers were reflective or if the design of the modules did not elicit reflective responses. Recommendation: The content of the modules should be analyzed to develop tasks that will elicit more reflective responses. (Comments on ESS course)

Additionally, the final project, classroom application of sphere analysis, was important in evaluating course effectiveness. Reflection in the dialogue of most students was clearly differentiated and could provide a means for the development of appropriate intervention. For example "Lynn" (see Table 11.2) was more reflective than "MaryAnn," who focused on technical queries (see Table 11.4). The level of reflectivity could be used to evaluate student performance as well as offer course facilitators a way of developing intervention strategies.

Table 11.4. Comparison of Classroom Application of Volcanoes

MaryAnn's Classroom Application of Volcanoes	Lynn's Classroom Application of Volcanoes
I would divide my classes up into 4 groups. I would have one group work on volcano research; learning about where they are located, how often they erupt, etc. I would have another group research _____ Mountain (which is where we live) and the mountains that we are near to see if they are tied into volcanic formation and its origins. I would have a third group devising a theory by using inquiry learning and a demonstration of how our particular mountains were formed and over what time frame. The fourth group would be in charge of designing a replica of group three's ideas.	I would start with a KWL to determine what the students know and want to know about volcanoes. This will lead to a discussion about what the students know about volcanoes and allow me to address any misconceptions. We would redo the groups we used for the deforestation activities. I would start the children researching in the library for information about volcanoes and their effects on the various spheres. Once the information was collected, we would be ready to begin. My children would then make brochures to entice people to come and visit a true volcano. For example, the most chosen one is always Hawaii. They act as a travel agent "selling the idea of a volcanic vacation." The brochures are supposed to be 90% factual with a little fiction ☺. The other activities also include my reading the *Magic School Bus—Journey to the Interior of the Earth*. The journals for the week have the children take a trip into a volcano. The concluding activity has the spheres make three pages for their own magic school bus book. The experts on the topics make the pages that deal with their topic. They then put together and put their best pages together to make a *Magic School Bus Book*. The books would be graded for content and sent back from the publisher (me) for corrections. I would give the children my grading rubric before we begin so they know what is expected!

CASE STUDY 2

In a federal report on the use of the online course in a southeastern university, Beer, McCurdy, Basinger, and Holmes (2002) reported similar barriers with the same online course in earth system science. Undergraduate students indicated difficulties with access to the course Web site because of a lack of adequate equipment and access to the Internet at home. To address this need, the instructor basically created another similar Web site on the local server. There were problems locating the appropriate place to post course assignments and the instructor used e-mail instead of the discussion board to avoid the confusion. While the educators in the research concluded the classroom applications of earth system science were effective in pre-K–12 classroom application, the geology instructor indicated the online environment might not be the best environment in which to learn the course content. One recommendation emerged from the research, which mirrored the investigation in Case Study 1: the researchers advocated allotting additional time to train students in the use of computers and in learning online. In conclusion, typical barriers were described (i.e., technical difficulties in navigating the Web site and unfamiliarity among students with the online learning environment) in both case studies. In both instances, the instructors and researchers concluded the course was effective in positively impacting student learning and most of the changes needed for course improvement simply involved restructuring the navigation.

CASE STUDY SUMMARY

As a result of the research in Case Studies 1 and 2, navigation in the Web site was changed to make the online learning environment more user friendly. Another outcome of the investigations involved change in the type of prompts used in the discussion board and some of the questions in the assignments. The lack of reflectivity in the dialogue unveiled by the model outlined in Case Study 1 was used to implement changes in the online course. The use of qualitative analysis of critical dialogue for reflectivity (research model for Case Study 1) was used to determine the effectiveness of an online course because true learning takes place in an environment rich with critical discourse (Hoadley, 2000; John-Steiner & Meehan, 2000). Therefore, to evaluate the effectiveness of an online course, one viable method for determining student learning is to conduct qualitative analysis of the reflectivity of the discourse. For example, as a result of these results, the course designers developed a series of questions designed to elicit more reflective responses to be used in subsequent semesters. The following questions are a sample of essential questions about air from week 15 of the course.

> Design your activities to elicit your students' questions, and to help them to think about the essential questions you have been addressing:
>
> - How do you study air?
> - How does air change?
> - What causes air to change?
> - How does water exist in air?
> - How does air help living things?
> - How do you know about weather? (Meyers et al., 2001)

While the research process was valuable in targeting areas of the course requiring improvement, the amount of time involved in the analysis was quite extensive. Additional research to develop a more efficient process should be centered on clarifying and classifying dialogue into appropriate categories.

RECOMMENDATIONS FOR THE DEVELOPMENT OF AN EFFECTIVE ONLINE UNDERGRADUATE SCIENCE COURSE

What recommendations can be made for use of online Web course designs for teaching science courses at the undergraduate level? Based on the review of research and investigations carried out and reported by the author, several areas of best practice should be incorporated in the design.

1. Ensure students have adequate access to the Internet and a computer before registering for the course (Cuban, Kirkpatrick, & Peck, 2001; Russett, 1994). Hold an initial face-to-face session to ascertain the technical skill of the students and address common technical questions (Russett, 1994).

2. Provide students with ongoing assessment of their progress and emphasize appropriate evaluation rubrics on the Web site at the beginning of the semester. Course facilitators should schedule periodic telephone or online chat office hours or face-to-face meetings to address needs of the participants (Riffell & Sibley, 2003).

3. Develop cohorts of students as close together geographically as possible to act as support groups. The ideal situation would be to establish a cohort of teachers involved in systemic improvement of the earth science curriculum in the same local school. As reported in research, student collaboration in online courses was important for student learning (DeSimone et al., 2000; Jiang & Ting, 1998).

4. Facilitators should examine the dialogue periodically for reflectivity and develop interventions when necessary. Changes should be made within the course, if possible.

5. Conduct an initial face-to-face meeting to assist in developing appropriate strategies for effective group collaborative learning. Periodically, use other modes of distance communication (teleconference, videoconference, or online chat session) to augment e-mail and the discussion board.

6. The use of qualitative analysis of course dialogue (e-mail and discussion board) for reflectivity is recommended to determine the effectiveness of an online course. Examine the dialogue periodically for reflectivity and develop interventions when necessary. True learning

takes place in an environment rich with critical discourse. Therefore, to evaluate the effectiveness of an online course, one viable
method for determining student learning is to conduct qualitative
analysis of the reflectivity of the discourse.

IMPLICATIONS AND FUTURE RESEARCH FOR REFORM IN UNDERGRADUATE ONLINE SCIENCE INSTRUCTION

Examination of the cited studies revealed the research base is still emerging. Earlier studies focused on primarily affective responses to the online
learning environment and on the development of the hardware structure
for online learning environments. As the body of the research continues to
grow, more studies have focused on development of exemplary software
materials, online pedagogy, and on factors related to student learning.
With the demands of tightened budgets, online learning environments
have been touted as cost saving. However, the barriers previously described
indicate course designers and instructors should address concerns during
planning and teaching in order to facilitate effective science learning in an
undergraduate online environment. Additional future research should be
conducted on the refinement of the model used to evaluate the online
course in Case Study 1 in order to develop appropriate protocols for
online course evaluation, addressing the need for creating and maintaining quality online science courses.

AUTHOR NOTE

This work was in part supported by the ESS-NOVA. Earth System Science—
NASA Opportunities for Visionary Academics (NOVA), a National Aeronautics
and Space Administration Project; and the National Center for Research
on Online Learning (NCOLR), a project funded by the U.S. Department
of Education, although the views expressed here are the author's only.

REFERENCES

Beer, G., McCurdy, M., Basinger, D., & Holmes, J. (2002). *Earth System Science Education Alliance Year-End Report.* Ruston: Louisiana Tech University.

Bohrer, G.J., Colbert, R., & Zide, M. (1998, February). *Professional development for
Bermudian educators.* Paper presented at the meeting of the Association of
Teacher Educators, Dallas, TX. (ERIC Document Reproduction Service No.
ED417140).

Borman, S. (2003). Nontraditional teaching. *Chemical & Engineering News, 81,* 45–47.

Bos, N., Krajcik, J., & Patrick, H. (1995). Telecommunications for teachers: Supporting reflection and collaboration. *Journal of Computers in Mathematics and Science Teaching, 14*(1/2), 187–202.

Carnevale, D. (2003, June 8). The virtual lab experiment: Some colleges use computer simulations to expand science offerings online. *The Chronicle of Higher Education, 49*(20), p. A20.

Chasteen, T.G. (2001). News from online: Teaching with chemical instrumentation on the web. *Journal of Chemical Education, 78,* 1144.

Cohen, D.J. (2002). *Syllabus finder: Searching 132,378 syllabi at the Center for History and New Media and over 100,000 syllabi via Google.* Retrieved May 22, 2003, from http://chnm.gmu.edu/tools/syllabi

Creanor, L. (2002). A tale of two courses: A comparative study of tutoring online. *Open Learning, 17,* 57–68.

Cuban, L., Kirkpatrick, H., & Peck, C. (2001). High access and low use of technologies in high school classrooms: Explaining an apparent paradox. *American Educational Research Journal, 38,* 813–834.

Denzin, N.K. (1978). *The research act: A theoretical introduction to sociological methods.* New York: McGraw-Hill.

DeSimone, C., Schmid, R., & Lou, Y. (2000, April). *A distance education course: A voyage using computer-mediated communication to support meaningful learning.* Paper presented at the annual meeting of the American Educational Research Association, New Orleans, LA.

Geisen, J.L., Wright, D., Sunal, D.W., & Sundberg, C. (2002, April). *Best practice in online instruction evaluation: Implications of the literature of online research design and instruction for higher education.* Paper presented at the annual meeting of the American Educational Research Association, New Orleans, LA.

Gonzalez, B.L., Arasasingham, R.D., & Wegner, P.A. (2003). *A cross-institutional analysis of the Web-assisted tools on visualization and proportional reasoning in general chemistry.* Retrieved May 24, 2003, from http://chemsrvr2.fullerton.edu/blg/ChemConf/ChemConf2003.pdf

Guy, F., & Lima, J. (1999). WorldClass system. *Educational Media International, 36*(1), 68–73.

Hall, B. (2001). *New technology definitions.* Retrieved February 11, 2003, from http://www.brandonhall.com/public/glossary/index.htm

Hall, D. (1997). Computer mediated communication in post-compulsory teacher education. *Open Learning,* 54–57.

Hargis, J. (2000, June). The self-regulated learner advantage: Learning science on the Internet. *Electronic Journal of Science Education, 4*(4), Article 1. Retrieved July 21, 2003, from http://unr.edu/homepage/crowther/ejse/hargis.html

Hedberg, J.G., & Brown, I. (2002). Understanding cross-cultural meaning through visual media. *Educational Media International, 39,* 23–30.

Hoadley, C. (2000). Teaching science through online, peer discussions: SpeakEasy in the Knowledge Integration Environment. *International Journal of Science Education, 22,* 839–857.

Jewett, F. (1998). *Course restructuring and the Instructional Development Initiative at Virginia Polytechnic Institute and State University: A benefit cost study.* Blacksburg: Virginia Polytechnic Institute and State University. (ERIC Document Reproduction Service No. ED423802).

Jiang, M., & Ting, E. (1998, April). *Course design, instruction, and students' online behaviors: A study of the instructional variables and students' perceptions of online learning.* Paper presented at the annual meeting of the American Educational Research Association, San Diego, CA. (ERIC Document Reproduction Service No. ED421970).

John-Steiner, V., & Meehan, T. (2000). Creativity and collaboration in knowledge construction. In C. Lee & P. Smagorinsky (Eds.), *Vygotskian perspectives on literacy research: Constructing meaning through collaborative inquiry.* Cambridge, England: Cambridge University Press.

Joyce, B., & Showers, B. (1995). *Student achievement through staff development: Fundamentals of school renewal* (2nd ed.). White Plains, NY: Longman.

Karr, C.L., Weck, B., Sunal, D.W., & Cook, T.M. (2003, Winter). Analysis of the effectiveness of online learning in a graduate engineering math course. *Journal of Interactive Online Learning, 1*(3). Retrieved July 21, 2003, from http://www.ncolr.org/jiol/archives/2003/winter/3/index.asp

Laine, P. (2003, Spring). Use of instructional technology as an integral part of a non-major science laboratory course: A new design. *Journal of Interactive Online Learning, 1*(4). Retrieved July 21, 2003, from http://www.ncolr.org/jiol/archives/2003/spring/1/index.asp

Lamy, M.-N., & Goodfellow, R. (1999). Language learning and technology. *Language Learning & Technology, 2*(2), 43–61.

Littlejohn, A., Suckling, C., Campbell, L., & McNicol, D. (2002). The amazingly patient tutor: Students' interactions with an online carbohydrate chemistry course. *British Journal of Educational Technology, 33,* 313–321.

Lundeberg, M., Bergland, M., Klyczek, K., & Hoffman, D. (2003, Winter). Windows to the world: Perspectives on case-based multimedia Web projects in science. *Journal of Interactive Online Learning, 1*(3). Retrieved July 21, 2003, from http://www.ncolr.org/jiol/archives/2003/winter/4/index.asp

Manteri, E.J. (2000). Using Internet class notes and PowerPoint in the physical geology lecture. *Journal of College Science Teaching,* 301–305.

McLellan, H. (1996, October). Information design via the Internet. In *VisionQuest: Journeys toward visual literacy. Selected readings from the Annual Conference of the International Visual Literacy Association* (28th), Cheyenne, WY. (ERIC Document Reproduction Service No. ED408942)

McNaught, C., Kenny, J., Kennedy, P., & Lord, R. (1999, October). *Developing and evaluating a university-wide online distributed learning system: The experience at RMIT University.* Educational Technology & Society, 2(4). Retrieved July 21, 2003, from http://ifets.ieee.org/periodical/vol_4_99/mcnaught.html

Mende, R. (1998, May). *Hypotheses for the virtual classroom: A case study.* Paper presented at the IT97 Conference: An Educational Technology Coordinating Committee Event, St. Catharines, Ontario, Canada. (ERIC Document Reproduction Service No. ED423919)

Meyers, R., Botti, J., Davis, H., Shay, E., Sebeck, J., Cox, K.T., et al. (2001). *Earth system science course.* Retrieved May 27, 2003, from Wheeling Jesuit University/

NASA Classroom of the Future Web site: http://www2.cet.edu/ete/hilk4/outline/week15.html#ess

Patton, M.Q. (1990). *Qualitative evaluation and research methods* (2nd ed.). London: Sage.

Riffell, S.K., & Sibley, D.H. (2003). Learning online. *Journal of College Science Teaching, 32,* 394–399.

Russett, J. (1994, March). *Telecommunications and preservice science teachers: The effects of using electronic mail and a directed exploration of Internet on attitudes.* Paper presented at the annual meeting of the National Association for Research in Science Teaching, Anaheim, CA. (ERIC Document Reproduction Service No. ED368571).

Schaeffer, E., Bhargava, T., Nash, J., Kernes, C., & Stocker, S. (2001, April). *Innovation from within the Box: Evaluation of online problem sets in a series of large lecture undergraduate science courses.* Paper presented at the annual meeting of the American Educational Research Association, Seattle, WA.

Slavin, R. (1990). *Cooperative learning: Theory, research and practice.* Englewood Cliffs, NJ: Prentice-Hall.

Stock-McIsaac, M. (2002). Online learning from an international perspective. *Educational Media International, 39,* 17–21.

Sunal, D.W., Sunal, C.S., Odell, M., & Sundberg, C.W. (2003). Research-supported best practices for developing courses for online learning. *Journal of Interactive Learning, 2*(1).

Sundberg, C.W., Mays, A., Sunal, D.W., Odell, M., & Ruchti, W. (2002, June). *Evidence for online course effectiveness: A qualitative design for evaluating a graduate science course for teachers.* Paper presented at the National Consortium for Online Learning Research, Tuscaloosa, AL.

Surbeck, E., Han, E., & Moyer, J. (1991). Assessing reflective responses in journals. *Educational Leadership, 48*(6), 25–27.

Tetter, T. (1997, October). *Teaching on the Internet. Meeting the challenges of electronic learning.* Paper presented at the fall conference of the Arkansas Association of Colleges of Teacher Education, Little Rock, AR. (ERIC Document Reproduction Service No. ED418957)

Thorpe, M. (2002). Rethinking learner support: the challenge of collaborative learning online. *Open Learning, 17,* 105–119.

Tiene, D. (2002). Addressing the global digital divide and its impact on educational opportunity. *Educational Media International, 39,* 211–222.

Van Dusen, G.C. (2000). Digital dilemma: Issues of access, cost, and quality in media-enhanced and distance education. *ASHE-ERIC Higher Education Reports, 27,* 1–120.

Wegner, S.B., Holloway, K., & Crader, A. (1997). *Utilizing a problem-based approach on the World Wide Web.* (ERIC Document Reproduction Service No. ED414262)

Wegner, S.B., Holloway, K.C., & Garton, E.M. (1999, November). The effects of Internet-based instruction on student learning. *Journal of Asynchronous Learning Networks, 3*(2), Article 7. Retrieved July 21, 2003, from http://www.aln.org/publications/jaln/v3n2/v3n2_wegner.asp

CHAPTER 12

PROFESSIONAL DEVELOPMENT OF UNIVERSITY SCIENCE FACULTY THROUGH ACTION RESEARCH

Carol Dianne Raubenheimer

ABSTRACT

Ideas are examined about action research held by college and university faculty involved in redesigning science courses to reflect reform-based concerns. The study (a) examines how faculty plan, gather, and analyze their action research data; (b) identifies outcomes from successful action research projects; and (c) concludes with implications for conducting professional development of university faculty through action research. Rubrics and content analysis were used on a diversity of products and processes involved in the action research process. Faculty teams from a national sample at 53 colleges and universities were investigated. The development of the Action Research Rubric (ARR) allowed quantification of approaches across action research plans. Final outcome reports and conference presentations of 25 action research projects were also analyzed. Data obtained from site visits to

Reform in Undergraduate Science Teaching for the 21st Century, pages 199–223

20 of the institutions were used to triangulate the findings from the analysis of action research projects. It was found that action research projects have produced a wealth of information about successful innovation in undergraduate courses, although many faculty are still novices in the implementation of action research projects. The majority used quantitative approaches, over qualitative designs. The outcome of these action research efforts has shown significant gains in student learning, as well as in faculty professional development in teaching, although the latter was less prevalent as a focus in the cases analyzed. Observational data undertaken during site evaluation visits support the idea that faculty shifted their practices away from traditional didactic lessons and that action research facilitated this process.

INTRODUCTION

Action research is a method for education practitioners to systematically and self-consciously examine classroom actions and outcomes, with the intention of improving their own practice. It begins with the practitioner generating a question about an aspect of his/her action and developing a systematic plan of action for gathering and analyzing data. This might include data collection through self or peer observation, interviews, and questionnaires with learners and other participants. It is an iterative process involving successive cycles of question generation, planning, action, observation, and reflection, with the latter being termed the four moments of action research (Hopkins, 1993; Zuber-Skerritt, 1992a). The difference between this approach and traditional research is that it does not end with data analysis, but leads to the generation of new questions and new action (Cross, 1998). Action research is concerned with results that are relevant to immediate practice and is a valid and valuable process in bringing about change and reform in education (Arhar, Holly, & Kasten, 2001).

LITERATURE REVIEW OF FACULTY ACTION RESEARCH PRACTICES

Previous published research with university faculty (Kember, 1998, 2000), as with K–12 teachers (Arhar, Holly & Kasten, 2001; Cochran-Smith & Lytle, 1992, 1999), has validated the effectiveness of action research in the development of expertise in teaching and for enhancing student learning and achievement. Action research is a powerful form of professional development for both teachers and educators in institutions of higher education (Elliot, 1991). Elliot further discussed how action research could be used to challenge the technical rationality that typically underpins government attempts at educational reform, through a process model to education (p. 135), in which teachers are empowered to bring about change in their specific contexts.

Zuber-Skerritt (1992a, 1992b) held that action research is particularly relevant for higher education because it focuses on improving teaching and professionalism in universities and colleges by focusing on the pedagogical aspects of higher education. There are numerous examples of faculty researching their own action, together with others, particularly in Australia, Hong Kong, the United Kingdom, and Israel (Burchell, 2000; Conway, Kember, Sivan, & Wu, 1994; Hadfield & Bennett, 1995; Kember, 1998, 2000; Kember, Lam, Yan, Yum, & Liu, 1997; Zoller, 1999; Zuber-Skerritt, 1992a, 1992b). From these accounts, action research is acknowledged as a powerful way to examine preconceptions, bring about curriculum change, improve instructional design and implementation, and enhance faculty development (Kember, 1998, 2000; Zuber-Skerritt, 1992a, 1992b).

For instance, action research was used to monitor the implementation of new courses at the Hong Kong Polytechnic (Conway et al., 1994) and indicated the value of faculty working together to implement new pedagogies. The successful implementation of action research projects requires a supportive environment and the active collaboration of various stakeholders (Bondy & Ross, 1998), but time and the perceptions of administrators and students can be roadblocks to the process. It is crucial that support mechanisms are in place to support the action research of practitioners within their own contexts. Kember (2000) concluded that action research could contribute to quality enhancement and quality assurance in higher education, provided that it is not linked to a positivist approach, but rather to interpretive or critical approaches.

In the US, classroom research has been advocated for many years as a mechanism for contributing to the scholarship of teaching in higher education (Cross, 1998; Cross & Angelo, 1989; Paulsen 1999, 2001). Classroom research has many of the same components as action research (Bondy & Ross, 1998), with the purpose being to improve teaching and learning, which means that the cycle from question formulation to changing practice must be completed (Cross, 1998).

Neither action research nor classroom research has been widely applied in higher education in the United States, except in some teacher education programs (Bondy & Ross, 1998; Ross & Bondy, 1996). In part this is because, historically scholarship has been synonymous with research publications and quantitative research (Kreber, 2000), and there has not been a focus on improved teaching practice for tenure and promotion review (Boyer, 1990), which could include feedback from action research. Thus, there are fewer examples of action research in institutions of higher education in the United States (Bondy & Ross, 1998). Despite this, Krockover, Adams, Eichinger, Nakhleh, and Shepardson (2001), working out of Purdue University, concluded efforts to implement action research models in higher education settings, though limited in number, support our belief that action research is an effective means of reforming teaching of introductory college and university science (p. 317).

Paulsen (2001) asserted that one of the main reasons why classroom research has not been widely applied in higher education is because of a confusion surrounding the terms "classroom assessment" and "classroom research." He added that the primary advocates of both approaches, namely Cross and Angelo (1989), have used the terms interchangeably, "thereby training faculty and faculty developers everywhere to think of classroom research in a somewhat narrow sense, as referring to the use of a set of classroom assessment techniques (CATs)" (p. 21). Cross (1998) has since made it clear that the two are separate processes, with classroom research being geared to understanding how students learn and how professors might enhance this process. Classroom research should be integrated with research and theories of learning (Cross & Steadman, 1996) and should not be an add-on activity but embedded in regular class sessions, engaging students as collaborators in the process (Cross, 1998). Classroom assessment techniques represent a range of strategies for assessing student outcomes, and can be used as one component in action research.

Carr and Kemmis (1986) categorized action research projects as serving one of three knowledge-constitutive interests, (a) technical, (b) practical, and (c) emancipatory, with each interest serving a particular purpose. These interests can be related to differing approaches to research, where different researchers emphasize different aspects of practice in their investigations and rely on different research methods and techniques that seem appropriate in the study of practice viewed from the particular perspective they adopt. Most faculty approach their research from their particular discipline and the accompanying paradigm (Huber, 1999). That is, different methodological perspectives in research on practice are related to different epistemological perspectives on the nature of practice (Kemmis & McTaggart, 2000) and it becomes possible to categorize different approaches to action research within one of the three knowledge-constitutive interests.

Keller (1998) contended that the research on and in American institutions of higher education has been dominated by "methodological monism" (p. 267) where the scientific method has predominated over other approaches. He argued for more research in higher education using other methods of inquiry, particularly pluralistic approaches. Kezar (2000) added that much of the critique of research in higher education points at the need to bridge the gap between research and practice, and suggests that action research provides one way to achieve this. Kember and McKay (1996) added that action research is "the mode of research associated with critical theory" (p. 531) and that it necessarily is concerned with social practice, is participatory, allows participants to decide topics, aims toward improvement, is cyclical and reflective, and involves systematic inquiry. This research considers the extent to which action research projects in the NOVA program served the critical interest.

AN INVESTIGATION OF FACULTY CONCEPTIONS AND PRACTICES OF ACTION RESEARCH IN A NATIONAL SAMPLE OF COLLEGES AND UNIVERSITIES

NASA Opportunities for Visionary Academics (NOVA) is a NASA-funded project aimed at improving science education at universities across the United States by introducing innovative science and mathematics classes for preservice teachers (Sunal et al., 2003). New courses that are developed utilize the NASA enterprises or themes as a major focus and adopt an inquiry approach to teaching science and mathematics. To become involved in NOVA activities, a team of three faculty members (one from education, one from science or mathematics, and an administrator) attends an initial training workshop. Thereafter they develop a grant proposal, which is submitted to NOVA for funding consideration. Beginning in 1998, as a regular part of the NOVA Phase I workshops, action research formed one component of the professional development process. All proposals to NOVA included an action research component. The education faculty member on the NOVA team usually took on the role of assisting science and mathematics faculty members in the development and use of action research plans in improving teaching, student learning, and course design.

Once funded, the team implemented the new science or math course, gathered action research data, and submitted a final report that included the results of the research. Each year, follow-up leadership development conferences are held for ongoing faculty development, additional project opportunities, and the presentation of action research projects to the NOVA network. Thus, NOVA is an example of a national professional development program for faculty to improve teaching effectiveness in a national network of 92 colleges and universities in 2003.

One of the main purposes of NOVA is to encourage faculty to adopt new approaches to teaching, assessment, and the evaluation of their work. NOVA promotes a cognitive apprenticeship approach, which involves people changing roles from teacher to learner within an action research model, involving cycles of planning, action, observation, and shared reflection (Sunal et al., 2001). This allows the teacher-learner to review the adequacy of traditional approaches and to make changes. Ownership of change is a central concept. Thus, NOVA has attempted to provide an alternative research framework to the dominant paradigm, and to bridge the gap between research and practice by promoting research on classroom action.

These cautions and elements of action research, cited above by Kember and McKay (1996), Kezar (2000), Keller (1998), Kemmis and McTaggart (2000), and Huber (1999), reflect the intentions of the NOVA program in promoting action research amongst teams of interdisciplinary faculty members. Analysis of approaches to action research adopted by faculty involved in NOVA at university campuses across the United States provides an opportunity to examine the extent to which action research has contributed to shifts in teaching and faculty research practice.

STUDY PURPOSE

The purpose of this study was to examine the ways in which faculty teaching content courses in science and mathematics engage in action research in their immediate contexts. Approaches to research typically reflect the disciplinary procedures within which faculty were trained (Huber, 1999), and this may be a factor for consideration in the implementation of action research projects, where the nature and assumptions implicit in action research, which is essentially a qualitative approach, may be foreign to science and mathematics professors. If action research is to become accepted as a valid approach for professional development, geared to understanding and changing practices, then it is important to identify faculty conceptions, how they engage in the process, and the value they derive from the outcome of such research initiatives.

Successful faculty development initiatives about teaching and learning are those that do not just provide educational information and input through workshops, but those that work toward changing the conceptions about the nature of teaching and learning held by faculty (Ho, 2000). The results of this study will enable the creation of more effective faculty development activities, relating to action research, in the future.

Thus, the main purpose of the study was (a) to consider faculty conceptions of action research as displayed through their action research plans and project reports; (b) to analyze how they planned, gathered, and analyzed their research data; (c) to review the approaches to action research used by faculty across diversity institutions; and (d) to examine the research outcomes from successful action research projects.

METHOD

Research Design

The research design is qualitative in nature, employing the analysis of key documents through the use of rubrics and content analysis (Merriam, 2001), namely the action research proposals submitted by institutions and final action research reports and conference presentations. All documents were acquired from the NOVA database housed at the Alabama Science Teaching and Learning Center at the University of Alabama.

Rubrics are commonly used as an alternative form of assessment in the scoring of open response questions (Taggart, Phifer, Nixon, & Wood, 2001). Action research proposals were scored using the Action Research Rubric (ARR) developed specifically for this purpose.

Typically content analysis is concerned with making frequency counts of the occurrence of each coding category within and across documents (Gall, Gall, & Borg, 2003), and this was done in this study. Documents were deductively coded against predetermined criteria and then frequency counts performed. However, in a qualitative design, content analysis differs from conventional content analysis in that there is a concern for generating meaning, for examining underlying assumptions and values implicit in the text, and for exploring theoretical relationships (Merriam, 2001). This study seeks to understand faculty approaches to action research in the context of the background literature already presented. Site visit data, including interviews and classroom observations of faculty, were used as supplementary data to verify findings from documentary analysis.

Subjects

The subjects of this research were science, mathematics, and education faculty from 53 institutions across the United States, representing all Carnegie institutional categories. The products analyzed were 53 action research proposals, 25 action research project presentations at Leadership Development Conferences, and classroom data gathered during site visits to 20 institutions. The 53 action research proposals represented all of those submitted after 1998, when this element became mandatory, until May 2002, when data was analyzed. Site visit data represented all site visits undertaken until May 2002.

Instrumentation and Data Gathered

Rubrics as a form of alternative assessment are common practice in educational settings and have been used to assess a diversity of products and processes including reading, writing, portfolios, research projects, as well as knowledge and skills in content disciplines (Taggart et al., 2001). They are useful tools for assessing restricted as well as extended performance tasks (Burke, 1999). Rubrics can be classified as holistic or as analytic, with holistic rubrics being used to rate a few skills without in-depth analysis (Wilson & Onwuegbuzie, 1999). In contrast, analytic rubrics are typically used with extended performance tasks because there are several criteria for assessing performance. Thus, when constructing an analytical rubric it is necessary to break the task down into smaller components and to rate each item on a scale.

Wilson and Onwuegbuzie (1999) created analytical rubrics for assessing graduate student performance in developing and presenting research plans. Their rubrics were used as the basis for developing a four- point rubric for scoring action research plans, where 1 = *novice,* 2 = *apprentice,* 3 = *proficient,* and 4 = *distinguished.* This is called the Action Research Rubric (ARR). The ARR was used to rate the action research plans of 53 institutions in the NOVA program. The ARR is an analytical rubric containing 14 items, with a total possible score of 56, which would assume a score of 4 on all 14 items. This allowed quantification of approaches across action research plans. Each institution received a total score and was categorized as (a) *novice-apprentice* (score of 14–27), (b) *apprentice-proficient* (score of 28–41), or *proficient-distinguished* (score of 42–56) (see Appendix).

Content analysis (Merriam, 2001), in which documents are coded against certain themes, was conducted on project plans and project reports. For action research plans this involved the coding of all proposals on several criteria, listed as (a) through (g) below. From these codes, a matrix was constructed for all 53 institutions that noted (a) the conception of action research articulated, (b) research questions or research focus, (c) research design, (d) proposed data collection methods, (e) proposed data sources, (f) proposed instrumentation, and (g) proposed data analysis.

Content analysis was also undertaken on the abstracts of papers presented at the 2001, 2002, and 2003 NOVA Leadership Development Conferences, in which papers were coded for research purpose and research method used. Descriptions of some exemplary action research projects are also provided.

Site visits were made to 20 of the participating institutions and included interviews with all NOVA team members, interviews with students, as well as an observation of a lesson. Classroom field notes were written and sessions rated using the ESTEEM instrument (Burry-Stock & Oxford, 1994).

Data obtained from site visits were used to triangulate the findings from the analysis of action research projects.

FINDINGS

A total of 53 action research plans were analyzed using the 4-point ARR. The mean score for the ARR was 36.07 (SD = 8.21). Eight institutions (15%) were categorized in the novice-apprenticeship range, 32 (60%) in the apprentice-proficient range and 13 (25%) in the proficient-distinguished range. So, at the proposal stage of action research plans, most institutions were still in the initial stages of developing action research plans because the majority of action research plans were categorized within the apprentice-proficient range. A quarter of the institutions had clear and well-developed action research plans that provided a comprehensive framework for meaningful data collection and analysis.

All 53 action research plans reviewed reflected the desire to move away from traditional lecture approaches to more interactive methods to enhance student learning. Action research conceptions were well articulated in 31 of these action research plans and the cyclical nature of change was clearly reflected. Some explanations given included "developing a best practice model," "analysis and improvement on an ongoing basis," "developing a plan for severe problems," and "authentic feedback to determine student strengths and weaknesses, for future improvements for improved student learning." This is consistent with a nontraditional inquiry approach to teaching where faculty are concerned to find better ways to enhance student learning. The purposes elaborated for these plans for action research were to enhance (a) student content knowledge, (b) student process skills, (c) student attitudes, and (d) student teaching outcomes. These were specifically stated in 50 action research plans, showing explicit concern for enhancing student learning.

The content analysis of the 53 action research plans revealed that most were weak on the data analysis component, with 39 action research plans (74%) making no mention about data analysis or simply stating that data would "be analyzed" or "compared" to other groups. Forty-one percent of institutions did not devise actual research questions or hypotheses, although in almost all cases these could be inferred. Many institutions (57%) did not specify the types of instruments that would be used to collect data, making broad statements about methods of data collection, such as by survey or questionnaire.

Quantitative approaches were represented in most of the action research plans, with the pretest-posttest design being the most common. Twenty-eight institutions (53%) suggested this approach. Another quantitative approach commonly advocated was that of experimental and control groups (30% of institutions), in some cases in a true quasi-experimental design or combined with a pretest-posttest design. In these instances the experimental NOVA course was to be compared to a matching or control class that was not being taught in the new way. Fourteen institutions (26%) made no mention of their design, while a few planned qualitative designs, such as case study, ethnography, or ongoing processes of data collection. Examples of better action research plans were made by two teams at large Midwest state universities, one with a strong emphasis on qualitative approaches, while a second had a strong quantitative design, indicating that strong action research plans could be either qualitative or quantitative in nature.

Typically quantitative approaches were usually to be combined with additional qualitative techniques that were designed to elicit more in-depth data, to provide a deeper understanding about undergraduate science teaching and learning. Sixty-two percent of institutions mentioned methods other than surveys to assess student outcomes. The range of techniques suggested for student assessment were wide and included journals, online tasks, performance tasks, rubrics, observation, interview, concept maps, quick writes, journals, portfolios, discussion groups, self-evaluation, peer review, and logs of class activities. This is an impressive array of qualitative techniques, which professors in science disciplines have embraced. These alternative approaches provide valuable ways to assess students' content and conceptual knowledge, skills, and attitudes, but as already noted there was insufficient discussion on how these would be used to gather data, and how that data would be analyzed.

An interesting, simple, and yet effective approach was suggested by a team of faculty at a medium-sized New York institution, where they planned to have students write anonymous "quick writes" (Cross & Steadman, 1996). In quick writes students would address key questions like "What have you learned this week?" "Are you having difficulty with a particular concept?" and "How can your instructor enhance your learning?" These were then collated each time by a different group of students who would collect, review, and summarize the answers. These results formed the basis of a discussion on the progress of the course.

Other innovative examples include the use of concept maps at a large California university. The concept map assessment comprised (a) a set of five key prompting questions about climate, (b) a reminder about the nature of hierarchical concept maps, and (c) a box with the word "climate" printed in it, with lots of white space below. Prompted by the key questions, students were asked to organize their preexisting knowledge about climate in a hierarchical concept map, starting from the labeled box. This was repeated after the students had completed the course, and there was a significant increase in content knowledge. Two people scored the concept maps for validity.

A team at a small Midwest state university used multiple assessment strategies for each unit, where each unit was assessed in a different way, giving students exposure to different methods, and instructors, multiple ways of assessing students. An interesting example was used by another small private Midwest university where students wrote weekly reflective journals using predetermined writing "stems" that included, "This week I learned…," "I learned this by…," "One thing I do not understand or have a problem with is…," and "I feel that…" These writing stems moved students away from the tendency to write purely descriptively and facilitated more reflective writing on areas in which the faculty were interested.

It is important to note that most institution faculty saw the students as the primary sources of data. There were only eight institution teams that mentioned faculty as a source of data, indicating that there is a strong focus on the students as the unit of analysis, rather than on the faculty member, teaching strategies, student-teacher interactions, or assessment methods. To this end, the bulk of the research questions focused on (a) student content knowledge, (b) student science and mathematic skills, (c) student attitudes toward science and mathematics, (d) student ability to integrate knowledge, (e) problem solving, and (f) teaching efficacy. Only 10 of the 54 institution faculty teams had an instructor-focused research question or a suggested focus that pertained to faculty and their teaching. A majority of these questions were not actualized in the methodology for data collection or analysis. In three cases action research was conceptualized as something for "future teachers" (not college professors) and was suggested as a component for students during their practicum or student teaching experiences following the science course.

The action research project being conducted at a medium-sized Midwest university is one example of an action research approach in which the instructors are focusing on their own teaching. The two instructors have co-constructed a "teaching portfolio" that includes personal reflections, teaching philosophies, methods and activities, student surveys, and suggested improvements for teaching. This is a potentially powerful tool that can lead to constructive reflection, course redesign, and changes in teaching.

A faculty team at a large California university focused on the role of scaffolding, the process whereby the instructor provides structured support to achieve student learning outcomes, as a vehicle for dealing with student misconceptions (see Chapters 6 and 7 for additional information on these types of student conceptions). The research was to include video and audiotapes of classrooms, transcriptions of the lessons, and discourse analysis for examining patterns of interaction in the classroom. Here the research focus was clearly on faculty, their classroom dynamics, and selected teaching strategies.

OUTCOME OF ACTION RESEARCH PROJECTS

Faculty who have engaged in action research projects have made presentations at recent national NOVA Leadership Development Conferences (LDCs). Nine papers, by different universities, were presented at the 2003 LDC, hosted at the Johnson Space Center. Five of these addressed students' attitudes and conceptual understanding in science or mathematics as a result of the new course, and all showed an increase in student performance in the NOVA courses. Of these, one reported an increase in faculty enthusiasm for teaching the course as a result of improved student interest and enhanced relationships between students and faculty. Using observation and assessment of student journals, another paper showed an increase in student perceptions of their teaching efficacy as a result of the NOVA course, with early hands-on laboratory teaching experiences being particularly important in increasing the confidence of preservice candidates. An additional case described the use of data from student assessments, in which it was found that students lacked the prerequisite skills to be successful in the course, particularly dimensional analysis (unit conversions). This resulted in the faculty modifying the course to the population of students for whom it was intended.

Only one institution specifically set out to examine the two professors' instruction, to find out if they were modeling instruction considered to be consistent with teaching science as inquiry. They utilized the Instructional Practices Rubric (Council of State Science Supervisors, 2001) to assess the extent to which students considered whether they were experiencing inquiry-based teaching. The results showed areas in which they were successful, particularly in how scientists study the natural world, and identified aspects for improvement, such as how to engage all learners in using analytical skills, mathematics, and technology.

There was one unusual paper presented, in which NASA resource materials had been utilized by an English professor in her Composition I course, for students to explore NASA personalities and career opportunities as objects for writing. This stimulated non-science majors' interest in NASA as a career possibility and went some way to demystify some science concepts. This was an example of how NOVA has "spin-off" effects on other disciplines and faculty.

At the 2002 LDC, held at the Goddard Space Center, 10 action research papers were presented. One presentation by a faculty team from a medium-sized Midwest university focused on investigating effective ways to teach science for improved student learning, highlighting the way in which faculty from different disciplines can work together productively. Another paper by a Midwestern state university highlighted the need for instructors to create a "friendly learning environment" for students with poor attitudes because attitudes to science are related to the desire to learn conceptually. All of the other eight papers reported positive quantitative and qualitative gains in student conceptual knowledge, attitudes to science, or science teaching efficacy as a result of the new teaching methodologies.

At the 2001 LDC six action research papers were presented, including papers describing student outcomes as well as those discussing new approaches to teaching and processes for faculty change. Four of these highlighted improvements were in student attitudes, content knowledge, process skills, and teaching efficacy. One institution discussed approaches to problem solving. A small west coast university team presented a unique project called Science Outreach, where education and science majors work together for eight weeks during the summer with students who have been homeschooled. Students are responsible for planning and implementing instruction. They reported that many of the science majors subsequently shifted their career goals to education. A large Plains state university described an action research model for education and science faculty to work together. This model, called "peer consultation," involves cycles of pre-observation discussions, written critique of observations (by the education faculty member), and written interactive reflections and follow-up between the two professors. This model provided valuable data for later analysis of action for improved teaching.

Of the 25 action research papers presented at the LDCs, 19 described improvements in student science or math content knowledge, attitudes about the subject, process skills, and teaching efficacy. While all of the new courses employed teaching approaches that were new and innovative for the science and mathematics professors, only four of the action research papers dealt directly with instructor-related issues, such as the classroom environment, modeling of good teaching, and use of new teaching strategies. Faculty did not easily perceive themselves as research subjects.

The studies of student outcomes mostly used a pretest-posttest design, and some made comparisons to traditional classes. These quantitative and qualitative projects show improved student outcomes as a result of the new courses. Clearly the action research projects have produced a valuable set of information about successful courses and improved student learning and show significant gains in student learning, as well as in faculty professional development.

EXEMPLARS

Four of the exemplary action research projects are those developed by faculty teams by two Far West universities and two Midwest universities. These are models of how action research can be used to improve instruction and to assess student outcomes.

A team from a large Far West university examined student attitudes to the revised NOVA geology course, their scientific reasoning, and their concepts about climate and compared data to four classes of traditional geology courses. To assess attitudes the faculty used a modified version of TOSRA (Test of Science-Related Attitudes; Fraser, 1981) which examines four different scales: (a) societal aspects of science, (b) science as inquiry, (c) nature and history of science, and (d) nature and study of climate. Data were analyzed using analysis of variance. Students in the innovative course had significantly higher scores on categories B (science as inquiry) and D (nature and study of climate) than did students in the regular, comparison classes. Thus, the innovative course-reform changes significantly impacted student attitudes to science as inquiry and to the nature and study of key climate concepts. Concept maps were used and scored quantitatively to assess conceptual knowledge and connections. To establish scoring validity, two researchers independently scored the concept maps. Analysis of student concept maps showed that students in the innovative course had significantly better connections between concepts than did their peers in the four other traditional classes (Dempsey, O'Sullivan, & White, 2001).

Faculty at a large Midwest university examined student attitudes using the Earth Science Values and Attitudes Instrument (Kern & Carpenter, 1984) to evaluate earth science attitudes, and STEBI B (Enoch & Riggs, 1990) to assess student science teaching efficacy. The course investigated was one section of a large introductory geology course. This section grouped education majors together. Data analysis showed that students taking the reformed or innovative course had significantly higher teaching efficacy than education majors in any of the other traditionally taught geology course sections. Students in the innovative course section were more confident to teach science in a school setting.

A positive correlation was also found between instructors' and students' perceptions toward geology and teaching geology in courses incorporating "best teaching practices." Students in the innovative course section displayed better attitudes to geology than did their counterparts in other sections, suggesting an increase in their perceived value and importance of geological events. Their positive attitude had significantly increased from the beginning to the end of the course, indicating that the shift was related to their achievement in the course (Goldston & Clement, 2001).

Faculty from a small Far West university examined changes made to the pedagogy and content in a freshmen integrated science course covering concepts in physics, chemistry, biology, astronomy, and geology. Three professors, one from science, one from education, and the faculty dean collaborated to redesign the course to make it more hands-on and participatory. Their objective was to "create science literate students not ones who could restate scientific material on exams" (Smelcer, 2002, p. 4). The purpose of the study was to consider whether these changes were effective in enhancing student understanding of science concepts and processes. The study sample size was 53 students. A pretest-posttest design was used in which the same test was completed by all students at the beginning and at the end of the semester. The test contained 75 true-false questions selected from the Project 2061 benchmarks for scientific literacy (Smelcer, 2002).

A paired-sample t test showed that there was a significant gain in content knowledge at the end of the semester, but that this did not depend on the student's gender, class standing, or major. Further analysis was conducted using a linear regression analysis to evaluate the prediction that the smaller the change in test scores, the higher the initial test score. A scatter plot of the two variables indicated that the two variables were related negatively, showing that students who start out with a higher initial test score have less of a change in the posttest score than students who start out with a low pretest score. These findings have implications for the level at which courses are introduced to meet all students' learning needs. In open-response questions several of the students who scored high in the pretest complained that the course was too easy and reviewed too much basic material. The challenge was to offer these students additional challenging tasks when they had completed assigned work, while not frustrating the lower scoring students who needed the basic review (Smelcer, 2002).

Content knowledge was assessed through pre-post tests designed by faculty at a small Midwest university. Using paired-sample t tests, data from 2000 and 2001 showed significant improvements in students' content knowledge on all topics covered, including physical science, earth science, chemistry, and space science concepts. To assess students' views on the nature of science, the VNOS instrument (Bell & Lederman, 2000) was used to compare students in the innovative reformed course to students in traditional science courses (level 2 physics and level 3 chemistry). Results from statistical analysis using an ANOVA showed no difference between the groups ($F = 0.616$, $p = 0.819$). In fact the students in the innovative course scored slightly higher than physics and chemistry majors, although the difference was not significant (Raubenheimer, 2002).

Students were also surveyed for perceptions about the NOVA course using a 4-point Likert-type questionnaire developed by the course presenters. Questions on the questionnaire focused on student course objectives and instructional strategies involving use of the learning cycle. Data were analyzed using one-sample t tests and comparing the student ratings to a test value of 2.5, which would indicate a neutral position. On the 14-item questionnaire 10 items were significantly above the test value, indicating a highly rated innovative course. Students particularly valued the hands-on, collaborative approach to learning science and the way in which science was related to every day experiences. They also felt better able to design and implement plans to investigate problems.

Based on some lower student perceptions in year 1, changes in methodology were made in year 2 of the course. In year 2, diagnostic tests were used as a tool for students to evaluate their own learning, whereas in year 1 the tests had simply been implemented as a research tool. More technology was integrated into the class, the textbook was changed, and more time was spent on explaining and elaborating the learning cycle. These were all aspects that had not been significantly different from the test value of 2.5 in year 1. The evaluation of perceptions, using the same questionnaire, indicated that all ratings by students were significantly higher than the test value in year 2. Clearly the revised methods had improved the course offering (Raubenheimer, 2002).

CLASSROOM OBSERVATIONS FROM THE NATIONAL SAMPLE OF COLLEGES AND UNIVERSITIES

Observational data, including student and faculty interviews and classroom observations, undertaken during site visits to 20 institutions, confirm that the national sample of faculty involved in action research has shifted its teaching practices. Only one lesson was observed (5%) where the faculty member was teaching a completely traditional didactic lesson, and this was a person who had not followed through on the action research plans. This instructor had an excellent understanding of the many ways in which children solve mathematical problems and wanted students (future teachers) to be aware of these and insisted they learn how to solve problems in these alternative, but very rigid, ways. Thus, the essence of constructivism, in which people solve problems in personal ways, was reduced to new forms of algorithms. During the interview this instructor stated she preferred student-centered instruction and this contradiction between espoused philosophy and actual practice needs further exploration.

Site-visit classroom observations found that 69% were primarily using inquiry-based lessons in which students were actively engaged in learning. These lessons were described by the observers as "an excellent learning cycle class involving an engagement activity, student discussion, mini-lecture, and a follow-up activity for application," or "the NOVA model was fully implemented," or "a fully developed lesson using innovative strategies and well integrated with the NASA enterprises." In all of these cases the interview data reported active faculty teams involved in design, planning, and implementation of science content courses. Faculty, in interview data, reported that the initial NOVA workshops had consolidated the teams and that NOVA had brought together faculty from different disciplines. Good collaboration has helped shift faculty from teacher-centered to student-centered approaches in teaching science content courses.

During faculty interviews many claimed that the new approaches are also being integrated into the other classes they teach, and into the classes of other faculty members who have observed their new strategies. Faculty also repeatedly stated that interaction between departments or disciplines, faculty collaboration, was a valuable form of professional development. Faculty from education were important team members, tending to be used as "consultants" or "sounding boards" providing "new" ideas on how to teach and assess student learning. For people who had only previously taught in a traditional format these were exciting opportunities for them to review their own teaching processes and to see the results of enhanced student outcomes using different teaching methods.

Within the remaining 32% of classes, one course was an online course and so the lesson was Internet based, with little opportunity for student interaction, although the instructor did answer questions from students. In 19% of the classes there were elements of student participation through the presentation of class projects, the construction of models, or the use of graphing calculators. A large portion of these sessions was dedicated to direct instruction without any student participation or interaction, indicating that some elements of student-centered approaches had been adopted, but that faculty were still experimenting with new approaches and shifting their views.

At one of these institutions two of the team members were new and had never attended a NOVA professional development session. They had only had tentative discussions with the education faculty member who was the only original NOVA team member. At another school, only one member had attended the NOVA workshop and had returned to promote this to other faculty and felt he was making progress. In both cases, the faculty had underdeveloped conceptions of action research. So, the absence of faculty teams experiencing a professional development program in the components of current science reforms in these cases has hindered progress in using new pedagogical approaches and in implementing action research projects.

Overall this data shows that the majority of faculty in the national sample that have undergone professional development through the NOVA program have shifted their practice to more student-centered approaches and that others are in transition. The data from classroom observations supports the notion that action research projects, together with other aspects of professional development, is a requirement for effective implementation of new reform models of teaching and learning in higher education.

SUMMARY AND IMPLICATIONS FOR FURTHER RESEARCH: PROFESSIONAL DEVELOPMENT OF UNIVERSITY SCIENCE FACULTY THROUGH ACTION RESEARCH

Approximately two thirds of the action research plans presented articulate statements about the role and purpose of action research. But, given that only a quarter of the action research plans were scored in the distinguished-to-proficient range, it is clear that additional work is needed to support faculty in developing (and implementing) comprehensive action research plans. There is a need to provide more input on instrumentation for data gathering and on data analysis, particularly with respect to qualitative techniques. In general faculty are comfortable with quantitative approaches, but less familiar with techniques for gathering and analyzing qualitative data. For instance, Kember and McKay (1996) noted that the use of control groups is unusual in action research groups, but this was the dominant conception of research design in this sample of action research projects. That faculty are more prone to using quantitative techniques suggests that it may be difficult to move faculty in science disciplines out of the research paradigm with which they are most familiar and comfortable.

The use of a broader range of methodological approaches, as promoted by Keller (1998) is likely to provide a richer and more in-depth platform for examining teaching and learning. Evidence in the action research plans and exemplary cases described shows that there is a move toward more diversity in methods in more than half of the cases, where a range of strategies have been used by science professors to assess student attitudes, perceptions, and conceptual knowledge. Evidence presented also shows that successful implementation of an inquiry approach to teaching was dependent on the successful functioning of interdisciplinary faculty teams. The inclusion of education faculty, who are likely to be more sympathetic to qualitative techniques, has aided in this emergence of alternative assessment strategies in science classes. Despite this positive move toward innovative ways of assessing students, there is less evidence on the extent to which alternative methods have been used to examine the effectiveness of particular approaches to teaching, to reflect on classroom relationships, to consider issues relating to classroom climate, to examine personal mental constructs and their impact on teaching, or even the impact of different assessment techniques. Research is primarily reserved for examining the impact on student learning and it might be argued that this has been done to justify the introduction of a new course and to counter any negative perceptions about innovation from other faculty or administrators.

There is a substantial body of research on the reciprocal relationship between students' prior experiences, their approaches to learning, their perceptions of the learning situation, and their learning outcomes (e.g., Prosser & Trigwell, 1999). For instance, students tend to adopt a "surface approach" to learning by memorizing facts if they perceive that the environment requires it. In contrast they will adopt a "deep approach" to learning if the teaching context demands that they make conceptual connections and that they strive for understanding. Prosser and Trigwell added that there is far less research into university teachers' conceptions of teaching, and "even less into their perceptions of the teaching context, their approaches to teaching, outcomes of teaching and relations between these aspects of the experience of teaching" (p. 21). However, it may be difficult to persuade faculty to engage in this kind of research because it is outside of their realm of familiarity. Schön (1995) described the paradoxical case of a professor at Massachusetts Institute of Technology who had the opportunity to research the impact of computer-aided instruction on student learning, but who was not able to develop a suitable research question. Schön concluded that introducing and legitimizing action research might prove equally difficult with other scholars in the disciplines who would be undertaking such research. More work is needed in developing the type of professional development that encourages faculty to move toward examining their own teaching practices and underlying views (Ho, 2000) within an action research mode that utilizes the pluralistic research methods suggested by Keller (1998).

Action research involves cycles of planning, acting, observing, and reflecting (Hopkins, 1993). The exemplary projects illustrate how faculty have been through these stages once, but there is little evidence of the continuation into additional cycles. Reflection necessitates the origination of new hypotheses and research questions for ongoing planning, action, observation, and reflection. Action research is an iterative process. Two cases out of 25 reports show how instructors modified instruction in the next course offering to account for any negative student perceptions and poor conceptual understanding in certain areas of the content, but other than this there is little evidence of the cyclical change implied in action research. Further data needs to be gathered to establish the extent to which ongoing cycles are being implemented. NOVA faculty should be encouraged to see change as an ongoing process of seeking improvement in teaching and learning, rather than as a task to complete as a funding/donor requirement.

That professors appear to be less able or willing to use alternative methods for personal evaluation and self-reflection, supports the analysis that there is a gap in higher education between research and practice (Kezar, 2000). It is possible that faculty go through different stages as they grapple with how to conduct action research. Initially they may find it easier to focus outside of themselves and on students. They may need far more support and input, such as through the involvement of a "critical friend" (Kember, 1998) or participant observer, to assist them in moving toward research, reflection, and analysis on their own practice. Kember (1998) described the role of the critical friend in the Action Learning Project, an interinstitutional project involving eight institutions in Hong Kong. Within the project there were a number of coordinators who worked with 10 to 12 teams (and there were more than 50 teams) in the role of "critical friend." Their role included that of (a) "rapport builder," (b) "coffee maker" (ongoing facilitator), (c) "mirror," (d) teaching consultant, (e) evaluation and research advisor, (f) writing consultant, and (g) matchmaker (putting people in touch with others). Within NOVA there are no "critical friends," other than the education team member, and she or he has other jobs and commitments to attend to and cannot dedicate the time needed to facilitate this in-depth process.

Professional development is not a one-time event and involves (a) a clear focus on learning and learners, (b) an emphasis on the individual and organizational change, (c) small changes guided by a grand vision, and (d) ongoing professional development that is procedurally embedded (Guskey, 2000). Action research is one way to achieve this and this research shows that most faculty in the national sample investigated here have a clear focus on learners and learning and have embarked upon change processes at the personal and institutional level by implementing these new courses and action research projects. The action research has enabled faculty to engage in thinking about student learning and to reflect upon their own teaching styles. All action research plans espouse the desire to move toward inquiry approaches in science teaching in higher education and this was observed in more than two thirds of the classroom observations. Thus, the action research projects have contributed to faculty development according to Guskey's definition.

The Carnegie Foundation for the Advancement of Teaching suggested three criteria for defining the scholarship of teaching and learning: (a) that teaching is deeply embedded in the discipline, (b) that it is an aspect of practice, and (c) that it is characterized by a transformational agenda (Hutchins, 2000). The majority of the action research plans, cases, and exemplars described in this article meet the first two criteria, although different action research projects are clearly in different stages of development and implementation. First, the action research designs proposed and used for inquiry are rooted in the scientific method as evidenced by the large number of action research plans that adopted a pretest-posttest design, together with control and experimental groups in some instances. Therefore, their inquiry is rooted in the particular tools of their discipline. Second, all of the cases involve faculty using new approaches to teaching and grappling to understand the impact on student learning, that is, the faculty are considering aspects of their teaching practice.

In considering the third criterion, it is important to consider the purpose of an action research plan. Within emancipatory action research, practitioners work together as a group and collectively identify problems and possible solutions (Carr & Kemmis, 1986). There is a concern for political change, consciousness raising, and the generation of new theories, as much as for practical improvements. Most of the action research plans in the national sample investigated are concerned with "practical" issues, like how to improve student attitudes and concepts, and can be located within the interpretive paradigm (Kember, 2000). These are not transformational agendas and so more research is needed, particularly on the exemplary action research projects that have produced results, to establish the extent to which action research projects have transformed personal and institutional practices. There is evidence from interviews with administrators in universities with innovative reform-based courses that this has happened, but these have not been the focus of action research agendas. So, action research projects may still be in transition toward transformational agendas. Despite this, the faculty from the national sample of universities and colleges involved in developing reform-based courses and conducting action research in the process can be considered to be making a substantive contribution to the scholarship of teaching and learning.

AUTHOR NOTE

This work was in part supported by NASA Opportunities for Visionary Academics (NOVA), a program funded by the National Aeronautics and Space Administration, although the views expressed here are the authors only.

APPENDIX

Action Research Rubric

Score	Explanation
1—novice	The item is not included.
2—apprentice	The item can be inferred from the text or there is evidence of intent that is not elaborated.
3—proficient	The item is stated in the text but is not elaborated or is explained in part.
4—distinguished	The item is fully elaborated in the text.

Scoring Dimensions

Action Research Plan	1	2	3	4
1. The rationale of the study is clearly presented.				
2. The purpose of the study is provided adequately.				
3. The action research cycle is evident (PAOR = Plan, act, observe, and reflect).				
4. There are clear research questions or hypotheses.				
5. The research methods are clearly stated.				
6. Research methods are appropriate to the questions.				
7. Instrumentation is described.				
8. States who the program participants are.				
9. Data sources are identified.				
10. Potential sample sizes are noted.				
11. Triangulation of data is proposed.				
12. States who will collect the data.				
13. Methods of data analysis are proposed.				
14. Methods of data analysis are appropriate.				

REFERENCES

Arhar, J.M., Holly, M.L., & Kasten, W.C. (2001). *Action research for teachers: Traveling the Yellow Brick Road.* Upper Saddle River, NJ: Merrill Prentice Hall.

Bell, R.L., & Lederman, N.G. (2000, April). *Decision making on science and technology issues: Do views of the nature of science really matter?* Paper presented at the annual meeting of the National Association for Research in Science Teaching, New Orleans, LA.

Bondy, E., & Ross, D. (1998). Teaching teams: Creating the context for faculty action research. *Innovative Higher Education, 22,* 231–249.

Boyer, E.L. (1990). *Scholarship reconsidered: Priorities of the professoriate.* Princeton, NJ: Carnegie Foundation for the Advancement of Teaching.

Burchell, H. (2000). Facilitating action research for curriculum development in higher education. *Innovations in Education and Training International, 37,* 263–269.

Burke, K. (1999). *How to assess authentic learning* (3rd ed.). Arlington Heights, IL: SkyLight Professional Development.

Burry-Stock, J.A., & Oxford, R.L. (1994). Expert Science Teaching Educational Evaluation Model (ESTEEM): Measuring excellence in science teaching for professional development. *Journal of Personnel Evaluation in Education, 8,* 267.

Carr, W., & Kemmis, S. (1986). *Becoming critical: Education, knowledge and action research.* London: Falmer Press.

Cochran-Smith, M., & Lytle, S.L. (1992). Communities for teacher research: Fringe or forefront. *American Journal of Education, 100,* 298–323.

Cochran-Smith, M., & Lytle, S.L. (1999). Relationship of knowledge and practice: Teacher learning in communities. *Review of Research in Education, 24,* 249–305.

Conway, R., Kember, D., Sivan, A., & Wu, M. (1994). Making departmental changes through action research, based on adult learning principles. *Higher Education, 28,* 265–282.

Council of State Science Supervisors. (2001). *Instructional Practices Rubric.* Hartford, CT: Author.

Cross, K.P. (1998). Classroom research: Implementing the scholarship of teaching. *New Directions for Teaching and Learning, 75,* 5–12.

Cross, K.P., & Angelo, T.A. (1989). Faculty members as classroom researchers. *Community Technical and Junior College Journal, 59*(5), 23–25.

Cross, K.P., & Steadman, M.H. (1996). *Classroom research: Implementing the scholarship of teaching.* San Francisco: Jossey-Bass.

Dempsey, D., O'Sullivan, K., & White, L. (2001, January). *Action research for SFSU's NASA-NOVA course: Planetary climate change.* Paper presented at the 4th annual NOVA Leadership Development Conference, Washington, DC.

Elliot, J. (1991). *Action research for educational change.* Buckingham, England: Open University Press.

Enochs, L.G., & Riggs, I.M. (1990). *Further development of an elementary Science Teaching Efficacy Belief Instrument: A preservice elementary scale.* East Lansing, MI: National Center for Research on Teacher Learning. (ERIC Document Reproduction Service No. ED319601).

Fraser, B.J. (1981). *TOSRA: Test of Science-Related Attitudes.* Victoria, Australia; Australian Council for Educational Research.

Gall, M.D., Gall, J.P., & Borg, W.R. (2003). *Educational research: An introduction.* Boston: Allyn and Bacon.

Goldston, D., & Clement, M. (2001, January). *Strengthening geology content courses for prospective elementary teachers.* Paper presented at the 4th annual NOVA Leadership Development Conference, Washington, DC.

222   C.D. RAUBENHEIMER

Guskey, T.R. (2000). *Evaluating professional development.* Thousand Oaks, CA: Corwin Press.

Hadfield, M., & Bennett, S. (1995). The action researcher as chameleon. *Educational Action Research, 3,* 323–335.

Ho, A.S.P. (2000). A conceptual change approach to staff development: A model for programme design. *International Journal for Academic Development, 5,* 30–41.

Hopkins, D. (1993). *A teacher's guide to classroom research* (2nd ed.). Buckingham, England: Open University Press.

Huber, M.T. (1999, September). *Disciplinary styles in the scholarship of teaching: Reflections on the Carnegie Academy for the Scholarship of Teaching and Leaning.* Paper presented at the 7th International Improving Student Learning Symposium, University of York, England.

Hutchins, P. (Ed.). (2000). *Opening lines: Approaches to the scholarship of teaching and learning.* Menlo Park: CA: Carnegie Foundation for the Advancement of Teaching.

Keller, G. (1998). Does higher education research need revisions? *Review of Higher Education, 21,* 267–278.

Kember, D. (1998). Action research: Towards an alternative framework for educational development. *Distance Education, 19,* 43–64.

Kember, D. (2000). *Action learning and action research.* London: Kogan Page.

Kember, D., & McKay, J. (1996). Action research into the quality of student learning: A paradigm for faculty development. *Journal of Higher Education, 67,* 528–54.

Kember, D., Lam, B., Yan, L., Yum, J.C.K., & Liu, S.B. (Eds.). (1997). *Case studies of improving teaching and learning from the Action Learning Project.* Hong Kong: Action Learning Project. Retrieved July 26, 2003, from http://celt.ust.hk/ideas/ar/pdf_files/alp1999/index.htm

Kemmis, S., & McTaggart, R. (2000). Participatory action research. In N.K. Denzin & Y. Lincoln (Eds.), *A handbook of qualitative research* (pp. 567–605). Thousand Oaks, CA: Sage.

Kern, E.L., & Carpenter, J.R. (1984). Enhancement of student values, interests and attitudes in earth science through a field-oriented approach. *Journal of Geological Education, 32,* 299–305.

Kezar, A. (2000). Higher education research at the millennium: Still trees without fruit? *Review of Higher Education, 23,* 443–468.

Kreber, C. (2000). Exploring the scholarship of teaching. *Journal of Higher Education, 71,* 476–495.

Krockover, G., Adams, P., Eichinger, D., Nakhleh, M., & Shepardson, D. (2001). Action-based research teams: Collaborating to improve science education. *Journal of College Science Teaching, 30,* 313–17.

Merriam, S.B. (2001). *Qualitative research and case study applications in education.* New York: Jossey-Bass.

Paulsen, M.P. (1999). How college students learn: Linking traditional educational research and contextual classroom research. *Journal of Staff, Program & Organizational Development, 16,* 63–71.

Paulsen, M.P. (2001). The relation between research and the scholarship of teaching. *New Directions for Teaching and Learning, 86,* 19–29.

Prosser, M., & Trigwell, K. (1999). *Understanding and learning in teaching: The experience in higher education.* Philadelphia: Society for Research into Higher Education.

Raubenheimer, C.D. (2002, March). *Student performance, perceptions of and attitudes to an integrated physical science course.* Paper presented at the 5th annual NOVA Leadership Development Conference, Baltimore.

Ross, D., & Bondy, E. (1996). The evolution of a college course through teacher educator action research. *Action in Teacher Education, 18,* 44–55.

Schön, D.A. (1995). The new scholarship requires a new epistemology. *Change, 27*(6), 26–34.

Smelcer, P. (2002, February). *Effectiveness of content and pedagogy change in SC 100: Integrated Science.* Paper presented at the 1st Northwest NOVA Cyber-Conference. Retrieved July 26, 2003, from http://nova.georgefox.edu/nwcc/arpapers/alaska-pacific.pdf

Sunal, D.W., Hodges, J., Sunal, C.S., Whitaker, K.W., Freeman, L.M., & Edwards, L. (2001). Teaching science in higher education: Faculty professional development and barriers to change. *School Science and Mathematics, 101,* 246–257.

Sunal, D., Staples, K., Odell, M., Freeman, M., Whitaker, K., Edwards, L., et al. (2003). *NOVA 1996–2003 Evaluation Report.* Tuscaloosa: University of Alabama, NOVA.

Taggart, G.L., Phifer, S.J., Nixon, J.A., & Wood, M. (Eds.). (2001). *Rubrics: A handbook for construction and use.* Lanham, MD: Scarecrow Press.

Wilson, V.A., & Onwuegbuzie, A.J. (1999, November). *Improving achievement and student satisfaction through criteria-based evaluation: Checklists and rubrics in educational research courses.* Paper presented at the annual meeting of the Mid-South Educational Research Association, Point Clear, AL.

Zoller, U. (1999). Scaling-up of higher-order cognitive skills-oriented college chemistry teaching: An action-oriented research. *Journal of Research in Science Teaching, 36,* 583–596.

Zuber-Skerritt, O. (1992a). *Professional development in higher education: A theoretical framework for action research.* London: Kogan Page.

Zuber-Skerritt, O. (1992b). *Action research in higher education: Examples and reflections.* London: Kogan Page.

CHAPTER 13

A CASE STUDY OF A NATIONAL UNDERGRADUATE SCIENCE REFORM EFFORT

Dennis W. Sunal, Christy MacKinnon, Carol Dianne Raubenheimer, Deborah A. McAllister, and Francis Gardner

ABSTRACT

Reforms in undergraduate science teaching have been successful. One reform program, sponsored by the National Aeronautics and Space Administration, has changed undergraduate science teaching on a national level, impacting hundreds of courses and thousands of students each semester. The reform program used the best of what had been learned through research on faculty professional development and in teaching and learning to enhance science literacy of undergraduate students in entry-level undergraduate courses. Collaboration between discipline and education faculties using a team approach was identified as an important factor in development and implementation of a course reform and resulted in a significant increase in type and expertise of research-supported teaching practices. The major student impact was an improvement in achievement in science understanding in areas related to the national standards for science. The results of reform efforts in higher education are important to professionals and an educated citizenry in a democratic society.

Reform in Undergraduate Science Teaching for the 21st Century, pages 225–239
Copyright © 2004 by Information Age Publishing

INTRODUCTION

Education in the United States faces significant challenges. Since the early 1980s a steady stream of reports has documented deficiencies in K–16 science, technology, engineering, and mathematics (STEM) education. Guidelines have been formulated to address the deficiencies, such as the *Nation at Risk* (National Commission on Excellence in Education, 1983), the *National Science Education Standards* (NSES) (National Research Council [NRC], 1996), *Benchmarks for Science Literacy, Project 2061* (American Association for the Advancement of Science, 1993), *Shaping the Future* (National Science Foundation, 1996), *Science Teacher Preparation in an Era of Standards-based Reform* (NRC, 1997), the *International Society for Technology in Education (ISTE) Standards* (Knezek, 2002), *Before It's Too Late* (U.S. Department of Education, 2001a), *Educating Teachers of Science, Mathematics and Technology* (NRC, 2001), *NASA Strategic Plan* and *NASA Implementation Plan for Education*, 1999–03, (National Aeronautics and Space Administration [NASA], 1999, 2000), and *No Child Left Behind* (U.S Department of Education, 2001b).

Although higher education science faculties are attempting to improve the effectiveness of undergraduate science courses, the process is slow and the results limited (Barinaga, 1991; Fedock, Zambo, & Cobern, 1996).

Many introductory undergraduate science courses, because of the ways they are taught, discourage significant numbers of students either from majoring in science or from taking additional science courses (Tobias, 1992). Sheila Tobias (1990) reported common features of courses turning off students include lack of relevance, passive student roles, emphasis on competition, and focus on algorithmic problem solving. The major concern for education majors in higher education STEM courses is even greater. The experiences for many education majors involve obscure connections between scientific ideas; lack of social, historical, and philosophical context; lack of meaningfulness in laboratory sections; and omission of important unifying themes (Floden, Gallagher, Wong, & Roseman, 1995).

As scientific research creates new knowledge and as educational research identifies more effective science teaching methods, faculty are under increasing pressure to implement more effective instruction. Floden et al. (1995) suggest that college science instructors must improve teaching quality, encourage student participation, increase literacy, and enhance 21st-century science conceptual knowledge and skills.

The key question that must be answered is, If K–12 reform in science teaching is to succeed, then what roles must be fulfilled by higher education faculty that promote relevant curriculum and instruction for undergraduate non-science majors, and more specifically, education majors?

A PROFESSIONAL DEVELOPMENT PROGRAM FOR HIGHER EDUCATION STEM FACULTY

NASA Opportunities for Visionary Academics (NOVA), a program of the National Aeronautics and Space Administration, is a professional development program with multiple strategies focused on assisting faculty to develop, implement, disseminate, and sustain innovative STEM curricula. The program was developed in response to the need to facilitate change in science teaching in higher education by providing assistance to faculty on a national basis. NOVA's goal, leading to the improvement of STEM literacy of future teachers, was to implement standards- and research-based change nationally in entry-level undergraduate STEM content in colleges and universities. A consortium of universities provides the leadership for NOVA: the University of Alabama, Fayetteville State University, and the University of Idaho. NOVA objectives are:

- Disseminate NASA's preservice education model nationally through 3-day workshops to a diverse population of higher education institutions, addressing critical concerns for equity and geographic distribution.
- Continue development of NASA's preservice education model aligned with NASA's Strategic Enterprises and the national standards and benchmarks for science, mathematics, and technology.
- Sustain the change process by mentoring workshop participants and collaborating with NOVA partner institutions (grant recipients).
- Increase the collaboration among the NOVA partner institutions by providing a forum to exchange innovative ideas for change in preservice education.
- Stimulate and conduct research on the effectiveness of NASA's preservice education model.

Since 1996, NOVA has invited the participation of higher education faculty concerned with the preparation of preservice teachers. Participation in NOVA includes several levels of opportunities for and commitment to enhanced knowledge and skills, through workshops, exemplary models, grants, mentoring, and collaboration within and between two- and four-year colleges and universities.

The various phases of the NOVA program consist of the following:

- Professional Development Workshop and Proposal Development (Phase I).
- Research and dissemination program (Phase II).
- NASA Field Center Program involving work with scientists, mathematicians, and engineers to develop NASA data-supported online course enhancements for NOVA courses (Phase III).

- Leadership Development Conference held each year for NOVA Network institutions, faculty, and administrators to foster national collaboration (LDC).
- Extensive evaluation including on-site assessment at NOVA universities and colleges (NOVA Evaluation).

EVALUATION OF THE IMPACT OF THE NOVA PROFESSIONAL DEVELOPMENT PROGRAM FOR SCIENCE FACULTY

Evaluation of NOVA activities from 1996 to 2003 included the gathering and analysis of extensive and diverse data. The evaluation involved data collected through such sources as surveys and interviews of faculty and students, faculty questionnaires, student course achievement outcomes, faculty research on student learning outcomes, on-site evaluation visits, focus group interviews, and comparisons of student learning outcomes among and between NOVA and traditional courses and institutions. The NOVA Evaluation was organized around 10 central points or questions. The questions presented in Table 13.1 involved characteristics of institutions, faculty, students programs, and courses, extent of implementation, dissemination, and sustaining of

Table 13.1.   Questions Guiding Evaluation

NOVA Evaluation Questions

1. What are the *characteristics* of the institutions, faculty, and students (gender, major, etc.) who matriculate in NOVA courses?

2. To what extent is the NOVA model being *disseminated and implemented* by other faculty at NOVA institutions or at other institutions?

3. How *congruent* is the NOVA program with recommendations found in major reports on the preparation of K-12 teachers?

4. What are the *participant reactions* to the NOVA professional development model?

5. How has the implementation of the NOVA model impacted *the collaborative work and organizational climate* of science, mathematics, engineering, and education faculties at NOVA institutions?

6. What new and modified *courses* are being offered as a result of NOVA?

7. To what extent does the NOVA course content *integrate data and information that are unique to NASA*, its fundamental questions, and its Enterprises?

8. What aspects of the NOVA professional development model are effective in *creating and sustaining* intended faculty knowledge and skills for reform action to occur?

9. What aspects of the NOVA professional development model *affect course disciplinary content, classroom pedagogy, and specific disciplinary pedagogy* related to student learning outcomes?

10. Does *scientific literacy among students* increase as a result of the faculty change through NOVA?

exemplary course models, congruency with national guidelines and reports, participant reactions, collaborative work and organizational climate in science departments, extent of integration of current research into course content, types of changes made in course content and pedagogy, and impact on scientific literacy among students. Additional details on results can be found in Sunal, Staples, et al. (2003). A summary of evaluation findings for each central question follows.

Question 1. Characteristics of NOVA Institutions, Faculty, and Students Involved in the Evaluation: Population and Sample

Since its inception, higher education institutions were invited to begin their participation in NOVA through a 3-day workshop and to continue the long-term process through involvement in professional development leading to systemic change. The program targeted underrepresented higher education institutions' faculty and students to facilitate change in STEM courses.

Between February 1996 and February 2003, 240 institution teams, located in 44 U.S. states, the District of Columbia, Puerto Rico, and the U.S. Virgin Islands, attended workshops (NOVA Phase I Program). Workshop participants were eligible to apply for grants of up to $34,000 to initiate change in STEM content and NOVA pedagogy.

Over the 7-year period, NOVA received 152 Phase I proposals for funding, of which 88 (58%) colleges and universities were funded. There are 308 multidisciplinary NOVA faculty (physics, chemistry, biology, geology, engineering, education, etc.) in these 88 institutions working collaboratively together, many for the first time, on improving student learning in higher education STEM courses.

As a result of Phase I funding, 154 new or substantially modified courses are being offered in the NOVA colleges and universities. In addition, a larger number of other existing courses have been indirectly affected in a number of ways. Over 54,000 undergraduate students have completed NOVA undergraduate STEM courses. The number grows by about 12,000 per year.

Meeting NOVA's first objective of disseminating its model to a diverse population, over 25% of the institutions in NOVA are minority institutions. As part of this effort NOVA partnered with the Minority University-Space Interdisciplinary Network (MU-SPIN) to provide workshops and collaboration with numerous Historically Black Colleges and Universities (HBCUs) through partnership in the NOVA Network.

The type of NOVA institution ranges from Doctoral/Research I Institutions to Associate of Arts Community Colleges, as assigned by the Carnegie Classification of Institutions of Higher Education. NOVA institutions include a greater number of master's degree and baccalaureate-granting institutions than other Carnegie types. Overall, the funded and nonfunded groups display similar trends in classification type and most have previously had little contact and use of current research data such as NASA databases or products for course planning or instruction. It is useful to note that a large majority of the NOVA BA and MA degree-granting institution faculty previously had not applied or been funded by government agencies in the past. These types of institutions provide the majority of science courses for preservice teachers in four-year institutions.

Question 2. Dissemination of the NOVA Model

Guidelines for the NOVA model of professional development in higher education were developed from an analysis of best practices research on faculty development in higher education and the use of a supporting framework of cognitive apprenticeship, focusing on faculty intervention and interactions (Sunal et al., 2001). The NOVA model focused on constructing, connecting, and collaborating, using the best of what has been learned through research on faculty professional development.

This professional development model content includes exemplary teaching and student inquiry strategies, methods of using technology to facilitate learning, considerations for teaching science to diverse undergraduates, and examples of successful innovative undergraduate course models. The framework supports interactive and collaborative learning. Learning outcomes are strengthened through interdisciplinary approaches to selection of content and the use of NASA research databases and materials to make connections to current research initiatives. Assessment is viewed as more of a learning process than a grading process. The content of faculty workshop learning activities, Web-enhanced distance learning, and continuing self-action in long-term follow-up at home institutions are all documented in literature-supported processes used in effective professional development. Developing plans to guide change, using action research in classrooms, working in collaborative teams, overcoming internal and external barriers to change, and funding innovative ideas are some of the skills developed in Phase I workshops to provide faculty with strategies for promoting and implementing change.

The change process is not a one-event activity. It is a systemic, long-term professional development and mentoring support system for faculty wishing to implement new courses or change existing courses. Workshops, written, video, and live descriptions of effective practice, expert and peer consultation and mentoring, Web distance learning and resources, funded course development, and involvement in a course development process are all elements in the change framework. Participation begins with application and participation in a Phase I workshop by a collaborative team of faculty and administrators, one each from science and education departments and one administrator (elicitation phase). It continues with a mentoring process during which team members are assisted in developing a proposal for change that will result in innovative teaching (reflection phase). As the proposal plan is implemented, mentoring continues, along with evaluation of ideas enacted (reconstruction phase). The conclusion of this action research process leads to a new level of understanding and individual professional development (Raubenheimer, 2003). The process, however, is a continuing cycle of activity. At this point faculty must redefine the problem (elicitation phase) and continue investigation by completing another round of the action research cycle (reflection phase and reconstruction phase). Part of the mentoring process is networking and site visitation by faculty who have implemented innovative changes in teaching at other institutions.

Dissemination of the reformed course models and classroom research results on the impact of course changes have occurred through many forms. Nationally, 23 Phase I workshops were held for 240 institution teams over 8 years. Over 200 NOVA-related publications in journals, reports, and presentations at international, national, and regional conferences have been documented. NOVA faculty teams of science and science education faculty repeatedly reported actively disseminating their research-supported best practice on teaching and learning in higher courses to colleagues within their own institution, local area institutions, and on Internet sites. The team-created institutional Web sites also provided information and resources for students to enhance learning outcomes. In addition, over 150 NOVA-related national, regional, and local workshops, consultancies, and other professional activities have been completed.

In over one half of NOVA institutions, non-NOVA faculty, in the same department and in other departments across the institution, demonstrated interest through the development of other clone and NOVA-like spin-off courses following the NOVA model. Most higher education administrators interviewed in this evaluation study reported faculty in their departments incorporating more hands-on and inquiry-based activities and NASA materials into other non-NOVA existing courses. They also reported faculty using the NOVA course and project as a springboard to grant writing for additional change in other or across departments and between institutions and school systems. For institutions that sent teams to the Phase I workshop but were not funded, whether or not they submitted a proposal to NOVA, over one third made significant changes in their courses based on the NOVA model. It is safe to say that the impact of NOVA professional development went far beyond the funded institutions.

Over the past 8 years, NOVA has evolved from development of innovative institution-based courses to a national dissemination network collaborating through a systemic change process and a set of STEM course models applicable to any higher education institution.

Question 3: Congruency of the NOVA Model with Recommendations found in Major Reports on Teacher Preparation

The NOVA model for Professional Development includes eight major elements involving professional development, systemic change, research, and dissemination aimed at changing how the sciences are taught in K–12 classrooms by influencing "how" content is taught at the college level. The NOVA model integrates the Mission of the NASA Enterprises with the National Science Education Standards (NSES), National Council of Teachers of Mathematics (NCTM) Principles and Standards, and other significant national guidelines to create reform in higher education STEM teaching and learning.

The model integrates content and pedagogy through the collaboration of faculty from science, mathematics, engineering, and education to develop exemplary STEM courses for preservice teachers as well as other majors. The guiding principle behind NOVA was to create a community of science faculty committed to continuous professional development and lifelong learning.

Question 4: Participant Reactions to the NOVA Professional Development Model

Participant reactions to the NOVA professional development process are positive. The Phase I workshops were evaluated as extremely useful in developing innovative pedagogy as well as necessary skills in technology, curriculum innovation, and writing successful proposals for funding change.

Reasons for participating in the NOVA program were varied. At some institutions, the initial interest in creating reformed courses, as expressed by administrators, was to introduce innovative teaching strategies and innovative STEM courses, and they contacted faculty who they thought might be interested. Change at other institutions was inspired by changes in state education requirements or content standards, so faculty needed help. Some faculty were using NOVA-like innovative methods in their courses, but wanted to validate their efforts and extend their knowledge to other innovations. Also, faculty members of the NOVA network recruited other new faculty. At most institutions, however, change began with one or two innovative individuals who wanted to improve the content knowledge of students and became more formal when the interdisciplinary team of science and science education faculty was formed, joined in many cases by an administrator, and attended a Phase I NOVA workshop.

NOVA science faculty increasingly gained confidence in their own teaching and expectations for high student outcomes. Faculty interview data identified professional development as a key part of the process. The science faculty team member(s) needed help with pedagogy; the education faculty member needed assistance in depth of content. A team approach was identified as a very important factor in development and implementation of a NOVA course. The longer faculty remained actively involved with the NOVA model the greater they rated their ability to plan, put in practice, and evaluate effective teaching practices in science courses in higher education.

NASA personnel, field center visits, and materials provided all members of the team with the organization and 21st-century STEM examples to create relevant course innovation. Team members worked together to provide the necessary instruction to ensure the success of the new course. Other factors shown to facilitate the change process for NOVA faculty included commitment to the project, belief (high efficacy) in the model, and monetary support for innovations.

Question 5: The Impact of the Professional Development Model on the Collaborative Work and Organizational Climate of Science, Mathematics, Engineering, and Education Faculties

Collaboration between STEM departments and the education faculty was greatly enhanced through participation in workshops and institutional implementation of the NOVA model. The team approach was found to be essential to innovative course development.

Collaboration between discipline and education faculties led to significant changes in student learning in the reformed courses. Comparison studies with traditional courses replaced by the reformed course show a significant increase in type and expertise of research-supported teaching practices. Teams that met regularly were found to have enhanced course change. Implementation of the NOVA model impacted the collaborative work of STEM department faculties. Examples of ways that teams have continued their work together include collaboration on other funding or grant opportunities, the development of new department programs, and professional dissemination.

Collaboration provided necessary breadth of expertise, experience, and sharing of the workload to address many barriers to the change process. Administrators were interested to see faculty initiating the process; their different viewpoints facilitated change. All faculty interviewed indicated that involvement in the NOVA Network facilitated the resolution of problems encountered. A campus climate that focused on student learning and the team's response to this climate were both important factors in facilitating change.

Question 6: The Effect of the NOVA Model on Higher Education Courses

A total of 154 reformed courses were developed and offered at the 88 NOVA institutions. The majority of the reformed science courses were in the discipline areas of biology, life science, life in space, or environmental science. The second most abundant was earth and space science; the third was physics, chemistry, and physical science courses. A large majority of the courses created or revised at NOVA institutions were at the entry level, freshman or sophomore levels. Several courses were developed as capstone courses to integrate basic science knowledge and enhance scientific literacy.

Many faculty and students reported that the NOVA course was a valuable learning model for the preservice teachers. Student attitudes were originally seen by faculty as barriers to change, but, as students became more familiar with the new methods, their acceptance became an important positive facilitator of change in the courses.

Question 7: The Impact of the NASA Mission, Data and Information, and Fundamental Questions on the NOVA Courses

Most of the NOVA professional development programs (Phase I, II, III and LDC) were held at, and utilized the resources of, NASA Field Centers (e.g., Johnson Space Center). Working at the Field Centers provided for NOVA Faculty extended on-site visits to facilities (e.g., mock-up of the international space station), interaction with NASA research personnel, and training in instructional resources and materials available at all 10 centers. The content for all of the NOVA courses include use of NASA information, data or materials over a range of science, mathematics, and technology content aligned with the standards found in the NSES, NCTM, and ISTE.

Many NOVA courses have multiple curriculum connections to the NASA Strategic Enterprises. Analyzed by institution, NOVA courses significantly connected over 300 times with the various NASA Strategic Enterprises. These connections were experienced by students in all of the NOVA courses, carried into K–12 classrooms by preservice students after completing the courses, shared between colleagues in discussions of course development, disseminated on institution Web sites, and presented at national conferences to peer faculty.

Question 8: Aspects of the Professional Development Model that Were Effective in Creating and Sustaining Change

A study of NOVA faculty teams found that the key criteria for successful implementation of course change was continuous faculty interaction, both within and between disciplines, administrative support, and a sense of common purpose and trust (Sunal et al., 2001). This suggests that unless structural mechanisms to support innovation are in place, attempts by faculty to initiate and successfully undertake change processes are likely to be thwarted with difficulties and even failure.

The key factor sustaining faculty coherence was the formation of teams and their maintenance through each phase of the NOVA process. NOVA facilitated team development addressing useful responses to critical barriers faced by collaborative teams. A majority of the teams reported that they would not have been successful without the content and skills developed in the professional development program. For faculty and course reform that meets the needs of preservice teachers in science, mathematics, and technology to occur in higher education, faculty development following the NOVA model was critical for successful changes to be planned, implemented, and sustained.

As a regular part of the NOVA Phase I workshops, action (practitioner) research forms one component of the professional development process. The outcome of these research efforts has demonstrated significant gains in student learning, as well as in faculty professional development. Extensive observational data gathered during on-site visits support the idea that faculty have significantly changed their practices. Many faculty reported that the reforms are also being integrated into the other classes they teach and into the classes of other department faculty members who have observed their new strategies.

Question 9: Impact of the NOVA Model on Classrooms and Students

NOVA courses address concerns emphasized in the national science and mathematics standards. Using findings from faculty action research presented at the past NOVA Leadership Development Conferences as a data source, it can be concluded that NOVA courses significantly improve student science content knowledge, attitudes toward science content, and science process skills (Raubenheimer, 2003). In addition, it was found that students had a higher efficacy to teach science or relate other subjects to science in a K–12 classroom setting (Sunal, Staples, et al., 2003).

One of the goals of the NOVA model focused on student learning using an inquiry strategy. Both NOVA Phase I proposals and action research reports reviewed reflected significant movement away from traditional lecture and memorization approaches to a greater focus on an inquiry approach to teaching, with interactive, student-centered methods to enhance student learning (Raubenheimer, 2003). The student consensus derived from focus group interviews at over 40 institutions was that their NOVA instructor "modeled how to teach."

Overall, from a variety of data sources, it was concluded that the typical reformed course emphasized construction of knowledge through inquiry, incorporating hands-on and technology-based activities. In-class cooperative student learning and cross-college faculty collaboration enhanced courses, which were more interactive than traditional courses, focusing on solving real-world STEM problems. The reformed courses were used as a model for developing additional courses in the team departments and for program review evidence to meet state higher education guidelines.

Several research studies conducted concerning the impact of NOVA courses on students concluded that (a) modeling, active engagement, project-based activities, and cooperative learning positively affected undergraduate education majors' science teaching efficacy, science content mastery, and teacher performance in K–12 science classroom instruction; (b) state standardized test scores of achievement in science among students of in-service teachers who have experienced a NOVA STEM course were significantly higher than students of teachers experienced only in traditional STEM courses in their undergraduate program; and (c) NOVA experiences resulted in a chain of increased development in STEM literacy among education majors and later to their students in classrooms (Staples, 2002).

Students reported that "understanding" was the goal of STEM teaching, not just covering content. Awareness of students' prior knowledge, ability to explain things on the students' level, and use of diverse ways of teaching were very important to the students. The instructor needed to have the ability to "make science come to life" and be relevant to the real world, and needed to have a strong knowledge of the content. The students reported that their NOVA experience made science less threatening, and they now had a better understanding of "what science is really about," and that "mathematics is something we use everyday" (Sunal, Staples, et al., 2003).

Question 10: Impact of the Professional Development Model on Science Literacy of Higher Education Students

Students in NOVA courses perceived that effective teaching in higher education STEM entry-level courses used an integrated inquiry and technology approach interactively involving students. Students experienced science as a process rather than being told about content. The integration of lecture and lab in the large majority of NOVA courses created a structure that promoted hands-on, experimental, inquiry learning (Sunal, 2003).

Overall, the key NOVA impact described by many NOVA institutions was that, after taking the NOVA courses, the students had shown an improvement in achievement in all areas of science understanding related to the national standards for science and mathematics (Sunal, Staples, et al., 2003). A majority of the institutions reported a significant increase in students' STEM content knowledge. Also, studies reported that the number of females and minorities interested in their NOVA courses increased and that science literacy increased significantly after completing the courses (Sunal, Staples, et al., 2003). The students reported that the model of instruction they experienced could be transferred to the K–12 classroom (Staples, 2002).

Two national studies on student understanding of the nature of science concluded that undergraduate students in NOVA reform-oriented classes demonstrated a significantly higher end-of-course growth and level of understanding as compared to students in other similar, "non-NOVA," courses (Sunal, Sunal, et al., 2003; Kallam & Kallam, 2003). These understandings and behaviors relate to students' ability to succeed in science-related tasks and careers, as well as their interest and anxiety about this content in their professional and personal lives.

SUMMARY

Reforms in undergraduate science teaching have been successful. In addition to reform being carried out through efforts of such organizations as the National Science Foundation and National Research Council, the National Aeronautics and Space Administration program, NASA Opportunities for Visionary Academics (NOVA) has lead national reform in undergraduate science teaching impacting hundreds of courses and thousands of students each semester. NOVA was developed in response to the need to facilitate change in science teaching in higher education by providing assistance to faculty on a national basis. The program used the best of what had been learned through research on faculty professional development and teaching to enhance science literacy of all undergraduate students with a focus on involvement of education majors in entry-level undergraduate courses.

The impact of the program resulted in science faculty increasingly gaining confidence in their own teaching and expectations for high student outcomes. Faculty interview data identified professional development as a key part of the process. Collaboration between discipline and education faculties using a team approach was identified as an important factor in development and implementation of a course reform and resulted in a significant increase in type and expertise of research-supported teaching practices. Teams that met regularly were found to have enhanced course change.

Observational data gathered during on-site visits support the idea that faculty have significantly changed their practices. Many faculty reported that the reforms are also being integrated into the other classes they teach and into the classes of other department faculty members who have observed their new strategies.

Overall, from a variety of data sources, it was concluded that the typical reformed course emphasized interactive construction of knowledge through inquiry, incorporating hands-on and technology-based activities, used in-class cooperative student learning, and focused more on solving real-world STEM problems. The reformed courses were used as a model for developing additional courses in the team departments and for program review evidence to meet state higher education guidelines. The major student impact, described by many NOVA institutions, was that after taking the NOVA courses, the students had shown an improvement in achievement in science understanding in areas related to the national standards for science.

Unless strong alternative experiences intervene, most students who experience "traditional" science courses may never master the important components of scientific literacy and may hold on to negative beliefs toward the importance and use of science in their everyday and public lives. The impact of NOVA and its effects in higher education are important to professionals and an educated citizenry in a democratic society.

AUTHOR NOTE

This work was in part supported by NASA Opportunities for Visionary Academics (NOVA), a program funded by the National Aeronautics and Space Administration, although the views expressed here are the authors' only.

REFERENCES

American Association for the Advancement of Science. (1993). *Benchmarks for scientific literacy, Project 2061.* New York: Oxford University Press.

Barinaga, M. (1991). Scientists educate the science educators. *Science,* 252(5009), 1061–1062.

Fedock, P.M., Zambo, R., & Cobern, W. (1996). The professional development of college science professors as science teacher educators. *Science Education,* 80(1), 5–19.

Floden, R., Gallagher, J., Wong, D., & Roseman, J. (1995, April). *Seminar on the goals of higher education.* Paper presented at the annual conference of the American Association for the Advancement of Science, Atlanta, GA.

Kallam, L., & Kallam, M. (2003, April). *Reforms in university courses on students' attitudes and beliefs regarding the nature of mathematics.* Paper presented at the annual conference of the American Educational Research Association, Chicago.

Knezek, D. (2002). *ISTE National Educational Technology Standards (NETS) and Performance Indicators: Educational technology foundations for all teachers.* Retrieved June 19, 2002, from the International Society for Technology in Education Web site: http://cnets.iste.org/teachers/t_stands.html

National Aeronautics and Space Administration. (1999). *NASA implementation plan for education, 1999–2003* (EP-1998-12-383-HQ). Washington, DC: Author.

National Aeronautics and Space Administration. (2000). *NASA strategic plan.* (NASA Policy Directive, NPD-1000.1B). Washington, DC: Author.

National Commission on Excellence in Education. (1983). *A nation at risk: The imperative for educational reform.* Washington, DC: U.S. Department of Education.

National Research Council. (1996). *National Science Education Standards.* Washington, DC: National Academy Press.

National Research Council. (1997). *Science teacher preparation in an era of standards-based reform.* Washington, DC: National Academy Press.

National Research Council. (2001). *Educating teachers of science, mathematics, and technology: New practices for the new millennium.* Washington, DC: National Academy Press.

National Science Foundation. (1996). *Shaping the future: New expectations for undergraduate education in science, mathematics, engineering, and technology* (Publication #NSF 96-139). Arlington, VA: Author.

Raubenheimer, C.D. (2003, April). *Faculty conceptions and practices of action research in undergraduate science courses.* Paper presented at the annual conference of the American Educational Research Association, Chicago.

Staples, K.A. (2002). *The effect of a nontraditional undergraduate science course on teacher and student performance in elementary science teaching.* Unpublished doctoral dissertation, University of Alabama, Tuscaloosa.

Sunal, D. (2003, April). *The effects of reform in university undergraduate science and mathematics courses.* Paper presented at the annual conference of the American Educational Research Association, Chicago.

Sunal, D., Bland, J., Sunal, C., Whitaker, K., Freeman, M., Edwards, L., et al. (2001). Teaching science in higher education: Faculty professional development and barriers to change. *School Science and Mathematics, 101,* 246–257.

Sunal, D., Staples, K., Odell, M., Freeman, M., Whitaker, K., Edwards, L., et al. (2003). *NOVA 1996–2003 evaluation report.* Tuscaloosa, AL: NOVA. Retrieved July 2003, from http://nova.ed.uidaho.edu/library/ViewLibrary.asp?Section=2

Sunal, D., Sunal, C., Whitaker, K., Odell, M., & MacKinnon, C. (2003, April). *The effect of standards-based reform in university courses on undergraduates' science knowledge.* Paper presented at the annual conference of the American Educational Research Association, Chicago.

Tobias, S. (1990). *They're not dumb, they're different: Stalking the second tier.* Tucson, AZ: Research Corporation.

Tobias, S. (1992). *Revitalizing undergraduate science: Some things work and most don't.* Tucson, AZ: Research Corporation.

U.S. Department of Education (2001a). *Before it's too late: A report to the nation from the National Commission on Mathematics and Science Teaching for the 21st Century.* Washington, DC: Author.

U.S. Department of Education. (2001b, August 21). *No child left behind..* Retrieved August 8, 2003, from http://www.ed.gov/offices/OESE/esea/nclb/titlepage.html

part II

PERSPECTIVES ON REFORM IN UNDERGRADUATE SCIENCE

This section examines from various perspectives the process of reform in developing science curriculum and the teaching of science at the undergraduate level. The perspectives vary from individuals who hold national leadership positions in professional science and science education organizations to higher education faculty sharing how they went about reforming course offerings in their various institutions. One chapter reports the perspective on an undergraduate student. The final chapter of the section examines a model for one-on-one collaboration for undergraduate science reform.

In Chapter 14, M. Jenice Goldston, Monica Clement, and Jacqueline Spears share the findings of a case study from a large university perspective that profiles collaboration between a scientist and a science educator to reform the teaching of geology to undergraduates. The study examined the geophysicist's personal practice theories as new curricular and pedagogical changes were incorporated into an undergraduate geology lecture course, and examined the nature of collaboration between the co-researchers in fostering the emergence of and changes in the geo-researcher's personal practice theories. Findings illuminate the complexity of the change process within the culture of higher education.

In Chapter 15, Frances Gardner, Jr., provides a description, from a small college perspective, of the development of a new course of study, Introduction to Life in Space. This interdisciplinary course targets non-science majors and potential preservice teacher candidates. The overarching objective for the course is the development of student skills necessary to transform students into more successful reflective thinkers through the use of technology. The final exam requires designing and conducting a scientific experiment.

In Chapter 16, using a perspective linking policy and research, Wayne Morgan points out that about one half of the undergraduates attending college in the United States are enrolled at a two-year college. He discusses appropriate approaches to teaching science for such a diverse population of students emphasizing the pedagogy championed in the National Science Education Standards. This also means that preservice teachers who begin their science training at two-year colleges will be likely to encounter instructors who will model the learning environments necessary to improve science education at all levels, promoting active learning and modeling science as inquiry.

Cheryl Mason and Steven Gilbert, in Chapter 17, use a science education research organization's perspective to address issues related to performance-based assessment in higher education. This approach has become increasingly common at the state and national level. The authors examine changing state and national policies that require institutions of higher learning to be accountable for student learning, rather than teaching. To facilitate these efforts, the authors recommend targeted research in relation to the development of individual and program-level assessments in the sciences at the postsecondary level, and the development of models for the involvement and support of faculty in reforming the undergraduate curriculum emphasizing science literacy. In addition, they recommend studies concerning the roles of critical thinking and metacognition, including the use of emerging technologies, as they impact undergraduate science education reform efforts.

George DeBoer, in Chapter 18, using a science education research organization perspective to discuss the reform of K–12 science education since the publication of several important documents by prestigious national organizations. He summarizes the vision in these documents concerning what students should know and be able to do to achieve science literacy by the time they graduated from college. The author examines the federal legislation in the past decade that has been supportive of this approach with its emphasis on accountability through testing, including the No Child Left Behind Act of 2001. The authors address questions such as whether higher education can move in the same direction as K–12 reform, whether the time is right for higher education to follow the path that K–12 is now on, and whether there are any intermediate steps that can begin to move higher education in that direction.

In chapter 19, Joe Herppert and April French use a scientist's perspective, gained during the reform of introductory undergraduate chemistry laboratory course, in an attempt to eliminate the distinction between the laboratory and the lecture. Also discussed is the context of the laboratory courses that were revised in their program, the instructional model used in the design of the experiments, and some of the research outcomes of the project. This chapter emphasizes the effects of reformed college-level science courses on preservice teachers. It is critically important that the nation address this challenge by both empowering existing science teachers and attracting students of the highest caliber into the profession.

In Chapter 20, Kimberly Staples reports the findings of a research study that identifies the effect of innovation in college science courses on university undergraduate students' perceptions of science and science teaching. The study focused on identifying specific innovative strategies that yield success in college science courses. The student perceptions were compared based on experiences in an innovative science course and a traditional science course. The results present strategies that improve student success in science.

In Chapter 21, M. Jenice Goldston, used a university science education researcher's perspective to describe a model that unfolds from the voices of two colleagues, a science educator and a geologist, from similar yet different academic cultures. The author addresses the issue that there are no guidelines that instruct one on how to develop collaborative working relationships across university departments, and describes a model, peer consultation, used as a professional development tool that utilizes collaboration and fosters reform in undergraduate science teaching practices.

CHAPTER 14

A LARGE UNIVERSITY PERSPECTIVE ON REFORM IN TEACHING UNDERGRADUATE SCIENCE

A Geologist's Personal Practice Theories and Pedagogical Change

M. Jenice Goldston, Monica Clement,
and Jacqueline Spears

ABSTRACT

This case study profiles collaboration between a geophysicist and a science educator to reform the teaching of geology to undergraduates. The research took place in a large Midwestern university with 50 undergraduates enrolled in Earth in Action. The study's focus was to explore (a) a university geophysicist's personal practice theories as new curricular and pedagogical changes were incorporated into an undergraduate geology lecture course, and (b) the nature of collaboration between the co-researchers in fostering the emergence of and changes in the geo-researcher's personal practice theories. The

Reform in Undergraduate Science Teaching for the 21st Century, pages 245–266
Copyright © 2004 by Information Age Publishing
All rights of reproduction in any form reserved.

geology course was modified to incorporate national science education standards and several "active learning" approaches. Primary method of data collection included classroom observations, personal essays, interviews, journals, and peer consultation. An interpretative approach and inductive analysis of data were used to discern personal practice theories and changes that took place during the course. Findings illuminate the complexity of the change process within the culture of higher education.

INTRODUCTION

As faculty in a large university we recognize that the climate and structure within institutions of higher education provide different issues, concerns, and barriers to reforming the way undergraduate students are taught compared to reform endeavors conducted within smaller institutions. For instance, large universities comprise distinct colleges, each with unique academic disciplines and cultures. Furthermore, large class enrollments, research faculty who specialize within narrowly defined subject areas, and the physical separation of the colleges provide a multilayered referent for viewing conflicts, issues, and decision making embedded within reform efforts.

The past decade has brought about an interest between university and college faculty in the use of innovative active learning approaches (Bonwell & Eison, 1991; Harris, 2001). Active learning is generally defined as any experience that engages the student beyond passive listening. Active learning methods range from simple think-pair-share to complex cooperative learning strategies. Those who have used active learning approaches have found them to be beneficial from sociopsychological stances which in turn may be related to student achievement (Hake, 1998; Sokoloff & Thorton, 1997; Wright et al., 1998). Despite the positive findings, there has been no widespread shift toward the use of active learning strategies in university classrooms. In fact, the predominant method of teaching in universities is still the "transmission" or lecture model.

Have you wondered why professors in large universities continue to teach as they do? If so consider the following: According to Harris (2001), there are many barriers that university faculty face when changing their teaching approaches. He concluded that lack of support, ability to cover content, limited knowledge of pedagogical techniques, and a strong research focus are among the obstacles. As an advocate of active learning, he encourages professors to find support, make small changes, and focus on student learning to improve their teaching. Though the suggestions are important, there is concern that these suggestions will not truly reform and sustain teaching practices at this level.

In this chapter, a geo-researcher from the Department of Geology and a science ed-researcher from the College of Education, both in a large Midwestern university, began an exploration to examine the geo-researcher's personal practice theories (PPT). Modifications that occur as "active learning practices" were incorporated into an introductory geology lecture designed for education majors.

THEORETICAL FOUNDATIONS

This research is embedded in three theoretical frameworks that include (a) teacher beliefs about pedagogy (personal practice theories), (b) reflective practices, and (c) teacher research (teacher as researcher). The frameworks merge into the complex dimensions of educational reform as we examine the personal practice theories and actions of a university instructor changing the way she teaches.

According to Carr and Kemmis (1983), "all practical activities are guided by some theory. Anybody engaged in the 'practice' of educating others must already possess some 'theory', idea, or notion of education which structures activities and informs decisions" (p. 110). Carr and Kemmis are referring to personal practice theories (PPTs) or personal practical theories (Cornett, 1990; Cornett, Yeotis, & Terwilliger, 1990; Sweeney, Bula, & Cornett, 2001). Personal practice theories are typically referred to as a logical, coherent set of beliefs that guide teaching. These beliefs are based on direct teaching experiences, selection of curriculum, and informal teaching experiences outside the schoolroom (Cornett, et al., 1990; Ross, Cornett, & McCutcheon, 1992; Sweeney et al., 2001; Tobin, McRobbie, & Anderson, 1997). In examining research on PPTs, one finds that "knowledge and beliefs are intricately intertwined and have an adaptive function" in teaching (Pajares, 1992, p. 325). Thus, a logical next step to building a nexus between teacher personal practice theories and their knowledge base is examining research findings on teacher beliefs.

The 1980s were dominated by research on teacher beliefs and cognition. For instance, Kagan (1990) operationalized teacher cognition as teacher self-reflections on beliefs and knowledge of teaching. Knowledge of teaching categories included knowledge of students, strategies for classroom management, and subject content knowledge. Others refer to the teacher knowledge base as one that includes (a) self knowledge, (b) subject matter knowledge, (c) pedagogical knowledge, (d) curriculum development knowledge, and (e) pedagogical content knowledge (Elbaz, 1983; Grossman, 1990; Shulman, 1986; Yeany, 1991). More recently Carlsen (1999), using a post-structural lens, synthesized the work of Shulman identifying five domains of teacher knowledge. He included (a) knowledge of general educational context, (b) knowledge of specific educational context, (c) general pedagogical knowledge, (d) subject matter knowledge, and (e) pedagogical content knowledge. Despite the constant revisions on what constitute a teacher knowledge base, teacher knowledge and personal practice theories are always seen as interconnected and influencing how one teaches. Research that explores reflective practices of teachers builds upon this teacher knowledge base.

The second framework connects personal practice theories to the change process. In general, a reflective approach refers to the implementation of reflection for personal and professional growth. "Reflection involves an individual in looking back over teaching experiences and engaging in internal dialogue about specific situations while questioning what is and what could be in the future" (Rosenthal, 1991). This is known as reflection-on-action. Another related reflective approach is "reflection-in-action" (Schön, 1983) which is a reflexive interchange between the individual and the event (MacKinnon, 1986). Through reflective practices, Schön's goal for professional growth is to move from what he called "technical rationality" to "practical rationality," where theory emerges from practice. Traditionally, theory informs practice. As used in this study, combined reflection "in and on action" through dialogue between the instructor and researcher exposed personal experiences that are used to determine the meaning of actions for them (Connelly & Clandinin, 1987). Thus, teacher reflections can become agents of purposeful change and an important tool for teachers examining their own practices through action research.

Action research, as a final framework, is known by many names that include: teacher research, classroom inquiry, teacher as researcher, classroom research, teacher inquiry, and interactive research. As discussed by Raubenheimer earlier in Chapter 12, the focus of action research by teachers generally includes seeking to understand, monitor, or evaluate some dimension of the educational setting with the goal of improving the educational environment. Whatever its purpose, action research is grounded in the personal experiences of the instructor and is expressed in their words and voice (Burnaford, Fischer, & Hobson, 2001). According to Grant and Zeichner (1984), action research incorporating reflection as a research tool leads to better decision making that leads to more effective teaching. Apple (1993) suggests that action research is best done by those most attached to the issues and strongly advocates teachers as researchers and change agents. Using the lenses of their experiences and histories, critical perspectives become a way to rethink knowledge and beliefs about teaching. Knowledge then becomes "transformative" (Freire, 1985) with change becoming valuable and meaningful. Thus, findings generated by teacher-researchers are powerful influences for improving their classroom practices (Cochran-Smith & Lytle, 1993; Flake, Kuhs, Donnelly, & Ebert, 1995; Goldston & Shroyer, 2000; Kincheloe, 1991; Schön, 1983).

PURPOSE

The purpose of this case study was to explore (a) the university geo-researcher's personal practice theories as curricular and pedagogical changes were implemented into an undergraduate geology lecture course, and (b) the nature of collaboration between the researchers in fostering the emergence of and changes in the geo-researcher's personal practice theories.

Contextual Setting

The study took place in a large Midwestern university with the co-researchers modifying GEOL 100, Earth in Action. This introductory course was reconceptualized to incorporate *active learning practices* and was restructured using the National Science Education Standards (National Research Council, 1996). One section of GEOL 100 was designated for undergraduate education majors and is the focus of this work. The following describes the participants and course changes.

Participants

Forty-eight undergraduates enrolled in Earth in Action (GEOL 100); 34 females and 14 males participated in the study. Preservice elementary education majors commonly take this course as part of the 12 hours of science content needed for the elementary education program. Education majors were encouraged to enroll in the modified course. Students enrolled in the course had not yet entered the elementary education program.

The geo-researcher was a geophysicist with a petroleum company for 6 years and entered teaching with a wide range of research experiences in the field of geology. She taught earth science for 6 years in a community college and currently teaches undergraduate geology in a large university where she is in her 4th year of teaching. The ed-researcher taught earth, life, and physical science in the public schools for 15 years. She currently is in her 8th year of teaching science education at the university.

Course Description

Original course. The original course, Earth in Action, was a typical "textbook course" that accommodated large undergraduate enrollments for lecture with student grades determined by three to four exams. The course topics included rocks, minerals, volcanism, rock cycle, plate tectonics, weathering and erosion, hydrologic cycle, earthquakes, and geological history.

Modified course.   One section of Earth in Action was modified for this study. Pedagogical changes focused on implementing active-learning, student-centered approaches. The modifications were based upon cognitive research findings that learning is an active process and new ideas are constructed by linking new information to what is already known and believed. As such, the modified course included field trips, student journals, cooperative teaming, guiding questions, open-ended inquiry, discussions about the nature of science, and process skills.

METHODOLOGY

The focus of this intrinsic case study (Stake, 1995) was to explore the geo-researcher's personal practice theories within the context of teaching an undergraduate geology course where active learning strategies were implemented. The study incorporated interpretative and naturalistic approaches (Lincoln & Guba, 1985; Stake, 1995). Prior to conducting the study, the co-researchers decided that the purpose was to explore prior personal practice theories as they intersected with pedagogical changes for the new course design. Methodological approaches, including naturalistic, interpretative techniques, were determined appropriate for the study with the ed-researcher assuming the role of participant observer. Direct interpretation and inductive analyses were used as advocated by Stake (1995) for case studies. In addition we included a "descriptive" step as suggested by Creswell (1998) to provide the details or "facts" about the case.

Data collection techniques were aligned to Cochran-Smith and Lytle's (1993) analytical framework for teacher research. The typology consists of empirical and conceptual research domains. The empirical research domain includes (a) classroom life accounts over time, (b) oral inquiries (examinations of classroom issues, context, experiences), and (c) classroom studies (interviews, observations exploring practice-based issues). The conceptual research domain includes essays about teachers' interpretations of assumptions or characteristics of the course. The following describes the data collection techniques used in this research.

Journals

The geo-researcher kept a journal as a place to record course "happenings," analyze teaching strategies and curriculum, and interpret and reflect upon the events of the university classroom. It was a "rethinking" place for making tangible emerging personal practice theories. In addition, undergraduates kept journals for the course. Journals were read and returned with feedback. Student journals influenced the geo-researcher's reflections and actions. For instance, Sue, an undergraduate wrote,

> Going to the spillway was a great experience—to actually get out and have "hands-on" application to what we were learning. I thought it was wonderful and it should be done again, but my only concern is that I felt kinda lost at times when I was there. I just felt like I didn't have a lot of guidance on what to do and I am totally new at this type of thing.

As a result of reading this entry, the geo-researcher modified the next field trip by providing both structured and unstructured activities. Without input she may not have recognized that some students were struggling with the open structure of the field experience. This kind of rethinking, reflective response to student journal entries occurred throughout the semester.

Oral Inquiries

Oral inquiries include data collection techniques that involve several individuals exploring events of the classroom and its environment. The co-researchers engaged in weekly oral inquiry through informal conversations, e-mail or phone. Conversations continued over a period of a year and provided the ed-researcher with data to enrich field notes and reflective memoranda. For instance, a conversation after a field trip to the spillway prompted the following question: "Why aren't the students curious about the geology of the spillway? They seemed totally bored." We began to discuss the lack of curiosity linked to the passive student role of "listener" perpetuated in traditional classrooms. Socialization and learned student roles in schools create expectations of being told what to do and they expect it. When the geo-researcher chose an open structure for the field trip and did not tell students exactly what to do, their lack of active engagement looked like boredom, but as seen in Sue's journal and in other journal entries, it was not.

Classroom Studies

A peer consultation model was used in exploring the geo-researcher's PPTs and providing insights on collaboration (Bernstein, 2000). The peer consultation model required (a) classroom visits by both researchers, (b) written classroom observations, and (c) exchange of written documents followed by responses. These steps constitute an iterative cycle for each classroom visit. Co-researchers conducted eight 50-min classroom observations during the semester. Peer consultation is an example of both classroom studies and oral inquiry used extensively in this work.

Essay

The geo-researcher wrote reflective essays throughout the semester in response to questions asked by the ed-researcher. Question examples included, Why have you taught geology as you have? or What barriers make changing the way you teach *difficult?* Essays drew heavily from the geo-researcher's experiences regarding teaching, learning, and the learning environment. Essays served as reflective tools used to uncover the personal practice theories.

EVOLUTIONARY STIRRINGS: BECOMING AWARE OF THE "TEACHING SELF"

My college teaching is modeled after my experiences in college classrooms—the lecture method. I teach this way because it is *easy,* once the format is worked out. It is organized and predictable. I know what will happen in class, I know what I will present and how I will present it, and I know what the exam over the material looks like. Very prescribed, orderly, neat, and tidy. I provide students with this orderly arrangement of information via the syllabus so that they know what they are supposed to do, when they are supposed to do it, and when they will be tested over the material. In lecture, I use many different ways to get information across, because I know students have different learning styles. I like to use the board a lot to draw diagrams, and I have students draw the diagrams too so that we draw as the concept is being presented. I also use overhead transparencies from the text; I bring in samples of rocks, minerals, and other things that will help students visualize what we cover in class. I am not rooted in the front of the class, but walk around freely during lecture. I want the students to know that I am with them, that I am accessible, and that I expect a certain involvement even from those who choose the middle and back of the class. I ask a lot of questions and use some active learning strategies (mostly think-pair-share) to engage the students and entice them to think more deeply about the material. (geo-researcher essay)

This passage, written early in the semester, describes the geo-researcher's teaching philosophy. From this essay and other data, eight personal practice theories initially surfaced. However as new data were concurrently collected and analyzed, the eight personal practice theories collapsed into three. The following three core PPTs emerged as the "handbook" for her teaching practices:

1. Teaching is neat, organized, and predictable = fixed curriculum.
2. Students are passive in lecture classes = silence means listening and learning.
3. Teacher is responsible for student learning = control of content coverage and time.

The core PPTs found in the geo-researcher's words and actions were not new to the ed-researcher. On the other hand, they were new to the geo-researcher. When asked to write her thoughts about teaching, she stated, "I could never, would never have thought to even do this." Even the idea of writing down her thoughts about teaching was new to her. Schön (1983) clarifies thoughts such as these by suggesting that we act in spontaneous, subconscious ways on a daily basis and these acts signify the presence of a knowledge base. However, when we are asked how or what we know that guides us to act in these ways we find that it is difficult to describe. Through reflective writing, the geologist's extant teaching self emerged and intuitive acts of teaching became tangible and clear.

The geo-researcher's teaching approach is ubiquitous, with a long history in university settings. Her PPTs are grounded in her experiences and a positivistic background in the sciences. The geo-researcher's PPTs focused on transmitting information, teaching effectively, and maintaining standards through control of content coverage. They are a logical set of beliefs that influence the way she selects curriculum and teaches. Freire (1971) refers to this as the "banking model," while others refer to it as the transmission model. The instructor deposits knowledge into the student where it is stored until withdrawal. Withdrawal is bringing forth the stored information upon demand. Embedded in the banking model is the view that the teacher is the primary giver of knowledge and students the passive receivers of knowledge (French, 1995). This model is objectivistic (Johnson, 1987) and suggests that students cannot create knowledge on their own and that teachers have power over them. From a critical perspective, these core PPTs represent the technical interest of critical theory whereby knowledge production is oriented toward control and management of the environment. This interest and the knowledge it produces is the focus of empirical scientific research (Carr & Kemmis, 1986; Giroux, 1980; Grundy, 1987; Habermas, 1972). The geo-researcher's core PPTs parallels this interest and its associated dominant research paradigm of the sciences.

Early in the course, the geo-researcher felt tension when implementing active learning strategies. One level of tension was the lack of experience in using a new approach, but a second level of tension appeared when actions needed to implement new strategies were not consistent with her core PPTs. This tension, according to the theory of conceptual change (though normally used when challenging content misconceptions), may set the stage for dissatisfaction when one's personal practice theories are challenged by alternate ways of thinking and acting (Posner, Strike, Hewson, & Gertzog, 1982). For the geo-researcher, there were "mighty struggles to make different decisions and carry out actions needed to implement new strategies" and escape her self-proclaimed "dungeon of ignorance."

EVOLUTIONARY BARRIERS: PERSONAL PRACTICE THEORIES—ESCAPING THE DUNGEON OF IGNORANCE

Despite the desire to teach for active learning, the geo-researcher struggled against her core PPTs to incorporate active learning strategies. The following episode takes the reader into the geology lecture classroom where the geo-researcher involves students in an active learning strategy known as expert teams. This was the first time the strategy was used in lecture.

Ed-researcher observation:

> The class is using an expert team "jigsaw" approach to learning about volcanism. Team members (5–6 students per team) are discussing material on volcanoes. One member from the team is putting relevant information on a chart on the board. The instructor is moving around in front of the room. Student teams are talking about what they have found and want to put on the board. Content information is recorded on a chart.... One student recorder at the board turns and looks to his team for support about the information to be placed in the [category] rift zone. The team communicates with him as he completes that portion of the chart.... A student from another group gets up and corrects recorded information in another portion of the chart. The instructor begins the discussion with "Something causes rocks to melt. What are your thoughts on this?...

> **Instructor:** Is anything missing from the chart?
> **Student 1:** Yes, The Mid-Atlantic Ridge should be under the rift zone, it isn't there.
> **Instructor:** (writes it in) OK, good. Now, let's consider.... What is magma like?
> **Student 1:** It flows slowly.
> **Student 2:** Low viscosity, like molasses.
> **Instructor:** What does that mean?
> **Student 2:** I think, about 1000 degrees C [low viscosity] to 1400 degrees C for [high viscosity]
> **Instructor:** OK, so temperature is related to viscosity. What else? Think about water and molasses? Which has high or low viscosity?

Ed-researcher reflective memo:

Several students responded en masse. It is hard to discern, but students seem to confuse high and low viscosity. I hear both correct and incorrect responses. From the student responses the instructor recognizes that the students think high viscosity is "fast flowing" and low viscosity is "slow flowing." She restates that high viscosity is like molasses and flows slowly while low viscosity is faster flowing.

Ed-researcher classroom observation:

Instructor: OK, so what else [is characteristic of magma]?

There was a long pause then another student states...

Student 3: Density?
Student 4: Something to do with silicon dioxide?
Instructor: [nodded] OK, look at the pie charts in the book regarding the composition of magma.

Ed-researcher reflective memos:

The instructor clarifies the difference in high and low viscosity in magma using the book. At this point the class dynamics changed (last 15 minutes of class). The geo-researcher is lecturing about various aspects of magma composition. Students are not asking questions. It was as if the return to the "text" for clarification was a semiotic cue to move back into a comfortable lecture format for both students and instructor.

Reflection after lecture:

I was worried about the *time*. I wanted to complete the table (chart) because it is big. It is full of students' thoughts and it would be virtually impossible to reproduce. So I started thinking about my time constraints. This is when I started to lecture at the students instead of engaging in discussion with them. I saw them fade. I saw it happen before my very eyes. I could see I was losing them. I need to get over the feeling of time constraint 'cause when I focus on that, the focus on student learning goes right out the window. When we are governed by *our* schedule, *our* agenda, we lose our students. [When I think like this, t]hey are nothing more to us than sponges absorbing what we have to say. I want engaged, active, participating, alive, interested students. Therefore, the rigidity of the lecture routine must go. (geo-researcher journal)

Peer consultation:

From an earlier conversation, you (geo-researcher) were worried that the expert student teams would slow you down in terms of the content that needed to be "covered." How does the idea of having the students "uncover" the content appeal to you? [Remember we discussed this earlier.] When I observed the lecture class with the expert teams, the students were actively engaged and student conversation in the teams was focused on the geology task. When I asked you after class, how the lesson went, you said that you moved from student-centered to lecture (teacher-centered) as the lesson progressed. I agree. The students were with you until the last 15 minutes of class. Given that, what might be the reason for this? Once you answer this, ask yourself, which approach produces the kinds of student interest and learning you talk about wanting. Which is more important, content coverage or student learning? Of course both are linked to time. (ed-researcher written reaction to classroom observation)

In this chronological data sequence, the use of active learning strategies challenged the geo-researcher's core PPTs. The ed-researcher's observations illustrated that student involvement in the activity was engaging and dynamic early in the class. However, after the chart discussion exposed misunderstandings about viscosity, the geo-researcher was at a "loss" as to what to do. She seemed to ask herself, "Do I address the students' confusion about viscosity in an active way or do I move on and leave them to depend on passive acceptance of the book definition?" Examination of data revealed that the core PPTs posed powerful obstacles that prevented her from addressing "student confusion." The obstacles and the associated PPTs which dominated her actions are discussed below (see previous PPT listing in the Evolutionary Stirrings section).

The obstacles to orchestrating new strategies appeared on two levels. The first level is one of skill. The geo-researcher's inability to facilitate the discussion in a direction *that she wanted* (PPT 3) prompted her to return to what she knew how to do well—lecture. On a second level, time constraints driven by the unpredictable direction of student discussions also influenced her return to the textbook (PPT 2). Third, the students moved to an active role through discussions that made completing the "agenda" difficult (PPT 2). Fourth, the unpredictability of student responses made the curriculum messy and unorganized and she was unsure of where it would lead (PPT 1). Last, dealing with incorrect responses meant that her "agenda" to cover certain material was challenged if she was to meaningfully address students' misunderstandings about viscosity (PPT 3).

It is clear even with attempts to separate the PPTs for analysis that they are connected and do not act in isolation. In this single teaching episode all core PPTs were challenged, but their strength and influence tenaciously dictated her actions to move back into her comfort zone.

EVOLUTIONARY TRANSITIONS: PERSONAL PRACTICE THEORIES

The geo-researcher's actions in the earlier passage attest to the power her core personal practice theories have over her teaching. However, as the semester wore on, questioning her ways of teaching and core PPTs became commonplace in her journal writings and discussions with the ed-researcher. For instance, her nascent recognition unfolds that knowledge constructed by learners is often different from what is transmitted during lecture.

> My prescribed, very organized way of presenting the material did not get spit back out at me like a carbon copy. The way the students understood the content was different [from what I presented]—different components had crept in there. I realized that there are other things going on in these students' heads besides what I had put in there (this is the way I thought at the time). (geo-researcher journal)

Her words evoke images of constructivism whereby individuals actively construct knowledge from experiences and interactions. Furthermore, she was aware that learners can and do construct interpretations different from those she intended. There was a note of dissatisfaction in her words. She was rethinking the way students learn while analyzing the benefits of the alternatives and new ways of viewing teaching and learning. Again she challenged herself to change.

> I think it is common for instructors to worry about covering the content when they adopt more active learning strategies. My previous syllabus for this course was laid out in a rigid, day-by-day accounting of lecture topics and exam dates. The content included detailed information about each topic. In the revised course, I moved to a more student-directed format through the use of guiding questions and group learning, including jigsaw. The immediate result was that the rigid syllabus was obsolete. The more interesting result was that I learned how rigid I was, and that I would have to struggle with this rigidity. (geo-researcher essay)

"Being provided with a new set of theoretical or conceptual lenses can be empowering ... but it also complicates one's life" (Prawat, 1992, p. 357). The geo-researcher voiced, as others before her, that covering the content is a predominant mission of lecture courses (Faust & Paulson, 1998). Changing the way one teaches is not easy, even when one is aware of and confronting one's own personal practice theories, let alone challenging the university's expectations. Using active learning practices placed greater demands on the geo-researcher. According to Cohen (1988, p. 255), those "who take this path work harder, concentrate more, and embrace larger pedagogical responsibilities than if they only assign the text chapter."

Facing and "embracing larger pedagogical responsibilities" continued to create tension, as well as new understanding as assessment and power surface in the geo-researcher's PPTs. As she refocused her efforts to teach in ways that centered on "students' efforts to understand," she struggled to assimilate current knowledge of how students learn while implementing newly acquired active learning pedagogies.

> The free flow of ideas and knowledge construction makes it impossible to "teach to the test"—which is what I used to do and which may be the source of all these time and control issues. This is the first time I have even thought this. Students have their own personal ways of constructing knowledge and if they are encouraged to make these constructions it is difficult for me to imagine how to assess that knowledge. The challenge is whether to be a sage on the stage, or guide on the side? I had a hard time letting go of that power position, sage on the stage. The theory that learners construct new knowledge upon a preexisting scaffolding of beliefs and knowledge suggests that learning must start from that scaffold. Student understanding must be the springboard for discussion and knowledge acquisition. This method requires the teacher to assume a role more like the guide on the side. The new role feels like it holds less power. (geo-researcher essay)

Ausabel's statement, "the most important single factor influencing learning is what the learner already knows" (as cited in Kyle & Shymansky, 1989, ¶ 2), comes to mind. Through readings and discussions the geo-researcher was more aware of the educational and psychological research base of how one learns and comes to know. These alternative epistemologies in teaching and learning made sense to her; however, this knowledge did not alleviate her need for control (PPT 3). She viewed lecturing as a way to control content and make assessments fair and easy to design. In addition she was unsure of how to assess in other ways. To ameliorate her dilemma, alternative assessment strategies were discussed. The geo-researcher then selected student-generated responses for exams to accommodate pedagogical changes used in the course. This action, along with her reflections and continued use of active learning strategies, suggests subtle shifts in her PPTs.

> I am pacing myself quite differently in the lecture class—not giving nearly as much detailed information, but opening up the material for more discussion. I need to get better at moderating discussion. In using more discussion I still feel a certain loss of control. (geo-researcher journal)

> Four students selected questions out of the basket at the beginning of lecture class. Student groups got together and looked up, and discussed information

on their particular question [fossil fuels]. Why I liked this [approach]: students were active, students were learning new things, students were teaching each other, students were reading the text, and students were asking very good questions. (geo-researcher journal)

Research on misconceptions through conceptual change strategies suggests that for change to occur in a learner—in this case, the geo-researcher—certain conditions must be met. First, an individual must be dissatisfied with existing prior notions (personal practice theories). This was accomplished by making the geo-researcher aware of her personal theories and opening them to examination through discussion and reflection. Second, an individual must have intelligible alternatives available that are plausible and useful in extending their understanding to new situations. This was done through exploring active learning strategies and assessments. Third, the alternatives must be more fruitful—in other words, what is gained by using the new alternative rather than the old one? (Posner et al., 1982; Smith, 1991). This last condition is the most difficult to discern. Are active strategies more fruitful to the geo-researcher? For the geo-researcher, the answer is "yes." However, it took more than reflections, discussions, and adopting active learning strategies to reach this point; it took feedback from other individuals affected by the pedagogical changes.

In undertaking practical social research, one must be aware that the focus of research must continually shift from a zoom lens that examines the details to a wide-angle lens to view the study more holistically. By doing so, it provides various entry points and perspectives representative of all the participants in the study. The tensions and changes in the course did not simply affect the geo-researcher. It affected her students. Their years of formal schooling and the internalized expectation of their roles had been challenged and as a result they strove to meet the changes needed for active learning. For instance, during a class using think-pair-share strategies, one student said, "Man, it's too early to think this hard!" To this student it was easier to sit and listen. However, this voice represented a minority perspective. On the other hand, a dominant perspective voiced by the students regarding pedagogical strategies was positive. Over half of the students wrote unsolicited journal statements similar to the following: "I think that by doing group work we have really benefitted. I feel it is very helpful to get into groups to help us to understand certain concepts better." In an interview by an outside evaluator, one student said, "The learning experiences have allowed me to find not only knowledge, but interest in a topic I normally would not care about. I liked the activity that we did today." Reading student journals had a strong influence on the geo-researcher's PPTs, she voiced. "I am just beginning to fathom how it [using active learning strategies] will work when I do let go. It feels free; it feels like a positive, true learning environment." The geo-researcher appeared to have met the three conditions necessary for conceptual change regarding teaching. In her own words, she found that active learning approaches provided her students with a "true learning environment."

PEDAGOGICAL SHIFTS THROUGH COLLABORATION: RETHINKING "EDUCATION" IN LARGE UNIVERSITY SETTINGS

The majority of this chapter focused on a geologist's endeavor to change the way she taught undergraduate geology. These changes, however, were not accomplished in isolation. In fact, peer consultation with its conversations, classroom visits, written observations and queries played a critical role in the change process as discussed below.

As part of the peer consultation model, the geo-researcher visited the ed-researcher's science methods class. During one visit, the methods students were introduced to the 5-E lesson plan format, a modified learning cycle (Bybee et al., 1989). The 5-E lesson plan focused on eliciting the students' ideas about the nature of science and scientific processes.

> You presented the "5E" model to the students on an overhead and talked about the "E's": Engage, Explore, Explain, Elaborate, and Evaluate. I like the 5E model. You engaged your students by telling them a story about NASA's need for their help. Students were hooked! You continued along the 5E path with the NOS [nature of science] phenomena and the tubes with string. I particularly liked the definite separation of observation and inference. One thing I particularly liked was that you had students go to the board. I like to do that too, and think I learn a lot by seeing how my students make a concrete drawing of their ideas. Having students then explain for their peers what observations they made, and what conclusions were drawn, and having them demonstrate what they observed was really good. "How can we improve our observations?" invited the students to think about what a "real" science perspective might be like, and to consider more detailed observations in the future. I was surprised at how many students came up with the idea of doing measurements, even though few of them did so when they could have. Throughout the lab session, you were very open to student input, very high-energy, creative, and flexible. I got ideas from this lab session on how I can link my lecture/lab with your methods course, and in fact I did start my lab today with an exploration of my students' ideas about science. I have never done that before and it was quite interesting. The information I get from observing is quite important to my own goals for the education students in my classes. (geo-researcher classroom observations)

Peer observation provided ideas for the geo-researcher. In fact, active learning strategies modeled in the science methods course were used later in the geology course. Peer consultation techniques such as visits, discussions, and written observations and reflections fostered change in both courses. For instance, after one classroom visit and the ensuing discussion regarding evolution between the researchers, the ed-researcher changed the way she addressed evolution in the methods class. For the geo-researcher, the ed-researcher's science methods class became a resource, a teaching lab. The observations provided concrete examples of how active learning strategies could unfold in a course. In particular, the geo-researcher's view that student discussion may not be valuable ("When students are in charge in the class, valuable time may be wasted and learning cannot occur if I am not in charge.... Class discussion isn't really a true learning experience") was challenged as she observed methods students interacting and constructing their understandings of investigative activities. In fact, the geo-researcher later stated, "Students' ideas are valuable and it is important that I am aware of how my students understand the information and that they help each other to understand by using their own words and descriptions." This is the complete reversal of her original PPT 2 that students who are silent are learning.

Reciprocal classroom visits also fostered trust and respect. Researchers shared their perspectives honestly in ways that prompted reflection about various episodes in teaching. Discussions were important, but the written reflections in journals and observations were most powerful as they made thoughts tangible and real. Peer consultation techniques converged to assist the geo-researcher in fostering the emergence of her PPTs and in challenging them. Peer observation of the methods class offered visual evidence of alternative ways of teaching, course connections, and new ideas. As a result of collaboration with the ed-researcher, the geo-researcher often voiced dissatisfaction with her PPTs and found the alternative approaches more intelligible, useful, and, by her own admission, "freeing."

CONCLUSIONS

Changing roles or trying to cultivate a dual role of geologist and education expert has been a struggle. It's been a tug of war in my psyche to devote my energies to education from geology/geophysics. Although I have been primarily an educator for 10 years, I find that the research institution's view of teaching as a field or scholarly endeavor is not good—and I even think I still harbor some feelings that education is a lesser field than geology. I am in the throes of metamorphosis ... and while I profess a desire to improve undergraduate education in geology, I have doubts and concerns about the validity of the work that she [ed-researcher] does in the field of education. (geo-researcher journal)

According to Harris (2001, p. 50), the university culture and its faculty face "strong disincentives to implement new techniques in their courses." These disincentives at a large university include lack of support from the departments, limited knowledge of alternative pedagogical techniques, the ability to cover the content, and the strong focus on research. In addition, there is a widespread perception that education and educational research does not hold an equal status to scientific research. This tacit, insidious barrier undermines collaborative attempts to change teaching at the university level. Moving beyond the aforementioned disincentives, "the nature of personal practice theories" dominated by the transmission model may be one of the greatest barriers to reform. What this study brings to light is that professional development strategies for university faculty suggested by Harris (2001) that include finding support, making incremental changes, focusing on student learning, and addressing student learning styles, though helpful, *might not be enough* to reform university teaching practices. In other words, change in teaching is complex, takes time, and science faculty are unlikely to embrace pedagogical change without undergoing major shifts in their own thinking about why they teach as they do. This study provides evidence that changing teaching practices is more complex than just adopting new pedagogies. Instead, through collaboration it became a process of uncovering the rationale of one's personal choice of practices and reexamining them in light of other options and current demands for higher achievement. As stated by the geo-researcher, "I can't imagine attempting these changes without a critical friend."

A recent report by the National Research Council (2001) calls for collaboration to prepare highly qualified science teachers. The report stated that "most instructors of new teachers—including postsecondary faculty in science, mathematics, technology and education—have not been able to provide the type of education that K–12 teachers need to succeed in their own classrooms" (p. 2). Furthermore, the document stated that those who teach grades K–8 do not have sufficient content knowledge to teach effectively. The solution is not having undergraduate education majors take more science content courses taught by the transmission mode. Rather, it is a call for science content specialists who teach undergraduates the "knowledge of the discipline" and science educators who prepare students on "how to teach science" to collaborate in reforming the "university culture of teaching."

IMPLICATIONS FOR REFORM IN UNDERGRADUATE SCIENCE IN A LARGE UNIVERSITY AND FUTURE RESEARCH

One implication gleaned from this study is that reforming undergraduate teaching in a large university is a multilayered process. Working with large class enrollments, overcoming the inertia of traditional lecture styles, seeking support from chairs and deans, challenging the idea of covering the content, and making time to experience alternative practices while juggling the demands of research are embedded issues that must be considered when engaging in teaching reform. Change is complex and occurs at both the level of the individual and the institution. At the individual level, reform in teaching should be more than simply adopting active learning strategies; rather "re-form" requires changing one's beliefs about the purpose of teaching. It takes time, persistence, and support. At the college or institutional level, reform requires professional development opportunities with sustained support from administration and experts who can provide teaching models and feedback, to foster personal reflection upon teaching decisions. Thus, we recommend partnerships between colleges of arts and sciences and colleges of education to foster and sustain reform endeavors. Furthermore, we recommend that university-wide efforts to support individuals who risk working across disciplines be explored and institutionalized. After all, teaching is one of the three fundamental missions of a university and seeking a balance among research, service, and teaching should be a university goal.

Further research regarding reform in large universities might address: (a) What type of professional support best fosters content specialists' reflections on teaching? (b) What are the perspectives of department chairs and deans on the *undergraduate learning experiences* offered within their college or department? (c) What are administrator's views on the level of undergraduate achievement within courses in their department or college? (d) What situations in a large university foster changes in undergraduate teaching? and (e) How does the use of active learning practices affect undergraduate achievement?

AUTHOR NOTE

This work was in part supported by NASA Opportunities for Visionary Academics (NOVA), a program funded by the National Aeronautics and Space Administration, although the views expressed here are the authors' only.

REFERENCES

Apple, M. (1993). *Official knowledge: Democratic education in a conservative age.* New York: Routledge.

Bernstein, D. (2000, June). *Peer review of teaching.* Presentation for University Teaching Scholar Lecture Series at Kansas State University, Manhattan.

Bonwell, C.C., & Eison, J.A. (1991). *Active learning: Creating excitement in the classroom* (ASHE-ERIC Higher Education Report No. 1). Washington, DC: Association for the Study of Higher Education.

Burnaford, G., Fischer, J., & Hobson, D. (2001). *Teachers doing research: The power of action through inquiry.* Mahwah, NJ: Erlbaum.

Bybee, R.W., Buchwald, C.E., Crissman, S., Heil, D.R., Kuerbis, P.J., Matsumoto, C., et al. (1989). *Science and technology education for the elementary years: Frameworks for curriculum and instruction.* Washington, DC: National Center for Improving Science Instruction.

Carlsen, W. (1999). Domains of teacher knowledge. In J. Newsome & N. Lederman (Eds.), *Examining pedagogical content knowledge* (pp. 133–144). Dortrecht, The Netherlands: Kluwer.

Carr, W., & Kemmis, S. (1983). *Becoming critical: Knowing through action research.* Geelong, Australia: Deakin University Press.

Carr, W., & Kemmis, S. (1986). *Becoming critical: Education, knowledge and action research.* Philadelphia: Falmer.

Cochran-Smith, M., & Lytle, S. (1993). *Inside outside: Teacher research and knowledge.* New York: Teachers College Press.

Cohen, D.K. (1988). Educational technology and school organization. In R.S. Nickerson & P.P. Zodhiates (Eds.), *Technology in education: Looking toward 2020* (pp. 231–264). Hillsdale, NJ: Erlbaum.

Connelly, F.M., & Clandinin, D.J. (1987). On narrative method, biography and narrative entities in the study of teaching. *Journal of Educational Thought, 21,* 130–139.

Cornett, J.W. (1990). Utilizing action research in graduate curriculum courses. *Theory Into Practice, 29,* 185–195.

Cornett, J.W., Yeotis, C., & Terwilliger, L. (1990). Teacher personal practical theories and their influence upon teacher curricular and instructional actions: A case study of a secondary science teacher. *Science Education, 74,* 517–529.

Creswell, J.W. (1998). *Qualitative inquiry and research design: Choosing among five traditions.* Thousand Oaks, CA: Sage.

Elbaz, F. (1983). *Teacher thinking: A study of practical knowledge.* New York: Nichols.

Faust, J., & Paulson, D. (1998). Active learning in the college classroom. *Journal on Excellence in College Teaching, 9*(2), 3–24.

Flake, C.L., Kuhs, T., Donnelly, A., & Ebert, C. (1995). Reinventing the role of teacher: Teacher as researcher. *Phi Delta Kappan, 76,* 405–408.

Freire, P. (1971). *Pedagogy of the oppressed* (2nd ed.). New York: Continuum.

Freire, P. (1985). *The politics of education: Culture, power and liberation.* Hadley, MA: Bergin & Garvey.

French, M.J. (1995). Exploring empowering strategies for teaching elementary science methods. *Teacher Education and Practice, 11*(1) 82–98.

Giroux, H.A. (1980). Critical theory and rationality in citizenship education. *Curriculum Inquiry, 10*(4), 329–366.

Goldston, M.J., & Shroyer, M. (2000). Teachers as researchers: Promoting effective science and mathematics teaching. *Teaching and Change, 7,* 327–342.

Grant, C., & Zeichner, K. (1984). On becoming a reflective teacher. In C. Grant (Ed), *Preparing for reflective teaching* (pp. 1–18). Boston: Allyn & Bacon.

Grossman, P.L. (1990). *The making of a teacher: Teacher knowledge and teacher education.* New York: Teachers College Press.

Grundy, S. (1987). *Curriculum: Product or praxis?* London: Falmer Press.

Habermas, J. (1972). *Knowledge and human interests.* Boston: Beacon.

Hake, R.R. (1998). Interactive engagement vs. traditional methods: A six thousand-student survey of mechanics test data for introductory physics courses. *American Journal of Physics, 66,* 63–75.

Harris, M. (2001). Strategies for implementing pedagogical changes by faculty at a research university. *Journal of Geoscience Education, 49*(1) 50–55.

Johnson, M. (1987). *The body in the mind.* Chicago: University of Chicago Press.

Kagan, D. (1990). Ways of evaluating teacher cognition: Inferences concerning the Goldilocks principle. *Review of Educational Research, 60,* 419–469.

Kincheloe, J. (1991). *Teachers as researchers: Qualitative path to empowerment.* New York: Falmer Press.

Kyle, B., & Shymansky, J. (1989, April 1). *Research matters—to the science teacher: Enhancing learning through conceptual change teaching* (No. 8902). Retrieved August 14, 2003, from the National Association for Research in Science Teaching Web site: http://www.educ.sfu.ca/narstsite/publications/research/concept.htm

Lincoln, Y.S., & Guba, E.G. (1985). *Naturalistic inquiry.* Beverly Hills, CA: Sage.

MacKinnon, A.M. (1986, April). *Detecting reflection-in-action in preservice elementary science teachers.* Paper presented at the annual meeting of the American Educational Research Association, San Francisco. (ERIC Document Reproduction Service No. ED276707)

National Research Council (1996). National Science Education Standards. Washington, DC: National Academy Press.

National Research Council. (2001). *Educating teachers of science, mathematics, and technology: New practices for the new millennium.* Washington DC: National Academy Press.

Pajares, F. (1992). Teachers' beliefs and education research: Cleaning up a messy construct. *Review of Educational Research, 62,* 307–332.

Posner, J., Strike, K., Hewson, P., & Gertzog, W. (1982). Accommodation of a scientific conception: Toward a theory of conceptual change. *Science Education, 66,* 211–227.

Prawat, R. (1992). Teachers' beliefs about teaching and learning: A constructivist perspective. *American Journal of Education, 5,* 355–389.

Rosenthal, D. (1991). A reflective approach to science methods courses for preservice teachers. *Journal of Science Teacher Education, 2*(1), 1–6.

Ross, E.W., Cornett, J.W., & McCutcheon, G. (1992). *Teacher personal theorizing: Connecting curriculum practice, theory and research.* Albany: State University of New York Press.

Schön, D.A. (1983). *The reflective practitioner: How professionals think in action.* New York: Basic Books.

Shulman, L.S. (1986). Those who understand: Knowledge growth in teaching. *Educational Researcher, 15,* 4–14.

Smith, E. (1991). A conceptual change model. In S. Glynn, R. Yeany, & B. Britton (Eds.), *The psychology of learning science* (pp. 43–63). Hillsdale, NJ: Erlbaum.

Sokoloff, D.R., & Thorton, R. (1997). Using interactive lecture demonstrations to create an active learning environment. *The Physics Teacher, 35,* 340–347.

Stake, R. (1995). *The art of case study research.* Thousand Oaks, CA: Sage.

Sweeney, A., Bula, O., & Cornett, J. (2001). The role of personal practice theories in the professional development of a beginning high school chemistry teacher. *Journal of Research in Science Teaching, 38,* 408–441.

Tobin, K.G., McRobbie, C.J., & Anderson, D. (1997). Dialectical constraints to the discursive practices of a high school physics community. *Journal of Research in Science Teaching, 34,* 491–507.

Wright, J.C., Millar, S.B., Dosciuk, S.A., Penberthy, D.L., Williams, P.H., & Wampold, B.E. (1998). A novel strategy for assessing the effects of curriculum reform on student competence. *Journal of Chemical Education, 75,* 986–992.

Yeany, R. (1991). Teacher knowledge bases: What are they? How do we affect them? Keynote address for the 1991 annual meeting of the Southeastern Association for the Education of Teachers in Science, February 8–9, 1991. In J. P. Prather (Ed.), *Effective interaction of science teachers, researchers and teacher educators* (SAETS Science Education Series, Monograph No. 1, pp. 1–7). Charlottesville, VA: Southeastern Association for the Education of Teachers of Science.

CHAPTER 15

A SMALL FOUR-YEAR COLLEGE PERSPECTIVE ON REFORM IN TEACHING UNDERGRADUATE SCIENCE

Francis Gardner, Jr.

ABSTRACT

Columbus State University is a senior unit of the University System of Georgia. While this provides stability administratively and functionally, it also reduces the autonomy in curriculum and institutional independence. The Introduction to Life in Space course was developed as one of several options for undergraduates to fulfill the University System Core Curriculum requirement in the sciences. All students must successfully complete at least one laboratory science course and one non-laboratory science course. Consequently, the course targeted the non-science major, and potential preservice teacher candidates who enter the university with a minimum science background and interest. The variety of issues that had to be considered during course development led to extensive discussion among members of the partnership team and subsequently other faculty. Some of the issues were developing a collaborative team; scope of the course content; balancing content, process skills, and pedagogy; and scheduling and room design requirements. The approval process became one of the major challenges to overcome in implementing the new course.

Reform in Undergraduate Science Teaching for the 21st Century, pages 267–284
Copyright © 2004 by Information Age Publishing

INTRODUCTION

- I definitely have a higher regard for science now. I never realized how much science we use in everyday life! Science is everywhere. I find the topic of Space to be very interesting in general and this class deepened that interest.
- I have learned a lot, and it still does not seem like enough. I guess that is confusing and intriguing at the same time?
- Yes! I have always loved math and hated science, but I now see that they go hand in hand. I feel more comfortable with science and not scared of it! I understand it more and the concepts. This has been a rewarding class.
- I have a somewhat better understanding but I'm not really interested in all of this. I think that's the key?

(Student course final evaluation comments for Introduction to Life in Space, spring semester, 2003)

FIRST STEPS: AWAKENING—THE BEGINNING OF THE CHANGE PROCESS

The student evaluative comments about the course Introduction to Life in Space are "typical" for the duration of the course at Columbus State University (CSU) in Columbus, Georgia. They reflect an awakening that has been shared by students and faculty that began with CSU faculty participation in the first workshop of NASA Opportunities for Visionary Academics (NOVA), a program of the National Aeronautics and Space Administration, conducted on the University of Alabama campus in early 1996. Five CSU representatives from the science and mathematics faculty attended this workshop to learn more about the NOVA program and its mission of creating interdisciplinary and inquiry-based undergraduate science, math, or technology courses for non-science majors (Sunal et al., 2003). Since CSU, as a member of the University System of Georgia, was in the initial planning stages for converting to a 16-week semester from the quarter system, a significant opportunity was to rethink science and mathematics instruction at CSU.

A partnership was formed between a professor of education, who serves as the chair of the Department of Curriculum and Instruction, and a professor of biology who directs the Science Education Outreach Center housed in the College of Science. Because semester conversion was going to drastically impact the entire university, the decision was made to develop a two-semester sequence of instruction that would be modeled after the courses designed by Trefil and Hazen (1995). With leadership from the biology professor a formal proposal was developed and subsequently funded by NOVA.

Fortunately, the national debate on instructional reform and related issues provided the focus for developing the new course. Lawson (1994) has proposed that understanding the nature of science and scientific reasoning processes, based on how humans learn, is much more critical to the development of a conceptual understanding of science than the typical trivial pursuit of factual memorization. Thus, the emerging course taking shape for us began to focus on the nature of science using guided inquiry while pursuing a theme centered on space exploration and environmental science. A student's understanding of the nature of scientific inquiry is enhanced by guided inquiry and open-ended approaches but they are problematic issues because many science teachers are ill prepared to use these instructional strategies (Sandoval & Morrison, 2003).

Preservice teachers were the intended audience, but because of small numbers that would not support offering the course each semester, non-science majors were added to the course. Interestingly, nearly 25% of the students who enrolled in the course have become science majors during the 4-year period the course has been offered.

COURSE DEVELOPMENT

Issues and Partial Answers

The variety of issues that had to be considered during course development led to extensive discussion among members of the partnership team and subsequently other faculty, particularly those on institutional curriculum committees. This included the scope of the semester course content, a struggle that continues because of the endless variety of pertinent information "out there." Should the course focus on a single subject or on integrated, multi-subject topics? What does "guided inquiry" really mean and what "balance" should be struck between information transfer, guided inquiry (potentially "cookbook labs"), open-ended inquiry, or a combination of these? How many credit hours should be offered and when should they be scheduled? This became a potential pitfall and critical issue since the way a class is scheduled impacts the ability to accomplish the stated goals and objectives of the course. Eventually we obtained support from deans and department chairs and the course was scheduled to meet three times per week for 2 hours each session awarding 4 semester hours of credit. This provided adequate time to set up and conduct experiments and allowed students to repeat laboratories (practice their motor skills) in the event they performed poorly the first time. We normally did not move on to new laboratory experiences until the majority of the class had acquired reasonable success with the new techniques. This allowed greater flexibility in conducting experiments in which the class could share data for comparative analysis not possible in limited small groups. For example, three groups could function as controls and three groups as treatments and the results are combined, thus allowing the use of appropriate statistical procedures for data analysis. A committed classroom was also a necessity allowing laboratories to extend over several class periods free from disruption from "outside" influences. Furthermore, the class was scheduled for midmorning to fit schedules of many non-science majors and working students. These two factors have been instrumental to the success of the course and have resulted in an intense demand for the course.

Balancing Content, Process Skills, and Pedagogy

The decision was made early on to offer an integrated course aligned with the National Science Education Standards (National Research Council [NRC], 1996) and focus on the theme of searching for life in the solar system incorporating NASA educational resources. A three-track approach was taken to develop the timeline of laboratories and related topics (see Table 15.1). The three tracks would, if at all possible, be run in nearly parallel fashion so a given class period might involve either one or a mix of two or three of the activities appropriate to a given track. In addition, the Special Projects Track was intentionally left open-ended as a place to introduce new content and activities each semester. This allows the incorporation of "state of the art" updates such as new launches by NASA, Solar System Exploration events, guest speakers, field trips, and so forth, to be correlated with the other two tracks. This flexibility requires students to take a long-term, holistic approach to the course since some early activities may not make much sense until near the end of the semester when everything begins to come together. This creates considerable anxiety in many grade-oriented students since evaluation for grading purposes does not begin in earnest until about one third to half way through the course. The approach was an intentional component of the course design in order to move the student toward a more introspective, independent, and self-reliant approach to learning. The students are expected to engage in all course activities and to self-monitor their individual progress as much as possible. It is at this time that they are given learning style inventories and other instructions and references to learning theory designed to help them become oriented to self-assessment as lifelong learners. In the end, as the evaluation analyses below demonstrate, most students, even though they struggle at first, manage to complete the course and consider the experience exceptional and very worthwhile.

**Table 15.1.   Introduction to Life in Space:
Timeline Overview and Instructional Tracks**

Track I	*NATURE OF SCIENCE and PROCESS SKILLS*

	WEEKS 1–5
Topics	Principles of Scientific Investigations • Development of Math, CBL, and Calculator Skills • Principles of Experimental Design (Candle Burning Experiment) • Descriptive Statistics (Coin Flipping & Normal Distributions) • Curve Fitting and Mathematical Modeling of Data

Track II	*SPACE and ENVIRONMENTAL SCIENCE (Content Information & Laboratories)*

	WEEKS 1–5
Topics	Structure of the Solar System • Behavior of Gases (P, V, T Relationships) • Energy Flow • Basic Properties of Water (Freezing and Melting Point, Density, Solvation) • Earth Conditions Necessary for Life • Extremophiles

Track III	*SPECIAL PROJECTS and ACTIVITIES*

	WEEKS 1–5
Topics	Introduction to NASA Enterprises • JPL Missions • Poster Presentation Group Work • Wisconsin Fast Plants (or Alternatives) with Gamma Irradiated Seeds • Field Trip to Coca-Cola Space Science Center

	ASSESSMENT PROCEDURES AND ACTIVITIES

Tasks	^ Pre-Test: Critical Thinking Skills First Student Evaluation ^ ^ Student Learning Style Inventory Online Discussion & Experiments with Light Properties ^ First Written Examination ^

Track I	*NATURE OF SCIENCE and PROCESS SKILLS*

	WEEKS 6–10
Topics	Experimental Design Exercises • Properties of Variables • Statistical Distributions and Inferential Statistics • Regression Lines/t-Tests • Research Methods in Scientific Investigations • Serial Dilutions/Toxicity Testing

Track II	*SPACE and ENVIRONMENTAL SCIENCE (Content Information & Laboratories)*

	WEEKS 6–10
Topics	Energy Flow in Ecosystems • Properties of Light • Energy Content of Foods and Fuels • Cell Respiration & Seed Germination • Habitable Zones in the Universe • Searching for Life "Out There" (Mars Exploration) • Lettuce Seed Germination Bioassay

Track III	*SPECIAL PROJECTS and ACTIVITIES*

	WEEKS 6–10
Topics	Continue Wisconsin Fast Plants (or Alternative Online Instructional Activities/Projects) • Periodic Sessions to Work on Poster Presentations or Special Projects • Research and Poster Preparations

**Table 15.1.   Introduction to Life in Space:
Timeline Overview and Instructional Tracks (Cont.)**

	ASSESSMENT PROCEDURES AND ACTIVITIES	
Tasks	^ Periodic Testing of Critical Thinking Skills	Student Evaluation ^
	^ Online Discussion & Experiments with Toxicology	
		Second Written Examination ^

Track I	NATURE OF SCIENCE and PROCESS SKILLS
	WEEKS 11–16
Topics	Completion of Class Data Analysis of Toxicity Testing • Discussion of Bioassay Applications to Space Exploration • Laboratory Examination: Use of Equipment • Preparation for Final Examination • Completion of Online Discussion Activities

Track II	SPACE and ENVIRONMENTAL SCIENCE (Content Information & Laboratories)
	WEEKS 11–16
Topics	Temperature Effects on Plants & Animals • Toxicity Studies: Bioassay of Environmentally Hazardous Materials • Microgravity Simulations • JPL Exploration of the Solar System • Stardust and Comets • Student Design of the Final Exam

Track III	SPECIAL PROJECTS and ACTIVITIES
	WEEKS 11–16
Topics	Finish Research on NASA Enterprises (or other NASA Venues/Projects) • Poster Presentations • Applications of Space Exploration • Final Laboratory Examination • Complete Final Exam Preparations • Open Labs

	ASSESSMENT PROCEDURES AND ACTIVITIES
Tasks	Laboratory Practical ^
	Post Test: Critical Thinking Skills ^
	Peer Evaluation of Poster Presentations ^
	Final Student Evaluation ^
	Final Examination ^

Curriculum Issues and Battles

Curriculum committees are responsible for insuring the integrity of course offerings sanctioned by a university. The approval process became the first major challenge to overcome in implementing our new course. Several revisions of the course proposal were scrutinized by departmental and college curriculum committees over a 2-year period. Some of the concerns for initial rejection of the proposal were sound and helped to guide the revision of the proposal to achieve a higher level of quality. The initial course proposal suffered from the same criticism often leveled at many college science courses, especially in the life sciences, that it was "a mile wide and an inch deep." The course design suffered from the same problem that student research proposals often do initially, of being too broad and unfocused. Consequently, this was a beneficial, though laborious process, forcing us to focus and refocus the purpose and content of the course several times. Once the target audience was identified as primarily non-science majors, it became easier to identify the goals and objectives of the course. The guiding principles for course development emerged from *The Myth of Scientific Literacy*, by Morris Shamos (1995).

The overarching course goal is to provide students with the scientific tools and experiences for them to understand that the true nature of science is tentative, investigative, and creative so that they recognize that "scientific" explanations must be evaluated according to the merits of the espousing person's scientific credentials and skills (i.e., their "expertise"), and the interpretation of the "facts" according to established principles and protocols. Everyone has the "right" to an opinion, but in science, opinions are only accepted if scientific thinking and testing back them. In other words, major course objectives are (a) to change student knowledge and attitudes about the nature, validity, and value of a rational and empirical scientific approach to problem solving; (b) to recognize the types of problems that science can solve; and (c) to establish some degree of self-confidence so students can recognize good science and qualified scientists when they see them. These outcomes, which became the centerpiece for the course, are eloquently summarized by Hurd (2002, p. 7):

> A standard is meaningless without reference to the student's interactions with life and living. Throughout the history of science education the sciences have been disconnected from the student's life. Learning to learn is an endeavor to introduce students to knowledge in a way that enables them to travel on their own throughout life as responsible citizens and productive workers.

Implementation

The most important conclusion that can be drawn from the science education research literature concerning what works is the power of "hands on/minds on" science teaching and learning (NRC, 2000). However, this catchy statement is an oversimplification of the complexity involved in incorporating these methods into the classroom. Students should *do science*, not just hear and read about it. It was decided that this would be the hallmark of our course.

Even though the final proposal to the curriculum committees passed muster, implementation became a confrontation with reality. We were still attempting to cover too much content and ended up refocused more narrowly on space exploration and environmental science as the content umbrella. The three-track approach allowed us flexibility in selecting content topics that are integrated across the science and mathematics disciplines. Additionally, because space exploration and environmental science are eclectic sciences and overlap extensively, they meet two of the important National Science Education Standards criteria for non-science majors: relevance and application to their daily lives. Consequently, each semester offers new opportunities for student projects that are current, pertinent, timely, and that demonstrate how math and science interact to impact students personally. The two subjects are also so broad that there is no possibility to "cover everything." Projects allow students to pursue individual topics can be pursued to a greater depth without giving in to the guilt trip of "failing to cover the field."

Space and Equipment Issues

Guided and open-ended inquiries require more committed space for fewer students, especially if class projects are scheduled for more than one period (Llewellyn, 2002). The administration was very supportive and provided a dedicated classroom in spite of the increasing competition for space in a growing university. Because our teaching approach involved laboratory activities nearly 70% of the time, equipment and supply availability became an issue. Fortunately, funding from NOVA, matching funds from the University and a private foundation, and other small public and private grants allowed the purchase of a sufficient number of graphing calculators, sensing probes, glassware and supplies, computers, and software to support a class size of 24 students (12 groups working in pairs). The acquired technology was critical for students to investigate scientific problems and to share data. It was recognized early on that we must demonstrate to students the absolute necessity for them to use applied technology for "doing good science" and help them develop the skills and confidence in using the instruments to directly address scientific questions, through experiments they design. We predicted that we would find significant changes in most students' science knowledge and skills perceptions, positive attitude changes, and personal comfort using science and mathematics tools and skills to solve problems. We did not seek to turn them into scientists but to enhance their understanding, appreciation, and respect for science and scientists. The following analysis of changes in student perceptions provides convincing evidence that we are making significant progress toward achieving these goals.

CONSEQUENCES AND OUTCOMES

Lessons Learned

Student satisfaction with a subject and classroom environment influences their attitudes and greatly impacts their success in learning (Fraser, 1994; Nolen, 2003). Consequently, an evaluation feedback approach was initiated early on. A questionnaire (How Are We Doing So Far?) was developed and administered three times during the semester at approximately 5-week intervals (Angelo, 1993; Doran, Chan, Tamir, & Lenhardt, 2002). Two sections of the questionnaire remained constant for each administration, while another portion reflected questions concerning specific activities, readings, videos, and laboratories or projects completed during the proceeding interval. The constant portion of the questionnaire was designed to monitor student comfort-discomfort with the course overall and clarity of instructions, as well as changes in students' perceptions of their own successes and difficulties with general categories of tasks and topics such as using the graphing calculators, statistical concepts, basics of experimental design, and mathematical applications.

Student Science Backgrounds

About 75% of the students enrolled in the course are non-science majors. A self-reporting questionnaire was administered the first class period in order to determine their level of science and mathematics preparation, and to develop a class profile with regard to previous math, science, and technology educational experiences. Approximately 39% of the students were freshmen, with the remaining class members divided between sophomore, junior, and senior level. Nearly a third of the students were liberal arts majors, with the remaining students evenly divided between majors in education, business, and the sciences. The ethnicity of the classes was surprisingly diverse. Nearly 21% of the students were African American, while 6% of the students were evenly divided between Hispanic and Asian students. A large number (38%) of the students enrolled in this course as juniors and seniors in spite of the fact that it is an elective course in the freshman-sophomore core curriculum. This reflected improper advising and problems of self-perception. Personal conversations with many of these students revealed that they consider themselves to have a "weak" background in science and mathematics, and an attitude that represents either a lack of confidence or pure dislike of science and math, or both. Surprisingly, this seems to be a similar situation for some science majors (41% were juniors or seniors) and non-science majors alike (36% were juniors or seniors). Ironically, the course content and especially the analytical skills taught are intended to help all majors as they enter their upper-level courses. Obviously, it is quite late in their academic careers to introduce them to skills that would help in their other upper-level courses. What is alarming is that even though a large proportion of the students had either secondary-level (75%) or college-level (13%) chemistry, they did not recall or understand the basic principles of the gas laws as evidenced on quizzes and laboratory activities. Similarly, their knowledge of basic principles of physics (i.e., laws of motion, gravity, electromagnetism, etc.) was substandard in spite of the fact that nearly 40% had secondary physics courses. Labs dealing with these phenomena initially gave them great difficulties conceptually. Generally, a "conceptual" approach was employed in most of the experiments and most applied mathematical (algebra) or statistical procedures associated with analyzing and recognizing definitive patterns in individual and class data. It is noteworthy that fewer than 5% of the students had been introduced formally to statistics or mathematical modeling. This was a real deficiency since we focused the course around data analyses using linear regression (applied, using the graphing calculator programs), descriptive statistics, and t tests. The basic philosophy employed here was to "learn to drive the car" and not worry about how the engine was built or how it operates. The emphasis was placed upon learning about the nature of science and how decisions about experimental design and conclusions were done using these analytical tools in order to evaluate the significance of experimental results. As indicated in Tables 15.2–15.4, significant improvements in student comfort and perceived understanding of the nature and processes of science occurred by the end of the semester. This is particularly important since 70% of the enrollment in this course was female (Kahle & Meece, 1994).

Are We Succeeding?

We designed a study to gather evidence that the approach being employed was favorably accepted by the students. Overall, these methods were having positive effects on student knowledge and perceptions about the nature of science and improving their "comfort," interpreted as *their* perceived success, with understanding and appreciating (valuing) science in ways that would be potentially relevant and beneficial to them in their future. An underlying assumption was that if the students were able to

Table 15.2.   Student Backgrounds in Mathematics and Science

Topic or Skill	Average Rating[a]	SD	n	Percent with High School Course	Percent with College Course
General Computer Skills	3.99	0.23	155		
Graphing Calculators	2.47	0.16	155		
Calculator Based Laboratory (CBL)	0.39	0.14	155		
Sensing Probes with CBLs	0.32	0.16	155		
Designing Experiments	1.16	0.35	155		
Internet Use	3.98	0.25	155		
PowerPoint Presentations	3.22	0.49	155		
International Space Station	1.31	0.23	155		
Search for Life In Space	1.36	0.24	155		
Physics				39.5	9.3
Chemistry				74.1	12.5
Astronomy				7.0	3.3
Biology				72.0	17.5
Earth Science				52.0	15.4
Ecology				12.2	5.9
Algebra				58.1	30.2
Geometry				83.7	8.5
Trigonometry				43.3	11.4
Statistics				4.5	2.6
Mathematical Modeling				4.5	4.2

Note: [a] Students ranked their skill or knowledge level on a scale from 0–5; 0 = *none*; 1 = *poor*; 2 = *fair*; 3= *average*; 4 = *above average*; 5 = *excellent*.

design, conduct, analyze, and develop valid conclusions based upon their own work, then they would gain an understanding and appreciation for science. This new understanding and appreciation would enhance the value placed on scientific processes because they would discover how to do science effectively. This, in turn, would be reflected in changes in their attitudes toward and comfort with the skills and tools required to do science and should be reflected in changes in the constant component of the evaluations described earlier.

The entire course was designed as a coordinated effort to systematically develop student abilities to design an experiment, conduct it, analyze the data, draw conclusions, and report the results. As an aside, this approach is completely different from that currently used in most introductory college science textbooks. Most textbooks give short shrift to the nature of science, presenting the scientific method in a few paragraphs as a completely linear process and quickly move into the content, seldom integrating "how science knows" into "what science knows." The textbooks are excellent references and encyclopedias but do not provide the necessary pedagogical methods to develop lifelong learners. Developing student ability "to do science" requires an entire semester of attention just as a beginning for most students. One component of this semester-long focus is a final examination process in which the student, under the instructor's tutelage, is guided through the inquiry process about a question that arises out of the student's interest during the course of the semester. Experiments and laboratories early on are "cookbooks" in nature to lay the groundwork of necessary skill development. Gradually, they are required to rely less and less on the detailed instructions from the instructor and the laboratory instructional materials, and to apply the techniques of inquiry.

During the final 3 weeks of the semester, the students engage in increasingly independent activities in which they bring all these skills to bear on a single question that is different for each student. The final examination requires them to complete an experiment they design by modifying previous ones conducted in class or, being even more creative, to generate entirely new questions while using one or more of the techniques they have learned, and to submit a packet of information they have accumulated as supporting evidence. In order to complete the final examination, the laboratory is open for supervised access to work out their design and modify their research, with interactive feedback from the instructor as required, during the final 3 weeks of the semester. Usually, valuable lessons are learned as they focus their question away from global projects, modify apparatus, build devices as required by discovery during preliminary trial and error sessions, and generally engage in the research processes characteristic of true hands-on/minds-on science.

Changes in Student Perceptions as Reflected in Periodic Student Evaluations

The student evaluations focused on two major issues: (a) overall student understanding of course goals and objectives, combined with their satisfaction in the methods being used; and (b) their perceived comfort with the science, mathematics, and technology being employed. These evaluations were administered at approximately 5-week intervals. Table 15.3 summarizes the changing distributions of student scores during this semester. Although these data represent one semester of the previous eight semesters, it is "typically representative" of previous student perceptions. A single

Table 15.3. Changes in Student Interest, Understanding, and Satisfaction With This Method of Instruction During the Semester

Response*	1	2	3	4	5	4+5	3+4+5
Q1: *On the scale below, rate your understanding of the overall goals and objectives of this course*							
1st Evaluation; Q1	0.0	27.3	50.0	18.2	9.1	27.3	77.3
2nd Evaluation; Q1	5.0	20.0	45.0	25.0	5.0	30.0	75.0
Final Evaluation; Q1	0.0	13.0	43.5	30.4	13.0	43.4	86.9
Q2: *Overall, how interesting did you find the course to this point?*							
1st Evaluation; Q2	0.0	9.1	40.9	31.8	18.2	50.0	90.9
2nd Evaluation; Q2	0.0	5.0	45.0	40.0	5.0	45.0	90.0
Final Evaluation; Q2	0.0	0.0	34.8	47.8	17.4	65.2	100.0
Q3: *Overall, how do you consider the approach that the instructor is using to teach this course?*							
1st Evaluation; Q3	0.0	4.5	40.9	45.4	9.1	54.5	95.4
2nd Evaluation; Q3	0.0	10.0	35.0	40.0	10.0	50.0	85.0
Final Evaluation; Q3	0.0	4.3	34.8	47.8	13.0	60.8	95.6

* Values are the percent responding
Q1: 1 = *Totally unclear;* 2 = *Somewhat unclear;* 3 = *Mostly Clear;* 4 = *Very Clear;* 5 = *Extremely Clear*
Q2: 1 = *Totally boring;* 2 = *Somewhat boring;* 3 = *Somewhat interesting;* 4 = *Very interesting;* 5 = *Extremely interesting*
Q3: 1 = *Useless;* 2 = *Not very useless;* 3 = *Somewhat useful;* 4 = *Very useful;* 5 = *Extremely useful*

semester is reported here because it is the first semester in which efforts have been made to systematically investigate possible relationships between student perceptions, performance, and learning styles.

More than 75% of the students had a clear (3 or above) understanding of what was expected of them during the entire semester. The fact that this increased to 87% in the final evaluation is a reflection of learning cycle approach (guided inquiry) that constitutes the primary method of instruction. It also indicates that because of the nature of the course where the three tracks begin to coalesce near the end of the semester as the students begin to converge their efforts on the final research project, the pieces of the puzzle begin to fall into place for them. The subject matter is inherently interesting and as indicated in the responses to Question 2, it would be quite difficult to reduce interest in space exploration topics. However, this is somewhat tempered by the fact that the math, science skills, and technology use create some real difficulties during this transition. The semester is truly a journey in which many different skills and topics have to be separately dealt with and then brought together by the end of the semester. It is also clear from their responses that the students consider the approach being used at least useful and 61% of them consider it very useful. It should be noted also that in eight semesters, the total number of students who drop the course is less than 3% and these have occurred in the first week of class. One must be careful though not to concentrate for too long on the science process skills learned without considering the content (space exploration and environmental science), which is a very interesting topic for study by the students. They begin to become discouraged and even grumpy if continual reference to and application of the process skill is not made. The relevance of the skills being learned must always be made apparent.

Furthermore, even though the math and science backgrounds of these students did not appear to be *considerably* weak, a large number of them were initially not comfortable with the technology, math, or science (50%, 50%, 64% respectively), and struggled throughout the semester. In spite of these challenges, though, their perceptions and attitudes toward the course goals and objectives remained high as evidenced in Table 15.3. Note in Table 15.4 that a significant number of students entered the course being familiar with the graphing calculators but with little experience

Table 15.4. Changes in Student Comfort with Technology, Science, and Mathematics Skills During the Semester

Response*	1	2	3	4	5	4+5	3+4+5	1+2
Q5a: *Rate your personal assessment of your comfort with using the Graphing Calculator.*								
1st Evaluation; Q5a	22.7	27.3	31.8	9.1	9.1	18.2	50.0	50.0
2nd Evaluation; Q5a	25.0	40.0	35.0	5.0	0.0	5.0	40.0	65.0
Final Evaluation; Q5a	26.1	26.1	30.4	4.3	13.0	17.3	47.7	52.2
Q5b: *Rate the degree of difficulty you are having with the math methods used.*								
1st Evaluation; Q5b	22.7	27.3	27.3	9.1	13.6	22.7	50.0	50.0
2nd Evaluation; Q5b	10.0	25.0	40.0	20.0	5.0	25.0	65.0	35.0
Final Evaluation; Q5b	17.4	17.4	34.8	26.1	4.3	30.4	65.2	34.8
Q5c: *Rate the degree of difficulty you are having with the statistical concepts used.*								
1st Evaluation; Q5c	9.1	22.7	40.9	13.6	9.1	22.7	63.6	31.8
2nd Evaluation; Q5c	10.0	15.0	60.0	10.0	5.0	15.0	75.0	25.0
Final Evaluation; Q5c	13.0	13.0	39.1	26.1	8.7	34.8	73.9	26.0
Q5d: *Rate the degree of difficulty you are having with the experimental design concepts used.*								
1st Evaluation; Q5d	0.0	36.4	27.3	27.3	9.1	36.4	63.7	36.4
2nd Evaluation; Q5d	0.0	25.0	45.0	25.0	5.0	30.0	75.0	25.0
Final Evaluation; Q5d	13.0	21.7	30.4	34.0	0.0	34.0	64.4	34.7

* Values are the percent responding
Q5a–5d: 1 = *Very easy;* 2 = *Easy;* 3 = *Moderately difficult;* 4 = *Difficult;* 5 = *Very difficult*

using sensing probes, applications of math and statistics to data analysis, and experimental design. The major mathematical procedure they were taught during the semester was the use of regression analysis (with the calculators and a computer program, Graphical Analysis) in which they were expected to interpret the results. Initially, their greatest difficulty was with linear algebra and line equations. Eventually, more than 90% of them mastered these skills at the minimum criterion level as exemplified by their performance on the final examination.

The application of statistics continued as a problem throughout the semester with more than 65% of the students still struggling even at the end of the semester. In addition to using regression analysis as a tool to determine correlations, they were also exposed to the use of descriptive statistics and two-sample t tests. All of these are programmed into the TI-83+ calculators and instruction was offered to teach them how to use these techniques to make decisions about scientific and statistical hypotheses. They were encouraged to "learn to drive the car and not worry about how the engine worked." This also implied that they would learn some of the rules for proper application and limits of these methods. This was also intended to become a "hook." By showing them how these tools are used we hoped to stir their desire to learn more and take a statistics course later in their academic career. Many majors now require such coursework. Once again the strategy reflects a learning cycle philosophy. Associated with their struggles in properly applying the math, technology, and statistics were their deficiencies in experimental design. The semester began with a candle-burning experiment that contained within it elements of the laboratory for the entire semester. It represented the nucleus around which experimental design and instruction about the nature of scientific investigation is built. It is noteworthy that on the final examination the pattern of difficulty with experimental design is completely shifted to the left in Table 15.4 and that more than 65% of the students rated their comfort with this skill as moderately difficult to very easy. This reflects gained self-confidence since during the interim between the first and final evaluations they were introduced to a number of difficult concepts and methods. The last evaluation indicates a similar comfort level with even more difficult concepts.

CONCLUSIONS

One of the goals of the course was to positively change student attitudes about science and mathematics, particularly those who typically do not like the subject, by enhancing their conceptual knowledge and inquiry skills. Examination of Table 15.3, Question 2 addresses this goal. Almost from the beginning of the course the majority (91%) of the students found the course at least "somewhat interesting." This wavered somewhat during the second 5-week period when the statistical and experimental design subjects were introduced (50% of those who found the course "very interesting" or "extremely interesting" on the first evaluation declined to 45% on the second evaluation) but significantly increased on the final evaluation (65% selected these choices on the final evaluation). Similar changes were seen in Question 3 in Table 15.3 indicating that the majority of the students found the method of conducting the course to be very useful (55% selected "very useful" or "extremely useful" on the first evaluation which dropped to 50% on the second and returned to 61% on the final evaluation). Overall, 96% of the students considered the approach used in this course at least "somewhat useful." Therefore, we conclude that the course goals were partially achieved. Question 1 in Table 15.3 suggests that we still need to improve how the goals and objectives are communicated to the students, especially early on. Also, we need to improve their conceptualization of the "big picture," especially with regard to space exploration and environmental science.

A second set of objectives was to improve student skills in science, mathematics, technology applications, and experimental design. The grading procedure for the course was based on "criterion referencing" that required any student receiving a letter grade in the course of "C" or above to demonstrate "minimum competency" with the skills outlined in Table 15.4. To date, we have been successful with more than 90% of the students completing these minimal requirements. However, the majority of the students do not find these tasks to be easy or very easy (about 35% on the final evaluation). It is noteworthy that even though, in general, students found the difficulty of these tasks increasing as the semester progressed, they never lost interest or declined in their approval of the methods of conducting the course and in fact, overall, their approval actually increased.

It is also clear from evaluating the final examinations that although many students had become adequate at applying the skills and methods, meeting minimum criteria to pass, their collective creativity and deeper understanding had just begun. How can we improve upon that? How can we improve upon the coordination and timing within the three tracks? One problem that all students have had is seeing the "big picture" in spite of multiple efforts to help them do so. Having the instructor do this for them only raises them to the rote memorization stage and not to the reasoning stages. Have critical thinking skills been improved? How can we modify our approaches to do so? Current studies are under way to find some of the answers.

RECOMMENDATIONS

This experience has convinced us that making the changes described above have been critical for both the students and the instructors. There is not a more meaningful description of the course that we set out to create than that of Arnold Arons (1978) in his Chautauqua Course nearly 20 years ago:

> Available laboratory is one in which the student must make some decisions of his own, profit from his mistakes, and retrace his steps if necessary. It must be one that allows the student to carry over his activities from day to day and does not compel him to "finish something within the confines of a straight jacket."

Finally, this journey has clearly been worthwhile. The instructors have gained new insights into what works for the majority of students. We have also become more closely tuned into what students struggle with. No matter how much they struggle with some of the concepts, the majority value tasks for which they can see a purpose, even when they have to struggle for a while before the light comes on. But, the light must come on at least some of the time and that is still the difficult part. For example, we often see linkages between concepts that they do not, indicating that the novice-expert gap is still not being bridged. So our search continues!

This search would not have been initiated without the support of NASA and this institution. The investment in the technology is also critical. Students must have time to repeatedly use equipment and redesign experiments in order to learn from their mistakes. It is also essential that the "competitive" atmosphere among students be removed as much as possible. Often they learn more from each other. Assembling class data is absolutely a must. It allows the use of comparative statistics and modeling that an individual or single group cannot do. All of this takes time! Most students initially complain about the group projects but overall rate them as beneficial. Consequently, class time has been given up for them to conduct their projects. This has been most fruitful and resulted in some excellent poster presentation sessions.

Typical of any investigation, more questions have been raised than answered. But one thing is clear: Teaching science for this instructor has regained its excitement and joy and I once again look forward with great anticipation to the next opportunity each semester brings to meet new students, new topics, and new challenges of learning for my students and myself. And I relish the chance to encourage my students to look to their future with the challenge. "Ad astra!"

REFERENCES

Angelo, T.A. (1993). *Classroom assessment techniques: A handbook for college teachers* (2nd ed.). San Francisco: Jossey-Bass.

Arons, A.B. (1978). Teaching science. In S.M. Cahn (Ed.), *Scholars who teach: The art of college teaching.* Chicago: Nelson-Hall.

Doran, R., Chan, F., Tamir, P., & Lenhardt, C. (2002). *Science educator's guide to laboratory assessment.* Washington, DC: NSTA Press.

Fraser, B.J. (1994). Research on classroom and school climate. In D. Gable (Ed), *Handbook of research on science teaching and learning* (pp. 493–541). New York: Macmillan.

Hurd, P.D. (2002). Modernizing science education. *Journal of Research in Science Teaching, 39*, 7.

Kahle, J.B., & Meece, J. (1994). Research on gender issues in the classroom. In D.L. Gabel (Ed.), *Handbook of research on science teaching and learning* (pp. 542–576). New York: Macmillian.

Lawson, A.E. (1994) Research on the acquisition of science knowledge: Epistemological foundations of cognitions. In D.L. Gabel (Ed.), *Handbook of research on science teaching and learning* (pp. 131–170). New York: Macmillian.

Llwewllyn, D. (2002). *Inquire within.* Thousand Oaks, CA: Corwin Press.

National Research Council. (1996). *National Science Education Standards.* Washington DC: National Academy Press.

National Research Council. (2000). *Inquiry and the National Science Education Standards.* Washington, DC: National Academy Press.

Nolen, S.B. (2003). Learning environment, motivation, and achievement in high school science. *Journal of Research in Science Teaching, 40*, 348.

Sandoval, W., & Morrison, K. (2003). High school students' ideas about theories and theory change after a biological inquiry unit. *Journal of Research in Science Teaching, 40*, 369–392.

Shamos, M. (1995). The *myth of scientific literacy.* New Brunswick, NJ: Rutgers University Press.

Sunal, D., Staples, K., Odell, M., Freeman, M., Whitaker, K., Edwards, L., et al. (2003). *NOVA 1996–2003 evaluation report.* Tuscaloosa, AL: NOVA. Retrieved July 2003, from http://nova.ed.uidaho.edu/library/ViewLibrary.asp?Section=2

Trefil, J., & Hazen, R.M. (1995). *The sciences: An integrated approach.* New York: Wiley.

CHAPTER 16

A TWO-YEAR COLLEGE PERSPECTIVE ON LINKING POLICY AND RESEARCH TO SUPPORT SCIENCE EDUCATION PROFESSIONAL PRACTICE

Wayne Morgan

ABSTRACT

Community colleges are a uniquely American institution. Nearly half of all undergraduates in the United States attend a community college. Tremendous pressure is put upon these institutions from accreditation agencies, state legislatures, four-year colleges and universities, K–12 districts, and the communities in which they are located. Two-year college faculty often find themselves facing difficult instructional decisions that balance the demand for rigor that comes frequently from four-year institutions and legislatures, with the demand for improved student learning that comes from accreditation agencies and the community. Fundamental to the mission of all two-year colleges is a commitment to helping students. Two-year college science fac-

Reform in Undergraduate Science Teaching for the 21st Century, pages 285–299
Copyright © 2004 by Information Age Publishing

ulty are among the most innovative in all of higher education. They continually strive to find instructional strategies that facilitate student learning, implement the types of inquiry advocated in the National Science Education Standards, and model the types of teaching techniques that will help preservice science teachers.

INTRODUCTION

Nearly half of the undergraduates attending college in the United States are enrolled at a two-year college. The two-year college represents the point of entry into higher education for minorities, women and nontraditional students. The diverse educational environment provides a significant challenge for science instructors. Balancing the need for rigor with the needs of the learner, science instructors at the two-year college must contend with myriad outside forces as they plan and execute courses in the life and physical sciences. Their innovative nature and commitment to learning enable the two-year college science faculty to implement the kind of pedagogy championed in the National Science Education Standards (National Research Council, 1996). This also means that preservice teachers who begin their science training at the two-year colleges will be likely to encounter instructors who will model the learning environments necessary to improve science education at all levels. It is necessary to delve into the background and challenges facing two-year colleges. This will set up the context for how two-year colleges respond in ways that promote active learning and model science as inquiry.

BACKGROUND ON TWO-YEAR COLLEGES

The two-year public college is a truly American institution. Starting in 1901 with a junior college devoted strictly to providing the first two years of course work that could eventually lead to a baccalaureate degree, the modern community college has evolved into a comprehensive institution where the diversity of courses runs the gambit from origami to organic chemistry. The nearly 1,200 two-year colleges in the United States provide educational opportunities for 44% of all undergraduates, 45% of all first-time college freshmen, 46% of black undergraduates, 55% of Hispanic undergraduates, 46% of Asian/Pacific Islander undergraduates, and 55% of Native American undergraduates (American Association of Community Colleges, 2003). Nearly 5.5 million students are currently enrolled at a two-year college. The number of students serviced jumps to 10 million if you include students enrolled in noncredit courses as well.

The mission of the two-year college is generally focused on four key areas: (a) providing transfer courses that contribute to both general education and courses within a major, (b) providing vocational and technical training leading to a terminal degree or certificate, (c) providing specialized education and training for business and industry, and (d) providing educational opportunities tailored to the needs and wants of the local community. Across the country local boards of trustees most frequently govern two-year colleges with oversight by some state agency. Balancing the diverse aims of the college with responsiveness to the local community would be daunting enough if the colleges were not also subject to state and federal policies, accrediting agency criteria, and sometimes tempestuous relationships with area K–12 school systems.

CHALLENGES FOR THE SCIENCE FACULTY

Two-year college science faculty find themselves in the position of balancing what they know about good instruction with the desire to make their courses the equivalent of their four-year college and university counterparts. Not only is it an issue of students transferring credits, it is a matter of professional integrity to try to maintain intellectual rigor in these introductory science courses. However, it is also increasingly important for two-year college faculty to provide instructional experiences that are truly meaningful for students and lead to real learning gains.

Currently the competition for students is intense among all types of postsecondary institutions. Each institution that makes contact with prospective students extols the quality of its instruction. Incoming students may equate quality instruction with the interactive types of experiences they have already had in high school or middle school science courses. When they arrive at the college campus, it is often an extreme disappointment to find that the science instructor spends the majority of the class time in lecture punctuated with some very limited discussion. This potential disconnect between the expectations for quality (meaningful) instruction and the actual experience in the college science classroom may be one of the major contributors to a serious retention problem (Poindexter, 2003).

Two-year college science faculty often find themselves at cross purposes when interacting with K–12 school systems in their area. Frequently school districts exert strong pressure on the two-year college to offer courses for college credit. This is especially true in many rural areas where it is not possible for advanced placement science courses to be offered in the high schools. Financially, this dual credit option can be very seductive for the two-year college. However, many science instructors oppose the awarding of college credit for courses offered in the high school even when the same textbook and syllabus is used. It is unfortunate that developing college-credit science courses offered at area high schools may be too effective in creating similar course environments. The inquiry-based approaches seen in science courses offered in the first two years of high school are abandoned in favor of a more "college-like" academic approach. Statewide science assessment is often done prior to students taking these college-credit courses so high schools generally do not see the less interactive, less inquiry-based pedagogy to be problematic. These dual-credit science offerings, along with advanced placement courses may actually stifle the development of process skills emphasized in the National Science Education Standards.

Although many two-year college science faculty are not trained as educators it is quite common for two-year colleges to devote a number of days each semester to professional development activities that address this gap in preparation for classroom teaching. Frequently, these professional development activities involve presentations that address pedagogy. Thus, the two-year college science instructor is exposed in a generic fashion to educational initiatives like collaborative learning, classroom assessment techniques, multiple intelligences, learning styles, and writing across the curriculum. Frequently, these initiatives appear to be the educational flavor of the week and have little impact on actual classroom instruction unless the administration makes implementation a priority.

Accrediting agencies have become more intrusive in the area of instruction. The Higher Learning Commission's criterion 3, student learning and effective teaching, is an example of the attention that is currently being paid to student learning in higher education. Criterion 3 states, "The organization provides evidence of student learning and teaching effectiveness that demonstrates it is fulfilling its educational mission" (Higher Learning Commission, 2003). No longer is it sufficient to merely document that a faculty is qualified to teach, an institution must be able to provide evidence that learning environments, based upon sound research into learning, are being created and that the result is an improvement in student learning.

State legislatures and higher education governing bodies are increasingly concerned about the quality of education provided by all institutions. The two-year college is often at the center of accountability debates in many states since state funding is often a significant portion of their budgets (Rifkin, 1998). Performance-based funding is one way that states have attempted to address the quality issue. Additionally, some states also tightly coordinate the general education offerings at two-year colleges so that there is a seamless transition when transferring to the four-year schools.

The two-year college finds itself as the access point to higher education for many students who traditionally have been underrepresented or have not attended college. Although diversity is to be celebrated it often makes the science classroom a challenging venue for learning. Not only can there be large differences in age there can be significant differences in preparation and aptitude. Ever increasing numbers of two-year college students are enrolled in remedial courses in mathematics and English (Jenkins & Borwell, 2002). It is not uncommon to find that a single science course may have students who are enrolled in remedial courses seated near a student who is enrolled in honors courses.

Among the diverse learners are preservice teachers. Ever increasing numbers of students are electing to enroll in science course work at two-year colleges. This means that many K–12 science teachers will receive their introductory science training outside of the college or university which will eventually credential them. It is incumbent upon the two-year college to facilitate preservice science teachers' learning of both declarative and procedural knowledge in science. More importantly it is essential that these preservice teachers experience science teaching that more closely models the pedagogy they will need to implement if they wish to help their students achieve the learning goals set forth in the National Science Education Standards. It is increasingly likely that preservice teachers will encounter interactive, inquiry-based pedagogy in their introductory science courses at the two-year college. Certainly, the converging influences of performance-based funding and accreditation are impacting classroom instruction in ways that align well with the changing emphases that will promote inquiry.

WAYS TO MEET THE SCIENCE INSTRUCTION CHALLENGES AT THE STATEWIDE LEVEL

What can be done to accommodate the complex situation facing science instructors at the two-year college? Three examples from a community college in central Kansas will serve to illustrate the types of solutions and collaborations that two-year colleges across the country have found to be successful in working with a wide variety of students.

The equivalence of course work undertaken at a two-year college and that undertaken at a four-year institution has been an issue of debate for many years in many states. In order to address the equivalence issue the Kansas Board of Regents, the governing body for all higher education in the state, convened faculty representatives from two- and four-year institutions across the state to develop a common set of course outcomes for many transfer courses. Included in the current list of courses for which common outcomes have been developed are several science courses: General Chemistry I and II, General Biology, and Physical Science (Kansas Board of Regents, 2003). This approach is designed to provide for seamless transfer of credit from the state's 19 community colleges to its 7 four-year colleges and universities. Once fully implemented the content of all courses for which the common outcomes have been established should be similar and the issue of equivalence will, in theory, be settled.

The Core Outcomes initiative in Kansas has the potential to impact science education in a far more substantive way than merely insuring the transfer of credit between institutions. The discussions between faculty members from diverse institutions that have occurred over the space of two academic years have been an invaluable experience. Arriving at consensus for the common science course outcomes has not been a trivial exercise. Serious, sometimes contentious, debate about what are essential concepts, skills, theories, and ideas within a course have been a part of the process. Biology and physics faculties have no current statewide meetings that enable them to interact and exchange ideas, so the Board of Regents-sponsored meeting provided a venue for very valuable interchange. Chemistry faculty in Kansas have convened annually for many years and have been able to use their annual meeting for further discussions initiated in the Core Outcomes project. Perhaps the most significant impact is that the development of a common set of outcomes now gives the two-year college faculty the freedom to provide innovative, learner-centered educational experiences in the classroom without fear that their courses will not be recognized. As long as a faculty member can demonstrate that students collectively have achieved these outcomes, there is no longer the need to merely emulate the style of teaching traditionally delivered at a typical four-year institution.

The development of statewide outcomes also meshes well with the mandates of accrediting bodies, such as the Higher Learning Commission. "An organization needs to be accountable to itself and to its constituencies, to be clear about what it intends students to know and to do, and to find ways of learning whether, as a result of the education provided, students actually know and can do (Higher Learning Commission, 2003). The development of a set of outcomes provides a foundation upon which to demonstrate student learning in the science courses. The Core Outcomes project, originally driven by legislative fiat, has enormous implications as performance-based funding goes into effect in the state. Legislators and regents alike are keen to determine if students are learning as a result of their educational experiences in our institutions of higher education. There can be no better beginning than establishing learning outcomes that can be assessed and that can lead to the improvement of learning.

WAYS TO MEET THE CHALLENGES IN THE CLASSROOM

Traditional pedagogy employed in many of the college science classrooms across this country has been devoted to training students as if they were technicians. A technician learns a great many facts about his or her field but has a minimum of context in which to place those facts. Protocols and algorithms predominate as the technician acquires a catalog of skills and techniques for future use. Science students have typically been required to buy expensive, encyclopedic textbooks and sit through lecture upon lecture where enormous quantities of information are presented to them. The textbook may include special applications' sections, but by and large the majority of textbooks go through chapter after chapter of scientific information with a minimum of context. It is deemed sufficient to merely state to the student occasionally that they will need to know this material for the next course in the sequence or in their major. The laboratory experience traditionally has been one of following directions and trying to interpret the language of the teaching assistant in charge. Students use a well-worn, perhaps anachronistic, set of techniques in lab and time-tested algorithms in lecture to "solve" problems that often have no relationship to their major or "the real world."

The scientific thinker, or emerging scientist, receives facts in a context that has relevance and meaning. Skills and techniques are developed as needed in response to problems in the laboratory or lecture. Scientific thinkers learn how to access factual information when it is needed. They also learn how to learn new concepts on their own. It is true that all of this happens for some students within the context of the traditional science course. Unfortunately, it is typically left to chance. Often it is akin to teaching someone to swim by throwing them in the pool. Some students will figure out a crude, but effective, set of strokes that allow them to survive the experience. A rare few even figure out from first principles or by watching the example of their "instructor" demonstrating the classic swimming strokes. However, there are some who fail to develop any effective strokes and drown. Most rarely develop any skills more sophisticated than treading water. In a similar manner, we throw students into science courses. Students are not coached in the ways of scientific thinking but are expected to figure it out by watching the exemplary practice of their instructor or by some other way.

Two-year college science instructors are among the most adaptive and innovative faculty in higher education when it comes to creating effective learning environments. An example from our course, Ecology/Environmental Problem Solving, may serve as an illustration.

AN EXAMPLE FROM ECOLOGY

Learning to recognize the complexity of issues concerning the environment and then applying scientific reasoning is often quite difficult for students in their first years of postsecondary study. In an effort to improve the problem-solving abilities of ecology students a major revision of the course was undertaken. The didactic portion of the course was significantly reduced and a number of problem-solving exercises were developed. It was decided early on that the best way to teach the problem-solving approaches was to place them in a realistic ecological context. One of the best examples of the problem-solving techniques, one in which students repeatedly acknowledge a change in how they view ecological problems, is the inductive pyramid.

The students begin the ecology exercise by reading a published account describing the discussions and conclusions from a global conference on the environment (Easterbrook, 1992; *Newsweek*, 1992). In this article a list of countries is identified as environmentally "good" while another list is identified as environmentally "bad." The terms "good" and "bad" in the context of this exercise come directly from the article. The task assigned to groups of students is to collect information about all of the countries and to use the inductive pyramid strategy to induce a set of principles or a theory that embodies what it means be environmentally "good" and also what it means to be environmentally "bad." Once each group has developed their two pyramids the results are presented to the entire class. After each group has made their presentation a whole-class discussion follows so that divergence in the inferences made can be reconciled and inductive reasoning errors can be identified.

The inductive pyramid (Clarke, 1989) is a technique designed to help students derive general principles or theories from a collection of scientific data or other information. The data or information is written along the bottom of a large piece of butcher paper. The data is examined for relationships and the relationships are described in writing in a tier above the data. Further relationships are sought out and the process is repeated until the most salient relationship among the members of the data set is found. The overarching principle or theory is written in the top tier of this pyramid.

During this week-long activity students must collaborate in order to research the countries, to determine what relevant information is to be included in the first tier of the pyramid, and to perform the inductive steps that lead them to formulation of a principle or theory about each group of countries. Typically the instructor will invite faculty members from the science department to come in and interact with the students as they begin to develop their pyramids. The interaction with the ecology instructor and other faculty members allows many errors in inductive reasoning to be addressed early on in the process. It is common for a faculty member and/or someone with environmental expertise from the community to be invited to the group presentations. This guided practice using the problem-solving technique in the context of the science knowledge being emphasized is a very effective approach for students to develop this skill (Taconis, Ferguson-Hessler, & Broekemp, 2001).

One of the assessment strategies incorporated into the course is the inclusion of questions on each exam concerning the students' perceptions and attitudes about the activities and topics within that unit. Several sample student comments will illustrate the typical kinds of response students have to the exercises. "I thought the projects [problem-solving tasks] were the best teaching tool, because the mind had to be exercised to accomplish them. One of the prime purposes, to me, of college is to teach people how to go past simplistic thinking and consider other ideas than their own." "The way you help us to discover things for ourselves rather than just give us a lot of data and no experience. Experience is the best teacher." It is typical for students to respond that the inductive pyramid has given them new insight into the complexity of environmental issues. Students often respond that they thought of issues in terms of absolutes until they completed the exercise and now they realized that issues are not necessarily cut and dried.

The ultimate measure of problem solving in the ecology course is the construction of a concept map (Mason, 1992) that illustrates the relationships between organisms throughout a food web when some environmental stress has been applied. When the webs constructed by students in the ecology course (Figure 16.1) have been compared to those in the second semester of the major's biology course (Figure 16.2), striking differences

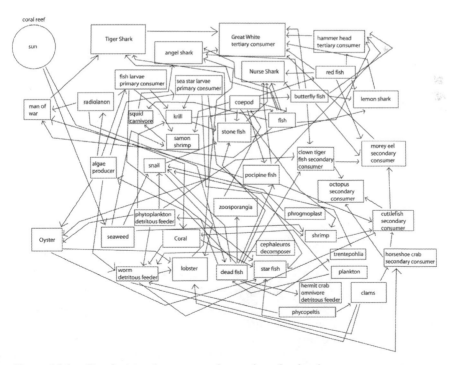

Figure 16.1.   Rendering of representative ecology food web.

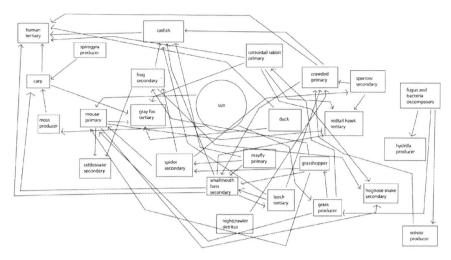

Figure 16.2.   Rendering of representative biology II food web.

are noted. Even though students in the second semester course have been exposed to a greater depth of information about how the organisms are connected or linked than the ecology students, the ecology students are generally much more sophisticated in identifying the relationships between organisms and how they would be affected by the environmental stress. The ecology class historically has been populated with a larger percentage of students with less science preparation in high school and few, if any, college science courses. Although not subjected to a rigorous scientific investigation, our conclusion has been that the explicit teaching of problem-solving methodology in context has produced significant learning in the ecology students as demonstrated by their consistently superior concept maps.[1] See Figures 16.1 and 16.2.

AN EXAMPLE FROM GENERAL CHEMISTRY

Problem solving in chemistry is often complicated by the necessity to convert from one representational system to another. Information about a chemical reaction may be presented verbally in a written narrative. In order to solve the problem, the student will be required to convert the verbal representation to some type of symbolic representation such as a chemical formula or structure. The ability to successfully solve chemistry problems is directly related to a student's facility in moving from one type of representation to another (Bodner & Domin, 1995). Changes have been made in both the lecture and laboratory at our institution in order to help students in General Chemistry become more adept at interconverting between types of representations and also to help understand them more deeply. Topics are typically introduced in lecture through a demonstration or hands-on activity that allows the student to encounter the chemistry principles at a macroscopic level first. Further development of the concepts requires the movement from the macroscopic to the molecular. In making that transition it is necessary to represent reactions and other chemical phenomenon with words, formulae, and structures. A conscious effort is made to illustrate the equivalent features of each type of representation while highlighting the unique information afforded through each as well.

In addition, examples of chemical phenomena are taken from contexts familiar, meaningful and/or engaging to the students. Consider an investigation taken from the general chemistry textbook *Chemistry* developed as a project of the American Chemical Society (American Chemical Society, 2003). Students are asked to observe what happens when samples of three different solutions are simultaneously applied to a display board. One sample is water, another methanol, and the third is acetone. The first thing the students must do is measure the time it takes for each of the liquids to evaporate from the display board. The students are then asked to explain at a molecular level what happened to each of the three liquids. This calls the students' attention to the molecular interactions that must be overcome in order for a liquid to evaporate. The discussion does not end there. The students are then required to put forward an explanation for why the acetone evaporates first and the water evaporates last. At this time the students are directed to construct models of the molecules and write out their structural formulas. The more robust explanations of the order of evaporation include an appropriate description of the types of intermolecular forces present in these liquids and how those intermolecular forces affect the molecules behavior. Starting with a simple phenomenon that is within the typical experience of the chemistry students, the activity culminates with the application of declarative knowledge about intermolecular forces and molecular structure to explain the behavior of the three liquids.

When the students are guided through problem-solving techniques that are presented within a pertinent science context there is a much greater opportunity for acquisition and transfer of those problem-solving skills (Taconis et al., 2001).

Changes have also been made in the laboratory to help students develop facility with chemical representation. Using small-scale laboratory techniques students are asked to make and record observations about chemical reactions. In several laboratory exercises students are asked to represent the macroscopic reactions symbolically. Further analysis of the reactions leads the students to other forms of chemical representations. The laboratory exercises are designed to help learners connect the symbolic language of chemistry with the observable phenomenon.

Consider an activity taken from a laboratory exercise to introduce students to acids and bases (Pfister, 2002; Thompson, 1990). Although the students may have conducted a titration in high school and certainly have been told about the neutralization of acids and bases, this simple activity serves to make the students' thought processes about the reactions more accessible to the instructor. In each of 10 wells on a 96-well plate the students will transfer 5 drops of 0.10 M hydrochloric acid. One drop of bromthymol blue indicator is then added to each of the wells. The student then adds one drop of 0.10 M sodium hydroxide to the first well, two drops to the second well, and so on until sodium hydroxide has been added to all the wells. The students are asked to record their observations about each well. They are also asked to write the appropriate equation for the reaction occurring in the wells. At this point the students are asked to identify the stoichiometric relationship between the hydrochloric acid and sodium hydroxide in this reaction and to explain how their experimental results support, or refute, that stoichiometry. The students' written responses render their ability to relate the observable phenomenon and the molecular-level reaction to the symbolic representation of the reaction. Students may be able to write and balance the equation but often fail to realize how the reaction of equal drops of acid and base of equal concentration provide experiment support for writing the reaction. Repeatedly students are asked to provide a molecular-level explanation and explain how it relates to the symbolic representation of the chemical phenomenon observed in the activities.

In these exercises the students are required to supply written descriptions and explanations of what they are observing. This approach also promotes a deeper understanding of what may be happening in a chemical reaction.[2] Rather than being a set of symbols to manipulate in some algorithmic fashion the reaction becomes more real for the student.

CONCLUSIONS

Two-year college science faculties face a serious challenge in instructional design and implementation. A growing body of research suggests that a pedagogy based predominantly on lecture and discussion may not be the most effective way to help science students learn and understand science. Many two-year college science instructors are in the vanguard of adopting innovative strategies for improving student learning. Scanning the programs from meetings of organizations like the Two-Year College Chemistry Consortium, an affiliate of the American Chemical Society, quickly gives evidence of the types of innovation occurring in two-year colleges across the country. Contextualizing the learning by using topics from food chemistry to forensic chemistry are just some of the innovations one can find among two-year college instructors. Perhaps it is easier for two-year college faculty to alter their learning environments since there are generally fewer faculty in a department to coordinate. The fact that some of the instructors teaching transfer courses may also be involved in vocational courses may also increase their flexibility. Vocational instructors must constantly adapt to the changing demands of the workplace. This often helps influence approaches taken in transfer courses at two-year colleges.

The heterogeneity of the two-year college student population makes it necessary for science faculty to be sensitive to student learning issues. The differences in age, experience, and ability found among the students enrolled in any given science course give the instructor only two choices. Focus on teaching and ignore the diversity of learners or focus on learning and do whatever is in your power to help students learn. In large part two-year science faculty choose the later approach.

Looming over science education at the two-year college is the issue of transfer of credit. Articulation of course work with four-year institutions often has a stifling effect on student learning in the two-year college. Until very recently emulating the four-year faculty meant extensive reliance on lecture and an undue focus on teaching rather than learning. The role of introductory science courses as gatekeepers for the major or for professional school admission has historically been one to produce a chilling effect. In the current accreditation environment, the emphasis has moved significantly away from teaching alone. Accrediting bodies expect demonstrations of student learning. All of higher education has to rethink the way they construct the learning environment. Two-year science faculty members are positioned to take advantage of the new priority given to learning. This is the validation many of those innovative instructors need for what they already had recognized as a serious issue. Two-year college science instructors typically know much about their students' abilities and progress. The outside request to prove that learning has occurred finally gives the two-year college instructors a chance to showcase what they do best.

The accountability issue is a significant one as more and more states move to performance-based funding for higher education (Rifkin, 1998). State governing bodies are increasingly interested in a seamless transition between two and four-year institutions as a measure of performance. Strategies range from simply using common course names and numbers to establishing a common set of core outcomes. It is clear that legislatures and governing bodies are becoming increasingly interested in what is happening in introductory and general education courses across all institutions of higher learning. Removing the pressure of trying to prove course equivalence will allow two-year college science faculty to focus on helping students learn.

Quite often it can be heard that the difference between two-year and four-year college faculty is whether their mission is teaching or research. Traditionally this dichotomy may have been accurate. The dichotomy in higher education today may be between a mission of learning and a mission of teaching regardless of the institution type. The challenging mix of students has made many in the two-year college come to grips with the need to focus on student learning. The situation is particularly acute since many of those science students are nontraditional and are unwilling to sit passively and accept a classroom environment where they feel they are not learning. The impetus for placing student learning in the center of our instructional design now comes from the converging interests of accrediting bodies, governing entities, and local stakeholders who often contribute significant tax dollars to two-year institutions.

The most significant problem on the horizon for science education at the two-year college may be developing research-based instructional design that leads to improved student learning. Two-year college instructors have had a long tradition of instructional innovation, however, tying those innovations to a research base does not always occur. Dissemination of science education research findings must be improved. Currently innovations in the science classroom are often the result of generic professional development opportunities that are given to the entire two-year college faculty. These educational strategies are frequently quite good, yet their efficacy in the science classroom may not be the same as in a different discipline. Opportunities must be given for science instructors to conveniently access sound research-based science content pedagogical strategies.

Projections indicate that soon a majority of undergraduates will be enrolled at a two-year college. The two-year college will continue to be the access point to higher education for minorities and at-risk students. With the culture of innovation and a growing focus on student learning, the science education of our students may be better in the future than ever before. Perhaps soon students will no longer lament that they "took" biology, chemistry, or physics, but they will say with confidence that they "learned" their science at a two-year college.

NOTES

1.   Joyce Selsor, the ecology instructor, has noted a qualitative difference in the webs over the past 10 years. Accuracy and depth of the relationships is typically superior with the ecology students webs.

2.   Interviews of beta testers using *Chemistry* at a meeting in Plano, Texas, repeatedly revealed a perception on the part of instructors that the critical thinking of their students had improved.

REFERENCES

American Association of Community Colleges. (2003). About community colleges: Fast facts. Retrieved August 8, 2003, from www.aacc.nche.edu/Content/NavigationMenu/AboutCommunityColleges/Fast_Facts1/Fast_Facts.htm

American Chemical Society. (2003). *Chemistry, gamma version.* New York: W. H. Freeman.

Bodner, G., & Domin, D. (1995). The role of representations in problem solving in chemistry. In D. Lavoie (Ed.), *Toward a cognitive-science perspective for scientific problem solving* (National Association for Research in Science Teaching Monograph No. 6). Manhattan, KS: Ag Press.

Clarke, J. (1989). Inductive towers: Letting students see how they think. *Journal of Reading, 33*(2), 86–95.

Easterbrook, G. (1992, June 1). A house of cards. *Newsweek, 119*(22), 24.

Higher Learning Commission. (2003). Restructured expectations: A transitional workbook. Chicago: North Central Association of Colleges and Schools.

Jenkins, D., & Borwell, K. (2002). State policies on community college remedial education: Findings from a national survey. Denver, CO: Education Commission of the States.

Kansas Board of Regents. (2003). Kansas Core Outcomes Project: Creating strong pathways to student success. Topeka, KS: Joint Council of Academic Officers.

Mason, C. (1992). Concept mapping: A tool to develop reflective science instruction. *Science Education, 76*(1), 51–63.

National Research Council (1996). *National Science Education Standards.* Washington, DC: National Academy Press.

Newsweek: Special Report (1992, June 1). Earth at the Summit. *Newsweek,* 119 (22), pp. 21–23, 29–33.

Pfister, R. (2003). *Chemistry I laboratory manual* (Spring 2003 ed.). Hutchinson, KS: Hutchinson Community College.

Poindexter, S. (2003). The case for holistic learning. *Change, 35*(1), 25–30.

Rifkin, T. (1998). *Improving articulation policy to increase transfer.* Denver, CO: Education Commission of the States.

Taconis, R., Ferguson-Hessler, M.G.M., & Broekemp, H. (2001). Teaching science problem solving: An overview of experimental work. *Journal of Research in Science Teaching, 38,* 441–468.

Thompson, S. (1990). *Chemtrek: Small-scale experiments for general chemistry.* Englewood Cliffs, NJ: Prentice-Hall.

CHAPTER 17

A SCIENCE EDUCATION RESEARCH ORGANIZATION'S PERSPECTIVE ON REFORM IN TEACHING UNDERGRADUATE SCIENCE

Cheryl L. Mason and Steven W. Gilbert

ABSTRACT

Over the past several decades, performance-based assessment in higher education has become increasingly common at the state and national level. Reliance on course credits alone to meet graduation requirements, a common practice today, is likely to erode in the coming decades. Driving this change are a number of factors, including prominent reform reports and professional accreditation standards that increasingly require institutions of higher learning to be accountable for student learning, rather than teaching. At present, most of these standards focus on broad indicators such as program or institutional pass rates, placement rates, employer reports, standardized test scores, and professional accreditation reviews. However, it is likely that regulators will call for more specific accountability measures in the future, that is, increased emphasis on the performance of students themselves, such

Reform in Undergraduate Science Teaching for the 21st Century, pages 301–316

as those required of teacher candidates by the National Science Teachers Association. To facilitate these efforts, more research is needed in relation to the development of individual and program-level assessments in the sciences at the postsecondary level, developing involvement and support of faculty, and reform of the undergraduate curriculum to include elements related to science literacy. In addition, studies should be made concerning the roles of critical thinking and metacognition, including the use of emerging technologies, as they impact undergraduate science education reform efforts.

INTRODUCTION

The National Association for Research in Science Teaching (NARST), the world's largest organization dedicated to research on science learning and teaching, is committed to improving science education through research at all levels of education. While much of the work that has been done pertains to the precollege level, it is clear that current efforts to reform education in the United States is not limited to the precollege level; rather such efforts can be expected to bring about changes in undergraduate science education as well. Many of the coming changes will focus attention on the need for more research into undergraduate science education, to assess the efficacy of many of the innovations that are being reported, and also to determine ways to systematically improve the delivery of science at that level (Colbeck, 2000). This is especially true of introductory courses, which often have a critical influence on student attitudes toward science as a career option and even on whether to take additional coursework in science. However, calls for accountability may also include upper-level courses and major programs.

The need for reform at all levels of education was brought to the public's attention through many reports in the 1980s, including *Time for Results* (National Governors' Association, 1986) and *A Nation at Risk* (National Commission on Excellence in Education, 1983). *A Nation at Risk* stated that average graduates of K–12 schools and institutions of higher education, at the time of the report, were not as well educated as graduates 25 to 35 years before, when fewer persons graduated, and concluded that "more and more young people emerge from high school ready neither for college nor for work" (p. 5).

As a consequence, the report called for new and higher standards for education and a commitment of schools and colleges to help students meet those standards, with the ultimate goal of creating a scientifically literate society. Intrinsic to this vision is the assumption that education should not focus only upon career goals, but should also add to the general quality of life through the extension of education beyond formal schooling into society. In order to achieve this goal, the report proposed the development of a continuum of learning committed to rigorous academic standards.

Since then, however, both the Secretary's Commission on Achieving Necessary Skills (SCANS) report (National Technical Information Service, 1991) and *Before It's Too Late* (U.S. Department of Education, 2002) have suggested that changes in science, technology, engineering and mathematics (STEM) education have been largely ignored over time. One of the reasons cited for this lack of change is the lack of scientifically based research to support both the allegations and suggested solutions (Shavelson & Towne, 2002).

PERFORMANCE ASSESSMENT

While change has not occurred rapidly in institutions, regional accrediting agencies have responded relatively quickly to the pressures generated by the numerous reports recommending reform. For example, *Goals for Education: Challenge 2000 of the Southern Regional Education Board* calls for regular assessments of the performances of undergraduate students, professors, and the higher education system overall (Creech, 2000). Creech, as well as Brakke and Brown (2002), insist that such assessments will be increasingly necessary in order to meet public demands for accountability, which in turn will be more essential in order to obtain funding and other kinds of public and governmental support.

Both the Southern Association of Colleges and Schools (Commission on Colleges, 2004) and the North Central Association require institutions to identify college-level competencies within the general educational core and provide evidence that graduates have attained those competencies. The Western Association of Schools and Colleges also requires that institutions verify the achievement of their stated program learning objectives (Accrediting Commission for Senior Colleges and Universities, 2001).

Despite these changes, performance assessment at the university levels tends to be specific to each institution. Creech (2000) has found that most states have not established minimum standards on the higher education indicators, and that there are few common measures across colleges and universities. He suggests that standards should be considered at least for the introductory courses where attitudes toward study in STEM subjects are developed. At the present time, the adequacy of college and university performance is largely determined from data such as program accreditation rates, results of internal and external reviews, employer assessments, student and alumni assessments, student evaluations of faculty, alumni assessments of preparation, performances on licensure and certification examinations, and percentages of graduates entering professional and graduate schools.

Within institutions, assessments of student learning and performance usually consist of general education grades or test scores, pass rates on certification and licensure exams, entrance test scores for graduate or professional schools, job placement rates, alumni assessment and student assessments of faculty instruction. Most of these assessments are rather limited in the range of abilities actually evaluated.

Despite all of this attention from accrediting bodies and state agencies, and the concurrent pressure for change in the quality of science education, many undergraduate institutions, including four-year colleges and comprehensive institutions, continue to focus primarily upon research dealing with scientific phenomena with little regard for the delivery of science content and the subsequent level of learning that content (Bransford, Brown, & Cocking, 1998; Daves, 2002; Mason, 1999). As a result, Goodstein echoes the opinion of many science educators when he states that our nation has some of the finest scientists in the world, but we also have some of the worst science education in the world (Daves, 2002). Whether or not this is demonstrably true, it is increasingly evident that this is the perception between many educators and policymakers alike.

UNDERGRADUATE SCIENCE EDUCATION REFORM

Reformist goals for science education have emerged from the work of the National Research Council (NRC). A major goal of science education, according to the NRC (Committee on Undergraduate Science Education, 1999) is that "institutions of higher education should provide diverse opportunities for all undergraduates to study science, mathematics, engineering and technology as practiced by scientists and engineers, and as early in their careers as possible" (p. 1). The NRC's vision for undergraduate education in STEM subjects is encompassed by the following recommendations. The emphases are the authors'.

1. Entry into the undergraduate program involves assessment of understanding of STEM subjects based on the recommendations of national K–12 science standards.
2. STEM courses, along with connections to society and human condition, are an integral part of the curriculum for all undergraduate students.
3. Colleges and universities *continually* and *systematically* evaluate the efficacy of STEM courses.

4. STEM faculties assume greater responsibility for the preservice and in-service education of K–12 teachers.
5. Postsecondary institutions promote innovative and *effective* undergraduate STEM teaching and learning.
6. Postsecondary institutions encourage graduate and postdoctoral students, along with faculty, to become skilled teachers for STEM courses as they acquire knowledge concerning effective teaching and learning.

Three words stand out in these vision statements and are essential for any assessment strategy: continuous, systematic, and effective. In order to create a database and environment for change, continuous and systematic assessment and evaluation strategies must be put into place and be regularly evaluated based on their ability to discriminate between effective and ineffective practices (NRC, 2003, p. viii).

Many science faculty members understand and accept the challenges faced in postsecondary science education and have made sincere efforts to change the current system. However, the lack of supportive research data often impedes efforts to implement systematic change. As a result, although demonstrative changes have been made in some undergraduate science programs, the reform efforts in other programs are best characterized as discontinuous, nonsystematic, and unable to effectively discriminate between effective and ineffective practices. This has a major effect on those entertaining the idea of becoming precollege teachers, since they tend to present science in the manner by which they were taught.

Laws and Hastings (2002) identified trends in the reformation of STEM teaching from more than 500 reports in 20 years. Changes include more emphasis on active, collaborative learning including experimentation and theory building; extended and problem-based research projects; more focus on depth rather the breadth of scientific knowledge; more emphasis on tenets of science and the relationship of science and mathematics to social issues; greater use of emerging technologies; and an increasing emphasis on the development of communication skills. These trends show that concern for effective teaching exists within the science community. Student learning, however, is the ultimate goal, rather than effective teaching per se.

Undergraduate Student Learning

Many indicators of teaching success are subjective and external, such as student evaluations of professors. Indicators of actual student learning may be regarded with suspicion, since the instructor has so little control over the background, quality, and habits of his or her students. Even so, more attention is being paid to undergraduate student learning than in the past, and research on teaching and learning in the universities is being given increasing consideration in both doctoral and faculty studies.

To support these efforts, more funding by agencies such as the National Science Foundation (NSF) and the National Institutes of Health (NIH) is dedicated to the improvement of undergraduate STEM education (Rothman & Narum, 1999). This concern is also reflected in the recent formulation of an NSF program that will fund Science of Learning Centers. The goal of these Centers is to integrate various research findings along a continuum from brain-based learning to classroom practice. These actions and activities will doubtlessly contribute to the improvement of undergraduate postsecondary education. Just as important, however, is the move to create coherent systems within university programs that define the graduates of that program based upon local, state, or national standards. Individual faculty efforts will have positive but limited effects if important advances are not institutionalized.

The impetus to develop better program-level performance assessment systems emerges from specialized accreditation requirements, but could certainly characterize university science programs. Standards for nursing education (Commission on Collegiate Nursing Education, 2002), engineering (Engineering Accreditation Commission, 2002) and science teacher education (National Science Teachers Association, 1998) are three of a large number of professions requiring that their students demonstrate specific competencies, rather than relying upon broad indicators of achievement such as grades and grade point averages.

Research Issues

Reform efforts have given rise to the need for more research on effective teaching, learning, and assessment at the postsecondary levels. Of particular interest is research on the quality of the assessments that are currently being used, the review of alternative ways of assessing students, and the development of new modes of assessment that are aligned with the practical needs of graduates. In particular, we might be concerned with the development of program-level assessment systems. The view that science education at the undergraduate level is primarily a matter of transmitting scientifically derived knowledge has given way to more of an interest in developing students as individuals who are literate, articulate, and capable of critical thinking and problem solving in the field.

Such a system is already required by certain accrediting organizations, including, in science teaching, the National Science Teachers Association.

The NSTA standards for science teacher preparation (NSTA, 1998) assume a relationship between the quality and nature of undergraduate preparation of K–12 teachers in science, and the quality and nature of their subsequent teaching. These standards are broadly based upon the ideal of science literacy. The goal of science literacy is a major recurring theme in all of the standards created since the publication of *Science for All Americans* (American Association for the Advancement of Science [AAAS], 1989) and the subsequent *Benchmarks for Science Literacy* (AAAS, 1993) including the *National Science Education Standards* (NRC, 1996). *Science for All Americans* (*SFAA*) bases its recommendations on a number of factors, including the utility of the knowledge and skills of science, the need to develop social responsibility, considerations of the intrinsic value of the knowledge, its philosophical value, and the need for childhood enrichment.

SFAA suggests that a science curriculum, in addition to fulfilling its traditional goals of developing scientific knowledge and laboratory skills, must also engage students in developing scientific habits of mind. Students should understand the values inherent in science, mathematics, and technology, and this should lead to the reinforcement of general societal values including curiosity, openness to new ideas, and skepticism. The social value of and attitudes toward science, mathematics, and technology should also be studied.

Among other things, the *NSTA Standards for Science Teacher Preparation* (NSTA, 1998) used in national accreditation of teacher education programs, reflect the concerns of *Benchmarks* (AAAS, 1993) and the *National Science Education Standards* (NRC, 1996) by requiring teacher candidates to demonstrate (a) a broad knowledge of science and of unifying concepts across science, (b) an ability to design and conduct inquiry-based experiments and apply mathematics to those studies, (c) a working knowledge of the nature of science, (d) an understanding of science's relationship to important social issues, and (e) how science teaching and learning can be contextualized in the local community.

In order for teachers at all levels to acquire the ability to address these issues and teach for science literacy as well as technical competence, they themselves must be scientifically literate. It follows, then, that there is a need for a broader conceptualization of the purposes of undergraduate education than has existed in the past. Lest it is thought that only teacher education is of concern here, consider the standards for engineering. Criteria for accrediting engineering programs now require that programs demonstrate that graduates can apply knowledge, design and conduct experiments, design a system to meet desired needs, function on multidisciplinary teams, identify and solve engineering problems, take professional and ethical responsibility, communicate effectively, understand engineering in a global and societal context, know contemporary issues, and create the ability to engage in lifelong learning (Engineering Accreditation Commission [EAC], 2002).

These specialized requirements are important for two main reasons. First, undergraduate STEM departments must help students develop the content background necessary for a number of professional areas that are increasingly asking for valid and reliable performance assessments. They must provide the required knowledge and skills, as well as validate student learning. These undergraduate programs must also address a professional audience beyond the classroom. In contrast to the days when only the course instructor needed to be satisfied with the assignments and assessments, it now becomes necessary to satisfy an audience of external reviewers, including policymakers.

Second, these new professional performance requirements set a precedent for educational practice that may well, and can be expected to, encompass general education and disciplinary majors programs, as well as the subject-specific professional programs. In addition to considering what should be included in the science curriculum, it is essential that faculty and graduate students also ask what is the best way to teach the subject, assess the results and evaluate the performances.

CATEGORICAL AREAS FOR RESEARCH

The remainder of this chapter will be devoted to a review of the categorical areas for research that are useful and relevant in the current reform effort. Performance assessment is a key issue in all of these areas, specifically as it is related to demonstrations of student and program achievement.

Content

Content traditionally refers to the major concepts, principles, theories, laws, and interrelationships of the field of study (NSTA, 1998). The development of conceptual knowledge has traditionally been the primary focus of most postsecondary science programs. While other concerns are making inroads, the development of content knowledge probably will always be a major concern for education at this level. However, studies have shown that science majors often have a poorly developed understanding of content when it is taught in a traditional lecture/laboratory manner (DiPasquale, Mason, & Kolkhorst, 2003; Mason, 1992, 1999).

Currently, most assessments are tests developed by instructors, sometimes with laboratory or other reports, and evaluation is typically expressed in terms of grades and grade point averages. New standards call for more emphasis on authentic and meaningful performances that rely on applications rather than rote recall.

Research is needed to determine what concepts are most important and how to articulate important concepts in a hierarchical manner. Content is frequently taught and assessed as if all concepts are equal. This is not the case, of course. Also important is to determine the best and most effective ways to assess content knowledge and, in particular, the ability of students to articulate ideas into meaningful principles and conceptual models.

Expression is also important: How can knowledge of different types be best expressed? Traditionally, most assessments are propositional—expressed in words and sentences or mathematically. In addition, assessments seek to elicit a right answer to the questions posed. Alternative forms of assessment, through imagery or pictorial representations, for example, may provide interesting insights into student knowledge that limited-response propositional items miss. Concept mapping and similar techniques might also be worth being broadly incorporated into an assessment system. Authentic projects and case studies also may be better measures of meaningful learning than traditional paper-and-pencil tests.

On a program level, it is also important that science units know what actual learning is taking place in courses under its control. Measures often are approximations at best. Few professors are held accountable for what they actually teach in their courses, and outcome measures are frequently not reviewed by anyone beyond the course professor, who awards a grade based on his or her own (unreviewed) criteria. Reform will require transparency in these processes and a significant research agenda for development with regard to content knowledge.

Design of Inquiry and Research

The National Research Council (NRC, 1996) envisions undergraduates who learn how to conduct scientific research, rather than just validating preexisting findings. Both engineering standards (EAC, 2002) and science teacher preparation standards (NSTA, 1998) require that candidates demonstrate some degree of proficiency and experience in research. The challenge is how to accomplish the goal of imparting knowledge of scientific research within a four-year program with limited supervisory resources. This goal is difficult to address, given the demands on the faculty members who will supervise such individual research projects.

College and university faculty members are increasingly including opportunities for students to design and conduct scientific research and experiments within their courses, but the overall impact on students of such efforts is not clear. In some areas, such as elementary science education and nursing education, the quantity of students allows for courses that are tailored to meet specific professional needs to be taught. Although instructors allude to the fact that these specialized courses are more successful than traditional courses in meeting the needs of these students, more rigorous research should be conducted.

Many science majors, and some nonmajors, will ultimately be employed in occupations requiring at least some research skills, or understanding of research. K–12 teachers are certainly called upon by the *National Science Education Standards* (NRC, 1996) and most state standards to teach at least some science content through inquiry. The degree of inquiry in K–12 science education progresses from exploration and the development of basic observation and inference skills in the lower grades to the full design of experiments and research in high schools.

As a consequence, undergraduate science preparation must provide science teacher candidates with knowledge and skills consistent with these goals. Observations during 15 years of NSTA reviews of science programs for the National Council for the Accreditation of Teacher Education have shown that, for the most part, this has often been relegated to greatly overworked science methods courses rather than in the content curriculum, and that most science programs provided undergraduates with few opportunities for true research. This situation appears to be changing as states have adopted national standards requiring undergraduate research experiences, but many questions need to be addressed regarding the nature of these experiences and their effects on program graduates.

History and Nature of Science

This is probably the area of inquiry that is least systematically developed in undergraduate STEM courses (Abd-el-Khalick & Lederman, 2000). Yet it is certainly reasonable to expect that a science major can distinguish and discuss the difference between scientific and nonscientific knowledge, and understand and communicate the historical basis for contemporary scientific knowledge and practices. The dimensions of this area of study might be conceptualized that science majors should (a) understand the historical and cultural development of science and the evolution of knowledge in their discipline; (b) understand the philosophical tenets, assumptions, goals, and values that distinguish science from technology and from other ways of knowing the world; and (c) communicate such knowledge to others and engage them in the critical analysis of false or doubtful assertions made in the name of science (NSTA, 2003).

Clearly such knowledge is essential in order for us to achieve the basic goal of science literacy. Given that even practicing scientists may have limited knowledge in this area, two broad research questions suggest themselves: How do we educate the professoriate? And how do we best infuse this additional study into higher education programs that are already packed with content? Research will be important to demonstrate both how to achieve the infusion of the nature of science into the college curriculum, and how to assess student learning so as to ensure that both majors and nonmajors alike understand and can communicate what science is really all about.

Issues and Applications

Skills related to the issues and applications of science are especially important to undergraduate students. Research has demonstrated that the study of issues and applications increases the relevance and appeal of science for all students, but in particular for members of groups underrepresented in the sciences. The study of applications of science in technological contexts such as engineering, medicine, public health, and biotechnology can help undergraduates decide on career goals and provide them with the broad perspectives desired in an undergraduate curriculum.

It is also important, however, that students of science are able to dissect and analyze science-related issues in a systematic way. Such skills are also important to the establishment of science literacy. Curriculum elements might be cross-curricular, with the social sciences, for example, and include case studies and techniques for analysis and decision making, including the analysis of risks, costs, and benefits of alternative solutions (NSTA, 2003).

Community

A number of accrediting bodies and state legislators are interested in developing more community involvement opportunities related to undergraduate education. Certainly the relationship of science to the community is of importance to both science and non-science majors. One of the major criticisms of science by those who study such things is the appearance that the sciences are seemingly aloof from community needs and values.

Undergraduates should know how science relates to the community, who important stakeholders are and what community resources are available to learn and promote science. In addition, science students should be able to communicate to the community and engage in projects that address issues of importance to the community in a scientific way (NSTA, 2003). More work on such "service learning" projects and their effects on students would be very beneficial.

The ability to communicate to the community is an important skill for scientists and for those in science-related fields. Research on how these skills can be systematically developed and demonstrated is important.

Data Processing Skills

Undergraduates in many programs may not have much of an opportunity to apply mathematics and data-processing skills, and may not have the ability upon completion of their programs to critically analyze such basic communications as journal articles in *Science* or *Nature*. In fact, science majors may complete some programs without familiarity with common journals and actual research products in their fields. The most common reason for undergraduates to avoid these journals may be their lack of skill in statistical analysis. Research on the integration of mathematics with science is needed to ascertain whether or not graduates can comprehend scientific arguments based on mathematics and statistics.

Integration

The integration of disciplines of the natural sciences (intradisciplinary) and of the natural sciences with other sciences or fields (interdisciplinary) is of increasing importance. Universities are establishing programs and departments dedicated to cross-disciplinary studies and research. Questions related to how the sciences can best be interrelated and of what value this can be to the undergraduate student need to be explored. All science majors could benefit from courses specifically designed to integrate knowledge across the sciences. Such courses could also provide the framework for understanding the unifying concepts identified earlier in this paper. Preservice teachers, especially elementary teachers, might well benefit from new configurations of subject matter that de-emphasize the individual disciplines and instead focus on patterns that cut across traditional boundaries. Considerable research has been reported in this area, but integration on a broad scale is more the exception than the rule in most programs.

Technology

A review of conference papers such as those collected and edited by Chambers (2000) shows that the technology of science instruction and learning is a field ripe for considerable research. Questions abound about the efficacy of computer-mediated instruction, the delivery of programs on the Internet (distance education), and the costs versus the benefits of various modes of instruction. The use of emerging technologies for the assessment and evaluation of students and programs also remains an open field for research.

Communication and Information

Skills in information processing and communications are areas of concern in contemporary science education, and questions of how to develop them in the context of science programs remain to be answered.

K–12 Science Education Reform

More and more students who have experienced the full impact of reforms in K–12 science education are moving into higher education. It is important to investigate whether the higher education community is building upon our knowledge of teaching and learning in order to provide them with an articulated, coherent educational experience in science, or instead is entrenched in traditional approaches to the teaching of STEM courses. This is an important question because it is clear from the research that the negative attitudes that teachers, especially general elementary teachers, may exhibit toward teaching science and mathematics are often due to their experiences in traditionally taught undergraduate STEM courses. Unfortunately, these experiences have been shown to impact the willingness and ability of elementary teachers and teachers who are non-science majors to teach science. On the other hand, there is evidence that integrated courses addressing science holistically and emphasizing inquiry and exploration tend to appeal to teacher candidates, resulting in more positive attitudes and more in-depth understanding of important concepts. Additional research needs to be conducted in this area and, in particular, the research should focus on introductory courses that impact the eventual learning and teaching of these teacher candidates. The extent to which colleges and universities as a whole have modified their programs over the past several years is basically unknown. As a result, there is a pressing need for more information on science programs concerning changes they are making, assessments they are developing, and data that they are gathering on student performance outcomes.

SUMMARY AND IMPLICATIONS FOR THE FUTURE

As professionals, university scientists and science educators ought to be asking many probing questions with regard to the design and delivery of undergraduate education. Some of these questions should involve the nature and design of the curriculum itself, which dates back to the 19th century. Is it necessary to stay with a system based on courses carrying credit hours? In this age of technology, are there ways to base academic credit on student performances tied to standards, rather than on the completion of courses with grades that are at best uncertain indicators of learning? Can learning tasks be structured for individuals or teams rather than structured within formal classes? What would such a system look like?

Other questions are related to the retention, tenure, and promotion system in higher education. Should institutions of higher education integrate research as part of their teaching mission with better indicators of student performances based on standards and criteria developed by the faculty as a whole? What would these indicators look like? How would the culture of the university science teaching community change? How can we make such changes happen?

Finally, how can the outcomes and achievements of students and programs be assessed, evaluated, and recognized? Can we develop a system in which it is possible to ensure that all students achieve set goals and objectives of the program, whether there are specific learning goals or broad goals such as the demonstration of creative research abilities?

The day is past when policymakers and the public will accept the performance of higher education units as a given; better ways need to be found to demonstrate success, and these demonstrations must be founded upon ongoing scientifically based research in the field.

REFERENCES

Abd-el-Khalick, F., & Lederman, N. G. (2000). Improving science teachers' conceptions of the nature of science: A critical review of the literature. *International Journal of Science Education, 22,* 655–701.

Accrediting Commission for Senior Colleges and Universities of the Western Association of Colleges and Schools. (2001). *Handbook of accreditation.* Alameda, CA: Author.

American Association for the Advancement of Science. (1989). *Science for all Americans.* Washington, DC: Author.

American Association for the Advancement of Science. (1993). *Benchmarks for science literacy.* Washington, DC: Author.

Brakke, D.F., & Brown, D.T. (2002). Assessment to improve student learning. *New directions for Higher Education, 119,* 119–122.

Bransford, J.D., Brown, A.L., & Cocking, R.D. (Eds.). (1998). *How people learn: Brain, mind, experience and school.* Washington, DC: National Academy of Sciences.

Chambers, J.A. (2000). *Selected papers from the 11th International Conference on College Teaching and Learning.* Jacksonville, FL: Center for the Advancement of Teaching and Learning.

Colbeck, C.L. (2000). Reshaping the forces that perpetuate the research-practice gap: Focus on new faculty. In A. Kezar & P. Eckel (Eds.), *Moving beyond the gap between research and practice in education* (Vol. 110, pp. 49–62). San Francisco: Jossey-Bass.

Commission on Colleges of the Southern Association of Colleges and Schools. (2004, January 6). *Principles of accreditation.* Retrieved February 27, 2004 from http://www.sacscoc.org/accrrevproj.asp

Commission on Collegiate Nursing Education. (2002). *Proposed standards for accreditation of baccalaureate and graduate nursing programs.* Retrieved July 21, 2003, from http://www.aacn.nche.edu/Accreditation/standards.htm

Committee on Undergraduate Science Education of the National Research Council. (1999). *Transforming undergraduate education in science, mathematics, engineering and technology.* Washington, DC: National Academy Press.

Creech, J.D. (2000). *Linking higher education performance indicators to goals.* Atlanta, GA: Southern Regional Education Board.

Daves, G.D., Jr. (2002). The national context for reform. *New Directions for Higher Education, 119,* 9–14.

DiPasquale, D.M., Mason, C.L., & Kolkhorst, F.W. (2003). Independent and critical thinking in an inquiry-based exercise physiology laboratory course. *Journal of College Science Teaching, 32,* 388–393.

Engineering Accreditation Commission. (2002). *Criteria for accrediting engineering programs.* Baltimore: Accreditation Board for Engineering and Technology.

Laws, P.W., & Hastings, N.B. (2002). Reforming science and mathematics teaching: FIPSE as a catalyst for change. *Change, 34,* 28–36.

Mason, C.L. (1992). Concept mapping: A tool to develop reflective science instruction. *Science Education, 76,* 51–63.

Mason, C.L. (1999). The TRIAD approach: A consensus for science teaching and learning. In J. Gess-Newsome & N. J. Lederman (Eds.), *Pedagogical content knowledge: Its role and usefulness in science teacher education* (pp. 277–299). Amsterdam: Kluwer Academic.

National Commission on Excellence in Education. (1983). *A nation at risk: The imperative for educational reform.* Washington, DC: U.S. Department of Education.

National Governors' Association. (1986). *Time for results: The governors' 1991 report on education.* Washington, DC: National Governors' Association Center for Policy Research and Analysis.

National Research Council. (1996). *National science education standards.* Washington, DC: National Academy Press.

National Research Council. (2003). *Evaluating and improving undergraduate teaching in science, technology, engineering and mathematics.* Washington, DC: National Academy Press.

National Science Teachers Association. (1998). *NSTA standards for science teacher preparation*. Retrieved July 21, 2003, from http://www.nvc.vt.edu/nsta-ncate/november98.htm

National Science Teachers Association. (2003). *Standards for science teacher preparation*. Retrieved July 21, 2003, from http://www.nvc.vt.edu/nsta-ncate/draftstandards.pdf

National Technical Information Service. (1991). *What work requires of schools.* Secretary's Commission on Achieving Necessary Skills (SCANS) report (Order No. PB92-146711INZ). Springfield, VA: U.S. Department of Commerce.

Rothman, F.G., & Narum, J.L. (1999). *Then, now and in the next decade: A commentary on strengthening undergraduate science, mathematics, engineering and technology education*. Washington, DC: Project Kaleidoscope.

Shavelson, R.J., & Towne, L. (Eds.). (2002). *Scientific research in education.* Washington, DC: National Academy Press.

U.S. Department of Education. (2002). *Before it's too late: A report to the nation from the National Commission on Mathematics and Science Teaching for the 21st Century.* Washington, DC: Author.

CHAPTER 18

A SCIENCE RESEARCH ORGANIZATION PERSPECTIVE ON REFORM IN TEACHING UNDERGRADUATE SCIENCE

K–12 science education has been in a period of reform since the publication of several important documents by prestigious national organizations about a decade ago. Each of these laid out a vision for science literacy and specified what it is that students should know and be able to do to achieve science literacy by the time they graduated from college. In addition to its emphasis on a smaller number of important ideas (sometimes referred to as a "more or less" approach), the current reform movement also places importance on teaching for understanding as opposed to recall of knowledge alone. In this chapter, I will address the questions of whether higher education can move in the same direction, if the time is right for higher education to follow the path that K–12 is now on, and if there are any intermediate steps that can begin to move higher education in that direction.

Reform in Undergraduate Science Teaching for the 21st Century, pages 317–327

INTRODUCTION

K–12 science education has been in a period of reform since the publication of several important documents by prestigious national organizations about a decade ago. These include *Science for All Americans* and *Benchmarks for Science Literacy* from the American Association for the Advancement of Science in 1989 and 1993, respectively, and the *National Science Education Standards* by the National Research Council in 1996. Each of these publications laid out a vision for science literacy and specified what it is that students should know and be able to do to achieve science literacy by the time they graduated from college. In addition, federal legislation in the past decade has been supportive of this approach with its emphasis on accountability through testing. This influence is being felt more than ever with passage of the No Child Left Behind Act of 2001. In addition to its emphasis on a smaller number of important ideas (sometimes referred to a "less is more" approach), the current reform movement also places importance on teaching for understanding as opposed to recall of knowledge alone and the idea that pedagogy matters. In other words, there are principles of good teaching that should be applied by teachers and instructional materials developers to support the growth of student understanding of important ideas.

The questions that I will address in this chapter are whether higher education can move in the same direction, if the time is right for higher education to follow the path that K–12 is now on, and if there are any intermediate steps that can begin to move higher education in that direction. The cultures of higher education and K–12 education are of course different in important ways and it is naive to think that changes in higher education will come about easily or that such change is even desirable. And it needs to be recognized that the standards-based approach is neither fully understood nor fully accepted in K–12 education either.

There are two issues that prove to be stumbling blocks to K–12 education reform that will inevitably be a problem for higher education as well. The first is faculty autonomy. Although teachers in elementary science classrooms are generally happy to have explicit guidance in the organization of the science curriculum (when science is taught there at all), high school science teachers who have majored in the subject they are teaching are more reluctant to have the curriculum spelled out for them. They prefer a certain degree of freedom to deliver the curriculum the way they feel is best. The standards-based movement has, therefore, produced some discontent among high school teachers who wish to retain more autonomy than the new emphasis on testing and specified outcomes allows. But any reluctance on the part of subject-matter experts in high school to be part of a goals-based reform movement is multiplied many times over at the college level. College faculties are very concerned about their academic freedom. Standardization of any kind is difficult to achieve. Even for staff-taught courses with a common syllabus and common text or reading list, faculties generally want to retain considerable autonomy to treat the subject the way they feel is best.

The second issue is the failure of faculties to fully understand that specifying a core body of learning outcomes can be the first step in establishing a genuine science of pedagogy. They need to see that once core learning goals have been identified, then efforts can go into creating instructional materials that support student understanding of those ideas as well as professional development programs that are consistent with the reform agenda. And when clearly stated learning goals are available, assessment instruments can be designed that measure student understanding of specific ideas that can in turn be used in research on curriculum development and pedagogical practice. A research-based approach to teaching and learning enables faculties at all levels to become mindful of how their practices and the materials they use facilitate student learning.

APPLYING K–12 REFORM IN HIGHER EDUCATION

Project 2061, a long-term nationwide reform initiative of the American Association for the Advancement of Science, recently conducted a case study to determine the feasibility of incorporating some of the central ideas from the K–12 reform movement into the college curriculum (American Association for the Advancement of Science [AAAS], 2003b). What became very evident was that establishing content standards and initiating outcomes-based accountability in higher education will not be an easy task. One of the first difficulties is getting higher education faculty to make the commitment of time necessary to even begin to learn about incorporating content standards into their curricula. With little to compel university faculty to change what they are doing, reformers must rely on their powers of persuasion to convince faculty that participating in workshops or other programs is worth their time. Faculty autonomy often hinders the participation that administrators can mandate at lower grade levels.

Another issue that limits the amount of effort that university faculties are willing to devote to a consideration of their teaching is the predominant emphasis in higher education on scholarly activities. Many professors prefer to spend their time pursuing research in their discipline rather than addressing issues of teaching. Research brings recognition from national peers and the possibility of career advancement. Additionally, even when there is a core of individuals at a college or university who are committed to change, administratively it is difficult to generate the widespread support that is needed for change to take hold on a continuing basis. Any attempts to do so require incentives, recognition and support from upper-level administrators who make tenure and promotion decisions, a critical mass of faculty who can influence others, and the ability to recruit new faculty over time (AAAS, 2003a).

The support of the university administration is a key element in any reform efforts. Formal structures for implementing reform need to be established that are sufficiently collegial and at the same time intellectually engaging to win the support of university faculty. In addition, those who are in the position of making personnel decisions must support these activities and grant suitable recognition for them. Administrators can provide funding earmarked for workshops and other professional development activities that create opportunities for faculty to gather and exchange ideas. But to be sustainable, even more aggressive measures will probably have to be taken. Depending on how extensive the changes are, load reduction for faculty, scheduling of regular staff meetings to discuss both content and pedagogical issues, end-of-year retreats to discuss issues covering the entire program as well as to make revisions in individual courses, the provision of administrative staff for the program, and mechanisms for recruiting faculty on an ongoing basis may also be necessary.

HOW THE K–12 EXPERIENCE IS RELATED TO HIGHER EDUCATION REFORM

Even with the obvious difficulties of influencing what goes on in higher education classrooms, there are compelling reasons why college and university faculties should consider the approaches being taken by K–12 educators and study them to see what aspects of the current reform movement might be relevant to higher education. For one, many science courses at the college level still use a traditional lecture format in which faculty members have little interaction with students or awareness of what students are thinking until the exams are given. Even then, most faculty members do little more than grade the tests and move on. In general, university faculties are not formally prepared to teach. They teach their subject in the way they think is best, but they have no prescribed preparation for uncovering what their students are thinking or the best ways to make use of the knowledge that students already have. More important, they have little idea of what concepts their students have already been exposed to, what are the learning trajectories that develop in the K–12 years, and where that knowledge can be built upon at the next level. University faculties are often surprised to find that their students do not have the prerequisite knowledge to understand the concepts being presented to them, yet they have no way of knowing how to uncover those learning deficits or what to do about it.

One specific area where the need for articulation is evident is in the transition from high school to college. It is well known that many more students go to college than succeed. Many enter college thinking they have the requisite knowledge and skills to do college work only to find out that they are prepared for nothing more than remedial work in college. This is another area where improved assessment is needed and where clearly articulated learning goals can guide the development of those assessments. As Kirst and Venezia (2002) observe, the current system of assessments is a patchwork that sends students conflicting signals about where they stand in the K–16 continuum of education. High schools and colleges often use different assessment tools that measure different skills in admission and placement of students. These differences confuse students about what standards they should strive to meet and which skills to concentrate their efforts on. However, clearly stated learning goals can define what is assessed. Kirst and Venezia write:

> K–12 assessments that are aligned with higher education standards can provide clear signals and incentives. These assessments should be diagnostic in nature, and the results should include performance levels that indicate to students whether their scores meet or exceed the level for college preparation and placement without remediation. (p. 36)

One effort to address this need for coherence in the high school-to-college transition is Standards for Success, a consortium of 15 member universities of the Association of American Universities, funded by the Pew Charitable Trusts. Their charge is to develop content standards in various academic areas that will provide guidance on what knowledge and skills students must have to be successful in college. Examples of ideas that students should understand to be successful in college-level biology include the following:

- That all living systems are composed of cells, which are the fundamental units of life, and that organisms may be unicellular or multicellular.
- That within multicellular organisms there are different types of cells and that these cells perform different functions for the organism.
- That different types of organisms (plants vs. animals) have different cellular specializations suited for the organism's lifestyle. (Standards for Success, 2002)

Content standards such as these are one sign that university faculty recognize the importance of identifying what is at the heart of the courses they teach and are willing to begin the very difficult process of articulating those ideas.

HOW TO LEVERAGE CHANGE

If there is value in changing the way science is taught in higher education and at the same time reluctance on the part of higher education faculties to alter what they are doing, something is needed to leverage that change. In K–12 education, the power of persuasion has moved some schools and faculties toward standards-based reform, but it is the current federal legislation directing states to assess the progress of all students with respect to clearly stated learning goals that will ultimately bring about change.

In higher education, there are a number of indications that such leveraging may be on the horizon as well. For example, the Middle States Commission on Higher Education has been in the process of establishing an outcomes-based accountability model for some time, a model in which colleges and universities will be assessed based on how well they meet the learning goals they set. In a draft of their standards for accreditation, the Commission writes: "The design of specific courses, programs, and learning activities should be linked to clearly articulated goals for the specific programs of which they are part and to the overarching mission of the institution" (Middle States Commission on Higher Education, 2001, p. 27). The Commission also explains how an institution should establish learning goals and use them once established:

> The institution defines the degrees it offers both by identifying the expected student learning and by creating a coherent program of study (not simply a collection of courses) that leads to those desired outcomes. Curricular issues, generally falling within the responsibilities of the faculty, might address such elements as skill building and mastery of increasingly difficult subject matter along with general education and the learning expected in the specific field of study. Institutions should document the development and attainment of students relative to those intended outcomes. (p. 27)

Content standards, in turn, influence the nature of an institution's assessment tools. Once established, these standards then play a key role in assessing not only the work of the students but also that of professors, departments and the institution as a whole. Therefore, assessment tools should carefully reflect the chosen learning goals. Noting that standards often lead to greater emphasis on assessment, the Commission writes:

> Because the purpose for assessing student learning is to help students improve and to maintain academic quality, the assessment measures chosen

should be those that provide the students, faculty, and others with information about student learning that is specific and useful for assessing and enhancing quality. The mission of the institution provides focus and direction to its outcomes assessment plan, and the plan should show how the institution translates its mission into learning goals and objectives. In order to carry out meaningful assessment activities, institutions must articulate statements of expected student learning at the institutional, program, and individual course levels, although the level of specificity will be greater at the course level. (p. 42)

Other hopeful signs include the attempts of the National Science Foundation (NSF) to bridge the gap between K–12 and higher education. All of the major education funding at NSF now incorporates efforts to have the faculties from schools of education and arts and sciences, as well as K–12 and university science faculties, work collaboratively. NSF programs such as the Math and Science Partnership program, the Centers for Learning and Teaching program, and the G-K12 program, require funded projects to make such links. The current solicitation for the Math and Science Partnership program, for example, states in its guidelines that: "Higher education core partner organizations [must] commit to engaging mathematics, science and/or engineering faculty in activities that strengthen their teaching practices and their roles in K–20 mathematics and science education, including K–12 teacher preparation and professional development" (National Science Foundation, 2002). Recognizing that each partner in the elementary school-through-graduate school continuum has something important to contribute to the educational experience of our youth is the first step in creating the coherence that is needed for that experience to be successful.

Still another hopeful sign is the report of the National Research Council (2002) on the development of *BIO2010: Undergraduate Education to Prepare Biomedical Research Scientists* by the Committee on Undergraduate Biology Education to Prepare Research Scientists for the 21st Century. The report addresses the way that future research biologists should be educated through an integration of concepts from biology, chemistry, physics, mathematics and computer science, and engineering. Included in the report is a listing of "central themes" from each of the separate contributing disciplines. Many of these themes are listed topically (periodic table, orbitals and electronic configurations, nuclear chemistry), but in some of the content areas the central themes are surprisingly similar to the statements found in *Benchmarks for Science Literacy* and the *National Science Education Standards*. For example, under "Concepts in Biology," the committee lists the following among 18 similar statements:

- Most biological processes are controlled by multiple proteins, which assemble into modular units to carry out and coordinate complex functions.
- In multicellular organisms, cells divide and differentiate to form tissues, organs and organ systems with distinct functions. These differences arise primarily from changes in gene expression.
- Many diseases arise from disruption of cellular communication and coordination by infection, mutation, chemical insult, or trauma.
- Groups of organisms exist as species, which include interbreeding populations sharing a gene pool. (National Research Council, 2002, p. 25)

Under "Additional Quantitative Principles Useful to Biology Students," the committee lists the following among their central themes:

- Equilibria arise when a process (or several processes) rate of change is zero.
- Equilibria can be dynamic, so that a periodic pattern of system response may arise. This period pattern may be stable in that for some range of initial conditions the system approaches this period pattern.
- There are relatively few ways for system components to interact. Negative feedbacks arise through competitive and predator-prey type interactions, positive feedback through mutualistic or commensal ones. (p. 34)

This limited set of learning goals provides university-level educators with an excellent example of how clearly stated learning goals can be identified for a particular course. Perhaps the prestige of institutions such as the National Research Council, the National Science Foundation, and the American Association for the Advancement of Science, along with the authority of accrediting bodies, can move higher education along the same reform path that K–12 is moving.

WHAT ROLE CAN TEACHER EDUCATION PLAY?

Our discussion up to this point has focused on changes that are possible between university arts and sciences faculties. But education department faculties also have an important role to play, both in K–12 and higher education reform. Ideally education department faculties can be the bridge connecting K–12 and higher education arts and science faculties. Schools of education are in the best position to take on the responsibility of doing the deep thinking and hard work needed to develop a coherent educational program that links K–12 and higher education through clearly stated learning goals and effective practice.

Although the task is not an easy one, there are a number of things that can and should be done by school of education faculties:

- Engage university science faculties in an examination of the K–12 science curriculum, both to inform them about current expectations for K–12 students and to motivate them to consider ways to apply K–12 reform principles to their own teaching.
- Influence state legislatures and state education departments to accept the principles of the reform movement, both with regard to curriculum and assessment. At the present time practicing teachers often believe that they do not have the time to devote to methods that focus on student understanding when the tests their students take emphasize knowledge acquisition. The combination of an overstuffed curriculum and high-stakes accountability makes it difficult for teachers to consider doing anything other than teaching to the test they and their students will be measured by.
- Influence the kinds of textbooks that are adopted by schools and the way those textbooks are implemented. Doing so will require that in-service and preservice teacher education experiences focus on ways to identify textbooks that are consistent with the reform agenda and enable teachers to implement them as intended. At the present time the textbooks that are available to teachers are often inconsistent with the recommendations of the national standards documents both with respect to content and pedagogy (AAAS, 2003b). Whereas the various national organizations have provided clear statements of what is important for students to learn, they have not yet changed the way curriculum materials developers approach the writing of textbooks. In fact, for the most part textbooks are not aligned with national learning goals. Because curriculum materials are powerful determinants of what happens in classrooms, they present a significant opportunity for improving the way science is taught. Regardless of what happens in preservice teacher education classes, without materials that are aligned with important learning goals, it will be difficult to advance the reform agenda.

HEADING IN THE RIGHT DIRECTION

In many ways the tasks that lie ahead seem daunting. However, there are a number of approaches that can be taken to move the reform agenda forward. One way to familiarize university faculty with the school curriculum is the *Atlas of Science Literacy* recently published by AAAS (AAAS, 2001). The *Atlas* organizes the learning goals from *Benchmarks for Science Literacy* (and some from the *National Science Education Standards*) into strand maps. Strand maps show the connections between a small number of ideas across four grade bands. A map on heredity dealing with "DNA and Inherited Characteristics" includes three strands: mechanism of inheritance, sexual reproduction, and cells and development. Benchmark ideas are linked by arrows showing the horizontal and vertical relationships among them. The *Atlas* maps are informative and engaging to university faculty because they show the progression of ideas in a given content area and they can be used to show where the knowledge is heading at the next level. A useful project would be for university science faculty to extend the maps to the next level.

AAAS has also paid attention over the past several years to the quality of K–12 textbooks. AAAS is now developing a NSF-funded Center for Curriculum Materials in Science with partners at the University of Michigan, Michigan State University, and Northwestern University to support curriculum development research. The Center will also help prepare the next generation of leaders with expertise in curriculum materials research and development. Courses are being developed for doctoral students and for preservice and in-service teachers. These courses will focus on the design, selection, and use of high quality materials. A key aspect of the mission of the Center will be to engage university science faculty in all aspects of the work.

CONCLUSION

As already noted, this journey will not be easy. Not only does a compelling case need to be made for the value of reform along the lines indicated, but the culture of higher education needs to be changed so that the consideration of teaching carries greater weight than it does at present. For everyone who desires change, the reasons will be different. The telling moment for one professor came when her students in an upper-level class could not recall concepts she herself had taught them in their first year of college. When she saw that her glowing student evaluations were failing to translate into long-term retention of the key ideas, she realized she needed to change how she approached teaching (AAAS, 2003a). By demonstrating how traditional content and pedagogy often lets students down, reformers may be able to motivate faculty to change how and what they teach. Change is possible, but the challenges are great.

REFERENCES

American Association for the Advancement of Science. (1989). *Science for all Americans.* New York: Oxford University Press.

American Association for the Advancement of Science. (1993). *Benchmarks for science literacy.* New York: Oxford University Press.

American Association for the Advancement of Science. (2001). *Atlas of science literacy.* Washington, DC: Author.

American Association for the Advancement of Science. (2003a). *Maryland MacArthur project.* Manuscript in preparation.

American Association for the Advancement of Science. (2003b). *Project 2061 textbook evaluations.* Retrieved July 10, 2003, from http://www.project2061.org/research/textbook/default.htm

Kirst, M., & Venezia, A. (2002, April). Bridging the great divide between secondary schools and postsecondary education. In *Gathering momentum: Building the learning connection between schools and colleges.* Proceedings of the Learning Connection Conference (pp. 30–38). Washington, DC: Institute for Education Leadership and the National Center for Public Policy and Higher Education.

Middle States Commission on Higher Education. (2001, August). *Characteristics of excellence in higher education: Standards for accreditation.* Draft for discussion. Retrieved July 23, 2003, from http://www.msache.org/CHX.pdf

National Research Council. (1996). *National Science Education Standards.* Washington, DC: National Academy Press.

National Research Council. (2002). *BIO2010: Undergraduate education to prepare biomedical research scientists.* Washington, DC: National Academy Press.

National Science Foundation. (2002). *Math and science partnership program (MSP): Program solicitation for NSF 02-190.* Retrieved on July 10, 2003, from http://www.nsf.gov/pubs/2002/nsf02190/nsf02190.htm

No Child Left Behind Act of 2001, 20 U.S.C. § 6301 *et seq.* (2002).

Standards for Success. (2002, October). *Knowledge and skills for university success: Final draft of standards in English, Mathematics, Science, Social Studies, Second Languages.* Presented at a meeting of the American Association of University Presidents, Atlanta GA.

CHAPTER 19

SCIENTISTS' PERSPECTIVE ON REFORM IN TEACHING UNDERGRADUATE SCIENCE

Serving Preservice Science Teachers in the Context of a Heterogeneous Student Population

Joseph A. Heppert and April French

ABSTRACT

The purpose of this chapter is to provide some perspectives gained during the reform of introductory undergraduate chemistry laboratory experiments at the University of Kansas. The chapter includes a brief outline of recent research on the reform of undergraduate science courses, including efforts in biology, chemistry, physics and astronomy, and geology that have focused on both the lecture and laboratory components of the courses, and some efforts that have sought to eliminate the distinction between the laboratory and the lecture. Also discussed is the context of the laboratory courses that were revised in our program, the instructional model used in the design of the experiments, and some of the research outcomes of the project. This

Reform in Undergraduate Science Teaching for the 21st Century, pages 329–350
Copyright © 2004 by Information Age Publishing

chapter emphasizes the effects of reformed college-level science courses on preservice teachers for two reasons. First, much of the initial funding for this study was provided though a National Science Foundation (n.d.) award to engender systemic improvement in teacher preparation programs, including the introductory science courses that serve preservice teachers. Second, the authors believe that the current deficit of highly qualified science teachers is a threat to the youth of the nation and to our international preeminence in science and technology. It is critically important that the nation address this challenge by both empowering existing science teachers and attracting students of the highest caliber into the profession.

BACKGROUND

With the *Benchmarks for Science Literacy* (American Association for the Advancement of Science [AAAS], 1993) calling for students to learn more problem solving and higher order cognitive skills instead of focusing overwhelmingly on learning factual information, teachers at all levels need to engage learners in the pursuit of knowledge. Researchers have found that students who actively participate in college science courses tend to have more favorable opinions about science as well as a better understanding of the nature of science (Brown, 2000; Russell & French, 2002; Zebrowski, 2001). Russell and French, for instance, compared the participation level of students in a traditional laboratory setting with those in an inquiry-based lab. They found that female students in an inquiry-based lab had better learning experiences than those in a traditional laboratory, based on student attitudes and behavior toward manipulating and handling data and equipment, and watching or actively discussing the experiment. Female students seem to benefit more from being actively engaged while learning. This finding, coupled with U.S. Census data, which documents that approximately 75% of the nation's teachers are female, has strong implications for the way teachers are educated (U.S. Census Bureau, 2001).

Engaging students in K–12 or undergraduate classes is easier said than done, as most teachers tend to "teach as they were taught" (McDermott, Shaffer, & Constantinou, 2000, p. 412). Preservice teachers who were taught through lecture-based methods are more likely to lecture to their classes, according to McDermott, while those who engaged in inquiry-based methods are more likely to incorporate these methods into their classrooms. Faculty at the university level also face challenges of their own when trying to develop inquiry-based curriculum. Sunal et al. (2001) suggest that barriers to change at the university level are similar to those found in elementary and secondary education: (a) society believing "teaching is telling"; (b) the lack of ongoing professional development, follow-up, and monitoring of faculty; (c) few awards and rewards for teaching excellence; (d) feeling that lecturing is the only effective means to handle a 250-plus-size classroom; and (e) instructors' beliefs and expectations toward teaching and learning (p. 247). Sunal et al. also include a discussion of barriers perceived by faculty members at the classroom and content levels, including a feeling that students have weak backgrounds in content or a fear of science, personal resistance to change, a sense of being under qualified, a lack of connection to decision makers, concerns about tenure and promotion issues, and dissatisfaction with available curricular materials (2001, p. 247).

MODELING INQUIRY FOR PRESERVICE TEACHERS

Researchers in education and college-level science instructors are exploring numerous models to close the gap between the goals and objectives outlined in the Benchmarks and the National Science Standards (National Research Council [NRC], 1996) and the appropriateness of the instruction for students in the classroom. An emphasis is being placed on designing inquiry-based coursework so that preservice teachers will have an immersion in inquiry before they take responsibility for a K–12 classroom. Courses for preservice teachers are designed to not only teach content knowledge in a deep and meaningful way, but also to model the instructional strategies that will better enable future K–12 students to understand and apply the content as well. McDermott, Shaffer, and Constantinou believe that "the routine problem solving that characterizes most introductory courses does not help teachers develop the reasoning ability necessary for handling the unanticipated questions that are likely to arise in a classroom" (2000, p. 412). The reasoning, critical thinking, and problem solving skills that McDermott et al. (2000, p. 412) argue are essential for preservice teachers are the skills that the authors of the Benchmarks (AAAS, 1993) feel are valuable citizenship tools for the general public. Coble and Koballa view inquiry-based science education as a means for students to develop an understanding of "the tentative nature of science and to be able to judge the assertions made by both scientists and nonscientists" (1996, p. 459) in order for the general public to judge for themselves the value of scientific work.

Researchers have examined inquiry-based courses for preservice teachers in a number of different disciplines. Margaret Johnson at Mesa Community College in Mesa, Arizona, studied students in an online introductory biology course that incorporates hands-on, inquiry-based labs (Johnson, 2002). Students in this course were required to purchase kits that contained the lab manual and some supplies for the lab. Other supplies, such as ammonia and acetone (found in nail polish remover) were purchased from local stores. Each week of the 14-week course, students began a new lab in which they were to develop alternative hypotheses about the question posed for the week. Students then used the course bulletin board to describe how they designed an experiment to test their hypothesis and corresponded with other students and the instructor to elicit feedback on whether their experiment established a causal relationship between the variables under investigation. One lab had students investigate the factors that affect the rate of diffusion of substances into cells using the diffusion of anthocyanin in cabbage water into gelatin cubes as a model.

Labs such as this both illustrate inquiry teaching for preservice teachers and provide content that future teachers could bring into their K-12 classroom. Other researchers fostering inquiry in biology laboratories and lectures have created methods to illustrate topics such as population doubling (Schlenker & Schlenker, 2000), natural selection and evolution (Udovic, Morris, Dickman, Postlethwait, & Wetherwax, 2002), metabolic rate (DeGolier, 2002), and respiration (Glasson & McKenzie, 1998), just to name a few. Zollman (1990) designed and implemented a physics course for preservice elementary teachers using a learning cycle model that organizes all learning experiences around laboratory-based activities. This course explicitly models instruction after the best practices which teachers are expected to use in their classrooms.

Other areas of science have also focused on supporting preservice teachers in gaining confidence in their content knowledge while they model inquiry-based teaching strategies. A course in geoscience at Lake Superior State University focused on many aspects of geology, such as the dynamic geologic process that occur on Earth and how humans can affect the environment in a field-based and laboratory setting (Brown, Kelso, & Rexroad, 2001). In the unit, *From Sediment to Rocks*, students were asked to explain how rocks are broken down, how soil formed, and how surface features changed (p. 451). Students visited sites where sediments had weathered chemically and mechanically to view the phenomena in real-life situations. The observations students made at the site were then related to influences of rock type, climate, and climate change. Students become more confident with their ability to identify sedimentary processes, as well as develop language skills, as they base their conclusions on supporting results in their field book and defend their interpretations in oral presentations. Other inquiry-based courses in geoscience have covered topics such as geologic time (Thomas, 2001), site interpretation (Stefanich, 1979), and oceanography (Martin & Howell, 2001).

During the 1990s, the National Science Foundation's Divisions of Undergraduate Education and Chemistry co-funded a series of projects intended to bring about systemic reform of the introductory chemistry curriculum. Information about the materials developed though these projects are available on the dissemination Web site at the NSF-sponsored Multi-Initiative Dissemination (MID) Project (UC Regents, 2003). Each project adopted a different approach to the reform of traditional chemistry curricula. The Workshop Chemistry Project (Gosser & Roth, 1998) developed a "peer-led team learning" approach to instruction, in which peer leaders act as facilitators for group learning environments. This strategy has been widely disseminated, and workshop modules have been developed for related subjects including introductory biology and organic chemistry. The Molecular Science Project (Russell, Chapman, & Wegner, 1998) focused on developing network deliverable resources that facilitate student abilities to use scientific data to construct understanding, and to develop effective written communication skills. The New Traditions Consortium (Landis et al., 1998) developed a variety of strategies to foster student-centered learning in small and large classrooms, including the use of "concept tests" in large lecture settings and inquiry-based group learning exercises as an alternative for traditional lecture settings. The ChemLinks and Modular CHEM Consortia (Anthony, Mernitz, & Spencer, 1998) focused on the development of active learning-oriented curricular units that bring the excitement of current scientific research and relevance of everyday observations into the teaching of chemistry concepts at the introductory level.

Teaching chemistry through inquiry-based activities has also focused on helping students better understand microscopic phenomena after obtaining a deep conceptual understanding of the macroscopic. Nelson (2002) suggests that although chemists spend much of their time at the microscopic level, students have a difficult time understanding this viewpoint because it is not readily observable. Because of the heavy focus on the micro-level (the atomic and molecular levels, as well as the electronic and nuclear), students become disengaged from the macroscopic or bulk level, leaving them to just accept what they have been told. Nelson (p. 216) calls for the return to the teaching style of the 1960s, where chemistry was taught by gradually moving down from the macroscopic to the microscopic level through a historical viewpoint or by specifically addressing commonly held misconceptions regarding the subject matter being presented.

Research on inquiry-based instruction has also focused on the importance of student participation in undergraduate research (Hutchison & Atwood, 2002) or laboratories with research-oriented curriculum (Hanks & Wright, 2002). For professors at the college level who are not as inclined to make radical changes in the way they teach, McNelis (1998, p. 479) suggests the use of lecture aids to help engage students in an introductory organic chemistry course. These aids provide the reoccurring portions of chemical structures to help reduce note taking and increase the amount of active thought by students when learning about organic reaction mechanisms.

Although inquiry-based teaching methods require more time than traditional lecture methods, resulting in fewer topics being discussed, the overall benefit of instruction that focuses on understanding the tentative nature of science and how scientists reason, benefits students more even with a reduction of factual information. The much more general motive, to benefit all students taking science classes, has driven many researchers in chemistry, including Cooper (1996), Abraham and Pavelich (1991), and Wink, Gislason, and Kuehn (2000), to devise inquiry-based laboratory curricula for introductory courses. Given the difficulty of identifying many middle/secondary preservice teachers and the heterogeneity of many science courses, it is our belief that implementing research-based changes in the undergraduate curriculum designed to benefit all students will be the most compelling rationale for curricular change in science content programs.

ESTABLISHING A COMMON AGENDA FOR CLASSES THAT SERVE PRESERVICE TEACHERS

Who Are Preservice Teachers, and Are They Deliberately Hiding From Us?

At times, it may seem to scientists as if preservice science teachers, who are our own majors, are deliberately trying to obscure their interest in teaching as a career. In reality, it is quite common for preservice middle/secondary teachers to embark on preparation for a career in education fairly late in their undergraduate studies. The way that we structure science degree programs and communicate with our students about career options tends to exaggerate this outcome. Preservice elementary education majors often self-identify as education students prior to matriculation, so schools of education can usually identify and communicate with these students from the outset of their degree programs. Preservice middle/secondary teachers, in contrast, frequently begin to consider education after they have had the opportunity to explore a range of opportunities to apply their degrees. Consequently, many science majors only begin considering teaching careers in the 4th year of their degree programs or even after they have received their bachelor's degrees. Efforts to communicate with our majors could identify at least some students who would consider science teaching at an earlier stage in their college career. Most science programs do not seek information among their majors about potential interest in middle/secondary science teaching early in their college career. Moreover, relatively few programs have formalized mentoring programs for future teachers. Extremely few undergraduate programs organize formal teaching mentorships that allow undergraduate science majors to explore teaching at the K–12 or college levels as a potential career path. NSF's Graduate Teaching Fellows in the K–12 Schools (NSF, 2003) program is one approach that has provided formalized teaching opportunities for both graduate and undergraduate students.

In science programs, we must consider preservice teachers to be "our" students. Not in the territorial sense that preservice teachers are "our" students as opposed to being students of the education program. Rather, we must adopt a partnership and a sense of mission in educating these students that equals the commitment we make in the education of our own majors. After all, teachers prepare our future majors, introducing K–12 students to science at a stage in their lives that usually has more influence on their future career choices than their college years. If for no other reason, the small numbers of students who intend to study the physical sciences at colleges and universities, and the declining numbers of domestic students majoring in engineering, are an indication of the effort that we should be investing in the education of preservice teachers. Estimates by the National Commission on Mathematics and Science Teaching for the 21st Century (2000) suggest that the nation is not producing enough qualified mathematics and science teachers to meet demand: The total need for newly trained middle/secondary mathematics and science teachers will approach 240,000 over the next 10 years. A recent decision by the American Chemical Society to create a special membership and dues category for teachers is an example of an effort that can increase the stature and appeal of precollege science teaching professions. We should, however, adopt the attitude that preparing the number of exquisitely prepared science teachers, required to build on our national dominance in science, is likely to require more from scientists than merely a superficial reorganization of the structure of our undergraduate programs.

Expecting More Than Content Mastery From Courses for Preservice Teachers

As we noted previously, though the National Science Education Standards (NRC, 1996) and Benchmarks (AAAS, 1993) support the need for an excellent content background for every science teacher, they affirm that more than content knowledge is needed for training effective teachers. Both documents emphasize that teachers need also to understand the current state of knowledge about human cognition and learning, methods for classroom management and for constructing a curriculum, strategies for teaching exceptional learners and the growing fraction of students with limited English skills, the social and historical context of the subjects they teach, applications of educational technology, and current issues in educational policy and law. In many institutions, expert faculty members in schools of education cover these topics. Beyond these subjects, future teachers need to develop sufficient confidence in the content they teach so they can relinquish absolute control over content delivery, and instead encourage future students to participate in determining the pace and scope of the content that they investigate. Allowing these students to construct their own understanding of science by developing and implementing research projects is the essence of scientific inquiry. In order to readily accept this instructional model, future teachers must be immersed in both the theoretical foundation and practice of scientific inquiry. It is our belief that science and education programs must collaborate in the development of undergraduate curricula that will accomplish this ambitious objective of relating science and scientific inquiry to science pedagogical models that support the implementation of the national science standards.

Curricular Issues in the Sciences

A discussion of the breath and depth of science curriculum is complicated by the fact that college-level science curricula are instructed in many different learning environments. Moreover, an increasing chorus from public and private sectors is contributing to the discussion about what should be taught in science courses and how student achievement should be assessed. State and local political leaders, who play an integral part in funding two-year and four-year college and university education, are increasingly demanding that courses offered in these environments have sufficient overlap in content that credit can transfer among the institutions in a state's education system. This policy is intended to alleviate the frustration among parents and students when courses taken at one state institution cannot be transferred to another. This has led to efforts in some states, including Kansas, to develop standards for "core competencies"—descriptions of knowledge, skills, and experiences that students will encounter if they take a corresponding course at any state institution (Joint Council of Academic Officers, 2003). Other states, including Texas, have emphasized the extent of this alignment by adopting a common numbering system for core courses that students might take in advanced placement, two-year college, state college or university environments. This trend toward mandated alignment of course content raises several issues for scientists involved in the education of teachers.

The first issue is quality control among the institutions offering the related courses. A committee of faculty from state supported two-year and four-year institutions in Kansas confronted several facets of this issue as we developed core competencies for a one-semester non-science major's chemistry course and a two-semester introductory chemistry sequence for science majors. Though committee members were relieved to find roughly a 90% commonality among the concepts listed in the syllabi for related courses offered at different institutions, coordinators of two-year college science programs expressed specific concerns about quality control of course sections taught by adjunct faculty at satellite sites (these included both formal satellite campuses and high school classrooms). The coordinators were committed to assessing the quality of these satellite sections by developing common finals for the courses. So, in the committee's experience, the drive toward the alignment of content in core courses can become an important contributor for providing quality science courses for preservice educators.

Another issue that raises more concern is the scope of the knowledge, skills, experiences, and attitudes—above and beyond factual knowledge of scientific concepts—which preservice teachers and science majors should gain from science content courses. As has been noted earlier in this chapter, teachers today are expected to be expert communicators, practiced facilitators of active learning environments, research mentors and sources of scientific resource materials (AAAS, 1993; NRC, 1996). Elementary and secondary preservice teachers cannot adopt these desirable professional roles if they have not observed the behaviors modeled by their instructors in college science courses and had adequate opportunity to practice the roles during their preservice preparation and field experiences. It is unreasonable to expect that the modeling of these science content pedagogical practices should be the sole responsibility of education programs. Preservice teachers spend an average of one or, at best, two semesters in science methods courses, and generally a maximum of one semester in student teaching. This period of time is completely inadequate to cover the basic information needed to understand instruction in a K–12 classroom, while simultaneously providing the sustained modeling of science instruction pedagogy that preservice teachers need to acclimate to inquiry-based and active-learning-based teaching strategies. Moreover, what modeling of best practices that can be accomplished during a one or two semester methods course seems likely to be completely overwhelmed if a contradictory approach is modeled in the 10 or more semester-long science courses taken by preservice teachers.

Consider that many of the key teaching skills outlined above are the functional equivalent of asking teachers to become consultants or research managers. Scientists in academics, industry, and government laboratories also employ these skills in the course of their professional activities, so modeling the same skills in science classrooms will benefit both future teachers and science majors. The fact that college-level science content instruction provides an important model for early career science teachers at the middle/secondary level further strengthens the case for modeling these skills in science content courses (Fetters, 1998; McGinnis & Parker, 1999; NRC, 1996, pp. 60–61).

It is very important to recognize that modeling active-learning strategies, group work, scientific communication, and inquiry-based instructional methods in science courses as an approach to serving preservice teachers only accomplishes half of the goal. Scientists who design and implement the standards-based objectives must be convinced that these strategies should be accorded equal status with content-based objectives because this will accrue benefits for *all* of the students taking these courses. In their classic study of why an unusually high proportion of *science majors* switch to non-science majors before completing college, Seymour and Hewitt (1997) uncovered substantial evidence that student understanding of the nature of both science careers and the scientific process is formed by the experiences they have in college science courses. Their study indicated that poor teaching, misperceptions about the nature of science careers, and a lack of obvious connection between science and socially relevant issues are common factors that dissuade many otherwise successful students from pursuing science majors. This and related data (Tobias, 1992) make a compelling case that all students taking science would benefit from infusing the collaborative, interdisciplinary, and problem-oriented nature of much modern scientific research into undergraduate science courses.

Finally, as more departments and institutions develop distinct science courses and curricula tailored to the needs of preservice teachers, we engender both enormous opportunity for progress and equally enormous challenges. Designing science courses that provide teachers with an exquisite background in scientific content and effective modeling of diverse instructional strategies is much more challenging than designing a course that accomplishes one or the other of these goals. Reviewers sometimes criticize grant writers for proposing such a unified course design because they believe that the challenge of accomplishing both objectives in one course as too great. This perspective fails to recognize what seems to be an essential truth: Teaching scientific content in one course and covering the theory behind research-based models for best practices in science teaching in another course will only perpetuate our current difficulty with educating teachers who are really prepared to *apply* research-based best practices in K–12 science instruction.

A second, and purely logistical concern with the increasing number of science courses specifically designed to serve preservice teachers, is that these courses must be accepted as fulfilling institutional and professional standards for degree requirements and accreditation. This requires ongoing collaboration among educational institutions, accrediting agencies, and course designers. These interactions should confront the fact that preservice preparation programs must address multiple objectives—including providing an adequate content background, an understanding of human cognitive development, a knowledge of instructional models, practice in classroom instruction and an understanding of the needs of exceptional learners—all of which are extremely time-intensive. A final issue with such courses is that course designers must come to the realization that they are not always able to control all aspects of instruction, even in their own courses. For example, a major challenge faced in implementing a redesign of our courses is that teaching assistants are largely responsible for chemistry laboratory contact hours.

Who Controls the Nature of Instruction Anyway?

Who, indeed! The answer to this question seems to depend both on the nature of the institution where the instruction occurs, and on how much care is taken in establishing a consensus for course objectives among the instructors. Ultimately, the factors that most strongly influence the nature of instruction seem to be the identities of the instructor of record and instructors who meet the laboratory and review sessions for the course. Many of our colleagues at two-year colleges or smaller four-year colleges have the advantage of meeting both the lecture and laboratory sections of their courses. This arrangement provides them with flexibility in establishing the instructional strategies that are used in the course and in establishing connections between the lecture and laboratory components of the class. In contrast, lecture courses at larger institutions, which can include 150 students or more, often rely on teaching assistants (TAs) as the primary instructors of small sections such as laboratory and problem solving sessions. The effectiveness of small-group problem-solving activities and inquiry-based laboratory experiences in these courses depends on the ability of teaching assistants to adopt roles as problem-solving facilitators and research coordinators.

One example of the central role of teaching assistants unfolded early in the implementation of a set of inquiry-based experiments for our science majors' chemistry sequence. An experienced, intelligent, and very well liked teaching assistant made the conscious decision to modify the laboratories, organized around a 5-E learning cycle model, into standard cookbook-style laboratories. He felt that, based on his experiences, cookbook-style experiments were more rigorous than inquiry experiments and that he could more effectively teach using a traditional instructional style. While this individual's efforts to circumvent inquiry-based instruction were an extreme case, the response of graduate teaching assistants to the role of becoming a facilitator of learning in the laboratory is extremely uneven.

Some teaching assistants rapidly acclimate to this role and grow in confidence throughout the semester-long class, while others continue to struggle throughout the semester. Our experience suggests that self-confidence with the content material introduced in the laboratory is a major factor in the comfort of teaching assistants with a less rigid, inquiry-based instructional environment in the laboratory. Adequate preparation for and familiarity with the experiment is probably another factor that influences TA performance: Student rankings for TAs teaching two laboratory sections are almost uniformly one-half point higher on a 5-point Likert scale for the second of the two laboratory sections. Furthermore, student attitude can influence on TA behavior. We have received repeated reports of students confronting TAs with statements like, "Don't tell me how to figure out what I need, just tell me what I need to do so that I can finish this experiment and leave." Teaching assistants, who are almost all advanced undergraduate or graduate students in chemistry, feel a similar pressure to take short cuts around time-intensive aspects of inquiry-based instruction in order to ease demands on their own schedules.

Robinson (2001) recognized the crucial role of teaching assistants in instruction during the implementation of our experiments, and designed and implemented a formalized program at our institution for preparing teaching assistants for inquiry-based laboratories. This program has prepared five generations of graduate students for teaching experiences through an examination of literature on instruction, participating in inquiry-based experiments and group learning activities, and observation-based evaluations of laboratory instruction.

DESIGN OF AN INQUIRY-BASED LABORATORY CURRICULUM

The target course in our study was a two-semester introductory chemistry sequence (Chemistry 184 and 188) for science majors (*Course List*, 2003). In reality, many of the students who begin this type of introductory course as a science major do not complete a degree in science. Less than 18% of the college-bound population majors are in a scientific or technological discipline (UCLA, 1996), which is mirrored by the enrollment trends at the University of Kansas (Office of Institutional Research and Planning, 1997). Embedded among the enrollees in this two-semester sequence is a small population of students identified as future teachers and a somewhat larger group who will eventually choose to pursue careers in middle/secondary education. Four percent of students in introductory chemistry courses at the University of Kansas (KU) self-identify as chemistry or secondary education majors (see Figure 19.1). In order to function as effec-

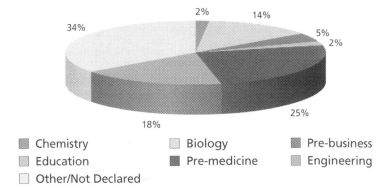

Figure 19.1.   Distribution of majors in introductory chemistry courses at the University of Kansas.

tive teachers, these individuals need a thorough exposure to the goals, methods, ethics, and scope of modern scientific discovery during their university careers. In the university environment, introductory science courses are the natural vehicles for fostering this literacy.

Our initial interest in revising our introductory course offerings began as a discussion of the role of those courses for chemistry majors, preservice teachers, and the broader population of undergraduate students at the university. Only 32% percent of students enrolled in these courses plan careers in allied scientific disciplines (Office of Institutional Research and Planning, 1997), yet, national studies and our own experience indicate that introductory science and technology programs at the university level generally strive to prepare students for subsequent courses in that major (Tobias, 1992). The multiple missions addressed by these courses have prompted the KU chemistry program to undertake an ongoing self-study of undergraduate course offerings. The goals of this planning are to increase scientific literacy in the undergraduate population, and to create a suite of introductory courses that better serve the needs of the diverse constituencies in our courses.

The William and Flora Hewlett Foundation Fund initially provided funding for this introductory chemistry course development project for General Education. Later funding was obtained as part of a comprehensive effort in a National Science Foundation Collaborative for Excellence in Teacher Preparation award to update undergraduate science courses taken by preservice teachers. The objectives of our project were to:

- Develop a new year-long introductory chemistry laboratory curriculum based on constructivist pedagogy.
- Establish cooperative learning teams for student participation in laboratories.

- Integrate network and Internet technologies into the new curriculum.
- Mentor the faculty and graduate assistants who will implement the new teaching pedagogy.
- Evaluate the effectiveness of these strategies in content delivery and in improving student retention and use of scientific principles.

Recasting the pedagogy employed in these introductory courses was a central objective of the project. Studies of cognitive processes suggest a large portion of the student population can benefit from curricula based on a constructivist model (Lawson, Abraham, Renner, 1989; Papert, 1990). Constructivist approaches to science education actively engage students in the collection and interpretation of scientific data, ultimately leading the students to construct their own understanding of the fundamental scientific principles underlying the investigation. Our curricula were built using a five-step learning cycle model (5-E) (Bybee, 1985):

1. Engagement: Develop student interest through the demonstration of a chemical reaction or process.
2. Exploration: Challenge students to explore and observe related systems.
3. Explanation: Develop hypotheses from student observations and examine the hypotheses in a broader context.
4. Elaboration: Test and extend the hypotheses through additional experiments.
5. Evaluation: Challenge students to draw conclusions from their observations and data.

These strategies naturally blur the lines between the traditional lecture and laboratory settings, because this pedagogy relies on the construction of concepts from experimentation rather than formal tutorial content delivery. In some notable instances at other institutions, the laboratory has become the entire course (Laws, 1991; Zollman, 1990). This new pedagogy changes the activities of the students, the role of the graduate teaching assistant (GTA) instructor, and the utilization of the laboratory space.

Contrasting models for the content and layout of a traditional laboratory experiment with that conceived for the constructivist, inquiry-based laboratory experience are shown in Figure 19.2. The traditional laboratory relies on providing students with a detailed experimental path to obtain data, which will then be manipulated in a predefined fashion (Halloun, 1985). Instructors are responsible for teaching students the exact protocols required to achieve a desired result. In this instructional model, the laboratory environment becomes a setting for students to obtain individual practice in certain laboratory protocols and to verify some well-established

Traditional Laboratory Pedagogy		**Inquiry Laboratory Pedagogy**
Complete Pre-laboratory Pages Answer questions about the laboratory	Pre-laboratory Activities	**E-1 Review Online Resource Materials** Background reading, solution chemistry, computer-graded exercises, electronic structure, ionic equations, initial explorations
Laboratory Quiz Model computations using the Nernst equation **Laboratory Lecture** Describe experimental procedures, Nernst equation, experimental write-up **Laboratory Experiment** Collect data using pre-made solutions, record on data sheets	Laboratory Activities	**E-2 Exploration of Central Concepts of the Laboratory** Explore conditions favoring electrolytic and voltaic reactions, determine requirements for an electro-chemical cell effect of activity on EMF **E-3 Breakout Discussion of Results** Discussion of results from inquiry teams, examination of how ion concentration affects EMF, development of the Nernst equation **E-4 Experiments that Amplify Concepts Discussed in Breakout** Preparation of solutions and examination of EMF as a function of ion activity **Final Laboratory Debriefing**
Complete Laboratory Report Pages Answer questions about data collected	Post-laboratory Activities	**E-5 Complete and Submit Laboratory Report** Write-up of protocols developed, data analysis, development of conclusions from observations

Figure 19.2.   Contrasting models for a laboratory to investigate the effect of ionic concentration on electromotive force.

scientific principles. The inquiry-based model, in contrast, is an environment where few definite bounds are placed on the actions of the student. An initial objective is stated and certain equipment is provided, but the inquiry team must cooperate in setting objectives and establishing protocols for the experiments. In this setting, the GTA initially plays the role of a consultant, assisting students in assembling materials for implementing their protocols. During the "breakout" session, the GTA draws together the results obtained by various inquiry groups, setting them into the broader context of an underlying principle: the relationship between reactant concentration and reaction driving force. The inquiry teams are then charged with performing another series of experiments to extend their understanding of the underlying principle. Finally, inquiry teams meet for a final discussion of experimental results. Laboratory write-ups from these sessions contain a mixture of hypothesis, experiment, data analysis, and conclusions, and may contain data and results shared among the inquiry teams. In this constructivist-based instructional pedagogy, the laboratory has become a site for discussion, conceptualization, investigation, skills development, and evaluation. This latter model is a more accurate representation of the actual scientific process.

Developing opportunities for cooperative learning experiences is a second principal priority of the project. Just as inquiry- and problem-based laboratory experiences can develop higher order cognitive skills (Udovic et al., 2002), cooperative learning projects can help students learn to integrate a range of viewpoints in reaching conclusions, and disseminate tasks among students with varying talents and perspectives (Barnett, Barab, & Hay, 2001; Henderson & Buising, 2001). This structure also provides the students with a more accurate model of the cooperation and specialization inherent in the modern scientific research experience. Each laboratory section is divided into five inquiry teams composed of four students. Inquiry teams are encouraged to cooperate in completing each inquiry-based laboratory experience. Discussions of experimental protocols, experiments and the evaluation of results, and preparation of a written report outlining the results and conclusions drawn from the experiments are jointly conducted. A component of each student's grade is determined by a peer evaluation of the extent of student partnership in each experiment.

OBSERVATIONS AND OUTCOMES

We applied a range of assessment strategies to gauge student learning, self-efficacy and attitude (Ellis, Heppert, Robinson, & Omar, in press) to the new curriculum. An American Chemical Society nationally normed examination was used to assess student content understanding (ACS, 1997). Though we reduced the overall number of topics covered in the chemistry laboratory, students scored on average at the 61st percentile nationally. This result indicates that the modified laboratory course still allowed students to assimilate content at a level competitive with their peers.

Student self-efficacy in chemistry was evaluated in the laboratory using a pre-, mid-, and post-survey developed for this study. The results showed a statistically significant increase in student self-efficacy over the year-long course sequence (Ellis et al., in press). Student attitude to the new experiments was measured using surveys and focus group interviews. Though students found the new experiments interesting, they continued to express the sense that the experiments were somewhat disconnected from material taught in the lecture portion of the course. This illustrates the strength of strategies that eliminate the separation between the lecture and laboratory experiences, such as the introductory physics course developed by Zollman (1990). Students indicated that a 4-week, self-directed water quality assessment inquiry project had two positive effects: (a) it reinforced the need for skills and concept development that had proceeded the extended unit and (b) it increased student understanding of the relevance of chemistry to their professional and social concerns.

The nature of the environment in the laboratory was evaluated through direct observation using the Inquiry Laboratory Assessment (Mason et al., 2003). Mid-semester and late-semester observations noted much greater student cooperation on implementing laboratory experiments and a far greater proportion of scientific discussion in the laboratory. Studies of student collaboration using the Collective Effort Collaborative Assessment Technique (CECAT) instrument (Walker & Angelo, 1998) indicated that student concern with social loafing in laboratory duties was a serious concern for less than 15% of students, and more than 85% of students felt that their inquiry teams were effective. Students expressed their greatest continuing concern about how their personal contributions to the cooperative laboratory projects are being assessed. In spite of some persistent concern about peer evaluation of effort, we have continued to use this method to define a component of the grade for each experiment.

Two additional noteworthy issues arose during the implementation of the new laboratory format. Our initial pilot introduced the second semester laboratories after the students had completed a series of cookbook-style laboratories during the first semester. This format led students to complain about the change in experimental format, and to the erroneous conclusion that they were required to do more work by completing a full-fledged laboratory report. We have found that the implementation of the experiments in successive years, in which all of the experiments adopt an inquiry format, have proceeded with much less student dissatisfaction. A second issue that became apparent through our in-laboratory observations and interviews with teaching assistants is that both students and teaching assistants strive to dive directly into attempts to run experiments in an effort to shorten the time required for the inquiry laboratory. This reduces the positive impact of the experiments for the students, who need adequate discussion time to assimilate the concepts discovered in the "Explore" phase of the learning cycle, and must carefully connect their knowledge of this concept to a plan for additional investigations in the "Elaborate" phase of the experiment. We have continued to work with teaching assistants to expand their abilities as facilitators of discussion and planning in inquiry-based laboratories.

CONCLUSIONS

Our efforts to redefine the pedagogical models used in introductory college chemistry laboratory instruction have resulted in positive outcomes for students, our graduate teaching assistants and our department. Among these positive outcomes, students

- Score, on average, above the 61st percentile on a nationally standardized examination for introductory chemistry.
- Recognize that an accumulation of chemical knowledge and skills allows them to approach posing complex questions and suggesting methods to test those hypotheses.
- Are more focused on discussions of scientific issues during the laboratory.
- Come to enjoy working in inquiry teams.
- Enjoy devising and implementing their own experimental protocols.

Moreover, our department has become more focused on developing additional advanced undergraduate courses that challenge students to become effective problem solvers.

We still face challenges in the implementation of effective inquiry-based courses for all science majors. These include:

- Convincing students that the content, skills, and experiences included in the courses are relevant to their future careers.
- Providing the mentoring needed to make teaching assistants into effective instructors.
- Making the coordination of the lecture and laboratory a seamless experience for our students.

It is too early to examine whether the new inquiry-based laboratory sequence has had an impact on the practice of preservice teachers. Many pre-secondary education students in the initial pilot tests of these courses are just now embarking on their professional practice. However, the experiments developed during this process have had an unanticipated impact on both preservice and in-service physical science teachers. We have adapted several of the new inquiry experiments that emphasize solution chemistry—experiments illustrating concepts such as concentration, equilibrium, acid-base chemistry and Beer's law relationships—for use in professional development workshops with middle/secondary teachers. These workshops have focused both on adapting the inquiry activities so that they can be presented at a level appropriate for middle/secondary students, and on providing university-level content enhancement related to these experiments for the benefit of the teachers enrolled in the workshop. Teacher response to this strategy for content enhancement has been so positive that we are currently piloting a summer university-level physics course using this model to frame the curriculum.

Our examination of the pedagogy of undergraduate courses in chemistry is currently taking several directions. In the revised introductory laboratories described above, we are developing strategies to support the planning and discussion involving students and teaching assistants that are crucial to the success of the 5-E learning cycle instructional model. Our analytical chemistry division has developed a problem-based learning approach for their senior-level instrumental analysis course (Department of Chemistry, n.d.), and has begun a related redesign of their junior-level analytical chemistry course. Organic chemists have also begun to examine methods to introduce more inquiry-based experiments into the sophomore-level organic chemistry laboratory.

AUTHOR NOTE

The authors acknowledge support by the William and Flora Hewlett Foundation Fund for General Education, and the NSF Division of Undergraduate Education NSF-CETP 9876676. By funding this project, NSF does not necessarily endorse the conclusions of this study. The authors further wish to acknowledge Professors Janet Robinson, Cynthia Larive, Robert Carlson, and Brian Larid from the Department of Chemistry, who participated in the development and adaptation of the laboratory curriculum, and Professor Jim Ellis from the Department of Teaching and Leadership, who directed the evaluation of the project.

REFERENCES

Abraham, M.R., & Pavelich, M.J. (1991). *Inquiries into chemistry*. Long Grove, IL: Waveland Press.

American Association for the Advancement of Science. (1993). *Benchmarks for scientific literacy—Project 2061*. New York: Oxford University Press.

American Chemical Society, Examinations Institute. (1997). *Form 1: Examination for general chemistry*. Milwaukee: University of Wisconsin-Milwaukee, Chemistry Department.

Anthony, S., Mernitz, H., & Spencer, B. (1998). The ChemLinks and ModularCHEM consortia: Using active and context-based learning to teach students how chemistry is actually done. *Journal of Chemical Education, 75*, 322–324.

Barnett, M., Barab, S.A., & Hay, K.E. (2001). The virtual solar system project. *Journal of College Science Teaching, 30,* 300–304.

Brown, F. (2000). The effect of an inquiry-oriented environmental science course on preservice elementary teachers' attitudes about science. *Journal of Elementary Science Education, 12*(2), 1–6.

Brown, L.M., Kelso, P.R., & Rexroad, C.B. (2001). Introductory geology for elementary education majors utilizing a constructivist approach. *Journal of Geoscience Education, 49,* 450–453.

Bybee, R.W. (Ed.). (1985). *Science, technology and society.* Washington, DC: National Science Teachers Association.

Coble, C.R., & Koballa, T.R., Jr. (1996). Science education. In J. Sikula, T.J. Buttery, & E. Guyton (Eds.), *Handbook of research on teacher education* (pp. 459–484). New York: Macmillan.

Cooper, M.M. (1996). *Cooperative chemistry laboratory manual.* New York: McGraw-Hill.

Course list. (2003, March 26). Retrieved July 23, 2003, from the University of Kansas, Department of Chemistry Web site: http://www.chem.ku.edu/Courses/

DeGolier, T. (2002). Using a guided-inquiry approach for investigating metabolic rate in mice. *The American Biology Teacher, 64,* 449–454.

Department of Chemistry. (n.d.). *Chemistry 636.* Retrieved July 23, 2003, from the University of Kansas, Department of Chemistry Web site: http://www.chem.ku.edu/GWilsonGroup/Chem_636/Packet/default.asp

Ellis, J.D., Heppert, J.A., Robinson, J.B., & Omar, M.H. (in press). Development and validation of a college chemistry efficacy beliefs instrument. *Journal of Science Teacher Education.*

Fetters, M.K. (1998, January). *Pushing the comfort zone: Confronting the perceptions of teaching and classroom culture.* Paper presented at the annual international conference of the Association for the Education of Teachers in Science, Minneapolis, MN. Retrieved July 23, 2003, from http://www.ed.psu.edu/CI/Journals/1998AETS/s1_6_fetters.rtf

Glasson, G.E., & McKenzie, W.L. (1998). Investigative learning in undergraduate freshman biology laboratories: A pilot project at Virginia Tech—New roles for students and teachers in an experimental design laboratory. *Journal of College Science Teaching, 27,* 189–193.

Gosser, D.K., & Roth, V. (1998). The Workshop Chemistry Project: Peer-led team learning. *Journal of Chemical Education, 75,* 185–187.

Halloun, I.A. (1985). The initial state of college physics students. *American Journal of Physics, 53,* 1043–1055.

Hanks, T.W., & Wright, L.L. (2002). Techniques in chemistry: The centerpiece of a research-oriented curriculum. *Journal of Chemical Education, 79,* 1127–1130.

Henderson, L., & Buising, C. (2001). A research-based molecular biology laboratory: Turning novice researchers into practicing scientists. *Journal of College Science Teaching, 30,* 322–327.

Hutchison, A.R., & Atwood, D.A. (2002). Research with first- and second-year undergraduates: A new model for undergraduate inquiry at research universities. *Journal of Chemical Education, 79,* 125–126.

Johnson, M. (2002). Introductory biology online: Assessing outcomes of two student populations. *Journal of College Science Teaching, 31,* 312–317.

Joint Council of Academic Officers. (2003, Spring). *Kansas Core Outcomes Project.* Retrieved July 23, 2003, from http://www.kansasregents.org/download/aca_affairs/core2003.pdf

Landis, C.R., Peace, G.E., Scharberg, M.A., Branz, S., Spencer, J., Ricci, R.W., et al. (1998). The New Traditions Consortium: Shifting from a faculty-centered paradigm to a student-centered paradigm. *Journal of Chemical Education, 75,* 741–744.

Laws, P. (1991, December). Calculus based physics without lectures. *Physics Today,* pp. 24–31.

Lawson, A.E., Abraham, M.R., & Renner, J.W. (1989). *A theory of instruction: Using the learning cycle to teach science concepts and thinking skills* (Monograph No. 1). Manhattan: Kansas State University, National Association for Research in Science Teaching.

Martin, E.E., & Howell, P.D. (2001). Active inquiry, Web-based oceanography exercises. *Journal of Geoscience Education, 49,* 158–165.

Mason, S.L., Robinson, J.B., Heppert, J.A., Wolfer, A., Ellis, J.D., & Doyle, R. (2003). *Using cooperative teams in inquiry: General chemistry laboratory investigations.* Manuscript in preparation.

McDermott, L.C., Shaffer, P.S., & Constantinou, C.P. (2000). Preparing teachers to teach physics and physical science by inquiry. *Physics Education, 35,* 411-416.

McGinnis, J.R., & Parker, C. (1999, April). *Teacher candidates' attitudes and beliefs of subject matter and pedagogy measured throughout their reform-based mathematics and science teacher preparation program.* Paper presented at the annual meeting of the American Educational Research Association, Montréal, Québec, Canada. Retrieved July 23, 2003, from http://www.towson.edu/csme/mctp/Research/AERA99Paper.htm

McNelis, B.J. (1998). Mechanism templates: Lecture aids for effective presentation of mechanism in introductory organic chemistry. *Journal of Chemical Education, 75,* 479–481.

National Commission on Mathematics and Science Teaching for the 21st Century. (2000). *Before it's too late: A report to the nation from The National Commission on Mathematics and Science Teaching for the 21st Century.* Jessup, MD: U.S. Department of Education, Education Publications Center.

National Research Council. (1996). *National Science Education Standards.* Washington, DC: National Academy Press.

National Science Foundation, Division of Undergraduate Education. (n.d.). *Directory of NSF Collaboratives for Excellence in Teacher Preparation.* Arlington, VA: Author. Retrieved July 23, 2003, from http://www.ehr.nsf.gov/ehr/due/awards/cetp.asp

National Science Foundation. (2003, January 28). *NSF Graduate Teaching Fellows in K–12 Education (GK–12).* Retrieved July 23, 2003, from http://www.nsf.gov/home/crssprgm/gk12/start.htm

Nelson, P.G. (2002). Teaching chemistry progressively: From substances, to atoms and molecules, to electrons and nuclei. *Chemistry Education: Research and Practice in Europe, 3,* 215–228.

Office of Institutional Research and Planning. (1997). *List of majors.* Lawrence: University of Kansas.

Papert, S. (1990). Introduction. In I. Harel (Ed.). *Constructionist learning* (pp. 1–8). Cambridge, MA: MIT Media Laboratory.

Robinson, J. (2001). New GTAs facilitate active learning in chemistry laboratories: Promoting GTA learning through formative assessment and peer review. *Journal of Graduate Teaching Assistant Development, 7*(3), 147–162.

Russell, A.A., Chapman, O.L., & Wegner, P.A. (1998). Molecular science: Network-deliverable curricula. *Journal of Chemical Education, 75,* 578–579.

Russell, C.P., & French, D.P. (2002). Factors affecting participation in traditional and inquiry-based laboratories. *Journal of College Science Teaching, 31,* 225–229.

Schlenker, R.M., & Schlenker, K.R. (2000). Integrating science, mathematics, and sociology in an inquiry-based study of changing population density. *Science Activities, 36*(4), 16–19.

Seymour, E., & Hewitt, N. (1997). *Talking about leaving: Why undergraduates leave the sciences.* Boulder, CO: Westview.

Stefanich, G.P. (1979). Eliciting inquiry in earth science. *Journal of Geological Education, 27,* 111–113.

Sunal, D.W., Sunal, C.S., Whitaker, K.W., Freeman, L.M., Hodges, J., Edwards, L., et al. (2001). Teaching science in higher education: Faculty professional development and barriers to change. *School Science and Mathematics, 101,* 246–258.

Thomas, R.C. (2001). Learning geologic time in the field. *Journal of Geoscience Education, 49,* 18–21.

Tobias, S. (1992). *Revitalizing undergraduate science: Why some things work and most don't.* Tucson, AZ: Research Corporation.

U.S. Census Bureau (2001). No. 593. Employed civilians by occupation, sex, race, and Hispanic origin. In *Statistical abstract of the United States: 2001* (pp. 380–382). Retrieved July 23, 2003, from http://www.census.gov/prod/2002pubs/01statab/labor.pdf

UC Regents. (2003, July 8). *Multi-Initiative Dissemination Project/NSF 2000–2004 Instructional Change in Chemistry.* Retrieved July 23, 2003, from http://www.cchem.berkeley.edu/~midp/index.html?main.html&1

UCLA, Higher Education Research Institute. (1996). *Weighted national norms for all freshmen, fall 1996.* Los Angeles, CA: Author.

Udovic, D., Morris, D., Dickman, A., Postlethwait, J., & Wetherwax, P. (2002). Workshop biology: Demonstrating the effectiveness of active learning in an introductory biology course. *BioScience, 52,* 272–281.

Walker, C.J., & Angelo T. (1998). Collective effort classroom assessment technique: Promoting high performance in student teams. *New directions for teaching and learning, 75,* 101–112.

Wink, D.J., Gislason, S.F., & Kuehn, J.E. (2000). Working with chemistry: A laboratory inquiry program. New York. W.H. Freeman.

Zebrowski, E., Jr. (2001). Natural disasters: A fascinating approach to scientific inquiry. *Journal of College Science Teaching, 30,* 376–381.

Zollman, D. (1990). Learning cycles for a large-enrollment class. *Physics Teacher, 8*(1), 20–25.

CHAPTER 20

A UNIVERSITY STUDENT'S PERSPECTIVE ON REFORM IN TEACHING UNDERGRADUATE SCIENCE

Kimberly A. Staples

ABSTRACT

The call for reform in science education provides the impetus for change in the way science is taught in K–16. However the vision for change in undergraduate science is influenced by faculty views of how science should be taught. The type and degree of change in science courses are inconsistent across content areas. In many U.S. colleges and universities today, faculty implement innovative strategies to improve science conceptual understanding and achievement. Success in undergraduate science courses weighs heavily on students' views of science, prior experience in science, and expectations of college science courses. Based on past course experiences, students enter science courses with opinions of the content, mode of delivery, and their anticipated performance. The purpose of this research is to identify the effect of innovation in college science courses on university students' perception of science and science teaching. The focus question for this research is, What are the specific innovative strategies that yield success in college sci-

Reform in Undergraduate Science Teaching for the 21st Century, pages 351–370

ence courses? In this study, student perceptions were compared based on experience in an innovative science course and a traditional science course. The results present strategies that improve student success in science.

INTRODUCTION

In *Before It's Too Late* (National Commission on Mathematics and Science Teaching for the 21st Century, 2000) the second major goal identified for improving science and mathematics teaching in the United States was to increase the number of science and mathematics teachers and improve the quality of their preparation. This mandate sends a direct message to undergraduate college science instructors of prospective classroom teachers. Teachers of K–12 science require a science learning environment that models the behaviors for students to construct knowledge rather than disseminate information. The research-based National Science Education Standards (National Research Council [NRC], 1996) do not advocate large lecture halls in which instructors lecture to students and subsequently verify concepts in the laboratory (Siebert & McIntosh, 2001). Traditional undergraduate content courses have consisted of experiments that students complete as verification activities to obtain the "right" answer (Crawford, 2001).

The major science reform effort in undergraduate college science courses focuses on the shift from science for the few to science for all. The improvement of student performance in science shifts course activities from teaching to learning (Seymour, 2001). Professional development models for undergraduate science, mathematics, and technology (SMET) education expose faculty to standards-based activities necessary for reform in content and pedagogy. The focus on collaboration between SMET faculty, science departments, universities, local schools, and teachers supports the reform initiatives set forth by professional societies of scientists and educators (American Association for the Advancement of Science [AAAS], 1993; NRC, 1996).

The National Science Education Standards (NRC, 1996) outlines the standards used to teach science content based on research in the cognitive sciences. The standards advocate science instruction that is inquiry-based, connected to pedagogy, and committed to lifelong learning. A constructivist learning approach provides the environment in which students can develop the declarative and procedural knowledge of science concepts. Students are provided opportunities to examine prior knowledge, experience disequilibrium with beliefs that oppose formerly held beliefs, and develop alternative concepts (Lawson, Abraham, & Renner, 1989). The effective implementation of a standards-based science course is mediated through a constructivist-learning environment.

Constructivist teaching practices have been recommended by reform initiatives for K–12 classrooms (AAAS, 1993). According to Fosnot (as cited in Cannon, 1997, p. 68), constructivism is "a theory about knowledge and learning"; it describes both what "knowing" is and how one "comes to know." Traditional college science courses that provide instruction in a lecture-style format do not provide student-centered instruction, opportunities for scientific inquiry that supports the construction of knowledge, or opportunities to reveal and change misconceptions (Siebert & McIntosh, 2001).

TRADITIONAL SCIENCE IN HIGHER EDUCATION

In a study by Light (2001) students rated 15 natural science courses low based on high grade competition. Students perceive good science courses as those with "modest" levels of grade competition. Surprisingly, the study found that course work was not a major factor that prevented student engagement in science. The lack of interaction among students, such as group work and the impersonal nature of instructors, affect student perceptions of college science courses.

The traditional science course does not emphasize student-centered, inquiry-oriented investigations in which students can construct their understanding by solving real-world problems. Undergraduate students' experiences in traditional college science classrooms are void of the inquiry and investigation skills they need to model in K–12 classrooms. They cannot deliver content in a manner that meets the academic needs of students (National Science Foundation [NSF], 1996). How a teacher knows what he or she knows is as equally important if not paramount to what the teacher knows. Therefore, a call for reform necessitates an examination of the delivery of science content to improve students' understanding and instruction of science. According to Treifel and Hazen (as cited in Mauldin & Lonney, 1999), "Many introductory science courses are geared toward science majors.... Introductory science courses designed for science majors fail to foster scientific understanding among nonscience majors.... Ultimately, this needless narrow approach to science education alienates most nonmajors" (p. 416). Traditional undergraduate science courses have served as "weeding out" mechanisms for science majors (Adams, 1990). Large class sizes make it difficult for the instructor to model and implement effective strategies for content courses. Barriers to change in undergraduate science courses directly relate to rewards associated with class size.

INNOVATIVE COLLEGE SCIENCE COURSES

The emergence of nontraditional content courses offers a real solution to the problem of developing competent teachers in the content areas. These courses focus on the appropriate content as well as on how a student learns the content that is consistent with improving the quality of science teacher preparation (National Commission, 2000). The design, implementation, and assessment of an effective content course that produces undergraduate students capable of appropriately teaching science in K–12 classrooms involves several important key elements. These important elements are habits of mind, scientific literacy, beliefs about the nature of science, and student views toward doing the inquiry process. In an effort to create reform in science, Stepans (as cited in Siebert & McIntosh, 2001) describes how education and science faculty team together to create science content courses that encourage conceptual understanding. Undergraduate students participate in laboratory activities, group projects, simulations, and field trips. The traditional lecture is de-emphasized. The students demonstrate improvement in attitudes toward science, science teaching, and content knowledge.

In a multiyear study funded by the National Science Foundation at Kansas State University, undergraduate students who were exposed to the learning cycle lesson format had a higher achievement than did students in traditional sections for that semester and the past 10 years (Stalheim-Smith & Scharmann, 1996). In a large nonmajor biology course, Ebert-May, Baldwin, and Burns (1999) found that cooperative learning, use of higher level questioning techniques, group work, position papers, and daily writing and speaking in class are effective strategies for science instruction. Students demonstrated an increase in biological literacy, scientific inquiry skills, reasoning, and critical thinking skills following the course. Also, Gili and Sokolove (2000) implemented active learning in an undergraduate biology course to improve student ability to produce insightful, thoughtful, and content-related questions. In contrast, students in a traditional biology course produced questions that were unchanged following the study.

According to Zelik and Bisard (2000), an effective science course addresses factual and structural misconceptions to produce conceptual change in introductory level science courses. During the study, one introductory astronomy course ($n = 181$) implemented cooperative learning groups while the other introductory astronomy course ($n = 101$) did not. The results of a pre–post misconception measure revealed that students in the treatment course had overall gains in conceptual change that were significantly higher than the control group.

In an investigation similar to Barab, Hay, Barnett, and Keating (2000), Greca and Moreira (2001) integrated physical, mathematical, and mental models to increase student construction of physical theories. Declarative and procedural knowledge examined through students' construction of mental models for physical models of theories support mental modeling. Students who do not construct mental models successfully rely on formulas to explain phenomena. They have difficulties making predictions of physical behaviors. Prior knowledge affects a student's ability to create mental models of everyday phenomena. This supports the cognitive base for using discourse to enhance students' ability to create mental models.

Dewey and Meyer (2000) investigated the effects of implementing real data in an active, hands-on learning environment in an introductory climatology course. Through the use of a climate software package, students compared laboratory results with experts and classmates. This method provides a shift from passive to active engagement in learning key concepts through hands-on data analysis. Student performance increased in ability to interpret results, apply interpretations to a research hypothesis, form conclusions, and present results to peers. The course size ($N = 35$) was conducive to implementing group work and facilitating a sense of community.

An inquiry-oriented approach to science instruction allows instructors to interact with students. French and Russell (2001) used a student-centered approach to improve construction of science concepts in a large mixed-majors biology course. Facilitators assisted instructors with students in groups during an inquiry-oriented course. The facilitators possessed a strong interest in teaching science or science education and had science pedagogical content knowledge. Concept maps used through multimedia facilitate group exercises. Survey results from students ($N = 588$) indicated that facilitators played a positive role during group work. The transition from a traditional to inquiry-based course provides the environment for integrating science concepts at various levels.

Innovative college science courses that reflect science reform create an inclusive environment for science learning. Group work, accessibility to the instructor, a constructivist approach to science instruction, and reduced class size facilitate science for all. McLaughlin and Thomas (1999) identified similar elements in nontraditional science content courses: (a) learning experiences framed within a context of content-specific pedagogy, (b) inquiry-based approaches to learning, (c) pedagogically oriented activities and assignments, and (d) sense of building a professional community. According to the work of Drugor (as cited in Siebert & McIntosh, 2001), an introductory biology course designed to promote a sense of community creates a learning environment where students feel important and comfortable. A sense of community is described in the National Science Education Standards (NRC, 1996). Teaching Standard E (p. 46) emphasizes collaboration among students, respect for diverse ideas, and a shared understanding of rules of scientific discourse.

The National Science Education Standards (NRC, 1996) emphasize active, inquiry-based learning. In a study by Ebert-May and Brewer (1997), biological literacy was compared between majors and nonmajors enrolled in a large ($N = 559$) biology course. The results of self-efficacy, a national biology test, and a process skills instrument were analyzed for the effect of nontraditional and traditional course sections. Results revealed that cooperative learning and questioning strategies improved students' understanding of content and process.

THE CONSTRUCTION OF KNOWLEDGE IN COLLEGE SCIENCE COURSES

The learning cycle is a three-phase inquiry approach that includes exploration, term introduction, and concept application. According to Lawson (2001), the learning cycle effectively helps students construct concepts and conceptual systems and develop reasoning patterns. In his work, Lawson illustrated how reasoning is used during concept construction with an "if/then/therefore" reasoning pattern. The researcher reported that students will not engage in constructing knowledge if they are simply told answers. However, if science instruction is more open-ended, students will use and improve their reasoning skills while exploring nature and using "if/then/therefore" reasoning to test their ideas.

During an action research study to improve reflection into classroom practice, Johnson (2000) used the Constructivist Learning Environment Survey (CLES) to evaluate the degree to which a college science classroom environment is consistent with constructivist perspectives. The results revealed that students' posttest scores increased significantly from the beginning of the semester. The students perceived that the current environment was more constructivist than were prior experiences in science courses. The CLES has been used to determine the level of dissatisfaction of students in traditional college science courses. The CLES is a 25-item, 5-choice, Likert-type survey with responses ranging from *almost always* to *almost never.* This instrument provides quantitative data on five important elements of a critical constructivist-learning environment for science as follows: (a) the degree of personal relevance to the students' studies, (b) the degree of student-shared control over their learning, (c) the degree to which students feel comfortable expressing concerns about their learning (critical voice), (d) the degree to which students are able to listen to and reflect on each others ideas to improve their understanding (negotiation), and (e) the extent to which science is viewed as ever changing (Taylor, Fraser, & Fisher, 1997). The CLES indicates the effect of the innovative course experience on student preference for learning in a constructivist environment. The alpha coefficient for reliability was 0.92 for individual student and 0.98 for class mean.

METHOD

The purpose of this research was to identify strategies that students perceive as effective in learning science. Undergraduate college students were surveyed to determine whether there was a differential effect as a result of the nontraditional course experience on their views of science, science teaching, and learning. The study included a comparison between undergraduate students enrolled in an innovative and traditional college science courses. The focus questions for the research were, In what environment do undergraduate students prefer to learn science? Which specific strategies are key to the construction of knowledge and understanding science content? Which specific strategies serve as deterrents to learning science? What characteristics do students prefer in college science instructors?

The study consisted of 247 undergraduate students divided into four groups. The first group of students was enrolled in an innovative science course. Within this group, 26 students were undergraduate education majors and 12 were non-education majors. These students experienced innovative science instructional strategies for undergraduate college science teaching. A constructivist-learning environment was implemented during this course. The instructor used the learning cycle lesson format for science instruction. Prior knowledge, student engagement, application of concepts, project-based assignments, and alternative assessment were integral components of the innovative science course experience.

The second group consisted of 174 undergraduate students enrolled in a traditional science course. Within this group, 16 students were education majors and 158 were non-education majors. These students experienced science lessons in a traditional format. The lecture portion of the course was separate from the laboratory. The students were not actively engaged during the lecture. Assessment consisted of weekly quizzes and multiple-choice exams. Both groups were first-year college students. The students enrolled in the traditional course did not have prior experience in an innovative college science course.

The third and fourth groups consisted of a stratified random sample of undergraduate students in their junior or senior year. The third group consisted of 17 undergraduate students who experienced the innovative science course. The fourth group consisted of 18 undergraduate juniors and seniors who did not experience the innovative science course.

COURSE DESCRIPTIONS

An Innovative College Science Course

The treatment in this study consisted of experience in an innovative undergraduate science course. The 15-week nontraditional science course met three times a week for 2-hr sessions in addition to times arranged outside regular class time. There was no differentiation between sessions. Full integration of science instruction and activities was the approach used to emphasize meaningful learning. The class with an enrollment size of 46 was divided into cooperative groups of four to five students. The students enrolled in this course were classified as education and non-education majors in their freshman and sophomore years of undergraduate study.

The course focused on problem solving, creative thinking, decision making, and higher order thinking. The primary goal of the course was to develop scientific literacy in undergraduate education majors planning to teach in elementary and middle schools.

The course content was organized into five science and engineering modules as follows: (a) Earth in Space; (b) The Dynamic Atmosphere; (c) Forces of Flight; (d) Designs for Flight; and (e) Guided Design Project, "Mission to Mars." The learning cycle instructional approach was in each module, following a sequence of exploration, invention, and expansion. The content coverage involved knowledge in multidisciplinary and interdisciplinary areas. The coverage included procedural knowledge, thought processes, and conceptual application through transfer to real-world relevant settings. Course activities included field trips to a local airport, weather station, and the U.S. Geological Survey, computer simulation problems, day and night sky observing sessions, teaching science lessons in local schools, wind tunnel testing, and monitoring real-time satellite images using the computer laboratory. Alternative assessment included open-ended questions on quizzes and examinations, constructing models of the solar system, using student-constructed models to solve problems, and doing project-based assignments.

A TRADITIONAL COLLEGE SCIENCE COURSE

The control in this study consisted of experience in a traditional science course. The 15-week traditional science course met three times a week for 50 min for the lecture. There were 20 laboratory sections. Each student was assigned to one of the laboratory sections that was held once a week for 1 hr 50 min. The major topics were The Study of Life, Living Cell, Continuity of Life, Evolution and Diversity of Life, and Living Environment. The course size was 381. The students enrolled in the control course were classified as education and non-education majors in their freshman and sophomore years of undergraduate study.

Assessment consisted of three 1-hr examinations during the regularly scheduled lecture periods and a final examination. The examinations contained 50 questions that covered the text and lecture material in a multiple-choice, matching, and true–false format. The major objectives of the course for students included (a) to gain an appreciation for the diversity of life in form and function; (b) to understand the scientific methods, execute experiments, and critically evaluate scientific information; and (c) to apply scientific knowledge gained in class to biological issues in everyday life.

INSTRUMENTATION

Data was collected on undergraduate students' preference of learning environment for science content, science learning and teaching. Quantitative and qualitative data was obtained through surveys, oral and written interviews. The undergraduate education majors in both groups were administered pre and posttest CLES surveys. At the end of the semester, eight students were randomly selected from the innovative and traditional courses to complete oral and written interviews. The interview protocols, Perception of Traditional Science Course and Perception of Innovative Science Course, consisted of 10 oral questions and 13 written responses. The interviews identified the effects of innovative science instruction on undergraduate students' beliefs about science, science teaching, and science learning.

RESULTS

Quantitative Data Analysis

This section describes the measures used for the quantitative data analysis. The Statistical Product and Service Solutions (SPSS) software, version 10.0 for Microsoft Windows, was used to analyze data from the CLES survey. An analysis of covariance (ANCOVA) test was used to quantitatively analyze the pretest and posttest surveys for all courses. The dependent variable for this test was experience in a nontraditional science course. The Wilk's lambda test indicated a significance of less than .05. An ANCOVA was conducted to determine the effect of course on the posttest scores of the CLES survey. The covariates were the CLES pretest scores. After significant adjustment by the covariates, the independent variable varied significantly by experience in course, posttest CLES, $F(3, 246) = 4.433, p < .05$. The results are presented in Table 20.1.

Table 20.1.   ANCOVA Summary Table for All Courses on Posttest CLES Scores

Source	Dependent Variable	SS	df	MS	F	p	η^2
Intercept	Post CLES	6299.623	1	6299.623	27.121	.000	.102
Pre CLES	Post CLES	8158.086	1	8158.086	35.122	.000	.128
COURSE	Post CLES	3088.991	3	1029.664	4.433	.005	.053
Error	Post CLES	55746.906	240	232.278			
Total	Post CLES	1798287.000	247				

Tukey HSD multiple comparisons were conducted to detect differences between experiences in courses. The quantitative results indicated that for the groups tested, the independent variable, experience in the nontraditional science course, had an effect on students' preference for a constructivist learning classroom environment. Innovative instruction based on national standards produced a significant effect on undergraduate students' perceptions of science teaching and learning.

The students enrolled in the traditional science course had a significantly lower ($M = 81.3$) preference for a constructivist learning environment for science than did students in courses 1 ($M = 89.5$), 3 ($M = 89.3$) and 4 ($M = 89.2$). The adjusted mean scores for the CLES are graphically represented in Figure 20.1.

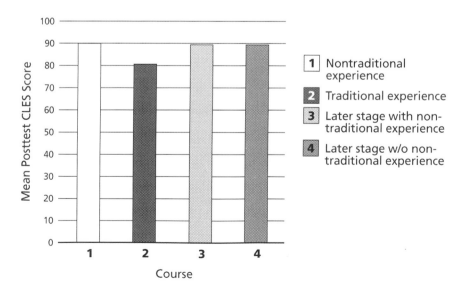

Figure 20.1.   Comparison of adjusted mean posttest scores on the CLES.

Qualitative Analysis

During the research study, oral and written interviews were completed at the end of the semester by undergraduate education majors in the innovative and traditional college science courses. Each student interviewed was randomly selected. The students were informed that participation in the research study was voluntary and would not affect their grades in the course. The data from the interviews were analyzed using a grounded theory methodology. The transactional system analysis was the approach used to identify actions and interactions with relationships that produce specific phenomena (Strauss & Corbin, 1991). Each interview was analyzed individually and compared between groups. For each group of students in the study, levels of actions related to science teaching and learning are identified.

Conditional paths reveal student activities referred to as "negotiations," which identify the effects of the innovative college science course. Each question was coded for concepts that form categories. The categories were used in the transactional system analysis to form levels of action during science teaching and learning. The qualitative analysis of data obtained from students in the study revealed eight levels of action based on interaction with science teaching and learning. The levels of action and interaction form a conditional matrix used to describe the negotiations that take place during science instruction. The categories and levels of action are listed in Table 20.2.

Table 20.2.   Levels of Actions Based on Experiences in College Science Courses

Category	Levels of Action
Personal Beliefs About Science:	1. Prior Experience
Perceptions of Science Learning:	2. Science Performance in the Course
	3. Science Instruction
	4. Change in Beliefs About Science
	5. Scientific Literacy
Beliefs About Science Teaching:	6. Belief About Teaching Science
	7. Ability to Teach Science
	8. Emphasis for Teaching Science

For each level, related concepts have been placed into categories to describe student engagement with science learning, science teaching, and science content. The levels of action build upon each other. As students

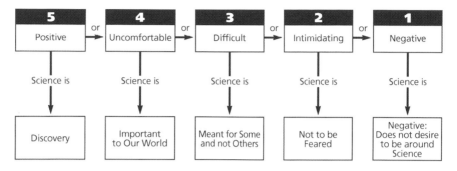

Figure 20.2.   Undergraduate students' beliefs about science for students in elementary classrooms.

interact in college science courses, their activities are negotiated based on conditions within each level. Therefore, the negotiations define how science will be transferred to students in the elementary classroom. The conditional matrix that represents the negotiations between levels of action is displayed in Figure 20.2. The arrows in the conditional matrix represent the negotiations that take place for the conditions within each level of action. As a result, the transactional analysis system can be explained as a bidirectional movement across levels of action. The lens through which undergraduate students view science defines the negotiations that take place across levels of action during science instruction. Their behaviors and perceptions of science teaching and learning during college science courses are based on personal beliefs about their ability to succeed.

Personal Beliefs About Science

An analysis of the data reveals that prior experience with science forms the foundation for interactions with science. Both groups of students in innovative and traditional courses used prior experiences with science to negotiate success across each level of action in college science courses. The undergraduates attached meaning to instruction and learning based on a personal belief about science. Their personal attitudes toward science represent the core of the conditional matrix. The undergraduate students who viewed science as positive were exposed to the innovative science course. One student stated, "Now I feel positive about science; I have seen ways to integrate science in my classroom" (post-semester interview with a student in the innovative course, December 7, 2001). None of the traditional science students held positive reactions to science. The positive responses were not associated with the innovative course. The levels of action for positive feelings about science were not self-centered. Therefore, the negotiation was centered on finding ways to integrate science into future roles as an elementary teacher of science.

> Our teacher gave lots of models, and we did a lot of creating models. Neat ideas to use illustrating concepts. If I were teaching a fourth grade lesson on seasons, I would use a globe and show how the earth is tilted and how this tilt combined with its movement around the sun effects the season. The best way to teach science in elementary school is hands-on, creating models, allowing kids to use their sense to discover. (post-semester interview with a nontraditional student, December 7, 2001)

Undergraduate students from both courses felt that science was challenging. The reasons for this view emanated from being inundated with science information during high school. Science was viewed as a vast body of knowledge. Those who felt science was challenging have a sense of agency with science but not to the same degree as students with positive views of science.

The perception of science as difficult came from an undergraduate education major in the traditional science course. Prior experiences did not provide a strong background in science. During high school the "reason" for science was never actualized. Therefore, the absence of a sense of the relevance of science created a barrier to learning. This level creates a reduced sense of agency with science.

> Science is hard for me because I don't have a strong background in science. In high school I didn't see the reason. Now it's easier to see why science and research affects my everyday life, such as disease. My instructor teaches stuff we would want to know about, past, present, and generations. Science is sometimes hard to understand, so I would emphasize the importance of science to my class and try to help them understand how important it is to daily lives. (interview with traditional student, December 3, 2001)

Clearly the student viewed science as relevant. However, a context for teaching science is not developed. Experience in the traditional science course produced the belief that science is important. However, the behavior required to teach science as important was not present. Undergraduate students who viewed science as intimidating described high school science teachers as less than effective. This view predisposed the students to high anxiety and fear of learning and teaching science. At this level the sense of agency with science is absent. Undergraduate students intimidated by science focused on survival mechanisms during college science courses. Prior experiences had produced the belief that science is not for all.

> I am scared of science. I did not have the best science teacher in high school. Right now, there is still fear when I think about science. In this course we did

a lot of group work. This helps you see other peoples' perspectives. They can help you out. (interview with nontraditional student, December 7, 2001)

Based on prior experiences students who hold negative feelings about science are totally disenfranchised from science teaching and learning. An undergraduate from each course communicated negative feelings about science. The desire not to be around science and not to teach science is indicative of the disenfranchisement from science. Students with negative feelings about science require instructional strategies that counter these beliefs. Students in this range negotiate activities in the course as validation for their deep-rooted beliefs about science.

I don't want to be around science. I'm only good at basic scientific theories. Science is the study of biology, astronomy, geology, physics, chemistry and engineering. I did not learn how to use the computer-simulated solar system programs because there was not a set of directions that told you how to use it. You have to figure them out. I couldn't apply knowledge during the first test because the questions were different from things in my notebook. (interview with nontraditional student, December 7, 2001)

Personal beliefs about science are central to negotiations that take place between each level of action during science teaching and learning. The instruction, learning, and science curriculum are internalized differently based on students' beliefs about science.

Perceptions of Science Learning

Science performance in the course.   The next level of action relates to students' perceptions of factors that influence their success in science courses. The first level of action was affected by personal beliefs about science. As a result, students negotiated events in each course as a support system for successfully completing the course. The difference between the negotiations was based on exposure to different instructional strategies.

The students in the lower range of personal self-efficacy in the innovative science course viewed "aids" as keys to their science success. Success in the course was dependent upon using open notebooks, notes, and having explicit directions for activities. Any form of exploration was viewed as a deterrent to learning science. The positive students viewed models, and created models to illustrate concepts as keys to their success in the course. The positive students did not interpret activities such as note taking and receiving step-by-step instructions as conditions required to insure success in the course.

Science instruction.   The students who experienced the innovative science course cited specific conditions as effective for learning. The students in the lower range of personal self-efficacy negotiated hands-on activities as an aid in understanding science. Visualizations, "re-teaching," models, and observations were internalized as aids for personal success in science. Group work was negotiated as a source of help for students in the lower range of beliefs about science. Positive beliefs about science were expressed by citing hands-on activities as a strategy for "figuring out" or inventing science concepts.

The traditional students with low personal science teaching efficacy cited notes on the board and PowerPoint presentations during the lecture as effective science instruction. The laboratory provided opportunities to visualize science concepts, making connections to the lecture. None of the traditional students negotiated group work as an effective strategy for learning science. Also, none of the traditional students interviewed cited the term "hands-on," even when describing laboratory activities.

> The most beneficial experience to learning science in this course was notes posted on the web page, and the instructor went into detail during lecture. This process helped me understand many things before the lecture, and then every question was answered in class. These notes helped me to be prepared for our lesson. (interview with a traditional student, December 3, 2001)

The traditional students continuously made reference to "teaching" science and the importance of "studying" science concepts. The students internalized strategies for teaching science that are not based on knowledge construction.

Change in Beliefs About Science

Modeling by the instructor and students were perceived as an instructional strategy that changed misconceptions about science concepts.

> The teacher modeled for us how the moon moves around the earth by holding a ball and moving around the light bulb. This helped us understand clearly how the moon phases operate. Modeling and student explanation was most beneficial to our learning science in this course. (interview with nontraditional student, December 7, 2001)

The students in the traditional course indicated that direct information provided by the instructor changed their beliefs about previously held science misconceptions. The instructor changed their perception of the relevance of science to everyday life, not an activity during the course.

Laboratory activities improved conceptual understanding only when they matched the lectures.

Science Literacy

The interviews revealed a direct interaction between negative views of science and science literacy. Students who felt disenfranchised from science also did not perceive themselves as scientifically literate. Overall, the students in the nontraditional science course felt scientifically literate. The ability to solve problems, to obtain necessary information to solve problems, and to communicate with a certain level of science knowledge, produce science literacy. None of the students in the traditional science course indicated they felt scientifically literate. The traditional science course focused on research related to current issues in science. The students negotiated that the inability to read research articles and a lack of basic knowledge contributed to poor science literacy.

Beliefs About Science Teaching

Effective science teaching.   The students in the innovative science course indicated that hands-on activities were the best ways to teach science. They transferred events in the course to future practices in elementary classrooms. Modeling, creating models, and using senses were cited as effective science teaching strategies for elementary students in the classroom. The negotiation for this level of action transfers events in the course to future events in the elementary classroom. The students had accepted, integrated, and identified the placement of effective science teaching strategies into their repertoire of scientific pedagogical skills.

> The hands-on strategy was great and group work. Hands-on really helps kids understand. Things that kids could relate to and use later in life are important to emphasize in science teaching. I think kids being able to apply what they learn is also important in science teaching. (interview with a nontraditional student, December 7, 2001)

The students in the traditional science course believed that step-by-step learning, experiments, and observations were effective ways to teach science. These students did not believe that they were scientifically literate. They negotiated that elements of teacher-centered instruction rather than those of student-centered instruction were effective strategies for science teaching. Students who felt intimidated by science expressed a desire to learn "simply" and "slowly" with models and visuals.

Ability to teach science.   Each student in the innovative course believed that he or she could teach science as a result of interactions with conditions in the course. Even students with negative beliefs about science had positive feelings about their ability to teach science. Those who believed science was hard and were disenfranchised from science felt a sense of ownership of science after completing the innovative science course.

> I think I can teach airplanes, metric conversions, and about stars. Before it was hard. However, the course has been interesting. Before I was not sure about phases of the moon, now I understand them. (interview with nontraditional student, December 7, 2001)

The students in the traditional course believed they could teach specific science concepts. But, they were limited in their view of what they could teach. The attitudes they expressed about the ability to teach science did not go beyond the topics covered in the course.

Emphasis for teaching science.   The innovative students negotiated science pedagogical knowledge and science content as areas for emphasis in teaching science. Applications, inquiry skills, the scientific method, teacher modeling of good behavior, discovery, and observation were elements they plan to emphasize in science teaching. The content included seasons, moon phases, forces, motion, conversions, and science applicable to everyday life.

The students in the traditional science course stated that observations, deductive skills, hypothesis testing, and scientific research should be emphasized during science teaching. Content such as growth, the solar system, and reproduction should be emphasized. Specific activities and content experienced during the traditional science course were negotiated as direct activities to emphasize in science teaching.

GOALS FOR SCIENCE TEACHING

Overall, both groups of students connected the goal of science for elementary classroom students to their personal beliefs about science. The fundamental difference was the negotiation that takes place across levels of action during the courses. The pedagogy and science content identified as important in science teaching varied between the innovative and traditional science courses. The students' goals for science teaching represent the outermost level of the conditional matrix. The desired outcomes for science teaching are presented in Figure 20.2.

Personal beliefs about science teaching and learning are affected by prior experiences with science instruction. Undergraduates exposed to innovative science instruction adopt constructivist classroom practices for personal science success and future classroom behaviors. By contrast, students in traditional science courses with low personal beliefs about science do not experience science content within a context of pedagogy.

SUMMARY

Efforts toward reform in science education and teaching focus on student-centered classrooms where students are active participants in the learning process. The goal of science teaching in classrooms today includes creating a society of individuals capable of making science-based decisions with regard to political, economic, and societal issues. Innovative science courses can influence or even change students' beliefs about science and alternative conceptions. Undergraduates' beliefs about science teaching and learning are affected by instructional strategies described in the National Science Education Standards (NRC, 1996) and by the American Association for the Advancement of Science (1990).

Innovative instructional strategies have multiple images in science content courses. Constructivism should be the foundation for strategies implemented in innovative science courses. The activities perceived by undergraduates as effective include modeling, active engagement, discussions, exposure to interdisciplinary science content, and collaborative work. Traditional science courses do not foster an environment for undergraduates to successfully connect science with everyday experiences. The nature of science is not supported in a traditional approach to teaching. The absence of inquiry-based investigations disconnects students with the nature of science. Undergraduates perceive the connection between concepts and the real world as an effective strategy toward science learning. This can be achieved through relevant, minds-on applications of science content. Modeling and inquiry-based instruction help students construct scientific knowledge.

Based on the literature and this study, a recommendation is made for college science courses to address science standards in the curriculum. The effect of science standards in introductory college science courses would be a focus for further research into reform in higher education.

REFERENCES

Adams, D. (1990). Science education for non-majors: The goal is literacy, the method is separate courses. *The Bulletin of Science, Technology, and Society, 10,* 125–129.

American Association for the Advancement of Science. (1990). *Science for all Americans: Project 2061* (2nd ed.). New York: Oxford University Press.

American Association for the Advancement of Science. (1993). *Benchmarks for science literacy.* New York: Oxford University Press.

Barab, S., Hay, K., Barnett, M., & Keating, T. (2000). Virtual solar system project: Building understanding through model building. *Journal of Research in Science Teaching, 37,* 719–756.

Cannon, J.R. (1997). The constructivist learning environment survey: May help halt student exodus from college science courses. *Journal of College Science Teaching, 27,* 67–71.

Crawford, B. (2001). Embracing the essence of inquiry: New roles for science teachers. *Journal of Research in Science Teaching, 37,* 916–937.

Dewey, K., & Meyer, S. (2000). Active learning in introductory climatology. *Journal of College Science Teaching, 29,* 265–278.

Ebert-May, D., & Brewer, C. (1997). Innovation in large lectures: Teaching for active learning. *Bioscience, 47,* 601–608.

Ebert-May, D., Baldwin, J., & Burns, D. (1999). The development of a college biology self-efficacy instrument for nonmajors. *Science Education, 83,* 397–408

French, D., & Russell, C. (2001). The lecture facilitator: Sorcerer's apprentice. *Journal of College Science Teaching, 31,* 116–121.

Gili, M., & Sokolove, P. (2000). Can undergraduate biology students learn to ask higher level questions? *Journal of Research in Science Teaching, 37,* 854–870.

Greca, M., & Moreira, M. (2001). Mental, physical, and mathematical models in the teaching and learning of physics. *Science Education, 86,* 106–121.

Johnson, K. (2000). Constructive evaluations. *The Science Teacher, 67,* 38–41.

Lawson, A. (2001). Using the learning cycle to teach biology concepts and reasoning patterns. *Journal of Biology Education, 35,* 165–169.

Lawson, A., Abraham, M., & Renner, J. (1989). A theory of instruction: Using the learning cycle to teach concepts and thinking skills. *National Association of Research in Science Teaching Monograph, 1.*

Light, R.J. (2001). *Making the most of college.* Cambridge, MA: Harvard University Press.

Mauldin, R., & Lonney, L. (1999). Scientific reasoning for nonscience majors: Ronald N. Giere's approach. *Journal of College Science Teaching, 28,* 416–426.

McLaughlin, A., & Thomas, D. (1999). Making science relevant: The experiences of prospective elementary school teachers in an innovative science content course. *Journal of Science Teacher Education, 10,* 69–91.

National Commission on Mathematics and Science Teaching for the 21st Century. (2000). *Before it's too late.* Jessup, MD: Education Publications Center.

National Research Council. (1996). *National Science Education Standards.* Washington, DC: National Academic Press.

National Science Foundation. (1996). *Shaping the future: New expectations for undergraduate education in science, mathematics, engineering, and technology.* Arlington, VA: National Science Foundation.

Seymour, E. (2001). Tracking the processes of change in U.S. undergraduate education in science, mathematics, engineering, and technology. *Science Education, 86,* 79–105.

Siebert, D., & McIntosh, W. (2001). *College pathways to the Science Education Standards.* Arlington, VA: National Science Teachers Association Press.

Stalheim-Smith, A., & Scharmann, L. (1996). General biology: Creating a positive environment for elementary education majors. *Journal of Science Teacher Education, 7,* 169–178.

Strauss, A.L., & Corbin, J. (1991). *Basics of qualitative research: Grounded theory procedures and techniques.* Newbury Park, CA: Sage.

Taylor, P.B., Fraser, B., & Fisher, D. (1997). Monitoring constructivist learning classroom environments. *International Journal of Educational Research, 27,* 293–302.

Zelik, M., & Bisard, W. (2000) Conceptual change in introductory-level astronomy course: Tracking misconceptions to reveal which—and how much—concepts change. *Journal of College Science Teaching, 29,* 229–232.

CHAPTER 21

A UNIVERSITY SCIENCE EDUCATION RESEARCHER'S PERSPECTIVE ON REFORM IN TEACHING UNDERGRADUATE SCIENCE

M. Jenice Goldston

ABSTRACT

This narrative inquiry explores a professional development partnership between a geologist and a science educator created to reform undergraduate geology lecture and laboratory courses. The study took place in a large university. Fifty undergraduates enrolled in the geology lecture and 20 students concurrently enrolled in the laboratory participated in the study. Data included student interviews and journals, geologist's essays, field notes, and peer consultation documentation. Data analysis focused on elements within the data that fostered course reform and uncovered dimensions of the collaboration that enlightened and gave meaning to actions, events, and dialogue experienced by the researchers. Discussion of findings uncovers tacit barriers to collaboration as well as components that contribute to a successful collaboration within a competitive university culture. Findings include (a)

Reform in Undergraduate Science Teaching for the 21st Century, pages 371–387
Copyright © 2004 by Information Age Publishing
All rights of reproduction in any form reserved.

negotiating the different languages of the researcher's disciplines, (b) providing sustained systematic support from multiple venues, (c) writing observations and reflections as part of peer consultation visits, and (d) "listening to silent collaborators."

INTRODUCTION

According to McLaughlin (1993), when creating a partnership for professional development it is important to recognize that collaboration as a medium to cultivate change through building relationships is increased when it is accompanied by shared experimentation and challenging discourse. Even so, forming collaborations within a large university is no simple task—it must meet the needs of the participants whereby something is gained by each of them and the purpose of the collaborative is better served through synergy rather than individual expertise (Loucks-Horsley, Hewson, Love, & Stiles, 1998).

The model and ideas that unfold in this chapter come from the voices of two colleagues, a science educator and a geologist, from similar yet different academic cultures. We work in a large university where the arts and sciences are a separate college from education, both physically and politically. The level of research, teaching, and service contributions determines professional success in both colleges. Research is paramount in both colleges; however, there is more balance between teaching and research in the College of Education. There is no handbook that instructs one on how to develop collaborative working relationships across university departments. For that matter, as I found out early in my career, there are no guidelines and few rewards for developing collaborative partnerships within a department or college. It became apparent that within higher education the impetus behind success was competition. Competition within and between colleges for grant funding and research is paramount in a university culture. Given this, the partnership discussed in this chapter is not that common. It began ironically as a joint endeavor to compete for grant funding to improve undergraduate geology education. Both the geologist and science educator saw mutual benefits by joining forces that led to a research relationship exploring the use of active-learning strategies in geology courses.

One focus of this chapter is to describe the partnership using a model, peer consultation, as a professional development tool that utilizes collaboration to build relationships that foster reforming undergraduate teaching practices. Peer consultation, a model dependent upon collaboration (Bernstein, 2000) was used to enhance the professional development of the geologist in changing her teaching practices. The following background section discusses key findings on professional development practices and the role of collaboration set within the K–12 settings. This research base provides a starting point for understanding the tacit dilemmas faced by the university researchers upon developing a professional development partnership.

BACKGROUND

Reform in Undergraduate Sciences: A Need for Professional Development

Results from international studies such as the TIMSS report cite failures in our educational system to produce students who can compete in science and mathematics with their counterparts in other nations (Snyder, Hoffman, & Geddes, 1998). Unfortunately, responsibility for these failures typically falls upon K–12 teachers, their professional development, or their university preparation. In the last decade accountability for educational failures has found its way into postsecondary education. For instance, in highlighting a need to reform education across levels K–16, the National Research Council (NRC) states, "Most instructors of new science teachers—including postsecondary faculty in science, mathematics, engineering, technology and education—have not been able to provide the type of education K–12 teachers need to succeed in their own classrooms" (NRC, 2001, p. 2). Furthermore, they stated that those who teach in grades K–8 do not have sufficient content knowledge to teach effectively. The findings poignantly suggest that educational reform, professional development, and accountability should encompass the K–16 spectrum, which includes undergraduate instruction.

Professional Development

Within the educational field there exists a plethora of research on effective K–12 professional development in science teaching and the change process itself (Coble & Koballa, 1996; Darling-Hammond, Hudson, & Kirby, 1989; Fullan, 1991, 1993; Loucks-Horsley, 1995, 1996; Loucks-Horsley et al., 1998). "Professional development in essence refers to opportunities offered for educators to develop new knowledge, skills, approaches, and dispositions to improve their effectiveness in their classrooms and organizations" (Loucks-Horsley et al., 1998, p. 79). In short, professional development is about improving one's practice. According to Loucks-Horsley et al. (1998), the following is what we know about the participants involved in effective professional development: (a) What learners know influences their learning, (b) learners construct new knowledge, and (c) knowledge is constructed through the process of change. From another stance, effective professional development builds on what we know about teachers and teaching. When translated into the university setting and *applied* to college and university instructors, these would include that (a) the purpose of instruction is to facilitate learning, (b) teaching as a profession has its own specialized knowledge, and (c) the practice of teaching is complex. Using the research base (National Staff Development Council, 1994, 1995a, 1995b; NRC, 1996) Loucks-Horsley and colleagues (1998) identified several broad tenets that foster effective professional development. These professional development tenets foster collaboration and collegiality, promote experimentation and risk taking, draw their content from available knowledge bases, involve participants in decision making, provide time for planning, practicing, and reflecting, and provide leadership and sustained support. Research on professional development and reform has been conducted primarily within the K–12 setting. However, these research findings and the following section on the change process make available important referents for viewing the professional development partnership between the geologist and the science educator.

The Process of Change

According to Fullan (1991, 1993), change can be both individual and organizational. In both cases, change takes time and persistence, those undergoing change need support and assistance to change, most systems resist change, and change is complex. Furthermore, educational changes, no matter what the setting, require people to act and think in new ways. Change is progressive and occurs over time through active engagement with new ideas, real experiences, understandings, and dialogue. According to Guskey (1986) changes in the way one thinks is the end product of instructors' using a new practice and seeing their students benefitting from it. Others have suggested that change may not be linear and that ideas, attitudes and behaviors are mutually interactive. For instance, one's current ideas and thoughts influence choices made while the counterpoint is that one's reflections on the activities influence subsequent thoughts and decisions.

A related referent for viewing this study is to situate the professional development partnership of the geologist and science educator within the "culture" of the university. Some anthropologists define culture as set of beliefs, actions, and patterns of behavior shared among members of a group; others see it as a set of ideas, expectations, or rules for what is acceptable in a particular group. No matter how culture is defined, few would disagree that culture molds expectations and behaviors within groups, organizations, and institutions.

According to Hord and Boyd, "professional development activities contribute to a culture of collegiality, critical inquiry, and continuous improvement" (1995, p. 10). In this study the reverse is true, collaboration was established before professional development activities occurred. Generally speaking, collaboration is fostered through meeting the needs of the participants while equipping individuals with new skills and strategies. Collaboration is not simply a contextual variable that enhances individual change and growth—learning and change are linked to the social culture in which the individual participates. Accordingly, the culture of the university was a persistent influence on the collaboration between the geologist and the science educator as well as the change processes. Research findings indicate that professional development thrives where collaboration, experimentation, and challenging discourse are possible and welcome (Hord & Boyd, 1995; Little, 1993; Norris, 1994). These conditions were elements of the professional development partnership of this work. Thus, while collegiality and collaboration linked to professional development foster change, the culture of the institution itself either contributes or hinders the process. The following section provides the impetus for improving undergraduate science teaching through professional development.

Collaboration as a Professional Development Process for Reform in Undergraduate Science

In this extended example, the focus of professional development is on a university geologist collaborating with a science educator. Though research on professional development in science teaching is predominantly based upon K–12 teachers, it provides a guide for exploring how tenets of professional development operate in a setting other than public schools. Past science reform initiatives placed little emphasis on changing the way undergraduate science content courses are taught. Their focus was generally on content and curriculum issues. If reform is to have an impact system-wide (K–16) all stakeholders must be actively involved. Therefore it is important for science faculty who teach undergraduates the "knowledge and skills of the discipline" and for science educators who have "specialized knowledge in pedagogy" to become co-participants in the reform.

Setting and Methodology

This study took place in a large Midwestern university as a partnership between the co-researchers evolved to explore ways to reform undergraduate geology courses. The course participants included 50 undergraduate students in lecture and 20 (also enrolled in the lecture) undergraduates in geology laboratory. Data were collected in the fall semester. However, discussions and reflections between the geologist and science educator extended well beyond that time frame into the present.

The study, a narrative inquiry in nature, draws from critical educational science supported by Wilfred Carr and Stephen Kemmis (1986). This research is a form of self-reflective inquiry conducted by the researchers within the social culture of the university to improve the rationality of their own practices, understanding their own practices, and the situations where the practices occur. From the stance of Lincoln and Guba (1985), the methods focused on getting at the truth in a collaborative, naturalistic manner.

Data collected for the study included student interview transcripts (5 volunteers were interviewed 3 times during the semester), student journals, geologist's journals, peer consultation documents (12 documents written by each researcher after visiting each other's classes), field notes and memos, and essays by the geologist. Data were analyzed, coded, and categorized in an inductive manner in search of themes and discrepancies. The *focus* of the analysis was on elements found within the written data that uncovered dimensions of the collaboration that would enlighten and give meaning to actions, events, and dialogue experienced by the researchers throughout the semester. Data were coded independently. Researchers then compared their separate analyzes for categorical similarities and differences. Differences were discussed and negotiated for meaning between the researchers as themes related to the collaboration evolved.

The purpose of this narrative inquiry was to determine critical components in collaborating to change undergraduate teaching in higher education that hindered or fostered the success. Included are the voices of undergraduates that shared the experiences with the geologist. They, too, were collaborators within the change process.

The extended example begins with key heuristics to help guide those who wish to engage in collaboration across colleges within a university. This section is called, "Rules of Engagement" because what I discerned was that though both researchers taught and conducted research within the same university, the research and teaching cultures within which we work were very different and affected our interactions. These rules emerged as we analyzed the interactions, events, and dialogue that transpired in our attempt to identify what made our "collaboration across cultures" work.

Rules of Engagement

Rule 1:   Build shared understanding of the languages of the disciplines to create a mutual vision.   When we began the collaboration, it was apparent that we shared a common language during early conversations regarding *some* teaching elements and terms, but it was more apparent that there were differences in the way we spoke and thought about teaching. As a matter of fact, the geologist called the language of the science educator, "edu-speak jargon." This was not meant to be a derogatory statement, rather it was a wake-up call, implying that the language of education was very different from that of the earth sciences and she did not understand some of it. Similarly the science educator made it clear that "igneous petrology" and terms like "syn" were geo-speak jargon. As we recognized the language difficulties, we began to move past "saving face" and "talking past one another" to *asking questions and learning* from each other. As a result, a large part of our collaborative interaction was discussing the language and big ideas of each discipline. The discussions opened the door to a body of educational research previously unknown to the geologist. For the science educator, the discussions updated her knowledge of geology and the new research techniques and technology in the field. In addition, it illuminated both researchers' knowledge about the culture of research and teaching in their distinctive disciplines and departments. So prior to the study, the researchers engaged in substantial dialogue to understand how each operated in their unique university cultures. This became important, a necessity, in order to tackle the changes needed to reform the teaching of undergraduate geology. Thus, a nascent collaboration had begun between the researchers.

According to Loucks-Horsley et al. (1998), an important professional development experience must foster collegiality and collaboration. Moving past edu-speak and geo-speak jargon, the researchers, began to discuss ways to reform teaching in geological sciences. Through the discussion, other hidden, tacit barriers blocked the path to change. Collaboration brought forth unspoken barriers via discussion that otherwise may have remained submerged. For instance, the geologists' beliefs about the science educator's discipline emerged. The geologist voiced concern that her current education research activities held a "lesser status" than "geology research" with her geology colleagues (and herself). She was concerned that it would not be valued by departmental colleagues and expressed dismay regarding the value of educational research in general. When she discussed her ideas, her peers voiced statements such as, "Qualitative research isn't real research!" or "Journals are not data!" She continued to work at incorporating active learning but was unconvinced her colleagues valued her educational research. For instance, she said, "I have a lot of support from my department chair and the Dean but I still feel like a 'lone ranger' in this endeavor. The support is there verbally and the chair seems interested in 'active learning' but hasn't changed her teaching practices."

Given this perspective, the science educator, picking up on the disdain for journals, reminded the geologist that Darwin kept a descriptive *journal* of data that he later used to construct his ideas about natural selection and the theory of evolution. Perhaps more important, the educator introduced her to educational research important to reform so she could judge its value for herself. These interactions made clear the university's dominant ideology regarding its three missions of research, teaching, and service. The missions are viewed hierarchically and research is the supreme master.

To summarize, it was the early and ongoing discussions that allowed the researchers to recognize, understand and accept the demands and differences between the disciplines. Negotiating the language and cultural differences of the disciplines through collaboration cultivated a mutual respect and a shared vision for using "active learning" to reform undergraduate geology courses.

Rule 2: Systematic sustained support (S^3) is necessary for those involved in reform efforts. Despite the context, systematic, ongoing support is an important, well-documented element of reform (Fullan, 1991, 1993; Loucks-Horsley, 1997; Loucks-Horsley et al., 1998). The geologist had support on two levels. First, the department and dean supported her in conducting educational research and on a second level the science educator supported her. Both forms of support were critical; without the departmental support the geologist would not have "risked" doing educational research because her peers did not view it as real "research." Support from the science educator was also a critical component as the geologist began to change the way she taught geology to undergraduates.

To enhance the collaboration and support the geologist's reform in teaching undergraduate geology, a peer consultation model (Bernstein, 2000) was used as part of the professional development process. The model included ongoing written dialogue and face-to-face interactions through classroom visits. The purposes of the visits determined in advance by the partners ranged from gaining insights into course syllabi to observing pedagogical strategies implemented within the courses. Written documentation of observations, questions, and reflections were shared with the partner after each class visit or task was completed. As seen below the model includes several interactive parts (see Table 21.1).

When analyzing the progress of the collaboration to reform geology teaching, the value of using this model was evident on many fronts. First, it provided a systematic schedule for exploring teaching practices in both researchers' classes. For instance, the science educator could observe how new teaching strategies were implemented in the geologist's classroom,

Table 21.1.   Peer Consultation Model

Consultation Activities		Written Documents[a] (descriptions)	Source	
			Self	Partner
Scholarship of the Syllabus	Part 1	Exchange syllabi.	✓	✓
		Reflect on syllabus and its goals.	✓	
	Part 2	Partner feedback focused on the syllabus's goals and activities.		✓
Pre-Teaching Pre-classroom observation conference	Part 1	Partner conference focused on objectives and other selected topics.	✓	✓
		Select classroom observation dates.	✓	✓
Syn-Teaching Essence of the Classroom Practice (reciprocal classroom observations)	Part 1	Reflection on class session focused on objectives and goals.	✓	
	Part 2	Partner's feedback to classroom observation.		✓
Post-Teaching Evaluation of the Classroom Practice	Part 1	Student work samples and assessment.	✓	
	Part 2	Reflection on the selection and assessment of work sample to demonstrate student learning.	✓	✓
		Partner feedback on how assessment of student work aligns with objectives, etc.		

Note: Shaded areas represent an iterative cycle for each classroom observation.
[a] All of the documents in the model are written documents.

labs, and field excursions, while the geologist's observations of the science methods classroom provided insights into how strategies new to her actually work in the classroom. Second, the use of *written responses* by the researchers as a reflective tool was a valuable component of the model. For example, a written exchange from the geologist to the science educator regarding the science methods course syllabus posed important questions and highlighted areas for continued discussion:

> This [your syllabus] is a little intimidating to look at—the page with the four categories with objectives listed underneath! I like your goal of getting across the idea of science as a process versus only a set of facts or truths. My feeling is that they will learn science in their college-level courses as a set of facts or

truths. Do you address this in class? I also like that you approach this class with teaching students about the nature of science. In a way, I think we as scientists do not talk about the nature of science specifically nearly as much as we should. I feel this topic is very important and relevant. . . . I would like to share this syllabus and your comments with my colleagues here in the geology department. I think they would be stunned to learn that your course and expectations are so rigorous. (Geologist [peer consultation])

The comments from this excerpt began to illuminate beliefs and ideas that the geologist held about science education courses. Her words implicitly suggested that she expected the science methods courses to be less rigorous than what she and her colleagues required in geology. Furthermore, the written exchanges prompted questions and other topics for discussion (i.e., nature of science). Another example that prompted dialogue through writing was an observation written by the science educator during one of the geology field trips:

> The males were digging with the hammers and using their lenses. Females were watching them. Some of the other groups of females were talking and drawing the columns as part of their task. One incident observed: a couple of males and females were working beside each other. The females were drawing the stratigraphic column and what they thought they were seeing. As I listened, what seemed to throw them was the actual construction of the "geologic story of the region" that they needed to create. The females were trying to cajole the "story" from the males near them. It was as if by magic the males somehow knew more than they did. The males said laughingly that "telling" would be cheating. I think the students might need an example of a "geological story" to help them with this part of the task. You may have given them one but they didn't make the connection. Watching the female-male interactions in the field prompts questions about gender and the sciences. (Science educator [peer consultation])

These are only 2 snapshots of 12 lengthy exchanges from reciprocal visits made between the researchers that highlight the nature of the written self-reflections, syllabi reactions, classroom observations, responses to observations, and e-mails. The written exchanges and questions that were evoked in the writings moved "teacher thinking" from the intangible to the tangible. Words in print were written examples that were revisited and evaluated in a way that verbal interactions were not—written words captured the "essence of thought" in the moment. Visualization of one's own thoughts and seeing others' thoughts in writing takes teacher thinking to a richer, deeper level of understanding about who we are and what we do. Simply talking about how to change teaching was not enough; strategies needed to be modeled, discussed, and tried out for "best fit." Follow-up discussion and reflective *writing* as part of the peer consultation model was critical for the success of the reform efforts by the geologist. Thus, the peer consultation model served us well. It allowed us to continually revisit and research—to search again our thoughts in context.

In addition, another key component of the peer consultation that worked in changing the way geology was taught to undergraduates involved a *particular kind of partnership*. Some partnerships for peer consultation may not focus on reform in teaching; rather, they may simply focus on looking at some aspect of teaching for the purpose of modeling. For this purpose, the partner may be a person from within the same department or college. However, if *reform in teaching* is the purpose, then an individual from one's department or college will not suffice. They can provide written observations; however, it is unlikely that they have the expertise to offer support via research-based alternatives or to be able to model more appropriate pedagogical strategies for the specific content.

For this study, reform means a *sustained change regarding how one teaches and how students learn*. Therefore, it means more than adopting a new strategy or two. It means to rethink, "re-form," and reflect upon the most important constituent of the teaching process—the student. To accomplish this in the sciences, we recommend that the partners be paired such that one of the partners has expertise in science teaching and the other expertise in a specific science discipline. Though the disciplines may be different, the fact remains that teaching is an integral part of both. Thus, reforming undergraduate science courses cannot occur in a meaningful way without knowledge of research-based pedagogy. Partnering individuals from both colleges multiplies the effort to reform undergraduate courses due to the synergy of knowledge and experience from individuals in both content areas.

In review, research-based studies on reform advocate systematic support for those involved within the change process. This study's findings concurs with previous work, and found that in reforming undergraduate geology teaching, collaboration, as a process of professional development, supported and promoted change. A form of collaboration found particularly useful included partnering the geologist with a pedagogical expert and using a model for exchanging *written* reflections and classroom observations that evoked teaching beliefs and focused on students' learning.

Rule 3:   Listening to voices of the undergraduates as collaborators is critical for monitoring and sustaining changes in teaching.   To be an educator means to learn from your students while they learn from you. As collaborators, student voices were heard by reading their words in journals and through interviews. Most journal entries were free flowing and the science educator conducted the interviews. Journals were read every 3 to 4 weeks. Reading and reflecting upon students' views of the course events provided insight and direction as the geologist implemented new pedagogical strategies.

For most teacher educators this is intuitive. In fact, the science educator has persistently used journals and interviewed students to improve her own practices. On the other hand, journaling was a novel idea to the geologist who initially was unsure of its value. However, by incorporating the use of student journals in the courses she received valuable feedback from undergraduates that helped her monitor the way new strategies were influencing student learning.

The geologist wanted to provide opportunities for the undergraduates "to experience geology as a geologist." Her ideas were in line with Loucks-Horsley and colleagues' (1998) findings on inquiry immersion. Inquiry immersion in science is an opportunity to experience firsthand both science content and processes. Learning through inquiry immersion embedded within an authentic context, the undergraduates put inquiry skills into practice and experienced the process for themselves.

Though inquiry immersion is a professional development approach for teachers, this immersion experience was designed to help undergraduates better translate theory and content into meaningful understanding. Therefore, the geologist included several field excursions in both lecture and lab with an extensive 2-day field study in the laboratory course. Because of her background as a geophysicist in the field, she found working in the field a valuable learning experience. She expected the same sentiment from the students in the class. However, early in the course, the voices of the students suggested otherwise. For instance, one student stated, "I don't want to try and find a place to park when we return to campus [from the field site]," and another student wrote in a journal, "I am not *into* outdoor experiences." These comments were mild compared to comments regarding the longer 2-day experience where one student interviewed stated,

> Field experiences are more appropriate for 300 or 400 level classes.... I could see an excursion at that time. [But] when you're taking a class that's required … it seems strange to have us doing it!

Hearing comments such as these helped the instructor to recognize that the changes she was implementing do have an impact on her normally "silent collaborators" and to provide the best possible experiences she must consider their views and make appropriate adjustments. She revised her plans to address these issues and lowered the number of initial excursions but kept the 2-day intensive field experience. Without listening to the students she may have embarked upon experiences doomed to fail because of student resistance and the impracticality of doing "too much." Thus the undergraduate journals revealed their concerns and insights as well as provided a reflective tool for sustaining changes within the course as seen in the following example.

During a 2-day open-inquiry field excursion, teams of students (3–4 members), generated questions, collected data, and later in the laboratory analyzed data and made interpretations. As part of the process, two students wrote,

- There were 4 people in my group including myself. We had shovels and disposable cameras and a list of tasks to complete. We went to a sandbar by Portertown. We had these fancy GPS things to use. You punch in your coordinates and it would tell us where to dig the trench. The trench would be about 2 feet down before water could come up because we were right by the river. We looked and saw the different layers of sediment. We measured how far down the water base was—we took pictures of our measurements. We will use these later in our presentation. All the tasks … it would be too much to handle alone! (Student 1)
- There was a bridge where two rivers connected and there was some sediment placed up before the fork of the rivers so it wouldn't erode away where the railroad tracks were. There were all these little rocks on the shore. I said they didn't look like native rocks and the instructor asked how I could tell. I noticed a bunch of non-native rocks dumped up stream about 100 yards to probably stop erosion. Little chunks had probably broken off and washed downstream. I collected some and I am going to test them. Anytime, you get people together you get a lot of good ideas. The theory of the rocks washing downstream, they [team members] incorporated what they thought into it [the theory]. (Student 2)

This excerpt was representative of the overall positive disposition of the undergraduates regarding the geologist's use of cooperative teams and the open-ended immersion inquiry. Embedded in their words was what the students were learning. For instance, the aforementioned students were thinking as a geologist working in the field. The student profiles here reveal their understanding of the process of science, techniques used in science, questioning, testing, and measuring to name a few. For the geologist, the words of the undergraduates provided the impetus to continue or modify various strategies supporting the change process within the courses.

By the end of the semester many of the students who viewed the field experience with trepidation and disinterest voiced both interest and value in the geology experiences. For instance, a student reported, "I never really looked at soil. I didn't know there were so many layers to it. You never think, 'This is where it came from or this is how it happened.'" While reflecting on the geology experiences, another student said, "I was real surprised. I enjoyed the field experiences and enjoy geology."

In summary, as the geologist incorporated active learning strategies into her courses, she found that support from the science educator was important but she also needed to know how the students perceived the strategies. Without this input she seemed at a loss as to how to interpret her progress and the progress of the students in light of the changes she was implementing. Input from the students via journals and interview transcripts, provided an informational grounding needed to revise strategies and plans she had for the courses in a reflective manner.

In review, using collaboration as a professional development process uncovers unspoken details that influence and are resistant to reform initiatives at large universities. Our findings on the use of collaboration as part of the process of professional development to reform undergraduate teaching requires individuals to pay careful attention to developing collaborations with open dialogue that can overcome professional culture differences and the hierarchical stratification of the disciplines found across a campus. In collaborations that focus on improving instruction, matching content specialists and pedagogical specialists maximizes the potential for reform. Once the partnership is established, finding time to engage in written reflections, making classroom observations, and evoking teaching perceptions that can be reflected upon in light of the reform innovations are critical.

Sustained support for those undergoing the professional development process is imperative. In this study, sustained support over time assisted in the internalization and ownership of the new strategies. Last, hearing the "voices of the undergraduates" provided a unique kind of support for monitoring, sustaining, and determining the impact of the reform efforts by the instructor.

RECOMMENDATIONS: USE OF COLLABORATION AS A PROFESSIONAL DEVELOPMENT PROCESS FOR REFORM IN UNDERGRADUATE SCIENCE

According to Loucks-Horsley et al. (1998), "it is difficult if not impossible to teach in ways that one has not learned" (p. 1). During teaching reform, instructors need opportunities to inquire into salient questions of their subject discipline as well as questions about learning and pedagogy—in supportive, collegial communities. Clearly, collaboration that fosters collegiality is important to professional development and reform in teaching. If undergraduate course reform is a goal, based upon our findings, we recommend the following:

- Building partnerships to maximize opportunities to build trust and respect through ongoing conversation.
- Negotiating understanding of the professional language differences and professional work cultures.
- Finding collaborators who can create shared goals that benefit all.

- Devising or using a model such as peer consultation to set expectations and provide structure for the collaboration through writing as well as taped conversations.
- Sustaining support for individuals reforming courses.
- Listening to undergraduates as part of the reform effort and reflecting upon their words to modify or sustain innovations.

IMPLICATIONS: USE OF COLLABORATION AS A PROFESSIONAL DEVELOPMENT PROCESS AND FUTURE RESEARCH FOR REFORM IN UNDERGRADUATE SCIENCE

Although the focus of change in this study was on the geologist, professional development for the instructor over time will succeed only if there is a concurrent effort to change the system. The system may be a department, college, or a university. Research on "system thinking" suggests that individuals behave in ways determined by a system's underlying frameworks, such as incentives, disincentives, rules, and cultures. Thus, failure to change may not be due to the individual; instead it may be due to a "system failure" (Patterson, 1993).

Given this line of thought, a potential direction for research is to explore ways to initiate change at the various levels within a university system. For instance, what strategies, incentives, or rewards foster reforming courses to facilitate undergraduate learning of the sciences at the level of the department? Others might include studies that examine ways to influence policy at the university level to support collaboration across the disciplines. In addition, exploring the perceptions of university faculty regarding their discipline as well as other disciplines may shed light on the hierarchical stratification we found regarding research and teaching in the sciences and education. Our findings on professional development collaborations within a university suggests, as did Boyer (1997), that there is a challenge before us: to redefine and broaden scholarship to encompass the wide range of work done by the professoriate to meet the rapidly changing social demands of teaching, research, and service.

REFERENCES

Bernstein, D. (2000, June). *Peer review of teaching.* Presentation for University Teaching Scholar Lecture Series, Kansas State University, Manhattan.

Boyer, E. (1997). *Scholarship reconsidered: Priorities of the professoriate.* San Francisco: Jossey-Bass.

Carr, W., & Kemmis, S. (1986). *Becoming critical: Education, knowledge and action research.* London: Falmer Press.

Coble, C., & Koballa, T. (1996). Science education. In J. Sikula (Ed.), *Handbook of research on teacher education* (pp. 459–485). New York: Simon Schuster.

Darling-Hammond, L., Hudson, L., & Kirby, S. N. (1989). *Redesigning teacher education: Opening the door for new recruits to science and mathematics teaching.* Santa Monica, CA: RAND.

Fullan, M.G. (1991). *The new meanings of educational change.* New York: Teachers College Press.

Fullan, M.G. (1993). *Change forces: Probing the depths of educational reform.* Bristol, PA: Falmer Press.

Guskey, T. (1986). Staff development and the process of teacher change. *Educational Researcher, 15*(5), 5–12.

Hord, S. M., & Boyd, V. (1995). Professional development fuels a culture of continuous improvement. *Journal of Staff Development, 16*(1), 10–15.

Lincoln, Y., & Guba, E. (1985). *Naturalistic inquiry.* Beverly Hills, CA: Sage.

Little, J.W. (1993). Teachers' professional development in a climate of educational reform. *Educational Evaluation and Policy Analysis, 15*(2), 129–151.

Loucks-Horsley, S. (1995). Professional development and the learner-centered school. *Theory in Practice, 34*(4), 265–271.

Loucks-Horsley, S. (1996). Professional development for science education: A critical and immediate challenge. In R. W. Bybee (Ed.), *National standards and the science curriculum: Challenges, opportunities, and recommendations* (pp. 83–95). Dubuque, IA: Kendall-Hunt.

Loucks-Horsley, S. (1997). Teacher change, staff development, and systemic change: Reflections from the eye of a paradigm shift. In S.N. Friel & S.W. Bright (Eds.), *Reflection on our work: NSF teacher enhancement in K–6 mathematics* (pp. 133–150). Lanham, MA: University Press of America.

Loucks-Horsley, S., Hewson, P., Love, N., & Stiles, K. (1998). *Designing professional development for teachers of science and mathematics.* Thousand Oaks, CA: Corwin Press.

McLaughlin, M.W. (1993). What matters most in teachers' workplace context? In J.W. Little & M.W. McLaughlin (Eds.), *Teachers' work: Individuals, colleagues, and contexts* (pp. 79–103). New York: Teachers College Press.

National Research Council. (1996). *National Science Education Standards.* Washington DC: National Academy Press.

National Research Council. (2001). *Educating teachers of science, mathematics, and technology: New practices for the new millennium.* Washington D.C: National Academy Press.

National Staff Development Council. (1994). *Standards for staff development: Middle level.* Oxford, OH: Author.

National Staff Development Council. (1995a). *Standards for staff development: Elementary level.* Oxford, OH: Author.

National Staff Development Council. (1995b). *Standards for staff development: High school level.* Oxford, OH: Author.

Norris, J.H. (1994). What leaders need to know about school culture. *Journal of Staff Development, 15*(2), 2–5.

Patterson, J.L. (1993). *Leadership for tomorrow's schools.* Alexandria, VA: Association for Supervision and Curriculum Development.

Snyder, T.D., Hoffman, C.M., & Geddes, C.M. (1998). *Digest of educational statistics, 1997* (NCES 98-105). Retrieved August 17, 2003, from the National Center for Education Statistics Web site: http://nces.ed.gov/pubs/digest97/98015.pdf

part III

INNOVATIVE MODELS FOR REFORM IN UNDERGRADUATE SCIENCE

The third and final section of this volume expands upon the lessons from research in previous chapters, and provides real examples of reform in teaching undergraduate science and engineering courses. The following chapters reflect the models for inquiry and innovation occurring in the diverse institutions. These models represent a few of the innovative undergraduate science courses that have been created across the United States within the last decade. They showcase some of the unique elements of reform in progress in higher education. By utilizing real-world scenarios and problem-solving techniques, the students can view the concepts through the lens of a scientist or engineer. Pedagogically, the inquiry-based courses intertwine laboratory work with scientific content to create seamless learning experiences for the students. And through collaboration, the courses are developed and taught, and the students learn.

The first two chapters in this section reveal different approaches to developing biology courses for beginning college students. In Chapter 22, Bonnie McCormick and Christy MacKinnon, a science educator and a biologist, respectively, created a course based upon E. O. Wilson's book, *The Diversity of Life* (1992). The instructors created a hands-on course that incorporated the use of unique local attractions, such as the Botanical Gardens, for study of the biodiversity that exists in the area. The other two foci of the course were evolution and ecology.

Charlene Waggoner, Monika Schaffner, Kimberly Keller, and Julia McArthur created a biology course, Environment of Life, which has routinely been taught to over 600 students each semester. Chapter 23 provides details of their infrastructure for inquiry model in which students are taught how to use inquiry skills during laboratory sessions and then the amount of help given to the students decreases as they gain these skills. The model was also used to help teaching assistants gain experience in running inquiry laboratories.

Dorothy Gabel has been involved with teaching an Introduction to Scientific Inquiry chemistry course for 25 years. Approximately 250 preservice elementary teachers take this required course in the education department for science credit each semester. In Chapter 24, Gabel describes a study of the effectiveness of incorporating Play-Doh modeling activities into this course to help students in representation of matter in conjunction with the use symbolic formulas and equations.

In Chapter 25, Dean Zollman writes about a unique approach to teaching a large (100-student) introductory physics course using the learning cycle. By allowing open laboratory periods, instead of set laboratory times, the students complete exploration and expansion laboratory activities between classes. With three classes per week, the typical three-phase learning cycle model fits perfectly with explorations on Monday, inventions on Wednesday, and expansions on Friday. This chapter also describes the implications of using interactive multimedia technology (Personal Response System) in large lecture courses.

M. Jenice Goldston and Monica Clement, in Chapter 26, provide an example of a reformed introductory geology course. Geology 103 was redesigned to include authentic field experience that allowed the students to experience geology as geologists. The authors discuss findings from the qualitative and quantitative data collected from students enrolled in the reformed course.

The next three chapters explore how engineering courses can be developed for preservice teachers and other non-science/non-engineering majors. In Chapter 27, Scott Graves, Michael Odell, Tim Ewers and John Ophus describe the changes made in reforming an existing interdisciplinary science course, INTER 103: Integrated Science for Elementary Education Majors. The course introduces students to the nature of science and scientific inquiry through the approaches of science, technology, and society, and of Earth system science. The course succeeds in employing strategies to help education students re-envision science as a "way of knowing" that involves an ongoing process of fine-tuning perception, evaluating evidence, refining insight, and continuously applying self-reflection as a means of gauging their own reactions to learning as it occurs in the classroom and in the field.

In Chapter 28, Jeanelle Bland Day examines an introductory engineering/science course taught in the Department of Aerospace Engineering and Mechanics. AEM 120 was developed with the elementary- and middle-level teacher in mind. Key concepts in the course solar system involved astronomy, flight, and forces. The author describes how faculty collaborated to help bridge gaps in expertise and created a course for that improved scientific literacy in students.

In Chapter 29, a very different approach to teaching about engineering was described by William Jordan, Bill Elmore, and C. W. Sundberg. Using the theme "Our Material World," the authors developed Engineering Problem Solving for Future Teachers as a way to teach about mechanical, chemical, and physical behavior of materials incorporating engineering problem-solving techniques. The authors discuss the creation of the course, give examples of course activities, and describe findings from original research.

The final chapter, 30, by Chuck Karr and Cynthia Szymanski Sunal, details reforms involved during the development of a course titled Artificial Intelligence Systems in Science. Using fuzzy logic, neural networks, and genetic algorithms as the basis of inquiry, the students explored scientific concepts such as classification systems in biology, brain functions, genetics, and chemical processes. Findings from research conducted in the class and changes in course format during subsequent course offerings are discussed.

CHAPTER 22

A MODEL FOR REFORM IN TEACHING IN THE BIOLOGICAL SCIENCES

Changing the Culture of an Introductory Biology Course

Bonnie McCormick and Christy MacKinnon

ABSTRACT

In 1997, we began a project to change the way we taught our introductory biology course at a small, Catholic university in South Texas. Through collaboration with the university administration and with support of a national professional development group, we were able to change the learning environment from one that was teacher-centered to one that allowed the learners to construct their own knowledge in a rich and meaningful way. The changes we made were informed by teaching standards in national reform documents. The lecture and the laboratory sections of the course were combined so that instruction could be organized using a learning cycle approach. The learning environment facilitated student-directed learning, collaboration, and inquiry. We evaluated student learning and attitudes

Reform in Undergraduate Science Teaching for the 21st Century, pages 393–408

toward science using a pre and posttest design. Students in this reform-based course performed as well or better on traditional content measures. Students in the traditionally taught course had a significant decline in their attitudes toward science. Alternative assessments, including student portfolios were used to evaluate student learning in the reform-based course. Student reflections provided supporting evidence that students are meeting the instructional objectives of the course.

INTRODUCTION

In 1997, we began to seriously think about changing the way that we taught our introductory biology course that serves our majors, non-majors, and preservice teachers. We had no idea at the time that our dissatisfaction with the way we were teaching our students would lead to a significant conceptual change in the way we viewed ourselves as teachers and as scholars. In the six years of investigating and implementing best practice in college teaching, we have transformed our roles as professors from that of the expert dispensing knowledge to designers of learning experiences that actively engage students in learning the concepts and processes of biology.

The University of the Incarnate Word (UIW) is a Catholic, liberal arts university located in San Antonio, Texas. UIW is recognized as a Hispanic Serving Institution (HSI) and has approximately 2,200 full-time undergraduate students. Approximately 48% of the student body is Hispanic, 6% are African American and 30% are Caucasian. Our university was founded by the Sisters of Charity of the Incarnate Word as a college for women, and although we are now a coeducational institution, women still comprise 64% of the student body. A large proportion of the students attending UIW are the first member of their family to attend college.

Our collaboration was the result of a shared need to change the learning environment from a format that was dominated by delivery of content through lecture to one that actively engaged students in the learning process. Our professional development introduced us to research on constructivist theories of learning. This provided our theoretical framework to alter our classroom environment to promote active construction of knowledge by the learner. Most of the recent reform efforts of professional groups and government agencies concerned with science and science education have been based on constructivism (Siebert & McIntosh, 2001). Constructivism views learning as a "building process by active learners interacting with the physical and social world" (Fosnot, 1996, p. 30).

RESEARCH ON TEACHING BIOLOGY

If systemic reform in science education is to succeed, there needs to be increased articulation among all parts of the system, including postsecondary education (National Science Foundation [NSF], 1996). Postsecondary institutions train not only future teachers, scientists, and engineers, but also prepare future users of science in a workplace transformed by the advances of science and technology (NSF, 1996). The teaching standards set forth by the National Science Education Standards (National Research Council [NRC], 1996) call for the science curriculum to emphasize understanding, reasoning, and problem solving. In order for postsecondary science classes to be aligned with these teaching standards, the learning environment, including pedagogical methods, must change from a lecture-centered format to an active learning environment (NSF, 1996).

Most of the recent reform efforts of professional groups and government agencies concerned with science and science education are based on constructivism (Fosnot, 1996). Constructivism is a family of theories that views the learner as the creator rather than the passive receiver of knowledge. Learning occurs through activity that allows learners to discover and build their own understandings by interacting with the environment, by making sense of these experiences, and by integrating new knowledge with prior experience (Padilla, 1991). Constructivist theory as a referent for teaching methods is an accepted strategy for improving teaching practice (Tobin, Tippins, & Gallard, 1993) and is consistent with the type of classrooms envisioned in reform documents.

The role of the teacher in the constructivist model is to structure the learning environment so that learners take an active role in developing and integrating new insights into their existing cognitive structures. Instructional methods that are consistent with this approach seek to determine what the learner already knows and what experiences can build on that base. Learning cycles engage students in the kind of thinking that constructivists argue is necessary to promote learning based on student actions (Lord, 1997). Although there are many versions of the learning cycle, a common learning cycle model has three phases (Lawson, Abraham, & Renner, 1989). (Refer to Chapter 6 for additional information regarding the learning cycle.) The exploration phase engages the student through experience with the environment by exploring new materials and ideas. The concept introduction phase involves social transmission in developing a conceptual framework of the principle or concept introduced in the exploration phase. During the elaboration phase of the learning cycle, students apply and extend the concept to another situation or task.

Research shows that learning cycles can provide a framework for constructing a learner-centered environment. The Biological Sciences Curriculum Study (BSCS, 1993) developed a conceptual framework for organizing postsecondary biology curricula based on unifying principles and major concepts using a learning cycle approach. Comparison of BSCS-type laboratory instruction with traditional instructional methods has shown that this is an effective method of teaching in the liberal arts college setting (Hall & McCurdy, 1990). A study comparing knowledge gains of non-major biology students taught using a conceptual framework with biology majors in a content-intensive course found that the non-majors performed as well as the majors on a content posttest and had a greater gain from pretest results (Sundberg, Dini, & Li, 1994). End of the course evaluations in an environmental science course that used a learning cycle approach which integrated lecture and laboratory found that a majority of students preferred this format to the traditional lecture with a separate lab (Poole & Kidder, 1996). A comparison of students taught in two large sections of a non-majors introductory biology course found that students taught using a learning cycle approach performed at a significantly higher level on course examinations than students taught in a traditional lecture format (Lord, 1997).

In addition to improving achievement in biology, courses designed using learning cycles have been found to improve reasoning skills and attitudes in comparison studies with traditionally taught courses (Kincaid & Johnson, 1997). Studies of attitudes of biology students toward science indicate that nontraditional course design can improve student attitudes toward science (Sundberg et al., 1994). One of the goals of curricular reform in the biological sciences at the postsecondary level is to improve attitudes of all students toward science in general and biology in particular (Lawson et al., 1989; Sundberg et al., 1994). Society and the workplace are being shaped by advances in science and technology, "yet for many students, learning science involves negative feelings and attitudes which discourage further exposure to any scientific inquiry" (Gogolin & Swartz, 1992, p. 488). Improving affective outcomes is important because most of the cognitive learning is forgotten over time while the affective outcomes remain (Gogolin & Swartz, 1992, p. 501).

AN INVESTIGATION IN CREATING ACTIVELY ENGAGED STUDENTS IN AN UNDERGRADUATE BIOLOGY COURSE

Although we had completely revised the required laboratory curriculum for the introductory biology course in 1993, there were still problems with congruence between lecture and laboratory content. Topics were not always covered in the laboratory during the same week that the topic was covered in the lecture. Students in the laboratory section were likely to have different instructors in their lecture sections, so there was not always correspondence between the lecture and laboratory curriculum. After reviewing the recommendations of the Biological Sciences Curriculum Study (BSCS), the National Association of Biology Teachers (NABT), the National Science Foundation (NSF), and the Society of College Science Teachers (BSCS, 1993; Gottfried & Hoots, 1993; Halyard, 1993; NSF, 1989), it was decided that integrating the lecture and lab components of the course was a viable mechanism for providing a curriculum that engaged the students as active participants in the classroom experience.

In the spring of 1997, we attended a meeting of the National Association of Research in Science Teaching to learn how other science educators implemented change in college courses. It was at this meeting that we learned about NASA's Opportunities for Visionary Academics (NOVA). NOVA is a national initiative that uses constructivist practices to offer professional development to university faculty teams who are interested in changing specific science, mathematics, and engineering courses. Through this program, NOVA seeks to create a national model of best practice for preparing future classroom teachers (Hodges, 1999).

We were accepted to attend a 3-day NOVA faculty development workshop. NOVA provides funding to support the creation of innovative science or mathematics courses that incorporate constructivist-based principles. A requirement of NOVA is that the team includes an administrator from the institution, a content specialist, and a science educator. The graduate dean had worked with us on other science education initiatives and agreed to be a part of our team. As a result of attending the workshop, we applied for and received funding to redesign the introductory biology course.

Course Design

Our plan was to reorganize the structure of the course by combining the lecture and laboratory components of the course so that the class met twice a week for three hours. This format allows the course to be taught using a learner-centered approach that engages the students in the process of discovering biological principles. The major conceptual themes of the course are evolution, biodiversity, and ecology. The course design follows the recommendations of NSF (1996, p. 5). This was done by

- changing the focus of instruction from memorization of facts to the mastery of concepts and applications,
- changing the learning environment so that students take an active role in the process,

- providing opportunities to apply technological tools to solve problems, and
- assessing the abilities of students to reason and solve problems using scientific principles.

These reform goals are consistent with the university's vision of the role of science in the undergraduate curriculum.

Our goal was to create a learning environment that focused on student involvement in discussion, hands-on activities, inquiry-based lab experiences, and small collaborative-learning groups. Course activities were organized using a learning cycle approach. A common version of the learning cycle uses the engagement, term introduction, and concept application sequence suggested by Lawson et al. (1989). The learning cycle model was chosen because these approaches are consistent with constructivist theory and because the learning cycle model provides a referent for prospective teachers, who form a part of the course student body that is consistent with the teaching standards specified by the National Science Education Standards (NRC, 1996).

The learning cycle allows the students to be actively engaged in understanding biological concepts and principles by relating new knowledge to prior knowledge, explaining what is discovered, building a group consensus, and extending knowledge to other situations. We developed several modules for the NOVA Project that integrated biology content with National Aeronautics and Space Administration (NASA) research resources (McCormick & MacKinnon, n.d.). The first module, used in the first week of class, engages students through discussion of the question, "How could you recognize life on Mars?" Students work in small groups to develop a definition of life and a list of the requirements of living organisms.

During the concept application phase, students use ratios and proportions to compare the size of cell structures and different types of cells. An electron micrograph of the putative Mars meteorite fossil is downloaded from the NASA Web site (see Beck, 2003), measured, and compared to sizes of known cell structures. Students complete individual Web-based investigations to write a short paper evaluating the evidence for life on Mars. The Mars Meteorite Home Page on the NASA Web site (see Baalke, n.d.) serves as starting point for students to obtain information about the scientific debate about whether there is evidence of life on Mars.

Groups contribute to a class discussion of commonalties of living organisms and the requirements of maintaining life. The instructor facilitates the student-led discussion and introduces scientific terms for common language descriptions of characteristics students describe during the term introduction phase. Students compare their list of new terms to information in their book and consistently find that they have developed a similar list of criteria. By allowing the students to develop a set of criteria for recognizing life, students gain confidence in their abilities and instructors gain information about students' prior knowledge and conceptions.

Technology use was an integral part of the redesigned course. Students utilized Web-based resources that were accessed in the computer lab or from laptop computers in the classroom. The computers were also used for computer-based simulations, data acquisition from electronic databases, and data analysis. Graphing calculators were available for data analysis. Students were trained in the use of software and hardware on an as-needed basis. Collaboration of the students helped facilitate technology use for both the students and the instructors. Invariably there was someone in the room who was more proficient than the instructor in at least some aspect of technology use. The collaborative atmosphere produced by technology use and collaboration on activities encouraged the view that the participants in the course were a community of learners.

Our change in teaching methods also required an expansion of course objectives to include not only content objectives but also process, affective, social, and metacognitive goals. We drew our teaching objectives from the principles of effective teaching recommended by the American Association for the Advancement of Science (1990). These include starting with questions about nature, using a team approach, engaging the students actively, concentrating on collection and use of evidence, de-emphasizing the memorization of technical vocabulary, rewarding creativity, providing experience in using technological tools, promoting group learning, supporting the role of women and minorities, building on prior knowledge, and relating to the world outside of school.

Broadening the expectation for course outcomes also meant changing assessment procedures so that both students and faculty could receive feedback on the success of achieving course expectations (Straits & Wilke, 2002). Students are evaluated using a variety of methods. Each exam has a performance component that relates to course activities and laboratory experiences. Students may be given a case study to explain, or they may use their lab and activity materials to complete an open-note lab practical. The last exam is a Web-based research project on the characteristics of biomes and literature analysis of the human impact on biomes.

Evaluation of the Reform-Based Curriculum

Students submit reading reflections to a group discussion board or by e-mail and discuss what was most interesting, what was difficult to understand, and what new questions were raised. This gives the students a chance to express their views and gives the instructor a chance to interact with each student on a regular basis through a dialogue about their reflective writings. Portfolios provide another type of assessment and allow students to reflect on the learning experience (Biggs, 1996). The portfolio consists of five items that demonstrate how the student has met the course objectives with a written reflection on how the chosen item demonstrates that the course objectives were met.

Changing to a reform-based curriculum requires time, effort, and resources to develop and prepare course materials. Change may create resistance from colleagues who only practice didactic teaching methods. If reform is to be accepted and sustained, it is important to demonstrate that the reform-based curriculum is effective in accomplishing course objectives. Therefore an important component of our project to reform the introductory biology course was to develop an action research plan to evaluate the effectiveness of the course. Action research is carried out in the classroom of the researcher. Its purpose is to answer questions about new methods of teaching and learning. (Refer to Chapter 12 for additional information on action research in college classrooms.) In our evaluation plan we investigated three questions:

1. Is the reform-based course effective in student acquisition of content knowledge?
2. Does the instructional method affect students' attitudes toward science?
3. Are we meeting our teaching objectives?

Our research design was to compare content acquisition and attitude toward science between the revised biology course and traditionally taught sections using a pretest and posttest design. Since many university science faculty members are skeptical of constructivist learning theories, we felt that it was important to demonstrate that students in a reform-based course could learn the intended content as well as those in a traditionally taught class. Because fostering positive attitudes toward science is viewed as a legitimate goal of science teaching (Koballa, 1989), we also examined how the instructional method affected students' attitudes in the reform-based course and in the traditionally taught course. Students' attitudes toward science can affect loss of interest in science and retention in science programs (Seymour, 1995).

A traditional multiple-choice test with questions matched to course objectives was given to three sections of the reform-based biology course and three sections of the traditionally taught biology course at the beginning and end of the semester. Analysis of covariance (ANCOVA) found that students in the reform-based course performed as well as or better than the students in the traditionally taught sections (see Table 22.1). This finding suggests that the integrated lecture and laboratory format was as effective as traditional methods in student acquisition of traditionally taught content knowledge. Additional details of this analysis can be found in McCormick and MacKinnon (2000).

Table 22.1. Analysis of Covariance Between Pretest and Posttest Content Scores

Source	Adjusted SS	df	MS	F	Sig.
Regression on Pretest	3595.91	1	3595.91	24.18	$p < .000$
Adjusted Means	1151.90	2	575.95	3.87	$p < .025$
Adjusted Error	12789.20	86	148.71		
Total	17537.01	90			

Instructional Method	Pretest Mean	Posttest Mean	Adjusted Mean	Difference from Covariance
Reform Course	38.62	55.06	54.19	15.57
Control 1	37.09	52.75	52.80	15.71
Control 2	35.81	45.69	46.05	10.24

The Assessing Attitude toward Biology Survey (Sundberg, et al, 1994) pre and posttest was given to students in one traditionally taught section and one section of the reform-based course (McCormick, MacKinnon, & Jones, 1999). The attitude scores of students in the traditionally taught lecture were lower on all scales of the attitude survey. The scales measured the subjects' attitude toward institutional requirements in the sciences, science in everyday life, personal comfort with science, the power and limits of science, and science and religion. Although the students in the traditionally taught course had a higher mean pretest score on the survey, these students had a significant decline as measured by the total attitude score ($t = 2.78$, $p < .007$). There was no significant difference in attitude scores of students in the reform-based biology course (see Figure 22.1). The results of the attitude instrument suggest that the manner of instruction may be important in maintaining or improving attitudes of the students enrolled in the reform-based course.

Portfolios and student reflections were part of the assessment process in the reform-based course. Preparation of portfolios allows students to select some of the evidence that they have met course objectives (Biggs, 1996) and also provides feedback to us about whether we are succeeding in meeting our teaching objectives. Assessment portfolios have the advantage of focusing evidence for an explicit purpose, exposing students' beliefs about their learning, and requiring student reflection (Kruger & Wallace, 1996).

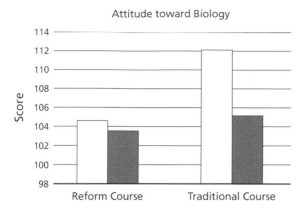

Figure 22.1.   Pre and posttest comparisons of total attitude scores. There is a significant difference in the pre and posttest mean in the traditional course ($p < .007$).

Students in the reform-based course selected five items to represent how the student achieved the course objectives. Each item is accompanied by a reflection that explains how the item of work helped the student achieve course objectives.

Students' reflections on the evidence presented in their portfolios demonstrate that students were able to achieve the expanded course objectives in the reform-based course. The format of the course was important to the students and some of the reflections included their feelings about the learning process. One student found the course format important not only to her learning process but also to her adjustment to college:

> In the beginning of the year I found many aspects of college life discouraging. I enjoyed coming to Diversity of Life because I felt like I learned as I went along because of our constant lab work. Each lab we did would help me access what I learned and would prove the ideas discussed. At home I could look through the labs and gather the information again until I felt that I had learned it. This lab I was particularly proud of because I earned a 100 on it, and at the beginning of the semester anything I did well on gave the encouragement to continue.

Assessment was a source of student reflection on positive aspects of the learning environment. Many students wrote that the exams were fun to take and that they learned from the assessment experience.

> When I got this test back and saw the grade I was so proud of myself. I have never been very good at biology and I actually was very proud of my grade. Another reason I chose this test because it was actually fun to take. I enjoyed the lab portion in which we got to move around and answer different questions about species of plants and animals.

Another student described how the first exam helped to build her confidence in her ability to learn and understand the concepts that were the focus of the class activities. She wrote, "This first test helped me settle my nerves and realize that I could make it through biology. I was also proud to see that I actually understood the Hardy-Weinberg Principle and recognized when it was violated in the case of the cichlids."

Many students included lab experiences that took place outside of class at the botanical gardens and zoo in their portfolios. One student wrote,

> One thing I loved about the class was we got to go see a lot of places. One place we went was the Botanical Gardens and we got to look at a bunch of trees. I know this cannot be that hard, but it takes a lot of observation to tell the difference between some trees.

Other students gained an appreciation for community resources for informal educational resources. "I loved this lab!" one student wrote.

> By going to the Botanical Center I learned that there are many beautiful plants right here in the very area that I live in. It is amazing how San Antonio provides this type of educational center to provide one with the opportunity to see the great beauty of this land.

Student comments also demonstrated metacognitive goals by their awareness of the learning process and pride in the completion of investigations. Concept mapping was mentioned by a student who wrote, "Making the concept map really enhanced my learning experience. This activity also gave me ideas on how to help my future students when it comes time for them to study these concepts." Others found that interaction with classmates was important to their learning. "All of the concepts, however, were learned easier due to the class interaction and participation." Some labs were chosen by the students because they fostered the collaborative learning environment of the classroom. "It helped me to develop a good relationship with my lab group." Another student reflecting on a lab completed at the zoo wrote, "I can honestly say that this was the most interesting lab that I have done. I got to know a lot of my classmates through labs like this, so after this lab, we knew each other a lot better."

A reflection on the process of science read, "This lab was unique unlike any other lab I had ever done in any other Biology course. Often you are asked to design an experiment. This was the first time I was then asked to perform an experiment." Another student shared her enthusiasm for participating in the process of science. "This lab was one of my favorites because we got to design it ourselves ... we got to test something we were curious about."

Some of the student reflections describe how students feel about sharing their learning. One student said, "When teachers give their students an assignment, the students often discover something new to the teacher. In essence, both the student and the teacher learn." After reflecting on her learning experience a student concluded by writing, "I even went home and explained what I had learned to my family." Some students wrote about how they might apply their knowledge in their chosen fields of study. "I am a political science major, and intend to go into public administration or law. Now, I can definitely make educated decisions about land use and management that protects the diversity of ecosystems" was the comment of a non-major.

These written reflections describe the rich learning experiences of diverse students. This is the type of learning envisioned in reform documents that call for student learning to have connections to their prior knowledge and to extend to the world outside of school (Siebert & McIntosh, 2001). Reflections also indicate future intent to use scientific knowledge to inform complex issues that society will face in the future. In addition, as students reflect on their own learning, they are taking steps toward becoming lifelong learners. These reflections provide valuable feedback to faculty as they work to refine the learning methods in this course and provide us with documentation of the value of the learning process to the student beyond the acquisition of traditional content knowledge.

SUMMARY: RECOMMENDATIONS FOR THE DEVELOPMENT OF AN EFFECTIVE COURSE IN THE BIOLOGICAL SCIENCES

The collaborative effort of the faculty involved in the NOVA University Network was an important aspect of the learner-centered course design and provided a forum to discuss and refine the activities, laboratories, and evaluation procedures so that these were consistent with the stated intent of creating a learner-centered environment. Designing an active-learning environment is time consuming and requires action research and creativity to implement. We found that it is far more than one person can do. In addition, having committed collaborators is critical to persistence when the work seems endless and colleagues are critical of the teaching philosophy used in the reform-based course. The support of a university administrator, affiliation with a nationally recognized support group (NOVA), funding from a prestigious source (NASA), and recognition as scholarship for tenure and promotion were critical elements in sustaining the project. This internal and external support provided credibility to our project to redesign the course curriculum among skeptical science faculty members.

Our major problem with expanding the course model to other sections of the course has been within the biology department. Some faculty members teaching in the combined lecture and lab format have only conformed to the time change. The teaching method has remained the same. There is a natural resistance to change and to teaching in a manner that is different than what science faculty have experienced through their years of education. There is also tension about the amount of work involved. Preparation and assessment time increases. There is less control over what happens during each class period when the lesson is focused on what the learner is doing rather than what the teacher is telling. Faculty typically have little formal training in learning theory or teaching methods so they need professional development to make the switch from a teacher-centered to a learner-centered paradigm.

In spite of the difficulties of making the switch to a new paradigm of teaching and learning, four faculty members are successfully teaching this course in the intended manner. Changing the focus of our efforts from teaching to student learning has transformed other courses as well. Our second introductory course for biology majors, our required genetics class, and several biology electives are being taught in this format. We have extended this format to our graduate biology courses for in-service teachers in our Master of Multidisciplinary Science program.

We find a deeper satisfaction in teaching learner-centered classes because the students share their experiences, feelings, and beliefs about the learning process that we would otherwise miss in a teacher-centered environment. Through alternative forms of assessment, we have a richer sense of how the students view their experience in our classroom and how they relate what they learn to their daily lives. It is possible that when we functioned as "tellers" of biology in teacher-centered environments that the students had rich learning experiences, but we had no way to know anything except what traditional content measures provided.

We have found that there is more satisfaction in focusing on designing instruction for active student engagement than preparing lectures, and there is more satisfaction in observing students actively applying their knowledge and skills to the learning process than watching passive students sit in a lecture hall. The shared experiences of the students and faculty members sustain change and promote additional improvements to the course materials and learning experiences. Our faculty reward system recognizes the extra time, research, and creativity required for the change. We are focused on learning, not teaching. Our focus is informed by the recommendations for reform and research on student learning. We are able to teach as scholars informed by our students' learning experiences.

IMPLICATIONS FOR REFORM IN UNDERGRADUATE BIOLOGY AND FUTURE RESEARCH

One important aspect of changing to a learner-centered environment is the process that takes place as faculty make the transition from a traditional, teacher-centered model to one that places the focus on the learner. Monitoring of teaching behaviors could provide valuable information on conditions that promote or hinder reform-minded instructors as they attempt to transform classrooms. Most formal teaching evaluations evaluate teacher-centered behaviors rather than learner-centered behaviors, making it difficult for faculty to judge reform-based learning environments. Research can contribute to an understanding of obstacles to change and elements that foster success in classrooms in transition. The role of collaboration in promoting and in sustaining change in reform-based undergraduate science classrooms could provide important information for faculty members considering change. Because we live in a society dominated by science and technology, the goal of science education reform is to promote science literacy for all students. That goal will not be achieved if science instruction results in classrooms that fail to provide knowledge of both the content and the process of science and fail to foster positive attitudes toward science through appropriate learning environments.

AUTHOR NOTE

This work was in part supported by NASA Opportunities for Visionary Academics (NOVA), a program funded by the National Aeronautics and Space Administration, although the views expressed here are the authors' only.

REFERENCES

American Association for the Advancement of Science. (1990). *Science for all Americans.* New York: Oxford University Press.

Baalke, R. (n.d.). *Mars Meteorite Home Page.* Retrieved August 13, 2003, from http://www.jpl.nasa.gov/snc/

Beck, B. (Ed.). (2003, August 7). *National Aeronautics and Space Administration.* Retrieved August 7, 2003, from http://www.nasa.gov

Biggs, J. (1996). Enhancing teaching through constructivist alignment. *Higher Education, 32,* 347–364.

Biological Sciences Curriculum Study. (1993). *Developing biological literacy.* Dubuque, IA: Kendall Hunt.

Fosnot, C.T. (1996). Constructivism: A psychological theory of learning. In C.T. Fosnot (Ed.), *Constructivism: Theory, perspectives, and practice* (pp. 8–33). New York: Teachers College Press.

Gogolin, L., & Swartz, F. (1992). A quantitative and qualitative inquiry into attitudes toward science of nonscience college students. *Journal of Research in Science Teaching, 29,* 487–504.

Gottfried, S., & Hoots, R. (1993). College biology teaching: A literature review, recommendations, & a research agenda. *The American Biology Teacher, 55,* 340–348.

Hall, D., & McCurdy, D. (1990). A comparison of a BSCS laboratory and a traditional laboratory on student achievement at two private liberal arts colleges. *Journal of College Science Teaching, 27,* 129–131.

Haylard, R.A. (1993). Introductory science courses: The SCST position statement. *Journal of College Science Teaching, 23,* 29 –31.

Hodges, J.B. (1999). *Factors associated with staff development process and the creation of innovative science courses in higher education.* Unpublished doctoral dissertation, University of Alabama, Tuscaloosa.

Kincaid, W.B., & Johnson, M.A. (1997). Our tortuous path to inquiry oriented instruction in an introductory biology course. In M. Caprio (Ed.), *From traditional approaches toward innovation* (pp. 61–66). Arlington, VA: Society for College Science Teachers.

Koballa, T.R. (1989). Changing and measuring attitudes in the science classroom (Research Matters to the Science Teacher, No. 8901). Athens, GA: The University of Georgia, National Association for Research in Science Teaching.

Kruger, B., & Wallace, J. (1996). Portfolio assessment: Possibilities and pointers for practice. *Australian Science Teachers Journal, 42*(1), 16–28.

Lawson, A.E., Abraham, M.R., & Renner, J.W. (1989). *A theory of instruction: Using the learning cycle to teach science concepts and thinking skills.* (NARST Monograph No. 1). Kansas State University: National Association for Research in Science Teaching.

Lord, T.R. (1997). A comparison between traditional and constructivist teaching in college biology. *Innovative Higher Education, 21,* 197–216.

McCormick, B.D., & MacKinnon, C.A. (2000, April). *Student perception of classroom environment in a learner-centered introductory biology class.* Paper presented at the annual conference of the National Association of Research in Science Teaching, New Orleans, LA.

McCormick, B., & MacKinnon, C. (n.d.). *NASA/NOVA.* Retrieved August 7, 2003, from http://www.uiwtx.edu/~mccormic/nova

McCormick, B.D., MacKinnon, C., & Jones, E.R.L. (1999, March). *Evaluation of attitude, achievement, and classroom environment in a learner-centered introductory biology class.* Paper presented at the annual meeting of the National Association of Research in Science Teaching, Boston, MA. (ERIC Document Reproduction Service No. ED453049).

National Research Council. (1996). *National Science Education Standards.* Washington, DC: National Academy Press.

National Science Foundation. (1989). *Report on the workshop on undergraduate biology education.* Arlington, VA: Author.

National Science Foundation. (1996). *Shaping the future: New expectations for undergraduate education in science, mathematics, engineering, and technology.* (Publication No. NSF 96-139) Arlington, VA: Author.

Padilla, M.J. (1991). Science activities, process skills, and thinking. In S.G.R. Yeany & B. Britton (Eds.), *Psychology of learning science* (pp. 205–217). Hillsdale, NJ: Erlbaum.

Poole, B.J., & Kidder, S.Q. (1996). Making connections in the undergraduate laboratory. *Journal of College Science Teaching, 26,* 34–36.

Seymour, E. (1995). Revisiting the "problem iceberg": Science, mathematics, and engineering students still chilled out. *Journal of College Science Teaching, 25,* 392–400.

Siebert, E.D., & McIntosh, W.J. (2001). *College pathways to the science education standards.* Arlington, VA: National Science Teachers Association Press.

Straits, W.J., & Wilke, R.R. (2002). Practical considerations for assessing inquiry-based instruction. *Journal of College Science Teaching, 31,* 432–435.

Sunal, D.W., Hodges, J.B., Sunal, C.S., Whitaker, K., Freeman, M., Edwards, L., et al. (2001). Teaching science in higher education: Faculty professional development and barriers to change. *School Science and Mathematics, 101,* 246–259.

Sundberg, M.D., Dini, M.L., & Li, E. (1994). Decreasing course content improves student comprehension of science and attitudes toward science in freshman biology. *Journal of Research in Science Teaching, 31,* 657–678.

Tobin, K., Tippins, D.J., & Gallard, A.J. (1993). Research on instructional strategies for teaching science. In D. Gabel (Ed.), *Handbook of research on science teaching and learning* (pp. 45–93). New York: Macmillan.

CHAPTER 23

A MODEL FOR REFORM IN TEACHING IN THE BIOLOGICAL SCIENCES

Infrastructure for Inquiry in an Introductory Biology Laboratory

Charlene Waggoner, Monika Schaffner, Kimberly L. Keller, and Julia McArthur

ABSTRACT

The Infrastructure for Inquiry model was developed to help both students and inexperienced teaching assistants succeed doing inquiry-based laboratory activities called for by educational reform documents. Students experiencing inquiry for the first time need support. The initial activities are very structured and highly modeled. The structure and modeling are reduced as the semester progresses until students are doing their own investigations. The program is designed to assess students not only on their content knowledge and skills, but also on the process of doing science. These assessments reward students for doing science and thinking reflectively about what they are doing. This demonstrates to the students that scientific thinking and

Reform in Undergraduate Science Teaching for the 21st Century, pages 409–424
Copyright © 2004 by Information Age Publishing

reflection are valued and important to this course. The teaching assistants are supported with detailed scripts and grading rubrics. The laboratory manual and assessment plan was developed for Biology 101, Environment of Life. Over the course of the project, 1,500 students took the course. Twelve graduate teaching assistants were involved in teaching the laboratories. To determine the impact of the curricular changes an external evaluation was conducted. Both qualitative and quantitative measures were used. Focus group data were used to evaluate student impressions of the course. A self-efficacy instrument was used to measure improved self-efficacy through a pre and a posttest. Most, but not all, students made gains in laboratory self-efficacy. Overall the program achieved its goal to improve confidence in students taking the introductory environmental science lab. However, variability in teaching from teaching assistants had an impact on the experience of some students.

INTRODUCTION

When inquiry is incorporated into the lab, the traditional methods of assessment, lab reports, lab practicals, and written exams do not directly assess the types of student learning being encouraged. In fact, the traditional assessment methods create an environment where the inquiry and teaching process are segregated from the "learning" (Costa & Kallick, 1995) The traditional assessments become steps that must be completed, but they are not seen as essential to memorizing the material that will be on the exam.

Because assessment drives instruction (Tobias & Raphael, 1995), it is critically important to both develop assessment methods designed to measure the inquiry skills of students as well as rigorously evaluate the developed assessment methods to determine their impact. For the assessment methods to succeed in the large-course format, structure must be in place for both the students and the teaching assistants responsible for the instruction of the labs. Providing this structure within the context of a large introductory course is the primary goal of the Inquiry Program developed at Bowling Green State University.

LITERATURE REVIEW: INFRASTRUCTURE FOR TEACHING BIOLOGY USING INQUIRY

The natural location for the inclusion of inquiry-based science instruction in university courses is the laboratory. There is a documented pattern of success using inquiry methods in undergraduate science courses. Stepans, Dyche, and Beiswenger, (1988) looked at two sections of an elementary education science content course. In one section, a learning cycle approach was the focus, and in the other section, lecture, demonstration, and recitation strategies were used. Both sections were taught the concept of buoyancy. Based on student interviews the learning cycle group achieved greater understanding of the terms used than the other group. Sundberg, (1997) reported on an inquiry-based laboratory that was designed for students to confront 10 common misconceptions in biology. To assess the effectiveness of this approach a content pre and posttest was administered to both the experimental group and to a control group of students from a remaining traditional biology section. "Average posttest scores for nonmajors taking the investigative laboratory were higher in virtually every category than the corresponding averages for their classmates who took a traditional laboratory" (pp. 147–148). Leonard (1983) compared an inquiry-based laboratory curriculum (Biological Sciences Curriculum Study-style) to a more traditional one. A multiple-choice test of biological concepts was used to compare the two different groups. Although there was no significant difference between the groups on the pretest, the group using the inquiry-based lab scored significantly higher on the posttest. Christianson and Fischer (1999) compared students from two traditional lecture courses with an inquiry teaching-based course. The students in the two settings were compared using a multiple-choice test focusing on the concepts of diffusion and osmosis. There were no significant differences observed on the pretest, but posttest results indicated that the students in the inquiry section developed a deeper understanding of diffusion and osmosis. Prince and Kelly (1997) describe the undergraduate introductory science experience at Hampshire College as problem-based and inquiry-based instruction. These experiences appear to have a strong impact on the graduates.

It can be concluded that undergraduate students who have not previously experienced inquiry-based instruction need support in their introductory courses through continuous formative-structured assessment. Only by providing this feedback can the students develop the inquiry skills necessary for enhanced conceptual learning of key science concepts.

In spite of the National Science Education Standards Inquiry Strand (National Research Council [NRC], 1996) being used as a guide for K–12 science instruction, students do not enter the university ready for science experimentation. Developing inquiry-based labs may not be the problem. There are a growing number of inquiry-based laboratory student manuals available (i.e., Dickey, 1995; Handelsman, Houser, & Kriegel, 1997; Morgan & Carter, 1999), as well as published laboratory activities in journals such as *Tested Studies for Laboratory Teaching* (Donahue, 2003). However, students must be prepared for an inquiry-based environment. Support is needed specifically in introductory courses. Formative-structured assessments can help students to develop the skills necessary for doing inquiry in a laboratory course.

Much work has been done at the K–12 level in terms of authentic assessment and portfolio assessment that can inform university instructors on ways for assessing scientific inquiry. Brown and Shavelson (1996) provide observational rubrics and performance scoring evaluations that are linked directly to inquiry activities and goals. Ostlund (1992) defines the process skills and six levels of achievement. Some other examples of assessment tools have been described by Ramig, Bailer, and Ramsey (1995) and Ruef (1998). Incline High School in Incline, Nevada, has a rubric for assessing inquiry and safety in K–12 (*Standards for Assessing Scientific Inquiry*, n.d.).

There are some examples of inquiry assessment at the university level. Heady, Coppola, and Titterington (2001) provide an excellent summary and examples of alternate assessment and inquiry-based assessment at the postsecondary level. Slater, Ryan, and Samson (1997), working with a college physics course concluded that the portfolio assessment does an equal job of conveying content knowledge and additionally contributes to student confidence in learning about physics. Brunkhorst (1996) describes using assessment strategies in geology that relate assessment to instruction through use of portfolios. There is a growing interest in higher education in developing assessment aligned with the goals of inquiry instruction.

Many of the described assessments are costly and time consuming, and require special expertise (Shavelson & Baxter, 1992). This expertise is often beyond the ability of inexperienced teaching assistants found in many large general education science courses (Ghosh, 1999). Garcia-Barbosa and Mascazine (1998) summarize the conditions for effective graduate teaching assistants in science courses. In their article, they provide a number of resources for the pedagogical development of graduate students and suggest that graduate students who are involved in exploring the pedagogy through various mechanisms develop enhanced teaching skills.

AN INVESTIGATION IN CREATING AN INFRASTRUCTURE FOR TEACHING BIOLOGY USING INQUIRY

Study Participants

Infrastructure for Inquiry was developed to help both students and inexperienced teaching assistants succeed in inquiry-based laboratory activities called for by educational reform documents (NRC, 1996). This model was developed for a large introductory course in environmental biology. Each semester between 640 and 820 students are enrolled in the course. There are between 26 and 28 lab sections each semester.

Infrastructure for Inquiry Model Investigated

Infrastructure for Inquiry was developed in response to the need to facilitate the success of both students and inexperienced graduate teaching assistants (GTAs) in inquiry-based laboratory activities. Students and instructors alike are provided with modeled examples of doing and teaching scientific inquiry. Assessments reward students for doing science and thinking reflectively about how they are doing it.

Infrastructure for Inquiry has two foci: supportive structures and linked assessments. To provide support for the students the initial activities are structured. During the semester the structure and modeling are gradually reduced until students are doing their own investigations. The program is designed to assess students not only on their content knowledge and skills, but also on the process of doing science. This demonstrates to the students that scientific thinking and reflection are valued and an important course goal. Infrastructure for Inquiry invites the teaching assistants into the learning process and builds a supportive community where pedagogy is discussed as frequently as survival strategies.

Infrastructure for Inquiry assessment tools are coupled to detailed grading rubrics that enable graduate teaching assistants (GTAs) to successfully evaluate student performance. The laboratory manual and assessment plan were developed for Biology 101, Environment of Life. Over the course of the project, 1,500 students took the course. Twelve graduate teaching assistants were involved in teaching the laboratories.

Existing inquiry-based laboratory activities were adopted or adapted for use in the laboratory manual. In general, students are given an opportunity to learn about a particular experimental method and are then asked to use that technique to answer a question of their own. Initial labs are highly structured with emphasis on particular elements such as experimental design or data analysis. Assessment is tied to the course goal of getting students to ask questions, formulate hypotheses, design and conduct experiments, analyze data, and report results.

Students must maintain a laboratory journal. They have pre-laboratory questions and skill points to determine their preparation and technical abilities. Face-to-face journal checks insure that the content of the journal is appropriate. There are written exams and group poster presentations. The heart of the assessment is the reflection page. At certain points throughout the semester, students are asked to select their best work and explain its strengths and weaknesses. They are then asked to explain how they will improve in the future. A final reflection asks them to review all of their experiments and to explain the strengths and weaknesses of the one they select. They are then asked to explain how this experiment demonstrates their growth throughout the semester. Student quotations come from this segment of the student work. Details of the assessment plan are described in the sections below.

Laboratory Journal

"Through lab reports and lab journals, I have been able to put my understanding into play and prove to myself and to others that I am learning and do understand these parts of doing science" (Student comment from final reflection).

While in the laboratory students keep a detailed laboratory journal that contains each section of the experimental process. In the lab manual, each portion of the experimental process is outlined in the Doing Science worksheet. The following is the section on experimental design (Waggoner & McArthur, 2002):

How did I design my experiment?

A carefully designed experiment should provide results that either support or disprove your hypothesis. There are many factors to consider when designing an experiment. These factors are either things the research can affect such as controls and independent variables or things that occur as a result of the choices made by the researcher, the dependent variable. Your hypothesis should easily tell you what is the independent variable and what is the dependent variable. "If I change the independent variable in this way then I expect the dependent variable to be affected in this way." A well-designed experiment pays attention to the evidence to be collected. It should be evidence that directly relates to the hypothesis.

Below is a copy of the rubric used to evaluate the experimental design of an investigation:

Experimental design rubric outline.

- Clearly identifies independent variable
- Clearly identifies dependent variable
- Clearly identifies controls
- Acknowledges necessary assumptions inherent in the experiment
- Includes enough controls to convince my peers of the validity or relevance of my results
- Specifically and directly addresses the research question
- Expected results will either support or disprove my hypothesis.

Table 23.1.   Factors in Experimental Design

Dependent variable. This is the thing that you are measuring. It is the one thing in the experiment that you cannot control. Changes in the dependent variable depend upon what you do when setting up the experiment.	How will you measure it?
Independent variable. The independent variable is the one factor in the experiment that the investigator chooses to change. The independent variable is the one thing that is different between the samples being studied. The choices made in selecting the independent variable and decisions about how to change it should be considered in light of how they address the research question.	How will you change it?
Controls. A well-designed experiment has as few variables as possible. There are many things that can be altered between samples in an experiment. A good experiment controls all of the factors that can be controlled, leaving the only possible explanation for the changes in the dependent variable the changes made by the researcher to the independent variable.	What will you do to make sure there are controls?
Assumptions. Often times there are factors in an experiment that are out of the researchers ability to control. Either they are issues where there is more than one interpretation or where it is impossible to clearly determine them. In these cases, a researcher must make assumptions. The assumptions should be clearly stated. Assumptions can affect the interpretation of results.	How do they impact interpretation of results?

What evidence will be obtained by this experimental design that will support or disprove my hypothesis?

Journal Checks

Students use the experimental design worksheet as a road map for designing and conducting their experiments. Their journals must reflect each element of the worksheet and periodic face-to-face checks of the journals are conducted. This format provides students with immediate feedback about the work they are doing and allows them in an initial low-stakes format to learn how to improve. The checks are done during the lab period. Each check is only a small part of the final grade. This ensures that a student having trouble getting started can still succeed in the course through the frequent intervention.

Reflection Pages

After looking at this experiment and then looking at my last experiment it is clear that I have become a more confident researcher. In the early experiments some of the aspects were provided through the lab manual, now none of the information is provided as I move comfortably through the experiments ... (Student comment from final reflection)

The heart of the assessment plan is a series of reflection pages. By completing these pages students reflect upon the process of doing science, demonstrate their understanding of the process of science, explain how they will improve their science knowledge and skills, and demonstrate their personal growth throughout the semester. Reflection pages allow students to take a few minutes to think about what they have been learning and doing in the laboratory. They cause the students to reflect about where they are and where they want to be in the future. The goal-setting component of the reflection allows the students to develop specific areas to work on during the remainder of the semester.

Reflection pages provide a unique insight into the student's thought processes. A student may be writing correct information in the lab journal yet be unable to discuss the strengths and weaknesses of the information in the reflection page. Each reflection allows the instructor to work directly with that student on that process. Furthermore, the long-term changes in the reflection pages demonstrate student growth. Often the final reflections will contain the "now I get it" comments.

Skill Points

> My lab skills have also improved, because at first I was a little timid in conducting experiments with my lab partners. Now I feel more confident in myself and in my abilities to contribute in lab groups. Overall Bio 101 has improved my curiosity in the Science field. Before taking this class I thought science was uninteresting, but now I know how much fun it really is. I am interested in taking more science courses in the future. (Student comment from final reflection)

The traditional laboratory practical is replaced with skill points. These are points given out by the teaching assistant after observing each student carry out a laboratory activity. It is slightly more time consuming, but it allows the instructor to directly assess the ability of each student to manipulate the materials and equipment. The teaching assistants use an established rubric to evaluate the skill points.

Pre-Lab Quizzes

> I admit that I came into this class with a cocky attitude that I already knew what I was doing in regards to forming an experiment. I was proven wrong. This experiment in particular made me take notice and relearn how to do an experiment the correct and scientific way. (Student comment from final reflection)

Each laboratory is accompanied by a series of pre-lab questions that are to be answered by the students before they get to the laboratory. On random days, they are allowed to use the sheet on which they have answered the questions to answer a short quiz. This provides incentive for the students to come to laboratory prepared. Students who come to lab ready to work and familiar with the laboratory activities are able to get more productive work done than students reading the lab for the first time in class.

Group Projects

> I have learned that it takes more than one person and one brain to come up with a thought out, intelligent plan for a lab. This semester has shown me how to go about not just doing the lab, but doing the journal checks and the reflection pages to make us think about the labs and the work we put into them. The posters and presentations that went along with them helped to see what others in the class were doing too. (Student comment from final reflection)

Students work on two group projects throughout the semester. For each project they fill out a group contract that indicates research question, e-mail, meeting times, and assigned activities. Each student and the instructor are given a copy. The initial research project has a defined experimental design, but the students ask their own question. They conduct the experiment, analyze the data, and report the results in a poster session.

Guidelines for preparing the poster are in the laboratory manual. Each student is responsible for standing by the poster for a 10-min period to answer questions. While one student answers questions, the remaining group members view the other posters. Students fill out a sheet asking for one question and provide one piece of constructive criticism for each poster. Students rotate through the presentations until everyone has seen every poster. At this time the students stand before their poster and the instructor chooses "reviewers" from the audience to ask a question. Each member of the group answers one question. Evaluation of posters and student performance is supported by rubrics. At the conclusion of the project each student is asked to fill out a group self-evaluation. The average rubric scores become part of each student's grade. As a group, students decide which poster represents the best report of the section.

Research Design for Infrastructure for Inquiry

The researchers wanted to know if student confidence in the process of doing science was affected by the changes in the laboratory structure. To determine the impact of the Infrastructure for Inquiry assessment tools, an external evaluation was conducted. Both qualitative and quantitative measures were collected. Focus-group data were used to evaluate the student impressions of the new course. During the 2000–2001 academic year, an instrument measuring students' self-reported confidence (self-efficacy) in biology laboratory activities was used as a pretest in the beginning of each semester and as a posttest at the end of each semester to measure quantitatively whether or not students gained confidence in scientific procedures. Qualitative data obtained from the students through focus-group interviews at the end of each semester was used to validate the quantitative results.

The quantitative analyses showed that there were significant gains in self-efficacy in each of the two cohorts, fall 2000 and spring 2001. The previous spring 2000 cohort was considered a pilot test of the program and only qualitative information was collected from this group. The principal investigators then used this information to improve and fine-tune the program for its full implementation during the 2000–2001 academic year.

Qualitative data collection involved two focus groups per semester. Samples of students participating in the course were asked the following five questions:

1. Describe some of the labs that you did. Include what you did in the lab and anything that stands out in your mind as a high point.
2. What didn't you like about the lab?
3. Do you think you learned anything that you can apply to everyday life? What was learned?
4. Discuss the features of the lab. Was there enough time?
5. Did you like the lab manual? Was the grading fair?

The qualitative analyses indicated that students overall enjoyed the hands-on learning involved with the laboratory activities. Several lab activities made impacts on students and their confidence in doing lab activities. Many of the problems students experienced with the labs were based upon interpersonal relationships; students frequently worked in groups and not all students communicated well or did their fair share of the work. Students felt that the laboratory did not follow the lecture and seemed to be two separate courses. A few students felt that their TAs were not properly prepared and did not always read journal entries.

Quantitative data were collected by means of pre and posttests on a self-efficacy instrument (Baldwin, Ebert-May, & Burns, 1999) modified for environmental science. Students in all sections of the modified laboratory answered 23 questions about their confidence in environmental biology laboratory activities with closed responses from 1 (*totally confident*) to 5 (*not at all confident*). A multivariate analysis of variance (MANOVA) was run on all variables for the pre and posttest scores of the combined cohorts, and a significant main effect occurred between pre and posttest ($p < 0.001$, $\eta^2 = 0.362$). ANOVAs of separate cohorts of pretest scores resulted in statistically significant differences among the lab sections ($p < 0.01$, $\eta^2 = 0.107$ for fall 2000; $p < 0.01$, $\eta^2 = 0.107$ for spring 2001). According to expectations, because students enter the class with differing degrees of ability, attitude, and interest, statistically significant differences were found among the lab sections at the pretest time. However, the effect-size measure for the pretest was very small so the variance in the self-efficacy scores was explained by being in a specific section and the difference is negligible. The posttest demonstrated that the difference among the course sections in the fall 2000 cohort was more pronounced ($p < 0.01$). This indicated that there were larger differences among some of the sections after students had taken the course than before, which was not expected. According to self-efficacy theory, if everyone had the same experience, there should be no group differences (Bandura, 1986). These differences among course sections were not significant in spring 2001 ($p = 0.273$). These 2001 posttest group scores may have resulted from varying levels of effective teaching by TAs during the fall, diminished in the spring creating a more consistent application of the Infrastructure for Inquiry model. There may also be individual class variations in experience and ability over time. Paired t tests for fall 2000 and spring 2001 were statistically significant between pre and posttests for combined sections; however, some sections were not significant (5 out of 18 sections for fall 2000 and 6 of 23 for spring 2001). In several cases, the same teaching assistant taught these sections. Most, but not all, students made gains in laboratory self-efficacy.

Summary of Investigation on Infrastructure for Inquiry

Overall the program achieved its goal to improve confidence, understanding and critical thinking in students taking the introductory environmental biology lab. The scaffolding built into the assessment used in the Infrastructure for Inquiry helps the students learn in a step-by-step manner. They also receive formative feedback all semester so by the end of the semester, they are able to design an experimental investigation. The reflection built into the assessments helps the students gain the meta-cognitive skills necessary to understand their growth.

Variability in teaching by TAs resulted in some students gaining more than others based on results from the self-efficacy instrument. Both the quantitative and qualitative data indicate that the teaching assistants were an extremely important component of the equation. The differences between teaching assistants came out very strongly in the focus groups. Some teaching assistants were clearly more able than others in implementing the assessments required for the course. Because teaching assistants are responsible for all the instruction in the labs, they should receive more intensive training to improve consistency among labs. Most of the teaching assistants assigned to this course are 1st-year students in the master's degree program, with little or no prior teaching experience. Helping them to understand the finer points of formative evaluation was a challenge. A training program should be established to ensure that the teaching assistants will be able to carry out the assessment strategies. There was a gain in students scores from the first semester to the second indicating that with experience the teaching assistants were able to improve students' confidence in laboratory science.

Currently, this work is expanding to other content areas. Colleagues in the physics department have developed a physics course designed for majors in middle childhood education with a specific focus on inquiry (Van Hook, Ballone, & McArthur, 2003). Overall, within the science and mathematics departments there has been an interest in improving undergraduate education (Ballone, Van Hook, Czerniak, & McArthur, 2002).

Continuing to study the impact of these inquiry-based courses is a priority. One study underway will compare the teaching performance of education majors enrolled in reform courses to students prepared in traditional science lecture sections.

SUMMARY: RECOMMENDATIONS FOR THE DEVELOPMENT OF AN INQUIRY-BASED COURSE IN ENVIRONMENTAL BIOLOGY

In science laboratories, as a component of undergraduate instruction, inquiry is a key learning outcome (NRC, 1996). Using inquiry-based activities in undergraduate science courses can produce measurable results in increased scores on content tests (Christianson & Fischer, 1999; Leonard, 1983; Stepans et al., 1988; Sundberg, 1997). In the case of preservice teachers, more positive attitudes toward science have also been observed. Hall (1992) looked at a biology class specifically designed for elementary teachers. This inquiry-based class was "influential in promoting positive attitudes toward science and science teaching among prospective elementary teachers" (p. 242). Crowther and Bonnstetter (1997) found an inquiry-based biology course for elementary teachers helped the students progress from "anxiety to empowerment in the teaching of science" (p. 15). This empowerment is extremely important for introductory courses. An implicit goal of many general education science courses is creating a positive attitude and understanding of science. The students in the non-major courses are not going on to take more content, so the specific content of one or two courses is not as important as the attitude toward science.

Students who have not had experience with inquiry need support specifically in introductory courses. Formative structured assessment is necessary to provide the students with the skills necessary for inquiry. In spite of the National Science Education Standards Inquiry Strand (NRC, 1996), students do not enter the university ready for science experimentation (Uno, 1988). To provide the students with an opportunity to grow and develop in their inquiry skills, such skills must be assessed in a systematic manner. Students learn to value what is assessed. If process skills are not assessed then students will not see inquiry as an important component in science.

Typically non-science university majors take two science courses to complete general education requirements. Therefore it is extremely important that students in these non-major courses are assessed on inquiry. It may be the only opportunity they have to receive meaningful feedback about scientific inquiry. Quotes from student journals in the study of Infrastructure for Inquiry demonstrate that they appreciated the emphasis on inquiry in the course:

> I plan on keeping my lab journal to use in other science classes.... This class has prepared me to think more scientifically and critically.... I like the idea that I actually understand what is going on in a science class.... I feel confident to explain to others about experiments and the results. I now feel prepared to take other lab courses because I comprehend what is being done in the lab.

IMPLICATIONS FOR REFORM IN UNDERGRADUATE BIOLOGY TEACHING AND FUTURE RESEARCH

In general, a supportive inquiry laboratory environment had a positive impact on most undergraduate students. Many students emerging from the course were able to verbalize a positive attitude toward inquiry and many of these non-majors began to refer to themselves as researchers, investigators, or scientists. The one factor that seemed to prevent some students from making these positive gains was the experience level of the teaching assistants. Further research needs to be done to address the pedagogical development of teaching assistants.

Another area that needs to be examined is the impact of inquiry on coverage of content. Anecdotally, the students in the new inquiry laboratory did as well on lecture exams as students taking the more traditional laboratory. Also, there is some anecdotal evidence that students transfer their inquiry skills to other courses, specifically a science methods course. However, further work needs to be done to quantify these observations.

ACKNOWLEDGMENTS

This work was supported by a grant from the National Science Foundation Division of Undergraduate Education (NSF DUE CCLI 9952421) and in part by NASA Opportunities for Visionary Academics (NOVA), funded by the National Aeronautics and Space Administration, although the views expressed here are the authors' only.

REFERENCES

Accongio, J.L., & Doran, R.L. (1993). *Classroom assessment: Key to reform in secondary science education.* Columbus, OH: ERIC Clearinghouse for Science, Mathematics, and Environmental Education. (ERIC Document Reproduction Service No. ED370774)

Baldwin, J., Ebert-May, D., & Burns, D. (1999). The development of a college biology self-efficacy instrument for nonmajors. *Science Education, 83,* 397–408.

Ballone, L., Van Hook, S., Czerniak, C., & McArthur, J. (2002, January). *Working together: Improving preservice teacher education by collaboration.* Paper presented at the annual meeting of the Association for the Education of Teachers of Science, Charlotte, NC.

Bandura, A. (1986). *Social foundations of thought and action: Social cognitive theory.* Englewood Cliffs, NJ: Prentice Hall.

Brunkhorst, B. (1996). Assessing student learning in undergraduate geology courses by correlating assessment with what we want to teach. *Journal of Geoscience Education, 44,* 373–378.

Brown, J., & Shavelson, R. (1996). *Assessing hands-on science: A teacher's guide to performance assessment.* Thousand Oaks, CA: Corwin Press.

Christianson, R., & Fisher, K.M. (1999). Comparison of student learning about diffusion and osmosis in constructivist and traditional classrooms. *International Journal of Science Education, 21,* 687–698.

Costas, A.L., & Kallick, B. (1990). *Assessment in the learning organization: Shifting the paradigm.* Alexandria, VA: Association for Supervision and Curriculum Development.

Crowther, D., & Bonnstetter, R. (1997, March). *Science experiences and attitudes of elementary education majors as they experience Biology 295: A multiple case study and substantive theory.* Paper presented at the international conference of the National Association for Research in Science Teaching, Chicago.

Dickey, J. (1995). *Laboratory investigations for biology.* Redwood City, CA: Benjamin/Cummings.

Donahue, M. (2003, June). *Tested studies for laboratory teaching.* Paper presented at the proceedings of the 25th Workshop/Conference of the Association for Biology Laboratory Education (ABLE), Las Vegas, Nevada.

Garcia-Barbosa, T., & Mascazine, J. (1998). *Guidelines for college science teaching assistants. ERIC Digest.* Columbus, OH: ERIC Clearinghouse for Science, Mathematics, and Environmental Education. (ERIC Document Reproduction Service No. ED433193)

Ghosh, R. (1999). The challenges of teaching large numbers of students in general education laboratory classes involving many graduate student assistants. *Bioscene, 25*, 7–11.

Hall, D. (1992). The influence of an innovative activity-centered biology program on attitudes toward science teaching among preservice elementary teachers. *Journal of College Science Teaching, 92*, 239–242.

Handelsman, J., Houser, B., & Kriegel, H. (1997). *Biology brought to life: A guide to teaching students to think like scientists.* Dubuque, IA: Times Mirror Higher Education Group.

Heady, J., Coppola, B., & Titterington, L. (2001). Assessment standards. In E. Siebert & W. McIntosh (Eds.), *College pathways to the science education standards* (pp. 57–63). Arlington, VA: NSTA Press.

Leonard, W. (1983). An experimental study of a BSCS-style laboratory approach for university general biology. *Journal of Research in Science Teaching, 20*, 807–813.

Morgan, J. G., & Carter, M. E. B. (1999). *Investigating biology* (3rd ed.). Menlo Park, CA: Benjamin Cummings.

National Research Council. (1996). *National Science Education Standards.* Washington, DC: National Academy Press.

Ostlund, K. (1992). Sizing up social skills. *Science Scope, 15*, 31–33.

Prince, G., & Kelly, N. (1997). Hampshire College as a model for progressive science education. In A. McNeal & C. D'Avanzo (Eds.), *Student-active science: Models of innovation in college science teaching* (pp. 189–200). New York: Saunders College Publishing.

Ramig, J., Bailer, J., & Ramsey, J. (1995). *Teaching science process skills.* Torrence, CA: Good Apple.

Ruef, K. (1998). *The private eye (5X) looking/thinking by analogy—A guide to developing the interdisciplinary mind.* Seattle, WA: The Private Eye Project.

Shavelson, R., & Baxter, G. (1992). What we've learned about assessing hands-on science. *Educational Leadership, 49*, 20–25.

Slater, T., Ryan, J., & Samson, S. (1997). Impact and dynamics of portfolio assessment and traditional assessment in a college physics course. *Journal of Research in Science Teaching, 34*, 255–271.

Standards for assessing scientific inquiry. (n.d.). Retrieved July 18, 2003, from http://www.inclinehs.org/science/standards.htm

Stepans, J., Dyche, S.E., & Beiswenger, R.E. (1988). The effect of two instructional methods in bringing about a conceptual change in the understanding of science concepts by prospective elementary teachers. *Science Education, 72*, 85–195.

Sundberg, M. (1997). Assessing the effectiveness of an investigative laboratory to confront common misconceptions in life sciences. In A. McNeal & C. D'Avanzo (Eds.), *Student active science: Models of innovation in college science teaching* (pp. 141–162). New York: Saunders College Publishing.

Tobias, S., & Raphael, R. (1995). In-class examinations in college science: New theory, new practice. *Journal of College Science Teaching, 24*, 242–244.

Uno, G. (1988). Teaching college and college-bound biology students. *The American Biology Teacher, 50*, 213–216.

Van Hook, S., Ballone, L., & McArthur, J. (2003, January). *Students' perceptions of a constructivist-based physics course.* Paper presented at the annual meeting of the Association for the Education of Teachers of Science, St. Louis, MO.

Waggoner, C., & McArthur, J. (2002). *Biology 101 laboratory.* Plymouth, MI: Hayden-McNeil.

CHAPTER 24

A MODEL FOR REFORM
IN TEACHING CHEMISTRY

With a Focus on Prospective
Elementary Teachers

Dorothy L. Gabel

ABSTRACT

The focus of this chapter is to provide the reader with a review of effective
teaching strategies that improve all students' understanding of chemistry, and
in particular, the chemistry background of prospective elementary teachers.
At the present time, the reform of chemistry teaching is becoming much
more widespread than in the past with the support of funding from the
National Science Foundation. Not only have new approaches to the teaching
of chemistry been supported, but also the dissemination of more interactive
models of teaching. The preparation of teachers, especially at the elementary
level, is very important because if children are turned off to chemistry in their
early years, they are less likely to select chemistry-related careers. This chapter
highlights the major teaching innovations that are being used throughout the
United States that are based on the NSF-funded projects and provides references to publications that describe the teaching practices and their effective-

Reform in Undergraduate Science Teaching for the 21st Century, pages 425–443
Copyright © 2004 by Information Age Publishing
All rights of reproduction in any form reserved.

ness. It also describes in more detail particular innovations that have been implemented at Indiana University in a chemistry-based course for prospective elementary teachers. This Introduction to Scientific Inquiry course is used as a model as one way to teach basic chemistry concepts using an inquiry approach as advocated in the National Science Education Standards.

INTRODUCTION

The chemistry background that elementary teachers need to teach science at the elementary level does not require the same depth that is commonly presented in a typical general introductory chemistry course. In addition, many concepts that are taught at the elementary level are not included in a typical chemistry course because instructors assume that prospective teachers understand the chemistry of everyday occurances such as burning. In addition, teachers are expected to teach science as "inquiry," a method not commonly used in introductory college chemistry courses. This chapter gives a short historical background of the changes in science instruction begun in the post-Sputnik era of the 1960s, and advocated again in the National Science Education Standards (National Research Council [NRC], 1996, 2000). It also gives a brief overview of changes that are occurring in chemistry courses today that will help make them more suitable in filling the needs of prospective elementary teachers.

BACKGROUND

If there is any truth to the adage that "teachers teach as they were taught," the reform of undergraduate science courses is a major component in bringing about change in science teaching and learning at the elementary and secondary levels. This will not occur until college-level science becomes more interactive and inquiry oriented.

In 1962, Joseph Schwab advocated teaching science as "enquiry." A dominant feature of many of the science educational books and programs produced in the post-Sputnik-era reform of the mid-60s was that they utilized an inquiry approach that was sometimes referred to as hands-on science. At the elementary level, the Science Curriculum Improvement Study (*Science Curriculum Improvement Study*, 1968) used the learning cycle approach and Science: A Process Approach (American Association for the Advancement of Science, 1967) organized the curriculum around the science process skills (such as making observations, inferences, predictions, measuring, classifying controlling variables, using models, etc.).

For the high school level, ChemStudy and the Chemical Bond Approach were developed (National Science Foundation [NSF], 1963a, 1963b). Although ChemStudy formed the basis of the chemistry textbooks on the market today (with much less content, however), it was the Chemical Bond Approach that more closely modeled what is advocated in the teaching of chemistry today. The final experiment in that text was a blank page (NSF, 1963b). Students were required to not only plan and do a scientific investigation, but also to design an experiment on a topic of their choice. How many chemistry teachers actually implemented this is unknown!

In 1996, the National Research Council in *National Science Education Standards* advocated (among other things) that science should be taught as inquiry at the elementary and secondary levels. In the special volume entitled, *Inquiry and the National Science Education Standards* (NRC, 2000), the essential features of classroom inquiry are described as follows:

> learners are engaged by scientifically oriented questions; learners give priority to evidence, which allows them to develop and evaluate explanations that address scientifically oriented questions; learners formulate explanations from evidence to address scientifically oriented questions; learners evaluate their explanations in light of alternative explanations, particularly those reflecting scientific understanding; and, learners communicate and justify their proposed explanations. (p. 25)

These features have variations according to the level of self-direction and the level of the direction given by the teacher or material. As described in the *Inquiry and the National Science Education Standards* (NRC, 2000, p. 29), the greatest level of inquiry occurs when the learner poses the question, determines what constitutes the evidence and collects it, formulates an explanation after summarizing evidence, independently examines other resources and forms the links to explanation, and forms reasonable and logical arguments to communicate explanations (see Table 24.1). These goals appear to exceed those advocated in the 60s in that more emphasis is placed on explanations and on student-generated problems to investigate.

If these are the goals that the nation is advocating for precollege students, then it is important that undergraduate and graduate students preparing to become teachers should have accomplished these by the end of their teacher preparation program. However, because the reform has not taken place universally at the precollege level, the work that must be done to bring prospective teachers to this level is a challenge to instructors teaching science courses and to science educators teaching methods classes. This is particularly true at universities that are preparing large numbers of teachers.

Table 24.1. Essential Features of Classroom Inquiry and Their Variations

Essential Feature	Variations			
1. Learner engages in scientifically oriented questions	Learner poses a question	Learner selects among questions, poses new questions	Learner sharpens or clarifies question provided by teacher, materials, or other source	Learner engages in question provided by teacher, materials, or other source
2. Learner gives priority to evidence in responding to questions	Learner determines what constitutes evidence and collects it	Learner directed to collect certain data	Learner given data and asked to analyze	Learner given data and told how to analyze
3. Learner formulates explanations from evidence	Learner formulates explanation after summarizing evidence	Learner guided in process of formulating explanations from evidence	Learner given possible ways to use evidence to formulate explanation	Learner provided with evidence
4. Learner connects explanations to scientific knowledge	Learner independently examines other resources and forms the links to explanations	Learner directed toward areas and sources of scientific knowledge	Learner given possible connections	
5. Learner communicates and justifies explanations	Learner forms reasonable and logical argument to communicate explanations	Learner coached in development of communication	Learner provided broad guidelines to sharpen communication	Learner given steps and procedures for communication

More _____ Amount of Learner Self-Direction _____ Less

Less _____ Amount of Direction from Teacher or Material _____ More

THEORETICAL BASE FOR TEACHING/LEARNING CHEMISTRY

Before examining effective strategies for the teaching of chemistry to prospective elementary teachers, it is important to briefly review how chemists view the physical world. The sciences can be understood and taught on three levels: the macroscopic or phenomena level, the particle level, and the symbolic level (Johnstone, 1993). A chemist with a mature understanding of chemistry integrates these three representations of matter in his/her long-term memory, and recalls them at will.

In considering the appropriate preparation of prospective elementary teachers of physical science and specifically chemistry, one must consider how children learn, the level of the subject matter that they are expected to teach, and the content that is appropriate for children at the elementary school level. The National Science Education Standards (NRC, 1996) and the American Chemical Society (Breslow, 1996) recommend that children in grades 1 to 6 be taught chemistry at the macroscopic level. For example, children can make observations and inferences about water and the changes it undergoes, how it can be frozen and melted, and how it can dissolve certain substances. This is not to say that elementary teachers should not be taught on a higher level and hence be expected to give explanations using atoms and molecules, and representing them using symbols and formulas. In fact, they need to study chemistry on this level so that their explanations to children on a lower level will be consistent with current knowledge of the chemical world.

In examining effective teaching strategies, it is also important to consider how people learn. A common theory in current use is an information-processing view of learning. In summary, once something is perceived by the senses, the information passes into short-term memory, and is eventually stored in long-term memory where concepts are linked together in what might be thought of as concept maps. The linkages can be correct (correspond to current scientific thought) or incorrect, that is, long-term memory may harbor information as misconceptions. Social constructivist theory maintains that social interaction between students and between teachers and students as well as self-reflection helps students make sense of the physical world. (For a summary of these ideas on the learning of chemistry, see Gabel, 1999a.) With the above in mind, it appears that if chemistry instruction includes learning by inquiry with opportunities for social interactions and focusing on the macroscopic level of phenomena, prospective elementary teachers should be well prepared to teach chemistry at the elementary school level.

Finally, however, the chemistry content of such a course must be considered. A cursory examination of elementary science texts shows that the content focuses on the common processes included in everyday life such as physical and chemical change, burning, decomposing, dissolving, melting, evaporating, freezing, and boiling. Also included are substances (elements and compounds), solutions, and heterogeneous mixtures. Middle and high school science instruction includes the same topics, but at a higher more abstract level using particles and symbols.

It must be remembered that, in addition to learning chemistry content in an interactive environment, and understanding the chemistry topics that they will teach, prospective elementary teachers also need laboratory experiences in science courses that model the various aspects of inquiry as recommended for children in the National Science Education Standards. This includes developing their own basic and integrated science skills, hypothesis testing, planning and conducting their own experiments, and giving explanations of their experimental findings.

EXISTING CHEMISTRY COURSES FOR PRESERVICE ELEMENTARY TEACHERS

An important question that institutions of higher education must answer is how to best meet the aforementioned requirements for prospective elementary teachers. Is it appropriate to require elementary education majors to enroll in an introductory chemistry course for all non-science majors, or should a special chemistry course for preservice teachers be offered? The University of Wisconsin at Madison offers a special companion course to their non-science chemistry course (Larson & Milddlecamp, 2003). California State Polytechnic University at Pomona created a special one-quarter chemistry course for their preservice teachers (Burke & Walton, 2002). Other institutions that have created special chemistry courses for their preservice teachers are Purdue University (Nakleh, personal communication, June 2003) and the University of Northern Colorado (Jones, personal communication, March, 2002). In many other institutions, undergraduate students will take a physical science course such as the one described by Jasien (1995). An alternative to these options is offered at Indiana University, and is described later in this chapter.

REFORM MOVEMENTS IN CHEMISTRY

What has appeared in the reform of general chemistry courses over the past few years has great promise for improving the instruction of all students taking chemistry, including prospective elementary teachers, even though those courses were not specifically designed with this in mind. At the national level, there is probably more innovation occurring in introductory chemistry courses at the present time than in all of the other sciences. This is due primarily to funding provided by the National Science Foundation in the 1990s through its Systemic Change Initiative for developing instructional change in chemistry. For the past few years, NSF has focused on the dissemination of these new courses through the funding of the Multi-Initiative Dissemination (MID) Project housed at the University of California in Berkeley. Leaders of five different NSF-funded-projects have combined their dissemination efforts by presenting joint one-and-a-half-day workshops around the country that introduce college chemistry instructors to new ways of teaching chemistry. Four mini-workshops (2 of the original 5 projects combined) are presented within the time frame of each workshop and common topics such as an introductory presentation on learning and another on evaluation are included. A grant funded in the spring of 2003 enables the continuation of these workshops for the 2003–2004 academic year. Descriptions of the four projects are given below.

ChemConnections (n.d.) ChemLinks Coalition and ModularChem Consortium (*Multi-Initiative Dissemination Project*, 2001) consist of the development of topical modules relating chemistry to everyday life to be used for the first two years of college chemistry. The 2- to 4-week modules start with relevant real world questions and develop the chemistry needed to answer them. Students model how chemistry is actually done, discover connections between chemistry, the other sciences, and the real world. To date, 13 modules have been developed. Each module features student-centered and active collaborative classroom activities and inquiry-based labs.

The Molecular Science Project (*Molecular Science Project*, 2001) consists of server-based instructional materials and techniques for the first two years of college chemistry that is network-based, so that it can be used in or out of class to provide stand-alone instruction. The materials involve peer collaboration, writing, critical thinking, and tutorials with visualizations of molecules, processes, and systems. Topics are taken from molecular life science, materials science, and environmental science.

The New Traditions program (*Establishing New Traditions*, 1997) has as its goal to facilitate cultural change in chemistry instruction at the college level by making it student centered rather than faculty centered. Included in the changes are the use of ConcepTests (Ellis et al., 2000) to supplement lectures, guided-inquiry and open-ended laboratory experiments, cooperative learning group activities (challenge problems and learning communities), the use of information technology/computer tools, and the use of new assessment techniques.

Peer-Led Team Learning (PLTL) consists of preparing students who have done well in chemistry to become peer leaders through intensive sessions with a faculty member on the chemistry content and pedagogy (Dreyfuss, 2003). The peer leader then becomes the leader of one or more weekly workshop groups of six to eight students enrolled in chemistry. This weekly meeting focuses on collaborative problem solving.

In summary, these new programs deviate quite remarkably from the way chemistry at the college level has traditionally been taught. All four programs involve collaborative learning to some degree. ChemConnections relates chemistry to the real world and hence may have greater relevance for prospective elementary teachers. It also has an emphasis on inquiry, as does the New Traditions program. Of the four innovations, it would appear that either ChemConnections or the New Traditions program would be an appropriate chemistry course for prospective elementary teachers in situations where no specialized course is available.

RESEARCH ON REFORM IN COLLEGE CHEMISTRY COURSES

Research on effective chemistry teaching in undergraduate, introductory courses for prospective elementary teachers, and even for general introductory courses, is difficult to replicate because there are many variables that are uncontrolled. However, some research has been reported on the effectiveness of the components of the NSF-funded programs discussed above.

Collaborative Learning

Collaborative learning, used in all of the aforementioned programs, is generally an effective teaching technique. Reviews of collaborative learning research by Gabel (1999a) and Bowen (2000) indicates that collaborative learning enhances science concept development and achievement. Although much of the research has been reported for high school science instruction, a meta-analysis of college science, mathematics, engineering and technology courses by the National Science Foundation's National Institute for Science Education (Springer, Stanne, & Donovan, 1999) indicates that college students generally perform better on standardized tests when taught with small-group methods, have better attitudes toward leaning, and retention in the course is greater. Specific studies related to chemistry instruction prior to 1998 can be found in Nurrenbern and Krupp (1995) and Nurrenbern and Robinson (1997). Other studies reporting the effectiveness of cooperative methods of instruction over the last few years include Wright et al. (1998) in analytical chemistry; Hagen (2000), Bradley, Ulrich, Jones, and Jones (2002), and Carpenter and McMillan (2003) in organic chemistry; Williamson and Rowe (2002) on problem solving; and Shibley and Zimmaro (2002) on laboratory performance in general chemistry. Although all studies do not show that there were differences in chemistry achievement, other benefits accrued, such as an increase in students' reasoning and communication skills, and in their enjoyment of the course.

ConcepTests

The use of ConcepTests in instruction is a specific interactive, collaborative teaching strategy that promotes student understanding, and their use is advocated in the New Traditions program. ConcepTests (*Project Galileo*, 2000) are short multiple-choice questions on a single concept that are used during instruction to let the student know whether he or she understands a concept rather than being used to determine a student's grade. The technique using this specific name was first introduced by Mazur (1997) in the teaching of physics at Harvard. The steps include:

1. The instructor (usually using multiple choice) poses a question.
2. Students are given time to think.
3. Students record their individual answers (optional).
4. Students convince their neighbors of the correct answer (peer instruction).
5. Students record their revised answers (optional).
6. Feedback is given to teacher: Answers are tallied.
7. The instructor explains correct answer.

In his study of physics achievement Mazur found that not only did conceptual understanding of physics increase, but students' scores on problem solving also increased.

In chemistry instruction at the University of Wisconsin, ConcepTests are widely used in both small and large classes and student response appears to be positive. The New Traditions project has produced a videotape illustrating how ConcepTests can be used in large lectures. More information about their use in chemistry instruction and a Web site (Ellis et al., 2000) containing chemistry ConcepTests items is available in Landis et al. (2000).

Peer-Led Team Learning

Peer-Led Team Learning (PLTL) also appears to be an effective teaching/learning strategy in an organic chemistry course. Lyle and Robinson

(2003) did a statistical evaluation of data obtained in the study by Tien, Roth, and Kampmeier (2002) who had used PLTL with a number of groups of students and compared their grades with students enrolled in the course prior to using PLTL as control groups over several years. They used the effect size corrected for SAT scores and determined that the number of students receiving a C in the course was significantly higher when PLTL was used.

Inquiry-Based Labs

Inquired-based science laboratory activities have been recommended for elementary and high school students by the National Science Education Standards (NRC, 2000). If elementary and secondary school teachers are required to teach using inquiry, they will need to experience inquiry-based laboratory in their science courses. Two research papers have been produced recently that utilize guided inquiry as advocated in the Chem-Connections and the New Traditions programs. Straumanis (2001) developed and delivered an organic chemistry course at the University of New Mexico using inquiry-based labs and compared student attrition and achievement to students taught in a traditional organic chemistry course. Attrition in the conventional course was 47% as compared to 12% in the guided-inquiry course. On the final exam (that consisted of American Chemical Society questions selected by the traditional course professor), there was no difference in test scores between the upper half of the students in each class. In the bottom half of the students in the traditional course, 47% withdrew and 5% got an F, as compared to students in the guided-inquiry course where 28% withdrew, 15% got a D, and 1% got an F. The guided-inquiry approach appeared to help some of the weaker students while not making a significant difference for the better students. Comments from both students and faculty who have used guided inquiry were quite positive.

Farrell, Moog, and Spencer (1999) provide another example of successful implementation of guided inquiry in a description of achievement in the general chemistry course at Franklin and Marshall College. Their guided-inquiry two-semester sequence of introductory chemistry has multiple sections of about 25 students per session. Over the past several years they have accumulated data so that comparisons of student grades for approximately 800 students can be made for students using the guided-inquiry approach from fall 1994 through spring 1997, versus a traditional lecture approach from fall 1990 through spring 1994. Data show that for the guided-inquiry group as compared to the lecture group, the number of A's and B's substantially increased in the guided-inquiry classes, the C's remained constant, and the number of D's and F's declined. An even greater tribute to its success was that in the 1997–1998 school year, every student who was enrolled in the first semester guided-inquiry course selected to take the second semester in a guided-inquiry course.

A MODEL TEACHER PREPARATION PROGRAM: INDIANA UNIVERSITY

The research relating improved prospective elementary teachers' understanding of chemistry at Indiana University has occurred primarily in the Introduction to Scientific Inquiry course that was developed and that has been taught by the author (with a cadre of graduate students) for the past 25 years. It is a required course for all students enrolled in the elementary education program (except for the 5% that test out).

Our current elementary teacher preparation program at Indiana University was established in 1991 as part of the Quality University Elementary Science Teaching (QUEST) Program with NSF funding. It was based on a previous program established in the 1970s that required all prospective elementary teachers to take four 3-credit-hour laboratory science courses (Introduction to Scientific Inquiry, Biology, Physical Science, and Earth Science). In addition, students take a 3-credit-hour science methods course with a 1-credit-hour field experience in a local school. In the late 1980s the elementary education program changed and all preservice elementary teachers were required to have an area of concentration. Students selecting the science option take in addition to the 12-hour general requirement, 15 additional hours of science including a 4-credit-hour integrated science course (funded partially by NOVA [NASA Opportunities for Visionary Academics]) that focuses on energy.

Introduction to Scientific Inquiry

The first required science course that all preservice elementary students take is called "Introduction to Scientific Inquiry." As indicated in the preface of the course textbook (*Introductory Science Skills*, written in 1984 and revised in 1993), the content was designed to help students

> understand the nature of scientific inquiry by involving (them) in "doing science" rather than merely reading about it and memorizing facts.... The content of the text emphasizes the science process skills, mathematical skills and the use of theories and models that are fundamental for learning the various science disciplines. (Gabel, 1993, p. v)

In the last unit, (chemistry) students "see how scientists use theories and models to explain the physical world in which we live." (p. v)

In the original planning of the QUEST science courses, the intent of the science educators was that this course would be taught in one of the science departments. However, because no science department wanted to develop or teach the course, and because the Indiana Commission of Higher Education insisted that it be a required course, rather than elective or remedial, the course was developed and is taught in the school of education for science credit.

At the present time, the course meets for 2 hours and 20 minutes twice a week. There are 11 sections of 24 students each fall, and 12 sections in the spring. The classes are taught by a faculty member and graduate assistants who meet each week in the fall semester for 90-minute sessions (and less frequently in spring) to review the content, materials, and procedures to be used the following week.

Modifications of the course have continued since the 1993 edition of the text, and published in the course supplement. The first factor stimulating revision was that the physical science course required for all elementary education majors and taught through the physics department increasingly emphasized physics and included very little chemistry content. In addition, a research study on children's understanding of burning (Gabel, Stockton, Monaghan, & MaKinster, 2001) conducted in our Saturday Science program (also established by QUEST) indicated that children had many naive conceptions on this topic. After surveying the chemistry content of elementary science textbook series, and desiring to make the course more conceptual, interactive, and inquiry based in accordance with the National Science Education Standards, three new components were added to the course over a 6-year period. These were the introduction of ConcepTests (CTs), the use of Chemical Applications (CAps), and using models as explanations (MAps).

ConcepTests (CTs)

In the Introduction to Scientific Inquiry course, ConcepTests are now used with the students on 15 occasions. Topics include observation, inference, and prediction; length and area; surface area and volume; variables and operational definitions; graphing; scientific notation; direct proportion and slope; direct proportion, inverse proportion, states of matter; substances; changes of state; mixtures; and chemical and physical change. No formal study has been done to determine their effectiveness; however, students have responded very favorably to this use of collaborative learning as indicated on student evaluations.

The procedure used is a modification of that described by Mazur (1997) in that each of the four students at a lab table responds to a ConcepTests question individually and selects an answer. The recorder collects students' responses to determine if there is consensus. The group leader then calls on group members to explain their answers and consensus is reached. The recorder then records the group consensus on a whiteboard. The process is repeated for other questions. The instructor then asks the six group leaders to hold up the white boards to see if there is class consensus on the answers, and then calls on the table spokesperson to give the reasoning of their group until class consensus is reached.

Chemistry Applications (CAps)

Chemistry Applications activities have been added to the course to (a) provide chemistry investigations related to everyday things that are replacements for those in the text that are chemistry-related, (b) introduce chemistry at the macroscopic level at the beginning of the course, and (c) gradually increase students' inquiry experiences throughout the course. This enables students to have sufficient time to conduct an independent science investigation at their residence during the last 6 weeks of the semester.

The general procedure of CAps is as follows: A question is presented, students individually answer the question, their group discusses the question, their group determines a plan for investigating/answering the question, the plan is carried out, and a group conclusion is drawn. Ten CAps are currently in use. These include Sinking and Floating, What Processes Occurred, Classifying Materials, Viscosity and Density, Burning a Candle under a Jar*, Estimating Particles, Increasing the Rate of Dissolving*, Explaining Floating and Sinking, Mixture or Substance, and Factors Affecting Evaporation Rate*. The three activities with asterisks are chemistry investigations that are designed to help students become increasingly independent in scientific inquiry. The last one is the final class project and should prepare the prospective elementary teachers to guide their students in conducting an inquiry-oriented science fair project.

Modeling Applications (MAps)

The use of MAps, having students make Play-Doh models representing atoms and molecules, is the most recent addition to the Introduction to Scientific Inquiry course. Prior to this, two-dimensional particle pictures were used to explain chemical phenomena and processes. The poster paper of Walters (2001) on "Teaching Chemistry Using Play-Doh" at the Gordon Conference on Chemistry Education made a convincing case for including this approach in instruction. Nine modeling activities are used in the course, seven of which include the use of PlayDoh as well as the drawing of particle pictures. These include Using Models to Explain . . . the States of Matter, the Separation of Liquids, Burning, Solutions, Density, Chemical Decomposition, Mixtures, Balancing Equations, and Using Models to Predict Evaporation Rate. Although no formal study was made about how students liked the use of PlayDoh modeling in the course, students appeared to be more interactive and involved when making the Play-Doh models than in drawing two-dimensional particle pictures.

Students begin making the PlayDoh models in the middle of the course after they have become familiar with the properties and processes of matter at the macroscopic level. During the last 3 weeks of the course the Play-Doh models are used in conjunction with representing matter using the symbolic level in writing chemical formulas and equations.

A study (Gabel, Hitt, & Yang, 2003) was conducted in spring 2002 on the effectiveness of using the Play-Doh models on course achievement by examining students' performance on a 50-item test over the final third of the course. Twelve sections of the course were taught using the text, CTs, CAps, and MAps as described above. However, the treatment group used the MAps that gave directions to make Play-Doh models on seven occasions, followed by making two-dimensional diagrams. The control group made only the two-dimensional diagrams. Three graduate students and one experienced professor taught sections of the control group, whereas one professor and two graduate students taught the experimental group sections. Unfortunately the data from one graduate student in the experimental group had to be deleted from the study because the last page of the test for that section was inadvertently missing.

The purpose of the study was threefold: to (a) determine the effectiveness of using Play-Doh to represent the particle nature of matter on chemistry achievement; (b) compare students' understanding on the macroscopic, particulate, and symbolic levels; and (c) examine students' understanding of pairs of chemistry concepts that are taught in the course and for which many students have misconceptions.

There were four forms of the test with variation in only six items that were randomly distributed to 237 students. KR–20 reliability varied from 0.75 to 0.86. Class means for the two-dimensional-picture-only group was 28.8 whereas class means for the Play-Doh group was 31.3. Comparisons between the two treatment groups using a t test was significant at the 0.02 level.

Other comparisons were made on students' understanding of the macroscopic, particulate, and symbolic levels of chemistry, and among the three concept-pairs: burning/decomposition; chemical/physical change, and melting/dissolving. This was done by comparing the percentage of students that correctly answered each of the last six test items that were randomly included in the four test forms. Findings indicated that, in general, students understood the concepts best at the macroscopic level (61.3%), as compared to the particle level (47.4%), and the symbolic level (37.0%). The only exception to this was that students understood the particulate level better than macroscopic level for melting/dissolving. Of the three concept pairs, melting and dissolving was the least understood, and physical/chemical change was the most understood. The conclusions in this subtest need to be cautiously interpreted because a major difficulty in writing test items on the three levels is that the macroscopic items in particular are very example oriented, and students may understand the concept using a different example than the one given in a particular test item. The results show that the macroscopic level is generally more understandable than the particle and symbolic levels, and this finding is supportive of the National Science Education Standards' (NRC, 1996) recommendation that chemistry instruction at the elementary level focus on the observable properties of matter and changes that it undergoes rather than on particles and symbols.

The conclusion of this study indicates that the use of Play-Doh facilitates an understanding of the chemistry that was taught in this particular course to a small degree. Because the overall test scores hover around 60% correct does not mean that the students have a good understanding of the chemistry that is taught in the course, but additional study of these basic concepts taught at the elementary school level is warranted.

GENERAL CONCLUSIONS

It appears that the reform of chemistry instruction in the past 8 years has been quite remarkable. At the present time there are excellent models for change in the teaching of chemistry, and these teaching strategies have been incorporated into a variety of different types of chemistry courses at different levels. It is particularly important that the science courses that preservice elementary teachers take in their teacher preparation program embody one or more of these reforms. It is desirable that prospective teachers are "turned on" to science so that they will learn the content that they will be teaching.

Perhaps of even greater importance, they need to have models of how to make the subject matter interesting. Instead of chemistry being driven by the memorization of content in a large lecture that students regurgitate, research supports the use of collaborative and inquiry-based learning, particularly guided inquiry for the teaching of college chemistry. Guided inquiry frequently makes use of models as explanations and this orientation is a particularly appropriate approach for prospective teachers even though the explanations expected to be given by children will be on a lower level.

The other important component in preparing prospective elementary teachers in chemistry is to make certain that they have a good understanding of the simple concepts and processes that they will teach. Unfortunately these may be overlooked in a general chemistry course or even in one especially designed for preservice teachers because it is assumed that they have this knowledge. Simple activities such as making observations about what happens when a pile of sugar and a pile of sulfur are heated with a Bunsen burner on a tin can lid are very instructive. To see that both melt at different temperatures—one burns and "disappears," but the other one turns black (or rather decomposes into carbon, which students think is burning), and then actually burns—is instructive. Prospective elementary teachers need to have a thorough understanding of these concepts at the macroscopic level, not merely to memorize them, but to give explanations at the macroscopic level as well as in terms of particles in case children ask why. They need to be able to answer questions such as one child asked on our Saturday Science Program, "If a flame produces water and carbon dioxide, how can water put out a flame?" They need to be able to make sense of science, that is to give explanations to children on their terms, to prevent unintentionally conveying alternative conceptions to their students.

The best chemistry instruction for prospective elementary teachers will need the input from both chemistry educators and science educators with a solid background in chemistry who are very familiar with the National Science Education Standards. Errors, such as indicating or inferring that the difference between a chemical and physical change is that physical changes are reversible whereas chemical changes are not, are still found in elementary textbooks and background materials for teachers. It is also important that teachers become familiar with common everyday reactions that can be used in elementary school science teaching that children can understand, rather than focusing instruction on the structure of the atom, which is no longer advocated as part of the elementary school curriculum.

FUTURE DIRECTIONS FOR RESEARCH

Research on teaching is very complex because teaching is a very complex human activity with many variables that are difficult to control. This is one of the problems with much of the research that has been reported in this chapter. Many times it is difficult to isolate the factors in the instruction that causes the change. Change might be due to the enthusiasm of doing something new, or finding something that fits a particular person's teaching style. This is why it is important to replicate studies on effective approaches to teaching, and to report them in detail. What may be effective in one set of circumstances may not be effective in another.

REFERENCES

American Association for the Advancement of Science. (1967). *Science: A process approach.* New York: Xerox Corp.

Bowen, C. (2000) A quantitative literature review of cooperative learning effects on high school and college chemistry achievement. *Journal of Chemical Education, 77,* 116.

Bradley, A.Z., Ulrick, S.M., Jones, M., Jr., & Jones, S.M. (2002). Teaching the sophomore Organic course without a lecture. Are you crazy? *Journal of Chemical Education, 79,* 514–519.

Breslow, R. (1996). *Chemistry today and tomorrow: The central, useful, and creative science.* Washington, DC: American Chemical Society.

Burke, B., & Walton, E. (2002). Modeling effective teaching and learning in chemistry. *Journal of Chemical Education, 79,* 155–156.

Carpenter, S.R., & McMillan, T. (2003). Incorporation of a cooperative learning technique in organic chemistry. *Journal of Chemical Education, 80,* 330–332.

ChemConnections: Systematic change initiatives in chemistry. (n.d.). Retrieved July 17, 2003, from http://chemlinks.beloit.edu/

Dreyfuss, A. E. (2003, July 7). *PLTL Workshop Project: Peer-Led Team Learning.* Retrieved July 17, 2003, from http://www.sci.ccny.cuny.edu/~chemwksp/index.html

Ellis, A.B., Cappellari, A., Lisensky, G.C., Lorenz, J.K., Meeker, K., Moore, D., et al. (2000, May 4). *ConcepTests.* Retrieved July 17, 2003, from University of Wisconsin–Madison, Chemistry Department Web site: http://www.chem.wisc.edu/~concept/

Establishing New Traditions: Revitalizing the chemistry curriculum. (1997). Retrieved July 17, 2003, from http://newtraditions.chem.wisc.edu/

Farrell, J.J., Moog, R.S., & Spencer, J.N. (1999). *Journal of Chemical Education, 76,* 570–574.

Gabel, D.L. (1993). *Introductory science skills* (2nd ed.). Prospect Hills, IL: Waveland Press.

Gabel, D.L. (1999a). Improving teaching and learning through chemistry education research: A look to the future. *Journal of Chemical Education, 76,* 548–553.

Gabel, D.L. (1999b). Science. In G. Cawelti (Ed.), *Handbook of research on improving student achievement* (pp. 156–179). Arlington, VA: Educational Research Service.

Gabel, D.L., Hitt, A., & Yang, L.L. (2003, March). *Integrating the macroscopic, particulate, and symbolic representations of matter.* Paper presented at the annual meeting of the National Association for Research in Science Teaching, Philadelphia.

Gabel, D.L., Stockton, J.D., Monaghan, D.L., & MaKinster, J.G. (2001). Changing children's conceptions of burning. *School Science and Mathematics, 101,* 439–451.

Hagen, J.P. (2000). Cooperative learning in Organic II. Increased retention on a commuter campus. *Journal of Chemical Education, 77,* 1441.

Jasien, P.G. (1995). A physical science discovery course for elementary teachers. *Journal of Chemical Education, 72,* 48.

Johnstone, A.H. (1993). The development of chemistry development in chemistry. *Journal of Chemical Education, 70,* 701–705.

Landis, C.R, Ellis, A.B., Lisensky, G.C., Lorenz, J.K., Meeker, K., & Wamser, C.C. (2000). *Chemistry concepts: A pathway to interactive classrooms.* Upper Saddle River, NJ: Prentice-Hall.

Larson, T., & Middlecamp, C.H. (2003). A companion course in general chemistry for pre-education students. *Journal of Chemical Education, 80,* 165–170.

Lyle, K.S., & Robinson, W.R. (2003). A statistical evaluation: Peer-led team learning in an organic chemistry course. *Journal of Chemical Education, 80,* 132–134.

Mazur, E. (1997). *Peer instruction: A user's manual.* Upper Saddle River, NJ: Prentice-Hall.

Molecular Science Project. (2001). Retrieved July 17, 2003, from http://www.molsci.ucla.edu/

Multi-Initiative Dissemination Project: Project descriptions and workshop methodology. (2001). Retrieved July 17, 2003, from http://www.cchem.berkeley.edu/~midp/index.html?main.html&1

National Research Council. (1996). *National Science Education Standards.* Washington, DC: National Academy Press.

National Research Council. (2000). *Inquiry and the National Science Education Standards.* Washington, DC: National Academy Press.

National Science Foundation. (1963a). *Chemistry, an experimental approach: Chemistry educational materials study.* San Francisco: W. H. Freeman.

National Science Foundation. (1963b). *Investigating chemical systems: Chemical bond approach.* St. Louis: McGraw Hill.

Nurrenbern, S.C., & Krupp, A. (1995). *Experiences in cooperative learning: A collection for chemistry teachers* (Publication 95-001). Madison: University of Wisconsin–Madison, Institute for Chemical Education.

Nurrenbern, S.C., & Robinson, W.R. (1997). Cooperative learning: A bibliography. *Journal of Chemical Education, 74,* 623.

Project Galileo: Your gateway to innovations in science education. (2000). Retrieved July 18, 2003, from http://galileo.harvard.edu/

Schwab, J.J. (1962). The teaching of science as enquiry. In J.J. Schwab & P.F. Brandwein (Eds.), *The teaching of science* (pp. 3–103). Cambridge, MA: Harvard University Press.

Science Curriculum Improvement Study. (1968). Chicago: Rand McNally.

Shibley, I.A., Jr., & Zimmaro, D.M. (2002). The influence of collaborative learning on student attitudes and performance in an introductory chemistry laboratory. *Journal of Chemical Education, 79,* 745–748.

Springer, L., Stanne, M. E., & Donovan, S.S. (1999). Effects of small-group learning on undergraduates in science, mathematics, engineering, and technology: A meta-analysis. *Review of EducationalResearch, 69,* 21–51.

Straumanis, A. (2001). *Guided inquiry: Organic chemistry.* Unpublished manuscript, Department of Chemistry, University of New Mexico, Albuquerque.

Tien, L.T., Roth, V., & Kampmeier, J.A. (2002). Implementation of a Peer-Led Team Learning instructional approach in an undergraduate organic chemistry course. *Journal of Research in Science Teaching, 39,* 606–632.

Walters, M. (2001, August). *Teaching chemistry using Play-Doh.* Poster session presented at the Gordon Conference, Mount Holyoke College, South Hadley, MA.

Williamson, V.M., & Rowe, M.W. (2002). Group problem-solving in college-level quantitative analysis: The good, the bad, and the ugly. *Journal of Chemical Education, 79,* 1131–1134.

Wright, J.C., Millar, S.B., Kosciuk, S.A., Penberthy, D.L., Williams, P.H., & Wampold, B.E. (1998). A novel strategy for assessing the effect of curriculum reform on student competence. *Journal of Chemical Education, 75,* 962–992.

CHAPTER 25

A MODEL FOR REFORM IN TEACHING PHYSICS

Large-Enrollment Physics Classes

Dean Zollman

ABSTRACT

Teachers at all levels require some knowledge of physics. A particularly large challenge is to provide appropriate higher education experiences for those future teachers who will be working with students in the first six to seven years of their schooling. We have developed an activity-based course for these teachers. The activities involve both traditional short experiments and technology-based ones. The university course is somewhat unique because the design allows for one faculty member to work with a relatively large number of students and yet maintain a student-centered environment.

INTRODUCTION

Today's research in learning indicates that the most appropriate way for children of all ages to learn science is through hands-on activities. Yet many

Reform in Undergraduate Science Teaching for the 21st Century, pages 445–458
Copyright © 2004 by Information Age Publishing
445

universities provide for future teachers' science courses that are primarily in the lecture format. Thus, future teachers learn a model of teaching that is not appropriate for them to use.

The most common way to offer hands-on activities is in small classes. Yet, the economic constraints on our departments precluded offering a large number of small classes to future elementary school teachers. We needed to find a way that provided strong instruction in physics and an appropriate role model for the teaching of science. The course needed to be designed so that the faculty load was similar to a traditional lecture-laboratory course for about 100 students.

Concepts of Physics at Kansas State University is an introductory-level physics course which serves students who are preparing to teach in elementary school. Each year approximately 100 students, mostly second- and third-year university students, enroll in Concepts of Physics. Their goal is to obtain a sufficient background in physics so that they can teach at the elementary school level.

REVIEW OF LITERATURE:
TEACHING UNDERGRADUATE PHYSICS

A significant body of research on student learning of physics exists. Much of this research was reviewed by McDermott and Redish (1999) and Tiberghien, Jossem, and Barojas (1998). More recent information can be found in the proceedings of the annual Physics Education Research Conferences (Gordon Research Conferences, 2003). No similar review of recent curriculum reform at the undergraduate level exists. Reform based on physics education research has been an ongoing process dating back to the early 1980s (Laws, 1991; Manogue et al., 2001; McDermott, 1991, 2001; Redish, 1994; Van Heuvelen, 1991; Wilson, 1994; Zollman, 1990).

A science instructional model used in a number of undergraduate institutions is the learning cycle. The learning cycle is derived from the intellectual development model of Jean Piaget and was first conceptualized as an instructional model in the early 1960s (Karplus, 1977). The learning cycle includes a sequence of three different types of activities (Zollman, 1997). The first activity, exploration, requires the student to explore a concept by performing a series of activities. Students are given a general goal, some equipment, and some general ideas about the concepts involved. They are asked to explore the concept experimentally, in as much detail as they can, and to relate it to other experiences they have had. The second phase of the learning cycle, concept introduction, provides a model or concept to explain observations of the exploration. Frequently, the concept-introduction stage is not an experimental activity but expository statement of concepts and principles. Following the concept introduction, the students move to concept application. Here, they use the concepts that were introduced and apply them to new situations. This application of the principles and concepts leads to further understanding of the theories and the models. The complete cycle has been used successfully to teach a wide variety of topics to students at all grade levels.

COURSE DESIGN

The instructional model used in an introductory physics course at Kansas State University is the learning cycle that was originally developed by Robert Karplus. Because of the emphasis on student-centered activities, a learning cycle class at most universities usually has an enrollment of less than 30 students. However, the economics associated with small class size has limited adoption of this method at many universities. To overcome this difficulty, we have adapted the learning cycle for a class of about 100 students with one faculty member assigned to it. During the past 25 years the course has evolved into one with an emphasis on the nature of science and on learning science by doing scientific activities.

The course is constructed of 15 activity-based units, each of which is one week long. An activity-based unit is a learning experience that focuses on a series of 8 to 10 short experiments performed by all of the students in the class. Thus, students perform a large number of experiments and these activities form the backbone of the course.

Each unit involves hands-on activities and is based on the learning cycle format. To adapt the learning cycle for a large-enrollment course taught by a single faculty member, we use a combination of activities completed in an open laboratory environment and large class meetings. This adaptation utilizes an open laboratory that is available to the students about 30 hours per week. The three parts of the learning cycle are as follows:

1. *Exploration.* This part of the learning cycle is a series of hands-on activities. Instruction sheets guide the students through a series of short activities. Students working alone or in small groups perform most of the exploration.

2. *Concept Introduction.* One large class meeting each week is devoted to this phase. Students are asked to describe their exploration observations and any related experiences. Using these observations, the instructor guides the students toward a model or a theory that can be used to explain the observations.

3. *Concept Application.* The final phase of the cycle again involves hands-on activities. Using the model developed in the large class

(concept-introduction phase), the students make predictions for new situations. Their predictions are tested experimentally. As in the exploration phase, instruction sheets are available. Additional applications and summaries are discussed in a large class meeting.

The teaching assistants, frequently students who have taken the course in previous years, act as proctors who help with occasional conceptual or equipment difficulties rather than as instructors in the course.

OPERATION OF THE COURSE

Each learning cycle begins on a Monday. At that time, the equipment for the exploration is available in the open laboratory. Students must complete all exploration activities before the class on Wednesday. The concepts that were explored are introduced in a lecture-discussion format during a 50-min class on Wednesday.

To assure that the students stay involved in the large class we use a classroom response system such as ClassTalk (Better Education Inc., n.d.) or the Personal Response System (Center for Enhanced Learning and Teaching, Hong Kong University of Science and Technology, 2002). With these systems the instructor poses a question. The students are encouraged to discuss their answer with others and then enter a response into a small handheld device. By posing three to five questions per class the instructor receives immediate feedback on students' understanding, and the students are motivated to be attentive throughout the class.

Following this class meeting, the application equipment is available in the open laboratory until class time on Friday. The class meets on Friday to ask questions about the week's work. In addition, the concepts introduced on Wednesday are applied to situations that, usually, lead to questions not easily answered with present knowledge. These applications introduce the exploration of the next cycle.

Each exploration and application is composed of six to eight short experiments called activities. The equipment for each activity is placed in the open laboratory at marked stations. The students are told what equipment is located at each station and presented with questions to answer about it. For example, an application activity on electrostatics states,

> At station EM-9 is a small Wimshurst Machine. By turning the crank you can charge the two spheres. A small aluminum ball is suspended between the spheres. Describe and explain the motion of the ball as you turn the crank.

The students are guided from station to station until they complete all activities (see Figure 25.1). For each activity they must write answers to all

Figure 25.1.   Traditional hands-on activities as well as interactive multimedia are used in the explorations and applications.

questions on their activity sheets. When they finish, the students leave the completed activity sheets in the laboratory.

To assure that students are adequately motivated to complete the activities, grades are assigned to each exploration and application. Explorations are graded on a satisfactory-unsatisfactory basis. To obtain a satisfactory, a student must try to answer each question. We do not grade on right or wrong answers—only attempts at exploring new phenomena.

The applications are graded on a scale of 0 to 8. In this case, grades are based on the students' abilities to use physics concepts in their explanations and to use those quantitative relationships presented in the class and text.

SAMPLE LEARNING CYCLE

As an example of a week's activity, consider the 1st of 5 weeks on the topic of energy. When students start the exploration, they have already studied kinematics, momentum, and forces. Thus, they begin by trying to explain the motion of a pendulum by using either conservation of momentum or Newton's laws. Then they look at several experiments involving motion and change of motion. (At this point the term energy has not been introduced.)

First, a toy car is rolled down an incline into an aluminum can. By releasing the car from several different locations on the incline, the students determine qualitatively the relationship between release location and damage to the can. The activity sheet then instructs the students to change the angle of the incline and repeat the experiments. (A similar experiment involves driving nails by dropping weights on them. They compare the distance the nail is driven for different release heights and different-sized weights.)

The exploration concludes with a station at which the cart and can are placed on a horizontal surface. The students are asked to make a dent in the can without lifting the cart or can from the table. Once they accomplish that, they are asked to do something to make a bigger dent. Most students decide to move the cart at a higher speed; a few think of adding mass to it.

After completing the exploration, the students express in writing any similarities they can see in the various observation activities. These statements will be in their own language since we have not yet introduced the vocabulary of energy-related concepts.

The concept introduction begins with a discussion of the difficulties involved in describing the motion of the pendulum and with the "exchange of something" that causes the pendulum to move fast at the bottom and slow at the top of its swing. The discussion is partially student-centered. The instructor leads off with a question, but the students do most of the talking. The discussion motivates a reason for introducing a new concept.

The general concepts of energy and gravitational potential energy are introduced. Students, by referring to their observations during the exploration, provide a list of variables upon which the potential energy depends. By recalling the nail-driving activities, they can also state the functional dependence of gravitational potential energy on mass and height.

A similar discussion occurs for kinetic energy. Most students will state that kinetic energy depends on speed. (However, none of the activities have enabled them to determine the functional dependence.) A few students will have discovered that adding mass to the cart will have an effect. Thus, with guidance from the instructor and frequent reference to their exploration activities, the students help construct the basic ideas of mechanical energy.

To conclude the introduction, we return to the pendulum and develop the idea of conservation of energy. With this material, the students are ready to begin the concept application.

The beginning of the application is simply a check to determine if the students can plug numbers correctly into the equations. After measuring their masses and walking speeds, they calculate their kinetic energies while walking and their change in gravitational potential energies when they move from the first to second floor of the physics building. For the next activity, they return to the nail driver, calculate its potential energy at several heights, and use conservation of energy to state its kinetic energy just before it hits the nail. Even though they have "learned" conservation of energy, many students reach a state of disequilibrium here. "How can I determine kinetic energy when I don't know the speed?" is a frequent question. Without the application, the students would not have noticed this problem in their learning until the next test. With this concrete example, they are able to address it at once.

A toy car with a loop-the-loop track is the equipment for the next activity. The students are asked to measure the height of the loop and predict the kinetic energy needed for a car to go through it. Using a photocell timer, the students determine the speed and calculate the kinetic energy at the bottom of the loop. When they compare the actual kinetic energy with their prediction, they find a significant discrepancy. The students are asked to speculate about the reason for differences between these two numbers and told that we will discuss it during Friday's class.

Next, the students drop a feather and a BB from the same height. The two objects have equal mass so they begin with the same potential energy (which the students calculate). Without making any measurements, all students notice that the two objects have different kinetic energies as they reach the floor. Again, they speculate about these differences.

Finally a two-hill "roller coaster" track is used. The students are asked to predict, and then experimentally determine, the point on the higher hill from which a steel ball must be released to roll over the lower hill. The experiment is repeated with a cork-covered ball of the same mass. They discuss the differences between these results and the results predicted by conservation of potential and kinetic energy.

After answering questions during Friday's class meeting, we continue looking at situations wherein the sum of kinetic and gravitation potential energies is not conserved. Particular attention is paid to the differences between the BB and the feather and between the bare steel and cork-covered balls. Because the students have studied friction, they speculate that it must be involved. A discussion of work by a frictional force prepares the students for the next activity: exploring thermal energy.

In this example of our adapted learning cycle, students used traditional laboratory equipment to perform all observations and measurements. However, other cycles include activities based on contemporary technology.

INTERACTIVE MULTIMEDIA IN EXPLORATIONS AND APPLICATIONS

In an exploration that introduces impulse, students are asked to view a short video sequence that shows a mannequin in a car that collides with a fixed wall (see Figure 25.2). Another mannequin would be in an identical

Figure 25.2.   A scene from Physics and Automobile Collisions that is used to explore Newton's laws.

collision but would have an airbag in the car. Students are asked to describe similarities and differences in the two events and what is different, and to speculate on why the airbag makes a difference in the amount of damage done to the mannequin. In general, students would make comments such as "the airbag is softer than the windshield" or "the mannequin sinks into the airbag but it bounces off the windshield." Their wording would not be in terms of Newton's laws of motion or impulse that are the concepts to be studied after this exploration.

This same video sequence could be used for an application activity. After the students have been introduced to the concepts of impulse and Newton's second law, they can be asked to look at the two video scenes and describe in terms of these concepts why the mannequin which is protected by the airbag receives less damage than the one which is not. Thus, an identical video sequence can be used in both an application and an exploration.

Video scenes such as these can also be used in situations where the students need to apply their knowledge of physics but for which there is no right answer. One of the favorites in this category is a video sequence from the second Mohammed Ali–Sonny Liston prizefight. In this sequence, Ali swings at Liston and Liston falls down. A voice-over narrator says that this punch was very controversial, and some people do not believe that Liston was actually hit. The students are asked to watch the video one frame at a time and try to determine if Liston was hit. However, they cannot just say, "Yes, I think so," or "No"; they must state their reasoning in terms of conservation of momentum, which is the concept being applied. Thus, they must talk about the momentum of Ali's fist before the interaction as well as the momentum of Liston's head. They must then look at the momenta of these two objects after the collision and come to some conclusion. However, because the camera angle does not allow for extremely careful measurement, no definitive answer can be determined. Thus, the students are required to come to their own conclusions and defend their conclusions based on the laws of physics. They are told that either yes or no is correct but that their reasoning is what really counts.

In other scenes careful measurement can be completed. Students are asked to apply the principle of conservation of energy to a pole-vaulter on the videodisc Physics of Sports (Zollman & Noble, 1988). They step through a pole vault sequence one frame at a time. Using VideoPoint software (Luetzelschwab & Laws, 2001) they measure the distance that the vaulter is moving during each frame before he leaves the ground. With this information they can calculate the vaulter's speed and then his kinetic energy. They then move to the frame at which he is highest above the ground and measure the distance to which he has ascended. Now, they can calculate the gravitational potential energy of the vaulter. The students must then determine if all of the gravitational potential energy that the vaulter obtained came from his kinetic energy when he was running on the ground. If not, they must speculate on where the remaining energy may have come from. Thus, this application is also an exploration for the next concept: the many forms of energy in addition to kinetic and gravitational potential energy.

Another example, which is partially quantitative and partially qualitative, involves the students analyzing the forces on a diver as she goes from a 3-m board into the water. The students are asked to look at the scene and then to go to several frames that have been preselected by the instructor. For each of these frames they are to state all of the forces acting on the diver and then the net force on the diver. This activity is particularly useful in bringing out some of the students' conceptions that are not consistent with the accepted way in which Newton's laws are applied. In particular, many students will state that as the diver is ascending, she must have a force in the up direction acting on her. The reason that the students give is she is moving up therefore there must be a force acting up. By the time the students come to this interactive video activity, they have already learned that a force does not necessarily need to be in the direction of motion. However, this idea is very firmly held by most students when they enter a physics class and is difficult to change. Thus, a "real-life" example helps bring it out so that the instructor can discuss it further. These and other similar activities are particularly helpful in this respect because they apply to real events with which the student can identify.

MULTIMEDIA IN THE LARGE-CLASS SETTING

The lecture response system allows the instructor to pose questions to the entire class during the large-class meetings, which are for the concept introduction and part of the applications. When the students respond, the responses are collected by the instructor's computer and sorted. Thus, the instructor and students can interact regularly even though the class is rather large.

An example of a large class meeting in which both multimedia and the classroom response system are used in combination is one in which students summarize work on the transfer of thermal energy. To begin the activity the class watches a scene from the film Wizard of Oz in which Dorothy throws water on the Wicked Witch of the West, and the witch melts. The instructor states that, because the witch melted, we should be able to determine the latent heat of fusion for her. The students are then asked if they have a sufficient amount of information to calculate this particular variable. The majority will answer that they do not, so they are asked to enter into their computers the variables that we need to know to make the calculation. From the students' responses the instructor can make a list of variables that the class will need to either estimate or measure. The instructor can also address variables that are listed by the students but are not necessary in this situation. Thus, the interactive system enables the instructor to uncover student misconceptions and address them in real time. For each of the variables that are needed the students are asked to enter in their computers an estimate based on the information on the screen. Thus, they estimate the mass of the witch, the temperature in the castle, the change in the water temperature, and the amount of water that struck the witch.

The estimates are made by looking at individual pictures in the film. The students can request any picture. For each variable the students enter an estimate into their computers. The instructor uses the values that were entered to determine an approximate average value for each variable. These averages are then used in the determination of the latent heat of fusion of the Wicked Witch of the West.

While on the surface this activity is a calculation of one number, the process that the students go through reviews most of the concepts related to thermal energy transfer. Further, the process shows students how we can use estimation as part of the scientific process. The students will also raise questions about thermal energy that we cannot account. For example, "steam" rises from the witch. That would indicate that she did not simply melt. Then, we must estimate if these contributions are significant. Thus, the overall result of the activity is an interactive discussion, which takes almost a full class period and includes a variety of topics related to the process of science as well as the concepts that are studied. The combination of interactive video and a class response system provides motivation and assures that all students are involved.

In all parts of the course, the multimedia and interactive classroom activities are not separated from other hands-on activities. Students move quickly and easily from an experiment involving standard laboratory materials to an interactive video station and then back to lab apparatus within a single class period. Likewise, in the large-class activities the instructor uses computer modeling, interactive video, and physics demonstrations together with the response system. Thus, the interactive media component of the course is fully integrated with all other aspects and provides a model for including these types of activities in their future teaching.

OUTCOMES OF LEARNING CYCLE IN THE PHYSICS COURSE

During the past 25 years, Concepts of Physics has been the primary course in physics for future elementary teachers at Kansas State University. When the course was first introduced, a study indicated that students in the course learned the content better than in a traditional mode. More recent studies have focused on the use of technology in the course (Escalada & Zollman, 1997). The studies show that the students' attitudes toward physics and toward the methods of learning as well as their learning of the content are very positive.

We find that students who complete the course frequently make statements such as, "I could not have succeeded in physics without the activities." Further, when they become student teachers or teachers, they return to borrow equipment or discuss how to teach certain topics. Thus, the goal to provide an appropriate role model for teaching science has been achieved.

SUMMARY: RECOMMENDATIONS FOR THE DEVELOPMENT OF AN EFFECTIVE COURSE IN PHYSICS

When this course began, it was based rather tightly on the learning cycle model. In the past 25 years, research has modified some underlying principles about ways to teach science (Erickson, 2000). Simultaneously, research in physics teaching and learning has vastly increased our knowledge of conceptions that students bring to a physics class and how to build on and modify those conceptions. This newly acquired knowledge about teaching and learning has been introduced into the basic structure of the course in a rather seamless manner. For example, explorations now frequently provide the beginnings for students to see how their naive conceptions may not be generalized. Applications will offer students the opportunity to investigate the value of newly acquired knowledge in a variety of different contexts and to transfer knowledge from one component of the course to another. Each of these additions to the course represents research that has developed since the course was begun.

We are able to modify the learning experiences to account for up-to-date research because the foundation for the course is very solid. The learning cycle was initially based on Piaget's model of intellectual development and provides a good classroom environment for any constructivist approach to teaching and learning. Because it has a solid base, the course structure was flexible enough to be able to be changed as knowledge about teaching and learning improved.

The primary conclusion is that a course of this nature needs to have a solid foundation based on contemporary ideas of teaching and learning. With that foundation an instructor can be in a position to make changes as needed and yet not need to modify the underlying operation of the course.

Transferability to other faculty is also a critical part of success. A good structure must survive the "enthusiasm curve" of the developer. In our case, the course was designed and implemented by the author of this chapter. A challenge was to convert the course from "Zollman's Course" to one which other faculty would feel comfortable teaching. This process began about seven years ago. Since then two other faculty, whose research areas are not physics education, have taught Concepts of Physics while the originator acted as his mentor. The transfer has been successful.

IMPLICATIONS FOR REFORM IN UNDERGRADUATE PHYSICS AND FUTURE RESEARCH

Reform in undergraduate physics teaching and learning will continue to be based on physics education research and on intuition of physics faculty. Both the research and the intuition are necessary components of this type of development. The research informs us of models for student thinking and applications to instruction. Intuition based on experience can help improve the process in areas where the research base is not yet developed.

Physics education research has broadened greatly in recent years. We have a solid body of research that helps us understand the naive conceptions that students bring to a physics class and a good understanding of which of these conceptions are most resistant to change by instruction.

Now the research is looking at underlying models that students use, how those models are cued by the contexts in which they are used and how information and thinking processes transfer within a physics course and from one course to another. This research should be able to feed back into the conceptual change work and perhaps lower the resistance to some of that change. Thus, future research will build on but go far beyond the conceptual change research that has served us so well.

AUTHOR NOTE

Development was with support of a grant from the U.S. National Science Foundation. Professors Mick O'Shea and Talat Rahman have taught the course in recent years. Additional information on ideas expressed in this paper can be found in other publications about this course (Zollman, 1990, 1993, 1994, 1996; Zollman & Fuller, 1995).

REFERENCES

Better Education Inc. (n.d.). Retrieved August 9, 2003, from http://www.bedu.com/

Center for Enhanced Learning and Teaching, Hong Kong University of Science and Technology. (2002).*Introduction: What is PRS?* Retrieved August 9, 2003, from http://celt.ust.hk/ideas/prs/intro.htm

Erickson, G. (2000). Research programmes and the student science learning literature. In R. Millar & J. Leach & J. Osborne (Eds.), *Improving science education: The contribution of research* (pp. 271–292). Buckingham, England: Open University Press.

Escalada, L., & Zollman, D. (1997). An investigation on the effects of using interactive video in a physics classroom on student learning and attitudes. *Journal of Research in Science Teaching, 34,* 467–490.

Gordon Research Conferences. (2003). *Conference Programs.* Retrieved August 15, 2003, from http://www.grc.uri.edu/04sched.htm.

Karplus, R. (1977). Science teaching and the development of reasoning. *Journal of Research in Science Teaching, 14,* 169.

Laws, P. (1991). Calculus-based physics without lectures. *Physics Today, 44*(12), 24–31.

Luetzelschwab, M., & Laws, P. (2001). Videopoint 2.5 [Software]. Lenox, MA: Lenox Softworks.

Manogue, C.A., Siemens, P.J., Tate, J., Browne, K., Niess, M.L., & Wolfer, A.J. (2001). Paradigms in physics: A new upper-division curriculum. *American Journal of Physics, 69,* 978–990.

McDermott, L.C. (1991). Millikan Lecture 1990: What we teach and what is learned—Closing the gap. *American Journal of Physics, 59,* 301–315.

McDermott, L.C. (2001). Oersted Medal Lecture 2001: Physics education research—The key to student learning. *American Journal of Physics, 69,* 1127–1137.

McDermott, L.C., & Redish, E.F. (1999). Resource letter: PER-1: Physics education research. *American Journal of Physics, 67,* 755–767.

Redish, E.F. (1994). Implications of cognitive studies for teaching physics. *American Journal of Physics, 62,* 796–803.

Tiberghien, A., Jossem, E.L., & Barojas, J. (Eds.). (1998). *Connecting research in physics education with teacher education.* Retrieved August 9, 2003, from the International Commission of Physics Education Web site: http://www.physics.ohio-state.edu/~jossem/ICPE/TOC.html

Van Heuvelen, A. (1991). Learning to think like a physicist: A review of research-based instructional strategies. *American Journal of Physics, 59,* 891–897.

Wilson, J.M. (1994). The CUPLE physics studio. *The Physics Teacher, 32,* 518–529.

Zollman, D. (1990). Learning cycles in a large enrollment class. *The Physics Teacher, 28,* 20–25.

Zollman, D. (1993). Interactive video activities for elementary education students. In J.V. Boettcher (Ed.), *101 success stories of information technology in higher education.* New York: McGraw Hill.

Zollman, D. (1994). Preparing future science teachers: The physics component of a new programme. *Physics Education, 29,* 271–275.

Zollman, D. (1996). Millikan Lecture 1995: Do they just sit there? Reflections on helping students learn physics. *American Journal of Physics, 64,* 114–119.

Zollman, D. (1997). Learning cycle physics. In E.F. Redish & J.S. Rigden (Eds.), *The changing role of physics departments in modern universities* (pp. 1137–1149). College Park, MD: American Institute of Physics.

Zollman, D., & Fuller, R. (1995). Teaching and learning physics with interactive video. *Physics Today, 47*(4), 41–47.

Zollman, D., & Noble, M.L. (1988). *Physics of sports* [videodisc]. Seattle, WA: Videodiscovery.

CHAPTER 26

A MODEL FOR REFORM IN TEACHING GEOLOGICAL LABORATORY SCIENCE

M. Jenice Goldston and Monica Clement

ABSTRACT

One approach to reform teaching in geological laboratory sciences is to align the course with the National Science Education Standards and use an authentic problem-based field study to assist undergraduates in developing scientific process skills and an understanding of scientific inquiry. Working in formal cooperative teams, undergraduates enrolled in a geology laboratory were immersed in field research—a central focus of the course. Undergraduates spent two days in the field generating questions, hypothesizing, and collecting data. Following the fieldwork, six laboratory sessions were used to analyze and interpret their findings to construct local "geological history." Research findings for this case study focused on the undergraduates' perspectives of the modified geology laboratory course. Qualitative results emerging from undergraduate student interviews, field observations, undergraduate student journals, and the geologist's journal suggest that (a) authentic field experiences provided opportunities for students to explore their notion of science and make connections by experiencing scientific work; (b) cooperative learning teams fostered discussion, exchange of ideas, and positive student interac-

Reform in Undergraduate Science Teaching for the 21st Century, pages 459–476
Copyright © 2004 by Information Age Publishing
459

tions; (c) the field experiences were viewed as motivational and useful, and they enhanced students' questioning and curiosity; and (d) undergraduate students reveal a favorable disposition toward pedagogical techniques and strategies used in the modified geology course and make connections between the strategies and applications for teaching children.

INTRODUCTION

Undoubtedly, the most predominant pedagogical approach used in introductory geology courses is lecture and the most prevalent activities used in laboratories are what are known as "verification" or "cookbook" labs (Scheritzer, 1995). In addition, laboratory activities are rarely connected to lecture topics. Science education research findings, while recognizing the value of traditional lectures as a time-honored approach, suggest that lecture and verification approaches are not always effective in fostering student learning. This is especially true when the focus is on fostering long-term retention, critical-thinking skills, problem solving, and developing the inquiry nature of the discipline. Many arts and science faculty are aware of and recognize a need to reform courses to more effectively teach their discipline, but they are often stymied by lack of experience and knowledge of what works. Thus, shaping intentions for reform into action is often difficult.

In describing a model for geology laboratory teaching, note that this example is only a small part of a larger collaborative research study. Accordingly, the chapter focuses on problem-based lab experiences and an extended, authentic field study which thrust undergraduates into the role of geologist and explored their perceptions of the experience. Pedagogical constructs that guided the modifications of the geology laboratory course included making laboratory experiences student centered, fostering inquiry representative of the discipline, emphasizing data collection, and interpreting evidences representative of the work of geologists. Simply put, the focus of the geology laboratory course shifted from looking at the changes within the course from an instructor's perspective to looking at it from a student's perspective.

LITERATURE REVIEW: TEACHING THE GEOLOGICAL SCIENCES

During the last 10 years, research in geoscience education has provided a rich array of studies that explore innovations focused on the improvement of undergraduate geology and earth science courses. In many studies, there is evidence that instructional approaches focused on constructivism improve student learning (Lawson & Thompson, 1988; Saunders, 1992; von Glasersfeld, 1989). Instruction guided by the general tenets of constructivism advocate that learners construct ideas and knowledge in active ways that move them toward more sophisticated knowledge linking prior knowledge to new information through assimilation and accommodation (Piaget, 1986). Pedagogical strategies that are compatible with constructivist teaching include active learning experiences such as cooperative learning, authentic experiences, problem-based learning, and many others (Beiersdorfer & Beiersdorfer, 1995; Carpenter et al., 1999; MacDonald & Korinek, 1995; Pinet, 1995; Smith & Hoersch, 1995). Many undergraduate course reform initiatives in the geological sciences aligned with the principles of constructivism. For instance, numerous descriptive studies have explored undergraduate teaching incorporating cooperative strategies that utilized both informal cooperative groups and formal cooperative groups. Informal cooperative groups are, in general, content-free ways to organize social interaction in the course while formal cooperative groups are more complex and teams work together over longer periods of time. For instance, some strategies such as Think-Pair-Share or Think-Pair-Square are considered informal and simple to incorporate into a course, while a problem-based team approach to field site research is more formal.

Pinet (1995) describes the benefits of using active learning with cooperative teams of undergraduates placed into companies to explore geological issues such as watershed pollution or erosion of farmland. He found that the undergraduates gained confidence in defining and solving problems and became more intellectually independent regarding the underlying geological processes. Most important, the formal cooperative experiences "provided opportunities to criticize arguments offered by others" (Pinet, 1995). MacDonald and Korinek (1995) found that incorporating cooperative writing activities enhanced student geology content, reduced student isolation, and fostered communication skills. Another study exploring a form of cooperative learning known as the jigsaw approach found that the geology undergraduates had an enhanced ability to argue fluently and demonstrate deeper levels of understanding (Tewksbury, 1995). It is important to point out that though there are numerous positive outcomes in the research related to cooperative learning strategies, these studies also point out difficulties using cooperative learning that range from the slow pace of content coverage to student resistance in participation.

Active learning strategies selected for this study were intended to transform passive occurrences into engaging relevant experiences utilizing cooperative teams in problem-based and authentic immersion inquiry. In the generic sense, problem-based learning is any learning that is a product of answering questions, interpreting observations, or understanding problem solving (Barrows, 1986). According to Barrows (1986) problem-based learning goals include (a) developing knowledge in the context of subsequent applications, (b) fostering critical thinking, (c) enhancing the motivation of the learners, and (d) fostering self-directed learning. A study by Smith and Hoersch (1995), found that problem-based learning encourages students to think critically and to reconcile mutually contradictory evidence. Others have also stated that problem-based activities build more highly developed cognitive skills (Bloom, 1956).

Anton Lawson (1999) is a proponent of authentic immersion in hands-on and minds-on lab and field experiences that are problem solving or inquiry in nature. He supports teaching that involves social interaction and provides students with many experiences to articulate their scientific reasoning, both oral and written. In this study, problem-based activities and authentic inquiries that provided opportunities for undergraduates to articulate their findings and understandings were central to the geology lab modifications.

In summary, over the last 10 years research findings reveal an influence of constructivist practices in reforming undergraduate science courses. This influence is seen in undergraduate geology courses where dominant modifications are active learning strategies that cross a diversity of contexts through discussion, writing tasks, projects, and presentations. Given this background, our constructivist model for teaching geology lab centers on an extended, authentic field study where the learners pursue their own generated questions. According to research findings with this model, learners gain more content knowledge and a broader view of the concepts and skills required of the discipline.

AN INVESTIGATION OF REFORM IN TEACHING THE GEOLOGICAL SCIENCES

Prior to starting the research, the co-researchers, a geologist and science educator, began modifying the course Geology Laboratory (GEOL 103) and aligning it with the National Science Education Standards (National Research Council [NRC], 1996). The alignment to standards was a totally new idea to the geologist, who perused the standards and found them relevant to what she taught in the geology laboratory. She voiced her observation that the labs were hands-on but not very motivating. As a geophysicist, she felt that having students "do geology" would be a more authentic experience. Thus, problem-based field experiences were developed for GEOL 103, including an extended, authentic field study to allow undergraduates to experience geology as a geologist.

Course Description

The following sections give an overview of the courses and the changes made within them. Two courses, Earth in Action (GEOL 100) and Geology Laboratory (GEOL 103), were reconceptualized to incorporate *constructivist teaching practices*, and *problem-based field experiences*. One section of GEOL 100 and one section of GEOL 103 were designated for undergraduate education majors. Though this chapter centers on the geology lab (GEOL 103) only, the brief description of changes in both courses gives the reader a broader, more complete contextual framework for interpreting the study.

Original Courses

The original course, Earth in Action, GEOL 100, was a common "textbook course" with large undergraduate enrollments for lecture; student grades were determined by 3 to 4 exams. The geology laboratory course focused on a set of 8 to 10 prescribed activities covering a range of geology topics such as minerals, rocks, maps, soils, and earthquakes. The laboratory did not have to be taken in conjunction with the lecture course; thus, geology laboratory was separate and unconnected to geology lecture. Students were graded on lab assignments and exams.

Modified Courses

One section each of GEOL 100 and GEOL 103 was modified for this study. Education majors were encouraged to enroll in both of the identified sections (lab and lecture). Changes in the courses focused on shifting from passive learning (teacher-directed) to active learning (student-centered) approaches. The modified GEOL 100–103 courses incorporated constructivist practices for teaching, national science education standards (NRC, 1996), technology, and problem-based field experiences.

The lab course shifted from prescriptive lab activities to more open-ended inquiries that placed students in the role of geologists attempting to solve questions centered upon local geology. There were three short field trips to various local geological sites. In addition, the lab course incorporated an extended 2-day authentic, problem-based field experience that incorporated GPS technology and team collaboration. Team questions were structured around various aspects of Kansas's geology, for instance initial focus questions for a river site included (a) What earth materials are present at the site? (b) Where did these earth materials originate? and (c) How did these earth materials get here? Teams then developed further question and strategies to ascertain appropriate data for their questions. Data collected at the field sites were analyzed during six follow-up lab sessions. The lab course was modified to include additional lab time to analyze student data. Data were synthesized from team reports to create a geological "story" of the sites explored. Table 26.1 illustrates the features and modifications of the laboratory changes.

As seen in Table 26.1, 20 undergraduate students (12 female, 8 male) were enrolled in the modified geology laboratory course. The geologist taught the students concurrently enrolled in the modified geology lab and geology lecture. Preservice elementary education majors commonly take this course as part of the 12 hours of science content needed for the elementary education program. Undergraduates were involved in inquiry labs that integrated topics on minerals, rocks, soils, and weathering, to name a few. In addition, as seen in the chart, the students in the revised geology laboratory formed teams and spent a weekend in the field exploring two sites to collect data. As part of the immersion experience, their task was to develop questions and create a geologic history of the sites by synthesizing the interpretation and results of the teams' research.

Table 26.1.   Original and Revised Geology Laboratory Course

Geology 103		
	Original Course	*Revised Course*
Class size	24 individuals	20 individuals
Students/semester (all sections)	approximately 360 students	approximately 360 students
Percentage of Ed. Majors	10–15%	75%
Teaching methods and experiences	Short lectures—minerals, volcanism, rocks, earthquakes, weathering, etc.	4–5 Problem-based inquiry labs.
		Student teams (4 members) for lab activities.
	Student partners.	Inquiry labs followed by team discussion of findings.
	8–10 verification labs.	
	Exams and lab reports.	2-day, on-site extended authentic field study with team-designed questions, data acquisition, analysis, interpretation, and presentation.
		Lab time set aside for team analysis and discussion of findings.
		Lab reports (individual)
		Presentation and team reports (field site findings).

METHODS

To address education majors' perspectives of the changes in the revised lab and the extended authentic field study, qualitative methods were utilized.

Qualitative Methods

Qualitative methods were used in ascertaining education majors' views of the pedagogical changes conducted within the geology laboratory. Qualitative data included laboratory and field observations conducted by the science educator, undergraduate journals, student interviews, and the geologist's journal. As part of the course modifications, the undergraduate journals were collected periodically, read, and returned to students with feedback from the geologist. Some journal entries were free response and others were written in response to teacher-structured questions. Students were interviewed by the science educator to gain insights into their perspectives of the changes being implemented into the course. Five students were randomly selected from a list of volunteers. Each student was interviewed 4 times for a total of 20 interviews. The semi-structured interviews were conducted at the beginning, middle, and end of the course. In addition, the science educator took field notes of the lab experiences and the geologist kept a reflective journal. The co-researchers, the geologist and science educator, met frequently to discuss and reflect upon their observations of the course "happenings." Last, because our research was part of a larger research grant, additional data came from an outside evaluator who administered a questionnaire and interviewed undergraduates.

Procedure for Qualitative Data Analysis

Data analysis for this case study began by inductively examining the various data from interviews, journals, and field notes. The raw data were read and carefully coded to uncover patterns that appeared to emerge. All the data were repeatedly examined to find patterns, themes, or singularities using codes and contextual analysis to support or reject emergent themes (Bogdan & Biklen, 1992; Lincoln & Guba, 1985). Coded data were analyzed using four forms of data analysis and interpretation advocated by Stakes (1995) for case studies. These include data clustering, direct interpretation, patterns, and naturalistic generalizations.

Multiple readings of the data independently allowed the researchers to generate codes to construct initial categories that were grouped for reexamination (data aggregation). As with any inductive process, the researchers continued to peruse the data, categorize and recategorize based upon the words of the participants. The pattern coding allowed the researchers to identify and negotiate emergent themes in the data (Miles & Huberman, 1994). Pattern coding arising from direct observations of the data proceeded toward interpretation or explanation (direct interpretation). The developing themes were tested for validity against the data by the continual search for affirmative or discrepant evidence. Themes began tentatively and were continually reconceptualized and negotiated between the researchers as required to best represent the evidences (naturalistic generalization). Triangulation of findings occurred using multiple data sources and the researcher's independent analysis of data. Two students were asked to verify the themes and our interpretations of them providing a member check and credibility for the interpretative themes (Lincoln & Guba, 1985; Patton, 1990).

FINDINGS

The process of change appears different based upon one's situational position in the social setting. The interpretive themes presented are based upon the education majors' perspectives as those that lived the events and experiences of the geology lab course. They are interpreted through the lens of the researchers who were also a part of the experiences. As the geo-researcher incorporated active learning into authentic experiences, it was important for her to hear the voices of the undergraduate collaborators as they gave a more complete picture of the events as viewed from those most affected by the course changes.

Data analyzed from the student journal entries, interview transcripts, researcher journals, and field notes formed categories identified by the co-researchers. These categories reflect events of the course that were associated with the changes being implemented by the geo-researcher. Core categories emerged that included (a) views of scientists and science, (b) active learning via groups, (c) fieldwork and field experiences, and (d) pedagogical models. The data within these categories representing the words of the undergraduates brought forth the following themes associated with the changes implemented within the geology laboratory.

Theme 1: Views of Scientists and Science

Authentic, problem-based experiences in the field provided opportunities for students to explore their notion of science and make connections by experiencing scientific work.

Interviews from the undergraduates revealed their ideas of science, scientists, and scientific work. When students were interviewed, their views of geology became explicit. The interview questions were designed to have them reflect on what they thought of scientific processes and science in general. An example of an interview question was, "Does what you are doing in the geology lab compare to what geologists do?" The undergraduate responses are represented in the following samples.

- We used the scientific process a lot [in geology lab]. I used to think of scientists as the men in the white lab coats, but going out and doing things in the field has shown me that science is very dynamic. (Student 1 [interview])
- I sensed that even as we worked [in the field] that we began to think scientifically. (Student 2 [interview])
- She [the instructor] challenges us to think on our own. It's good. You aren't doing a formulated thing. It is actually a challenge so you feel you are doing something important. (Student 3 [interview])

The experiences in the geology lab and in the field provided undergraduates opportunities to think and act as a geologist, and to solve questions that they generated and that did not have a textbook answer. From the interview data, the authentic inquiry experiences were something the undergraduate students saw as meaningful because of the vested intellectual interest each had in the challenge of searching for the solution.

While interview questions made the views of undergraduates regarding science explicit and in some cases revealed prior notions of science and scientists, views of science also found their way *unprompted* into undergraduates' journal reflections about field and lab experiences. Many comments indicate that the student was thinking "like a scientist," but it is not clear that they are aware of it themselves as seen in the following:

- Today's field trip was interesting. At first it was hard to determine the different rock types because some things had the same color and they were different. Some things were different colors yet they were the same type of rock. After we climbed up the rocks and dug into them a little bit we were able to see the different layers of rock type. Basically these were alternating layers of shale and limestone where we observed. This shows that millions of years ago the viewing area was an ocean. The different layers of rock show the different levels of the water throughout time. Each layer is made of a different composition because they were formed at different time periods while the area around the rocks was also changing. There are a few different ways that we determined the composition of the rocks. Basically we looked at their hardness, texture, and color. These helped us identify what the rock was and where it was in the rock cycle. It is hard to determine how old these rocks are. Really all we can do besides carbon dating the rock is tell how old it is in relation to the rocks around it. We know that the rocks on the bottom of the stratigraphic column are the oldest and as you move up the column the rocks are younger. (Student 4 [journal], Field trip to river/dam location)

- 3 horizons were obvious in one area of the field while the second spot seemed to only have A & B horizons. These sights were only 50 ft. apart. What causes this discrepancy between the two sites? Water flow? Sediment? Soil make-up? Or was the third horizon just not uncovered? (Student 5 [journal], Field trip to an area for soil study)

Thinking like a geologist (albeit a novice) is obvious in students' questions and reflections. It is not clear that they connect their thoughts, questions, or actions to those of novice scientists exploring geological phenomena. Unsolicited, these reflections were evidence for the geo-researcher that the changes in the lab were "working" in ways that are not always obvious. She voiced that these reflections "show the kind of scientific thinking I wanted to see." Reading the journals supported the course changes while providing insights that informed subsequent teaching.

After the 2-day field experience, the science educator observed teams of undergraduate students working in the lab analyzing and discussing *their* data. Watching the lab activity, the science educator was struck by the self-directed engagement of all the students. The science educator's field notes stated, "One group is busy working on the computers, two groups are discussing data in small teams, one team is analyzing rock samples under the microscopes, and another is looking at soil samples on filter paper. . . . I am struck by something—every single person is actively engaged. No one looks bored or looks like they would rather be elsewhere." Detailed field notes point to the undergraduates' use of scientific processes and questions to analyze and interpret data collected in the field as seen below.

- Team of women (Group A):
  A team of four women are working through the process of sorting the data collected, discussing the lay of land at the sites of collection, and synthesizing the data they collected. [They] used filters to examine the size of the sand grains and discussed the possible composition of the grains.

  Student a: Look at the sand types? What do you think about them?
  Student b: They look very similar. OK, so what do we do with that information? What does it mean?
  Student c: I am not sure (pause), but we could look at sediment collected from the other teams and compare it to ours.

- Another team of three women and one man (Group B):

    Student d: Let's look at the data and collection sites with maps on the computer.

    Student e: Look, the river has changed; it has shifted position. So, here's (points to the spot on the map) where we were standing and digging was once the river. How many years ago was that map made?

The field experiences and the work the undergraduates conducted in the laboratory represent an active engagement with geological concepts, processes, and scientific inquiry. The snapshots represent only glimpses of their interactions as the teams diligently generated questions and analyzed the data collected from the field to tell the "story of the geological site." They spent more than 6 weeks interpreting their findings which culminated in a presentation of their synthesized findings to preservice teachers in the College of Education. This presentation served to allow them to articulate their newly acquired knowledge and the processes needed to accomplish the task.

Theme 2: Active Learning via Groups

Active learning groups fostered discussion, exchange of ideas and positive student interactions in most cases.

The use of formal cooperative teams was central to the constructivist approach implemented into the geology laboratory. For the geologist, observing interactions and discerning the undergraduates' views of their interactions was important to understanding the pedagogical modifications implemented in the course. The undergraduates interviewed were asked specifically about team interactions; however, any comments on team work taken from journals were unsolicited. As such, the number of students (15 out of 20) who made comments regarding team work in their journal entries was astounding. From journal entries, data indicates that team work fostered choice, decision making, and new ideas, and was generally a positive experience as seen in the following:

- Any time you get people together you get a lot of good ideas … We got to assign our own tasks. It was hard to get anyone to step up and say "I'll do this." I had a good group, I was lucky. (Student 3 [interview])
- I liked the group activity that we did today. We had to come up with a way to tell the age of the Earth and of rocks without using absolute dating. It was interesting to hear everyone's ideas together. (Student 8 [journal])

Discrepant data suggests that not everyone was completely satisfied with work conducted in a cooperative group. In the following excerpts it is obvious that teams do not "just happen." In fact, the few comments that were negative about cooperative group learning were related to feelings of being isolated or having a "loner" in the team who did not share their knowledge with the group. What was interesting was that there were no comments suggesting the "free loader" symptom often associated with cooperative grouping. From the data collected on collaborative teams, excerpts such as these were rare.

- My group was OK, but they didn't include me in a lot of things. Especially "Jim" he just did stuff on his own. (Student 12 [journal])
- I really liked working with a group rather than lecture. It kind of bothers me when one person in the group doesn't participate at all, but rather does the work on his own while we sat confused. Eventually he tried to help us. (Student 14 [journal])

The following excerpts represent the overall nature of the undergraduate students' comments regarding active learning via groups.

- I think that by doing group work we have really benefitted. I feel that it is very helpful to get into groups to help us to understand certain concepts better. (Student 15 [journal])
- I think it helps getting classmates' feedback. You don't feel stupid telling the teacher something wrong. It is better to talk it out with a classmate. (Student 4 [interview])
- This active learning experience has allowed me to find not only knowledge, but interest in a topic I normally would not care about. This has been a great experience. (student interviewed by outside evaluator's interview)

In summary, according to the undergraduates in the courses, active learning in groups provided opportunities (a) for students to make decisions and learn from each other without depending on the instructor for answers, (b) to actively engage, (c) to test out their ideas within the safe environment of peers, and (c) for the instructor to stimulate interest, ideas, discussion, and understanding of concepts that might not be gained in a lecture format.

Theme 3: Immersion—Extended Authentic Field Experience

Initially all students did not embrace the authentic 2-day immersion field study required of the laboratory. Some students were uncomfortable with the open inquiry of the short field experiences and felt a loss as to what to do. However, in most cases the field experience was viewed as motivational and useful, and enhanced students' questioning and curiosity regarding geological structures.

The extended authentic field experience was designed for the course to enhance students integrated process skills (estimating, measuring, graphing, modeling, data collection, hypothesizing, testing, and interpreting) by providing them opportunities to learn them directly by interacting with the environment and each other. However, tradition dies slowly and this constructivist philosophy was not initially embraced by all, as seen in the following:

- I am less than enthusiastic about the field experiences. (Student 13 [journal])
- I didn't really want to do this (the field experience). But once we got out there, I enjoyed it and actually learned a lot. (Student 19 [interview])

Other undergraduates were simply surprised that field experiences were included in the course while still others simply expected to gain little from the extended field experiences as seen in these snapshots:

- I didn't think we would be doing all of what we are going to do. Such as actually going out into the field and doing research and getting a hands-on project.... I just basically assumed that we would be sitting behind a desk the entire time. I'm glad I was wrong. (Student 14 [journal])
- I thought the geology field trip was very interesting. I went into the day expecting nothing. As the day progressed, I realized that I was having fun and learning a lot.... It was neat to see how you could dig a trench in one spot and then move over several yards and dig another and they would be entirely different. Moreover, I was impressed by how much I learned (especially about how to read a map). It was fun to see them and you (instructor) on a more "personal" level. (Student 20 [journal])

The inquiry nature of the field studies was uncomfortable for many students at the beginning of the semester. However, by the end of the semester they had become more comfortable with the "lack of guidance." In addition, what the co-researchers found in the data was that the field studies fostered curiosity evident in the questions found in the undergraduates' journals, questions that might not otherwise have occurred. The following quotations are representative samples supporting this theme.

- [First reflection: field site] I felt kinda lost at times when I was out there. I just felt like I didn't have a lot of guidance on what to do and I am totally new at this type of thing.... [Last reflection] This (2-day field trip) was a great experience—to actually get out and have a "hands-on" appreciation of what we are learning. I thought it was a wonderful idea and it should be done again. (Student 12 [journal])

- There are many questions I want answered. Is there sandstone under the water? Was the water high enough at one point to cover the little island we were on? What is the rock above the second layer of shale? How can little weeds grow on shale? On the other side of the stream, is it sandstone? If it is then why isn't sandstone on this side? (Student 16 [journal])

It is clear that the undergraduates' dispositions toward the extended field trip initially fell along the continuum ranging from positive to negative. However, once the field study took place the undergraduates' dispositions toward it shifted positively and the students were surprised by what they learned and what they could find out on their own. It seems that the old adage, "experience is the best teacher" may explain the shifts in their ideas about field research. Once they did field research and realized they could do it, they shifted their views about research. In review, data analysis of the journals, interviews, observations, and the field notes revealed that problem-based inquiry field studies fostered opportunities that allowed undergraduates to interact with one another in alternative ways to actively construct knowledge, to have greater access to the instructor, and to experience scientific investigative processes firsthand.

Theme 4: Pedagogical Modeling

Undergraduate students reveal a favorable disposition toward pedagogical techniques and strategies used in the modified geology course and make connections between the strategies and applications for teaching children.

The old adage that "we teach as we have been taught" seems to be appropriate for beginning this section. The twist on the adage is that we do not have to teach the way we have been taught if we are aware of teaching, and willing to teach, in other ways. The geo-researcher began a journey to change the way she was teaching to enhance student achievement, inquiry skills, communication skills, motivation, and critical thinking, to name a few. As the new strategies (cooperative teams, problem-based field experiences, journaling, and open-inquiry approaches) were used, what was interesting were the unsolicited comments regarding the strategies found in the undergraduates' journals. Students wrote about their own learning styles and how some of the strategies modeled in the geology course fit their style. Ideas for teaching their own classes in the future, based on geology experiences, were also found in journal data. The following samples are representative of the undergraduates' writings:

- I definitely think it's important that teachers be able to apply what they learn to something that they can teach their students. I have especially enjoyed this class because we have learned things that are relevant. Teachers are not going to teach 4th graders the in-depth details of radioactive dating, but they might need to inform them about what it is. I believe we should focus on teaching teachers what they will need to know in order to teach students. Of course, we want them to be well-educated individuals, but we should emphasize the material they will be teaching. (Student 16 [journal])

- An idea … to present to my future class … ways to show crystallization. The object is to place a number of beads in a pan and vibrate it. When the beads are shaken up and then suddenly stopped it represents a rapid cooling effect. When this happens there is no bead arrangement and represents glass—a mass of unordered beads. When you slowly stop the vibration the beads end up ordered. This is an example of a crystal. This would be a great way to show my students that crystals need freedom to move and time to cool slowly for formation. (Student 15 [journal])

- The example that was used to show how rocks are deformed was an excellent hand-on experience to be able to demonstrate how the deformation does take place. The rubber band representing elastic deformation was a great way to help me understand that the stress is not permanent and could go back. The silly putty flowing and also being pulled apart quickly also was great in explaining ductile and brittle deformation. I really think that these visual examples are helpful in helping people to understand an idea. (Student 18 [journal])

- I do like the idea of a journal. I can write in this if I have any questions. Sometimes you can understand concepts better if you just write them down. (Student 9 [journal])

SUMMARY OF THE INVESTIGATION OF REFORM IN TEACHING THE GEOLOGICAL LABORATORY SCIENCES

The teaching example presented in this chapter is one model utilizing active learning in geology laboratory courses for undergraduates. Extended, authentic field experiences and problem-based inquiry labs and field exercises taught through cooperative groups fostered questioning, scientific process skills, and knowledge of geology content in ways that were seen as valuable to the undergraduates and the instructor. Undergraduates gained new insights about how they viewed science and scientists, and voiced a change in their perspectives after the experiences in the geology laboratory. To most undergraduates in this study, geology as a research discipline was demystified and more familiar to them as a result of the experience. Using formal cooperative groups over a long period of time helped the teams to gain insights into the skills each person brought to the team. Initially some groups were slower to form the characteristic of positive interdependence, but by the end of the semester all the teams were working together to synthesize the data they had collected to create their "geologic story." Last, many of the active strategies incorporated into the class evoked positive feedback via journals where students wrote about how the various techniques were helping them to learn geology.

RECOMMENDATIONS FOR BEST PRACTICE IN TEACHING GEOLOGICAL LABORATORY SCIENCES

We find that the advantages of using constructivist strategies, such as cooperative teams, an extended field study, and problem-based inquiry exercises in the lab and field, increased students' interest, questioning, engagement with geological processes, and knowledge of geology, all of which are supported in earlier research. Problem-based open inquiry used in the geology lab or during short field studies evoked multiple avenues from which the teams could work together to arrive at a solution. In many cases, the problems posed did not have a readily accessible textbook solution. This situation created an opportunity for undergraduates to gain intellectual independence by judging the evidence on their own. This created a confidence that does not come with being told or finding an answer in a book. The American Association for the Advancement of Science (AAAS, 1993) in Project 2061 advocates scientific literacy for all students. Scientific literacy involves knowing how to conduct scientific inquiry as well as understanding scientific inquiry. Given this perspective, we recommend including an extended problem-based field study into geology laboratory courses for undergraduates. We found it essential in nurturing the development of undergraduates' ways of knowing and doing geological research. The model, by its very nature, promotes the essence of scientific literacy as defined by AAAS (1989).

IMPLICATIONS AND FUTURE RESEARCH FOR TEACHING GEOLOGICAL LABORATORY SCIENCES

According to Loucks-Horsley, Schmidt, and Raizen (1989), one cannot become scientifically literate "by sitting in a college lecture hall." They point out that conceptual reasoning is inhibited when undergraduates are overwhelmed with terms, facts, and too many topics. We found that it was difficult to modify the geology lab course by randomly eliminating information traditionally taught. It is amazing how we become attached to unnecessary details. However, we finally limited topics by aligning those selected with the National Science Education Standards (NRC, 1996). By doing so there was flexible time built into the course, which was necessary for student data analysis following the extended field study. Therefore, our study implies that one important way to begin to reform teaching is to have university faculty carefully select conceptual themes and omit unnecessary facts. This can provide time for student teams to experience an extended team field study where they construct knowledge grounded in their findings and their inquiries. Reforming teaching methods based upon constructivist practices introduced by the American Association for the Advancement of Science (AAAS, 1989) have recently been shown to improve student achievement (Lawson et al., 2002). Though we did not examine student achievement as a variable of this study, there is clearly a need for such research across the sciences. Thus, there is a need for longitudinal examination of the impact and use of extended, authentic field research on the achievement of undergraduates in the sciences.

AUTHOR NOTE

This work was in part supported by NASA Opportunities for Visionary Academics (NOVA), a program funded by the National Aeronautics and Space Administration, although the views expressed here are the authors' only.

REFERENCES

American Association for the Advancement of Science. (1989). *Science for all Americans*. Washington DC: Author.

American Association for the Advancement of Science. (1993). *Benchmarks for science literacy: Project 2061*. New York: Oxford University Press.

Barrows, H. (1986). A taxonomy of problem-based learning methods. *Medical Evaluation, 20,* 481–486.

Beiersdorfer, R., & Beiersdorfer, S. (1995). Collaborative learning in an advanced environmental-geology course. *Journal of Geological Education, 43,* 346–351.

Bloom, B.S. (Ed.). (1956). *A taxonomy of educational objectives: Handbook 1. Cognitive domain*. New York: David McKay.

Bogdan, R.C., & Biklen, S.K. (1992). *Qualitative research for education: An introduction to theory and methods*. Boston: Allyn & Bacon.

Carpenter, J., Tolhurst, J., Day, E., Zenger, S., Barron, A., & Dozier, K. (1999). A constructivist approach to a high-enrollment undergraduate environmental education course. *Journal of Geoscience Education, 47*, 249–254.

Lawson, A. (1999). What should students learn about the nature of science and how should we teach it? *Journal of College Science Teaching, 28*, 401–411.

Lawson, A., & Thompson, A. (1988). Formal reasoning ability and misconception concerning genetics and natural selection. *Journal of Research in Science Teaching, 25*, 733–746.

Lawson, A., Benford, R., Bloom, I., Carlson, M., Falconer, K., Hestenes, D., et al. (2002). Evaluating college science and mathematics instruction: A reform effort that improves teaching skills. *Journal of College Science Teaching, 31*, 388–393.

Lincoln, Y.S., & Guba, E.G. (1985). *Naturalistic inquiry*. Beverly Hills, CA: Sage.

Loucks-Horsley, S., Schmidt, W.H., & Raizen, S.A. (1989). *Developing and supporting teachers for elementary school science education*. Andover, MA: National Center for Improving Science Education.

MacDonald, H., & Korinek, L. (1995). Cooperative learning activities in large entry-level geology courses. *Journal of Geological Education, 43*, 341–345.

Miles, M., & Huberman, A. (1994). *Qualitative data analysis: A sourcebook of new methods* (2nd ed.). Thousand Oaks, CA: Sage.

National Research Council. (1996). *National Science Education Standards*. Washington DC: National Academy Press.

Patton, M. (1990). *Qualitative evaluation and research methods*. Newbury Park, CA: Sage.

Piaget, J. (1986). *The construction of reality in the child*. New York: Ballantine.

Pinet, P. (1995). Rediscovering geologic principles by collaborative learning. *Journal of Geological Education, 43*, 371–376.

Saunders, W.L. (1992). The constructivist perspective: Implications and teaching strategies for science. *School Science and Mathematics, 92*, 136–141.

Scheritzer, J. (1995). The use of learning stations as a strategy for teaching concepts by active-learning methods. *Journal of Geological Education, 43*, 366–370.

Smith, D., & Hoersch, A. (1995). Problem-based learning in the undergraduate geology classroom. *Journal of Geological Education, 43*, 385–390.

Stakes, R. (1995). *The art of case study research*. Thousand Oaks, CA: Sage.

Tewksbury, B. (1995). Specific strategies for using the "jigsaw" technique for working in groups in non-lecture based courses. *Journal of Geological Education, 43*, 322–326.

von Glasersfeld, E. (1989). Constructivism in education. In T. Husen & T.N. Postlethwaite (Eds.). *The international encyclopedia of education: Research and studies* (Suppl. Vol. 1, pp. 162–163). New York: Pergamon Press.

CHAPTER 27

A MODEL FOR REFORM IN TEACHING INTEGRATED SCIENCE

Promoting Scientific Literacy Among Undergraduate Non-Science Majors

Scott Graves, Michael Odell, Timothy Ewers, and John Ophus

ABSTRACT

Faculty members from the University of Idaho's Colleges of Letters and Science and of Education joined in a cooperative effort to modify an existing interdisciplinary science course, INTER 103: Integrated Science for Elementary Education Majors. The course introduces students to the nature of science and scientific inquiry through the approaches of science, technology, and society, and of Earth system science. Classroom and field activities include developing data-gathering and interpretation skills, addressing alternative science conceptions (with remediation where necessary), and participating in an ongoing study of a local watershed. The course presents a progression of inquiry skills from scientific observation through experimentation. The students are involved in scientific research utilizing protocols

Reform in Undergraduate Science Teaching for the 21st Century, pages 477–491
Copyright © 2004 by Information Age Publishing
All rights of reproduction in any form reserved.

developed by the GLOBE program, an environmental monitoring program to examine inputs into the earth system and the interactions surrounding the biosphere, hydrosphere, atmosphere, and lithosphere (soils) around a watershed. Students must collect hydrology data, map the land cover at their study site, and use historical data to find changes over time. From the data, students in the course prepare scientific reports and participate in a scientific poster session. The goal is for the students to understand the process of doing science, improve inquiry skills, and promote scientific literacy. The course succeeds in employing strategies to help education students re-envision science as a "way of knowing" that involves an ongoing process of fine-tuning perception, evaluating evidence, refining insight, and continuously applying self-reflection as a means of gauging their own reactions to learning as it occurs in the classroom and in the field.

INTRODUCTION

Science is neither simply a philosophy nor a belief system. It is a combination of mental operations that has become increasingly the habit of educated peoples, a culture of illuminations hit upon by a fortunate turn of history that yielded the most effective way of learning about the real world ever conceived.

—E.O. Wilson (1999)

Among the primary audience for undergraduate non-major introductory science courses at the University of Idaho (UI) are prospective elementary teachers. Elementary teachers lay the academic foundation in mathematics and science for all students. Research from Trends in International Mathematics and Science Study (TIMMS), National Assessment of Educational Progress (NAEP) and other comprehensive studies indicates that prospective elementary teachers have weak skills in mathematics and science. Many have not taken a science course since their sophomore or junior year of high school; they are not proficient in either the individual disciplines, or their integration in a systemic approach to teaching and learning. In Idaho most have taken math through Algebra 2, but may not have taken a mathematics course since their junior year of high school (Graves, 1999; Klett, 1997). Upon entering the university, prospective elementary teachers are faced with choosing science courses from a broad array of topics in life, earth, space, and physical science. It is not surprising to find that elementary education majors typically enroll in life and earth science courses. They tend to avoid the physical sciences and this lack of preparation can be seen in test scores of K–12 students.

Shaping the Future, a report commissioned by the National Science Foundation (NSF) examined undergraduate science education and while the focus was on undergraduate science, technology, engineering, and mathematics (STEM) education, it was recognized as only one part of the STEM education continuum that runs from preschool through postgraduate work (NSF, 1996). The components of the continuum are interdependent. Undergraduate STEM education is dependent upon the students prepared by the K–12 system. The higher education system prepares future teachers for the K–12 system as well as other university faculty who prepare teachers. Therefore, the K–12 community and higher education must work collaboratively for each community's mutual benefit. K–12 science education has been undergoing a transformation as a result of the National Science Education Standards (National Research Council, 1996), Project 2061 Benchmarks (American Association for the Advancement of Science, 1993), and revised state standards. The standards put an emphasis on inquiry, active learning, and the integration of technology. Unfortunately, many university introductory courses are primarily taught through lecture, and as a result, new elementary teachers may not have experienced science taught in the manner the standards indicate. To sustain these reforms at the K–12 level, undergraduate education will also have to change, especially in teacher preparation.

It is the goal of the committee that produced *Shaping the Future* that "all students have access to supportive, excellent undergraduate education in science, mathematics, engineering, and technology, and all students learn these subjects by direct experience with the methods and processes of inquiry" (NSF, 1996, p. 1).

SCIENCE, TECHNOLOGY AND SOCIETY AND EARTH SYSTEM SCIENCE

Major foci of integrated science courses at the University of Idaho (and many other institutions) include science, technology, and society (STS) and Earth system science (ESS). These two broadly encompassing themes have a significant history of success and provide contemporary relevance in programs of study. The STS focus is well justified in programs of study and in research in teaching and learning dating back to the mid-1970s (Lisowski, 1985).

Education in the 21st century impresses upon an increasingly connected global society, the necessity of addressing complex issues and challenges whose resolution involves innovative solutions that often employ advances in information technologies. Learners of all ages need more sophisticated problem-solving and decision-making skills and access to information to deal effectively with a wide variety of issues. The goals of Science-Technology-Society (STS) programs found in many colleges and universities, including the University of Idaho, are the development of logical, higher order thinking and reasoning skills in the context of science, technology, and society.

STS programs are designed to reflect national goals for education related to citizenship, problem solving, and higher order thinking. They are grounded in philosophical and pedagogical frameworks that promote the active, collaborative construction of knowledge, with historical reasoning providing launching points for informed and reasoned study of present and future issues. STS teaching strategies, course materials, and activities, comprise sound instructional models to help learners develop cognitive reasoning and skills necessary for participating in an evolving democratic citizenry.

STS programs are universally described as constituting a multi- and interdisciplinary inquiry. An exemplary STS degree program at Stanford University bears the following description: "STS encourages students to internalize a comprehensive and systemic way of thinking about technology and science in society—not just in terms of politics and economics, but also in terms of ethical and cultural aspects" (Robert McGinn, Department of Management Science and Engineering, Stanford University, as cited in Sanford, 2001, ¶ 6). Through studying particular issues at the confluence of science and technology in a developing and evolving society, learners can begin to understand how lessons learned from the past and present can inform purposeful thinking and actions in the future.

Founded in the early 1970s, science, technology, and society (STS) programs of teaching and research devote considerable study to science and technology in society, in both historical and contemporary perspectives. STS undergraduate programs of study have existed around the country and internationally since the mid-1970s (in the United States at Stanford, MIT, Cornell, Penn State, North Carolina State, University of Michigan, Vassar College, etc.) and abroad (in England, Canada, Australia, the Netherlands, Norway, and Sweden). STS programs focus on the intersections of science and technology with ethics, aesthetics, pubic policy, politics, cultural change, economic development, history of science, history of technology, organizations, history of medicine, history of engineering, work, information, and material culture.

STS research and teaching are predicated on the firm belief that science and technology are two of the most potent forces for individual, societal, and global change in the contemporary era. "Understanding the natures, causes, and social consequences of scientific and technological developments, how science and technology function in different societies, and how social forces attempt to shape and control these forces to serve diverse, often conflicting interests ... requires study beyond the purview of any single conventional academic discipline" (Leland Stanford Junior University, 2002).

In 1978, the National Science Teachers Association (NSTA) recognized the value of broadening its science and technology discourse toward viewing science as a discipline concerned with the study of the interaction and impact of science on society (NSTA, 1978). The dominance of technology in society today reaffirms these philosophical orientations and necessitates that today's students achieve a level of scientific and technological literacy that will help them deal with science-related societal issues for the improvement of their own lives and the advancement of society.

The National Science Foundation offered specific recommendations for K–12 instruction related to STS in 1982, recommending a 3rd-year elective course in STS for students in grades 11 and 12. The National Science Board followed suit in 1983 suggesting a 2-year required STS sequence for grades 9 and 10 and a science curriculum in grades 9 to 11 be structured around the interactions of science and technology.

The STS movement has seen a dramatic growth in international programs. Even before the turn of the millennium, more than 91 STS-related journals and newsletters with an international audience were in circulation (de la Mothe, 1983). Journals such as the *Science, Technology and Society Curriculum Newsletter* of Lehigh University (Cutcliffe, 1986), and the *S-STS Reporter* (Roy, 1985) are representative of these serials, providing information, insights, and suggestions for classroom instruction. Lisowski's 1985 review of the literature notes that "infusion within established programs, integration between disciplines and establishment of complete courses have been occurring with significant levels of success" (Conclusions section, ¶ 2).

Among the STS programs at major universities, K–12 school curricula, and in individual college classes, modular units of study can be found on a wide variety of STS topics. At the University of Idaho, the integrated science course employs an STS approach with a content focus on Earth system science (ESS).

Earth System Science

NASA's Earth Sciences Enterprise, in conjunction with other science organizations and many university researchers, has pioneered the early development of the new discipline of Earth system science. NASA's Mission to Planet Earth (MTPE) program mandate, and its current Earth Sciences Enterprise is to provide for a global observational capability ensuring that the data collected bear on the scientific and societal challenges that attend ESS and global change. NASA's MTPE is "dedicated to understanding the total Earth system and the effects of natural and human-induced changes on the global environment" (NASA, 2003). The Universities Space

Research Association described ESS in the following: "Earth system science views the Earth as a synergistic physical system of interrelated phenomena, governed by complex processes involving the geosphere, atmosphere, hydrosphere and biosphere" (Earth System Science Online, n.d., ¶ 1). Fundamental to an ESS approach is the need to emphasize relevant interactions among chemical, physical, biological, and dynamical processes that extend over spatial scales from microns to the size of planetary orbits, and over time scales of milliseconds to billions of years. In integrating the traditional science disciplines, the ESS approach has become widely accepted as a framework for learner inquiry in which they pose specific discipline-related and interdisciplinary questions that include a societal (part of STS) issues component. ESS forms the foundation of NASA's Earth science vision as well as the basis of the NSF geoscience long-range planning effort as part of the nation's global change research objectives. NASA's educational mandate is to help any and all educational institutions promote an informed society that understands and appreciates the interdependency of the Earth's physical climate system with all of life.

Integrated Science at the University of Idaho

In an effort to engage students preparing to become K–12 teachers in science, and develop better prepared teachers, concerned university faculty from the University of Idaho Colleges of Letters and Science and of Education began examining the undergraduate science curriculum required of students preparing to be elementary teachers. The core curriculum requires seven to eight credits of laboratory science. Most of these courses are general non-major courses such as Geology 101, Biology 100, Physics 101, and Chemistry 100. By and large these courses assisted students in learning science facts, but what was lacking were extended experiences in the methods and processes of scientific inquiry and an understanding of what scientists do on a day-to-day basis. It was not clear that the traditional core was preparing future teachers to understand how science can be used to make informed judgments about social and scientific matters, and how to communicate and work together to solve complex problems.

Many universities have also increased the requirements for non-majors and preservice teachers to enhance content preparation. Although the amount of content may be increased for prospective elementary educators, these students may still not get a holistic presentation of science as recommended by NSTA. If the nation's K–12 students are to be scientifically literate, teachers need to have the proper experiences to facilitate this. Significant growth in preservice teachers science content knowledge, increased efficacy and more positive attitudes toward science can be achieved by designing integrated science course experiences that stress a hands-on, minds-on approach in a collaborative atmosphere where doing science includes opportunities to demystify their previous, often negative experiences in science classrooms. These new course experiences are most beneficial where they provide students with the opportunity to identify any scientific misconceptions that they may have and to become aware of common misconceptions held by K–12 students (Graves, 1999).

To address these issues many universities including the University of Idaho have begun creating special "Integrated Science" courses for non-majors and preservice teachers to better meet their needs. The course embodies an STS and ESS approach. Integrated Science INTER 103 was developed by a team of University of Idaho scientists and educators to create an introductory general science course that focuses on the integrated nature of science and the process of science inquiry and content that is transferable to K–12 classrooms. Transferable, in this context, indicates that the content is aligned with the science standards for the state of Idaho and NSTA's *College Pathways to the Science Education Standards* (Siebert & McIntosh, 2001).

In a statement made in 1991, the Association of American Colleges and Universities called for a serious examination of the goals of colleges and universities with respect to the intellectual and ethical development of students. "In the final analysis, the real challenge of college, for students and faculty members alike, is empowering individuals to know that the world is far more complex than it first appears, and that they must make interpretive arguments and decision-judgments that entail real consequences for which they must take responsibility and from which they may not flee by disclaiming expertise" (AACU, 1991). Inferring from this statement that scientific literacy entails critical reflection and efficacy in deciding whether evidence is sufficient to warrant certain actions, actions that one has personal responsibility for, the integrated science course initiated at the University of Idaho was designed to instill confidence in science through remedying misconceptions, collaborating in problem solving, and promoting critical self-reflection on the process of learning science concepts.

Course Description for Integrated Science INTER 103

INTER103 focuses on the development of scientific literacy and areas of science in which we have found elementary educators to be underprepared. The course description is as follows:

INTER103 Integrated Science for Elementary Education Majors (4 cr). Scientific method, physics and chemistry of atoms and molecules, chemical energy and thermodynamics, electrical circuits, physics and biology of light, magnets and motors, geological evolution of the earth, forces shaping the earth, meteorology, fossil record and evolution, ecology, and topical issues in science. Two 3-hour class meetings each week. Prerequisites include Basic College Mathematics and declaration of major course of study in Elementary Education.

Classroom environment.   The course meets twice weekly for a total of 6 hours in the science education laboratory classroom in the College of Education. Teaching/facilitation is accomplished in a decidedly "constructivist" format with multiple content-expert facilitators who have specifically designed classroom pedagogy and instructional styles to focus on cooperative inquiry-based and student-centered learning. The physical classroom has six hexagonal tables for groups of six students each. At the front of the classroom is a standard scientific demonstration table; at the back are two wet lab stations that can accommodate two groups at a time. The classroom environment was developed in such a way to provide a model for science classrooms in K–12 schools. Around the perimeter of the classroom are 22 PCs. There are three multimedia stations in the classroom as well, one at the front for the instructor, and one Macintosh and one PC station in the back for students. These stations are equipped with additional peripherals such as VCRs, scanners, and color printers. These stations also have specialty software. The instructor station is also connected to a DVD/VCR, videodisc, Web Cam, and a SmartBoard and Projection system. The walls of the classroom have safety posters and information, posters illustrating concepts in the course, and bulletin boards where assigned groups in the course may post information.

Course structure and content.   The instructional approach utilized in the course is an STS-ESS approach. It incorporates actual field research based on the *College Pathways to the Science Education Standards* (Siebert & McIntosh, 2001), models inquiry and standards-based education, exemplifies the nature of science, and incorporates collaboration with peers and experts. Students are pre-assessed on a variety of attitudinal, content knowledge, and skills competencies during the first course meeting. (Descriptions of the various instruments and assessments appears later in this chapter.) Subsequent class sessions are designed around the specific areas of need as indicated by the results of this pretest. The instructors have developed a number of modules that include activities, media, and technology application, based on the content of earth, space, and physical science.

Course modules.   Based on the outcome of the initial assessments, the instructors decide which of the physics, chemistry, biology, or geology modules are most critical and how each of these modules will be integrated. For students who do not have the appropriate skills (as evidenced by very low scores on initial assessments), a course support Web site has been developed using the Idaho Virtual Campus system (University of Idaho, 2003). The support Web site is designed to enhance the live class and structure the students' out-of-class activities. Of particular use are a series of online modules specifically designed to remediate deficiencies in science process skills. Students are able to check out designated kits and complete any particular modules of need. This practice has allowed the course to proceed in a timely manner and not become consumed in remediation.

The instructors have adopted a STS-ESS approach in order to address the earth science and ecological components of the course in a holistic manner. The content is examined through the lens of events and interactions between the atmosphere, lithosphere, hydrosphere, and the biosphere, with humans as an active agent among the systems processes. In addition to being central to NASA's mission in ESS, remote sensing, systems concepts, and understanding models are common threads that run through all ESS module activities in the integrated sciences course. Among the major ESS themes, those that are addressed in the course include (a) system concepts and earth systems; (b) remote sensing; (c) hydrological cycle; (d) earth system history; (e) land use and land cover change; (f) earth energy budget; (g) human population-environment interactions; (h) economics, sustainability, and natural resources; (i) the sun and the solar system; and the processes and phenomena of (j) the geosphere, (k) the biosphere, and (l) the atmosphere.

Emphasizing an STS approach avoids the compartmentalization of teaching ESS by subdisciplines, and in all cases the topics focus on critical issues that bridge across the geosciences and other relevant disciplines. Through both the STS and ESS approaches the emphasis is on evolving and growing a personal knowledge base developed through acquisition, analysis and interpretation of global and local data. This ensures the development of holistic perspectives of Earth systems and introduces students to the underlying important issues and challenges adopted by the United Nations Conference on Environment and Development (UNCED) at the Earth Summit meeting in June 1992, in Rio de Janeiro, Brazil.

Throughout their investigations and in-class discussions, students are given opportunities to explore human-sphere-system interactions and human impacts on the physical environment, habitats, climate, and water quality. The particular *event* focus for the STS-ESS studies is land use practices in a local watershed with an evolving community context. The culminating experience for the ESS portion of the course is a semester-long field study of a local watershed (Paradise Creek, Latah County, Idaho), its resident flora and fauna, and the land cover and land use adjacent to a local stream system. Integrating technology, the watershed study is initiated with the examination of remote sensing images, air photos, contour maps, and existing ground data to begin investigating the local area.

The course also integrates components of the GLOBE Program (n.d.). GLOBE is a hands-on science and education program uniting students, teachers, and scientists in 102 countries to take environmental measurements, report their data online, and ultimately conduct student inquiry projects. These data are used in student research as well as by scientists around the world for their international research initiatives. To date, more than 9 million student measurements have been recorded on the GLOBE database from more than 13,000 schools. The GLOBE data collection protocols and data visualization tools from the Web site that the students use in the INTER 103 course help to demonstrate to the future teachers that it is possible to conduct long-term studies as a focus of science teaching, to facilitate student inquiry projects, and conduct through online support mechanisms similar studies with students in the future. All students receive GLOBE protocol training and certification during the course for each specific measurement taken.

The GLOBE investigation areas that are incorporated into INTER 103 include hydrology, land cover biology, and soils. To initiate the study, students are taken to a location along Paradise Creek and asked to make observations based on what they see. This particular location has a number of features that make it ideal for this type of activity. Paradise Creek flows adjacent to a highway and at this particular locale, the stream bank is eroding significantly. A water runoff pipe dumps directly into Paradise Creek at this locale, and above the opposite bank of the creek is a large wheat field. Students are asked to make field sketches and asked what information one would want to know if they wanted to determine the water quality. Students then generate questions that can be investigated through scientific inquiry. Teams of students are assigned a specific characteristic locale within the Paradise Creek watershed, which extends from a forested upland through agricultural areas and into suburban and urban settings, all within an area less than 50 square miles, and easily accessible from the university campus.

Weekly field activities include the collection of water-quality data (chemical assay and macro-invertebrate sampling), vegetative biometry (canopy and ground-cover surveys), as well as soil sampling. Teams also assess landscape features that have been human altered and research the nature and extent of human impacts. Students use calculator-based-lab (CBL) technologies, as well as traditional chemical assay methods to collect hydrology and soil fertility data. To analyze the data, they use the GLOBE Web site tools for graphing and visualization, as well as spreadsheets. The course Web site links to current reports on land use and its effects on the environment. The watershed study is conducted over the course of the semester, and students present a synthesis/summative poster in a class-long scientific poster session. Their poster presentations include written narrative overviews of their site-specific research, data collection strategies, data tables, interpretive graphs and charts, and maps of their specific site, with sampling locales identified, distribution of fauna and flora observed and ecosystem/habitat classifications based on the Modified UNESCO Classification Scheme (MUC) system. Most poster sessions also include a multimedia component (PowerPoint or Web pages). At the conclusion of the poster sessions students self-assess and peer critique their efforts and contributions to the team projects.

Other components of the course focus on basic physics and chemistry. Student demographic data and information regarding their course-taking patterns (course pre-surveys from INTER 103 and associated science methods courses) consistently indicate that most elementary education majors entering our program typically enrolled in a physical science course in high school. As much as possible, the instructors integrate chemistry and physics as it relates to the Paradise Creek watershed studies. Since this is not always possible, a portion of the course is dedicated to introductory chemistry and physics concepts. Currently, the course uses the Constructing Physics Understanding curriculum, a conceptual physics curriculum to facilitate physics topics, and also utilizes materials developed by the Indiana University QUEST program for the chemistry modules (CPU Project, 2000).

DISCUSSION: NATURE OF SCIENCE, SCIENCE LITERACY, AND SCIENCE PROCESS SKILLS

In establishing a research agenda for the INTER 103 course, the instructor/designers developed a suite of assessment instruments, tests, and performance criteria. Student demographic data and course-taking patterns are also collected. Throughout the course and in initial, formative, and summative assessments, the instructors use a variety of techniques including paper-and-pencil surveys, traditional objective tests, performances, projects, and electronic journals and portfolios. Students are pre-assessed before each module. Instructors look for misconceptions and specific areas needing attention. Students utilize assignment and assessment data to track their own progress. Their materials are managed using the IVC electronic portfolio tool (an online personal folio/journaling site associated with the class). Students are required to keep this journal up to date (following the course schedule and time line) and are required to complete reflective journal entries before and after each class session. This serves as a formative assessment for the instructors as well as a self-assessment for the students.

INTER 103 students were pre-assessed on science inquiry skills, attitudes, and understanding of the nature of science. The pre-assessments consisted of paper-and-pencil tests and one performance activity. The skills assessment examined students' knowledge and understanding of the science process skills used in scientific inquiry. Questions on the assessment looked at observation, inference, prediction, measurement, communication of data (graphs and tables), variables, estimation, large and small numbers, factor labeling, analysis of data, and experimentation. The tests were constructed from local resources and in personal consultation with the Institute for Mathematics, Interactive Technologies, and Science (IMITS) research expertise. For a majority of students, their posttest scores increased by a factor of 25 to 30% (IMITS Group, 2002).

The second instrument, created by Schoon and Boone (1998) examined students' alternative conceptions or misconceptions in science. Questions covered a variety of basic concepts including in the physical and biological as well as earth and space sciences. Data indicated that students held misconceptions primarily in the space and physical sciences. At the end of the course, many, if not most of the students had completely remedied their earlier misconstrued ideas, but this was a process that necessitated multiple individual and group discussions, demonstration, and personal reflection time (IMITS Group, 2002).

The third instrument asked students to design and perform a simple experiment, such as a test of strength for paper towels. Students were provided with materials to investigate a simple hypothesis (in this case, a selection of different brands of paper towel, standardized weights, sponges, buckets, twine or rubber bands, and water). Other performance tests included deriving the significance of pi from an open-ended inquiry focused on a selection of circular objects (with tape measures, rulers, string, and calculators). In this test, students were encouraged to analyze their data using charts of principle geometric components they decided to measure, such as diameter and circumference. In yet another performance test, the hypothesis given was that the mass of a liquid is directly proportional to its volume, and students were engaged in determining measurability and provableness of this assertion. This performance was also used as a check for the results of the skills test. In all cases, students made significant progress toward becoming more comfortable with approaching these types of problems, reasoning from principles in the sciences, and constructing and defending hypotheses, analytical procedures, and findings.

In a study of the INTER 103 and science methods courses taken in sequence by non-science majors, Graves (1999) found that those students who had participated in the INTER 103 course and then went on to take the science methods course (6 out of 71 participants) obtained exceptionally high scores on a measure of reflective judgment (Reasoning About Current

Issues instrument). This RCI instrument is based on the seminal work of King and Kitchener on cognitive development and reflective judgment theory (King & Kitchener, 1994). These same INTER 103 alumni also outscored their peers on measures of science efficacy, attitudes toward science, and openness to constructivist learning environments (Graves, 1999).

In the most recent iteration of the INTER 103 course, students also completed the Views on the Nature of Science survey (VNOS) form C (Lederman, Abd-El-Khalick, Bell, & Schwartz, 2002), which is designed to assess their understanding of science as a human endeavor, a process, and an evolving knowledge base. VNOS data were also collected in the introductory geology, physics, and chemistry classes for non-majors as a control. Recent work in progress by Ophus (IMITS Group, personal communication) suggests that the INTER 103 students achieved significantly higher scores (statistically) on the VNOS than did traditional science discipline students.

Based on the results of the assessments accumulated over 5 years, INTER 103 instructors continue to fine tune the course to improve elementary majors' inquiry skills and understanding of the nature of science. These students conduct an in-depth research project and experience science taught in an integrated holistic fashion focusing on concepts, not single disciplines. The course structure and content, as well as facilitation schema are continually updated, with the instructor/designers' belief that standards-based, inquiry-focused, and learner-centered science experiences can positively influence students' knowledge, attitudes, and future teaching methodology.

From these findings, as well as in instructors teaching notes and reflections, we strongly believe that students that have progressed through the INTER 103 course and have gone on to the science methods course consistently show a higher degree of readiness for teaching, with well-developed science teaching efficacy, as well as better attitudes and significantly advanced perspectives on constructivist approaches to teaching and learning in the sciences.

REFERENCES

American Association for the Advancement of Science. (1993). *Benchmarks for science literacy: Project 2061.* New York: Oxford University Press.

(2000). *Constructing physics understanding.* Retrieved August 18, 2003, from http://cpucips.sdsu.edu/web/CPU/default.html

Association of American Colleges and Universities. (1991). The challenge of connecting learning. Washington, DC: AAC&U Press, p. 4.

Cutcliffe, S. (Ed.). (1986). *Science, Technology, and Society Newsletter, 46.* Bethlehem, PA: Lehigh University.

de la Mothe, J. (1983). *Unity and diversity in STS curricula.* Quebec, Canada: Science, Mathematics and Environmental Education Clearinghouse. (ERIC Document Reproduction Service No. ED230431)

Earth System Science Online. (n.d.). *What is Earth system science?* Retrieved July 12, 2003, from the Universities Space Research Association Web site: http://www.usra.edu/esse/essonline/whatis.html

GLOBE Program. (n.d.). An exciting, worldwide, hands-on education and science program. Retrieved August 16, 2003, from http://www.globe.gov

Graves, S. M. (1999). *Alternative conceptions in science and science teaching self-efficacy.* Unpublished doctoral dissertation, University of Idaho, Moscow.

IMITS Group: Institute for Mathematics, Interactive Technologies and Science Education, University of Idaho, College of Education. (2002). Internal papers, personal communications (Graves, Klett, Odell, Ewers, Ophus).

King, P. M., & Kitchener, K. S. (1994). *The development of reflective judgment in adolescence and adulthood.* San Francisco: Jossey-Bass.

Klett, M. (1997). *The effect of alternative clinical teaching experiences on preservice teachers' self-efficacy.* Unpublished doctoral dissertation, University of Idaho, Moscow.

Lederman, N.G., Abd-El-Khalick, F., Bell, R. L., & Schwartz, R. (2002). Views of nature of science questionnaire: Toward valid and meaningful assessment of learner's conceptions of nature of science. *Journal of Research in Science Teaching, 39,* 497–521.

Leland Stanford Junior University. (2002). *Science, technology and society: Majoring in STS: FAQs and answers: 2002–2003.* Retrieved July 12, 2003, from http://www.stanford.edu/group/STS/stsfaq.html

Lisowski, M. (1985). *Science-technology-society in the science curriculum* (ERIC/SMEAC Special Digest No. 2). Columbus, OH: ERIC Clearinghouse for Science Mathematics and Environmental Education. Retrieved July 12, 2003, from http://www.ericfacility.net/ericdigests/ed274513.html

NASA. (2003). *Earth Science Enterprise.* Retrieved July 12, 2003, from http://www.earth.nasa.gov/Introduction/what.html

NASA. (1986, May). Earth system science—A closer view. Report of the Earth System Science Committee NASA Advisory Council. Washington, DC: National Aeronautics and Space Administration.

National Research Council. (1996). *National Science Education Standards.* Washington, DC: National Academy Press.

National Science Foundation (1996). *Shaping the future: New expectations for undergraduate education in science, mathematics, engineering, and technology* (NSF Publication No. 96-139). Washington, DC: Author.

National Science Teachers Association (NSTA). (1978). Science education: Accomplishments and needs, a working paper. Columbus, OH: ERIC Clearinghouse for Science, Mathematics, and Environmental Education (ERIC Document Reproduction Service No. 171151).

Roy, R. (1985). S-STS Reporter. University Park: Pennsylvania State University.

Sanford, J. (2001, August 22). Science, technology and society program survives, thrives. *Stanford Report.* Retrieved July 12, 2003, from http://www.stanford.edu/dept/news/report/news/august22/sts-822.html

Schoon, K.J., and Boone, W.J., 1998, Self-efficacy and alternative conceptions of science of preservice elementary teachers, *Science Education, 82,* 553–568.

Siebert, E.D., & McIntosh, W.J. (Eds.). (2001) *College pathways to the science education standards.* Arlington, VA: National Science Teachers Association Press.

University of Idaho. (2003). *Idaho Virtual Campus.* Retrieved July 12, 2003, from http://ivccourses.ed.uidaho.edu

Wilson, E. O. (1999). *Consilience: The unity of knowledge.* New York: Knopf.

CHAPTER 28

A MODEL FOR REFORM IN TEACHING IN ENGINEERING AND TECHNOLOGY

Creating Links Among Disciplines for Increased Scientific Literacy

Jeanelle Bland Day

ABSTRACT

The institutionally-imposed segregation of higher education faculty into the specific fields within departments of science, mathematics, engineering, and education has created a barrier to interdisciplinary collaboration. This separation has created difficulties in providing appropriate pedagogy and discipline knowledge in science course work. The lack of expertise in faculty was demonstrated in several key areas including course experiences leading to meaningfully understanding the elements of scientific literacy, developing positive dispositions toward science, making interconnections between the science disciplines, and understanding prior knowledge that freshmen bring into college science courses. This chapter describes how faculty in the College of Engineering and College of Education at the University of Alabama

Reform in Undergraduate Science Teaching for the 21st Century, pages 493–510
Copyright © 2004 by Information Age Publishing
493

bridged the gap and created an aerospace engineering course for preservice teachers to improve scientific literacy.

INTRODUCTION

Nearly 20 years ago, a report by the National Commission on Excellence in Education, *A Nation at Risk* (1983), warned that our educational system was on the brink of failure. The report estimated that more than 20 million American adults were functionally illiterate and average K–12 achievement test scores had declined steadily over a 15-year period (National Commission on Excellence in Education [NCEE], 1983). High schools in recent years have raised graduation standards and colleges have continued to pressure high schools by raising entrance requirements for specific courses (National Science Foundation [NSF], 1996a).

Higher education faculty in science, mathematics, engineering, and technology (SME&T), despite raising college entrance requirements, still blame high school teachers for sending underprepared students to college. Higher education faculty have now begun to realize that "teachers and principals in the K–12 system are all people who have been educated at the undergraduate level, mostly in situations in which SME&T programs have not taken seriously enough their vital part of the responsibility for the quality of America's teachers" (NSF, 1996a, p. 35).

Not only are SME&T faculty responsible for educating America's teachers, they are becoming responsible for educating increasing numbers of college students (NSF, 1996a). According to the report *Shaping the Future: New Expectations for Undergraduate Education in Science, Mathematics, Engineering, and Technology* (NSF, 1996a), undergraduate SME&T education is described as follows:

> Undergraduate SME&T education is the linchpin of the entire SME&T education enterprise—for it is at the undergraduate level that prospective K–12 teachers are educated, that most of the technical work force is prepared, and that future educators and professional practitioners in science, mathematics, and engineering learn their fields and, in many cases, prepare for more specialized work in graduate school. (p. 1)

As the workplace demands higher-skilled workers, and low-paying, low-skilled jobs disappear in the global marketplace, SME&T faculty will have to provide opportunities for more students to succeed in these areas of study since college has become a rite of passage for students in recent years (NSF, 1996a, p. 34). Sixty-seven percent of female and 60% of male high school graduates attend college within a few months of graduating (NSF, 1996a). With increasingly more diverse populations on college campuses than ever before, institutions will be "forced to think more about how individuals learn . . . and recognize that there are differences in learning styles which profoundly affect achievement" (NSF, 1996a, p. 28), which in turn will affect how teaching should be approached.

THE VALUE OF USING CONSTRUCTIVIST TEACHING METHODS IN COLLEGE CLASSROOMS

To more effectively teach students in lower-level courses, college faculty must have an understanding of how and why students learn (Clough & Kauffman, 1999). Due to current national trends in curriculum standards at the K–12 level that stress the use of such teaching techniques as cooperative grouping, authentic assessment, integrated curricula, and technology, incoming students will have been exposed to learning experiences that vastly differ from those commonly used in the past (National Center for History in the Schools, 1996; National Council for the Social Studies, 1994; National Council of Teachers of English and International Reading Association, 1996; National Council of Teachers of Mathematics, 1989; National Research Council [NRC], 1996). The prevailing theory of teaching in the area of science is the constructivist learning theory (CLT), which Clough and Kauffman suggest has five major elements:

> First, learning is an active process! Even when students sit passively in a lecture, for learning to occur they must be mentally active, selectively taking in and attending to information, and connecting and comparing it to prior knowledge and additional incoming information in an attempt to make sense of what is being received. Second, because the incoming sensory input is primarily organized by the individual receiving the stimuli, the meaning intended by the teacher is often not communicated in an intact manner. Third, knowledge that a student brings to current instruction may help or hinder the creation of meaning similar to that intended by the teacher. Fourth, students' prior knowledge that is at odds with the intended learning is, at times, incredibly resistant to change. That is, in attempting to make sense of instruction, students interpret and sometimes modify incoming stimuli so that it fits (i.e., connects) to what is already known. Fifth, as the number of links connected to new learning increases, the likelihood of long-term and meaningful learning increases.

The report *Shaping the Future: New Expectations for Undergraduate Education in Science, Mathematics, Engineering, and Technology* (NSF, 1996a) recommends that instructors in the targeted fields make their courses inquiry based (i.e., constructivist in nature). It is also suggested that the faculty use instructional pedagogy that helps students develop communication, critical thinking, and collaborative skills.

According to King (1993), college students do not "spontaneously engage in active learning" (p. 31). King asserts that students should be provided with adequate opportunities to engage in this type of learning through the implementation of learning strategies that purposefully encourage students to engage in active learning. Active learning may take place through the use of peer interactions such as mentoring or teaching, cooperative learning, and use of alternative assessment methods, such as the use of concept mapping (Clough & Kauffman, 1999; King, 1993). Perhaps the most important aspect of CLT is making concrete experiences available to students before abstract ideas are presented, thus providing a foundation upon which abstractions may be linked to prior knowledge.

REFORM IN UNDERGRADUATE SCIENCE AND ENGINEERING COURSES

A review of literature related to reform in undergraduate science and engineering courses has revealed that it has been occurring in small pockets throughout the country in certain courses, but not in complete colleges of engineering or sciences. The *Shaping the Future: Strategies for Revitalizing Undergraduate Education* (NSF, 1996b) document provided summaries from 44 institutions representing two-year colleges, four-year colleges, and universities that described their plans for improving undergraduate SME&T courses. A review of these summaries indicated that the institutions planned to concentrate on four broad areas in their reform efforts. In promoting changes in the curricula and university culture or environment, changes involved (a) use of faculty, students, and administrators to cooperatively plan institutional reform; (b) integration of discipline-based studies into thematic curricula; and (c) support of interdisciplinary instruction in the areas of mathematics and science.

In the area of using new teaching models, the summaries suggest that the institutions want to incorporate active learning and cooperative learning into module-based courses designed for non-science majors. The institutions planned to support faculty development by (a) expanding mentoring opportunities, (b) offering faculty forums where faculty could share ideas and materials, and (c) preparing faculty to work in collaborative, interdisciplinary settings. The institutions also planned to incorporate student-centered activities into the reform efforts by involving undergraduates in mentoring activities with faculty members, offering opportunities for undergraduate research, and improving training of graduate teaching assistants (NSF, 1996b).

For reforms to be effective at the institutional level, there must be a clear purpose about the intentions of the reform. The institutions should have powerful visions about teaching and learning, and there must also be buy-in of all involved stakeholders including faculty and administration (Olson, 1994). Slavin, when interviewed by Olson (1994) suggested that reform efforts are fruitless unless there is at least 80% of the faculty in agreement about changes to be made. This would hold true for efforts to reorganize an entire college, or to just create innovative courses addressing recommendations in the National Science Education Standards (NRC, 1996) in a university department.

In departments that have typically had science educators embedded within them such as chemistry, physics, and biology, innovations in courses have been documented. For instance, McIntosh (1996) redesigned a college physics course to coincide with the National Science Education Standards (NRC, 1996). Traditional laboratory exercises were replaced with open-ended investigations, lectures were diminished, and the use of technology became a major focus of the course.

A similar, yet more complex course was created by Bazler and Charles (1993) at Lehigh University. Using guidelines recommended in *Project 2061: Science for All Americans* (American Association for the Advancement of Science, 1989) the authors revamped two courses. Faculty from engineering and education worked together to create and team-teach courses that focused upon interdisciplinary science themes. The researchers discovered that the team approach to teaching created an awareness of each professor's shortcomings, and allowed for reflection on the teaching process (Bazler & Charles, 1993).

At Rollins College, a science literacy course for non-majors was created (Gregory, 1992). The course, taught during an interim term, centered around two themes: (a) science and the media, and (b) science and scientists. The students were required to read two newspapers each day, and a weekly news magazine. The students collected all articles related to science and wrote summaries and comments about the articles. Class sessions were reserved for discussions of current scientific topics related to the readings. Through the use of course evaluations, the authors discovered that the students' attitudes toward science had been positively affected by the course. A follow-up study of the students in the course a year after course completion revealed that the course had a lasting impact on the selected readings of the students. All respondents reported that they were still reading news articles related to science more frequently than before taking the course. One student commented that the course had persuaded him to take additional science courses during the year following the course.

In a study by Weld, Rogers, and Heard (1999), semester-length ecology and animal behavior courses at the University of Iowa were changed to incorporate independent research projects into the courses rather than laboratories. Student teams of four to five students ($N = 80$) worked cooperatively to carry out the research. In post-course surveys, the researchers discovered that the students showed gains in three areas: (a) knowledge in the way that nature works, (b) appreciation for the methodology of science, and (c) insight on possible careers in science.

Finally, in a study by Frey (1997), students ($N = 28$) were given cooperative group, home-study assignments in an introductory chemistry course. Each cooperative activity incorporated course material equivalent to approximately three textbook chapters. In a post-course survey, it was found that 94% of the students felt that the activities were worthwhile. In addition, 73% of the students wanted these types of activities to be incorporated into future science courses, and 67% of the students responded favorably to working in cooperative groups.

These are but a few examples of reformed courses at the undergraduate level, led by individual reformers in selected departments on college campuses. Based upon suggested recommendations for improvement in undergraduate education (NSF, 1996b), universities and colleges are falling short of these expectations. Even though the efforts to improve science and engineering courses have worked in selected courses, the reform lacks breadth. In studies such as the one by Frey (1997), students report that they want interesting and interactive activities to be incorporated into their courses. The research, however, fails to give clear examples where changes have been implemented throughout entire programs of study.

AN INVESTIGATION IN DEVELOPING AND TEACHING REFORMED UNDERGRADUATE SCIENCE COURSE IN THE COLLEGE OF ENGINEERING

The Course Design Process

Because typical elementary majors and non-science majors usually opt for biological sciences and earth sciences for their natural science electives, at the University of Alabama faculty in the College of Engineering and the College of Education decided to collaborate and create a new space science/physical sciences course, Aerospace Science for Educators, for preservice education students. Few students in education take physics or astronomy courses, and the physical sciences tend to be the weakest areas of science within the sciences in elementary classrooms. To combat the problem, in 1992, a science educator, Dennis Sunal, and two aerospace engineers, L. Michael Freeman and Kevin Whitaker, responded to a National Aeronautics and Space Administration Request for Proposals for preservice education courses. Receiving funding, the faculty set out to create an Aerospace Science for Educators course, which would satisfy a natural sciences requirement for preservice educators, particularly at the elementary and middle school level. The collaboration created a unique opportunity for the faculty members to share ideas about content and pedagogy.

Using CLT as the pedagogical framework for the course, the faculty began their course planning with a list of 15 topics to cover within the one-semester course. As laboratory activities were added, due to the course having integrated rather than separate laboratory times, the list was finally reduced to just three broad topics, each with its own module: Orientation to Space, Force and Motion, and Atmospheres. The number of key concepts covered in the course modules allow focus on depth over breadth. Sample concepts in the orientation to space module include sky observation methods and instruments; basic analytical geometry concepts; coordinate systems for locating objects on the earth, in the solar system, and the celestial sphere; orbital motion and observing artificial satellites; properties and relationships of bodies in the solar system; procedures for interplanetary navigation; earth's orientation in space; and location predication of bodies in the solar system. In the module Force and Motion, sample concepts developed involve Newton's laws of motion; physics of propulsion systems; the principles of flight design in atmospheres including Bernoulli's Principle; flight in space including minimum energy orbits; and measurement and mathematical analysis involving units of pressure, force, centripetal force, velocity, and acceleration. In the Atmospheres module, climate and weather are developed in relationship to the other course modules. Atmospheres are a natural fit with topics in other modules including properties and structure of atmospheres, atmospheric pressure systems, climate, measuring and forecasting weather conditions, properties of weather effects, and monitoring (remote sensing) Earth's weather and surface properties and conditions from space. The content of the course was aligned with the Benchmarks for Scientific Literacy (Project 2061) and with NASA's Strategic Enterprises (see Figure 28.1). There is no single textbook for the course; instead, readings are obtained from sources such as instructor-designed readings, journal articles, and other books. All activities and readings are found on the course Web site of the Alabama Virtual Campus.

To more closely integrate lecture topics and activities, the course meets 3 days per week for 2-hour blocks of time. The 4-semester-hour course integrates the laboratory into the lecture portion allowing the students to complete hands-on activities that relate to current topics while having the course instructor, not teaching assistants, available to answer questions. For instance, in the Orientation to Space module, students are required to attend three of five nighttime observation sessions, in which they must con-

| NASA Strategic Goals | AEM 120 Course Content | Benchmarks for Science Literacy, Project 2061 |

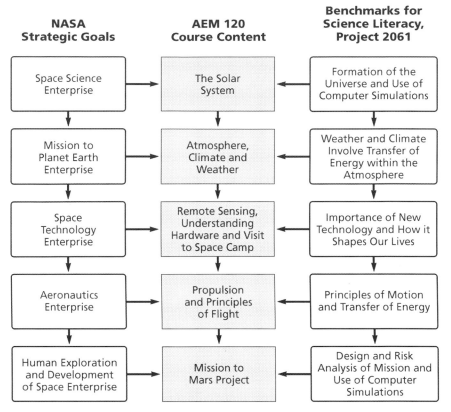

Figure 28.1.   NASA's Strategic Enterprises.

duct their own research (completed through the use of computer programs) on determining the positions of all of the major bodies within the solar system (Sunal, 1984). This research is completed by constructing a three-dimensional model of the solar system oriented to local stars in nearby space. This model, the Model of Nearby Space, allows prediction of events in the solar system and planning for trips between planets in the solar system. While this activity takes about 1 week in class to complete, the students use the model in nighttime observation sessions from that point forward.

A major semester-long project, Mission to Mars, involves planning the science components for a trip from Earth to Mars. Students use astronomy software and mathematical analysis to determine the date and time of launch from and to Earth with only the angle of separation of the two planets, and the number of days of the trip being given (changed every semester). This planning includes launch from Earth, interplanetary navigation, landing on Mars, and outlining a significant research program from a Martian base camp. The instructor takes the role of a facilitator and monitor, answering questions and guiding students during these sessions.

The course learning activities in each of the modules are based upon the use of the learning cycle. (See Chapter 5 for additional information on the use of the learning cycle in undergraduate science.) For the pedagogy used within the course, instead of reliance upon the traditional lecture as the primary mode of instruction in the course (Sunal, 1998), interactive experiences and related abstractions are connected with prior knowledge through the use of numerous carefully designed active learning activities. A common version of the learning cycle involves the use of three phases, which are exploration, invention, and expansion (Sunal & Sunal 2003; see Figure 28.2). During the exploration phase, students use activities to elicit prior knowledge and to ask questions. Discrepant events are used to gain attention of the students and produce questions (Appleton, 1997; Chiappetta, 1997; Sunal & Sunal, 2003). These activities provide the concrete experiences to which Clough and Kauffman (1999) refer.

The invention phase involves students in debating, looking for information, and discussing the activities completed during the exploration phase in an effort to learn more about a specific concept. The expansion phase gives students another opportunity to actively engage in applying and transferring the ideas just learned through the use of different problem-oriented activities. Throughout the learning cycle, the teacher serves as a facilitator in the learning process by using student questions to guide the lesson (Yager, 1991).

The course concepts are taught using the learning cycle, and are arranged as a series of activities that engage the learner during the course. Throughout the course, cooperative groups are used so that the students are interactive participants in the learning process. Cooperative groups are

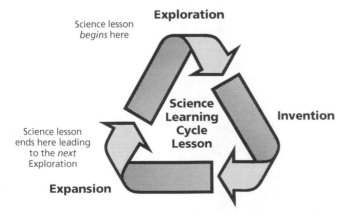

Figure 28.2.   The learning cycle.

also used for the semester-long guided design project, Mission to Mars. In this project, the students must cooperatively plan an entire mission to Mars from start to finish. The course offers numerous laboratory and field experiences to develop and apply basic concepts including numerous day and night sky observing sessions, visit to an astronomical observatory and viewing objects using real-time digital technology, daily use of computer simulations and Internet Web resources, a field trip to the U.S. Space and Rocket Center and the NASA research laboratories at Marshall Space Flight Center in Huntsville, Alabama. A field trip to the local airport allows students to get a close-up and hands-on look at several airplane and wing designs. Students also plan and conduct research in low-speed wind tunnels, and measure specific impulse of model rocket engines in the College of Engineering laboratories.

Student assessment is different than for typical college science courses, with students being graded on more than just a few tests and a final examination. The students in Aerospace Engineering and Mechanics (AEM) 120, Aerospace Science for Educators, are graded on an electronic portfolio that includes evidence from the Mission to Mars group project, three regular examinations, a final examination, work completed at all field trips and out-of-classroom laboratory sessions, and in-classroom individual and group activities. The open-book examinations do not contain the typical lower-level, factual, multiple-choice questions. The questions require the students to use analysis, evaluation, and synthesis skills, as well as mathematical relationships to correctly answer each question. For instance, on the Orientation to Space module examination, students are required to use their Model of Nearby Space, and a Star and Planet locator to determine the location of the sun on the ecliptic at various times throughout the year. They also have to determine the principal planets and constellations visible in an evening sky from a location on another planet in the solar system.

In its 12th year, the AEM 120 course has been taught to more than 950 preservice teachers and non-science majors. Throughout the course, the instructors have been from the fields of aerospace engineering and science education. The course has always been team taught with a major instructor assisted by other faculty. For the past 5 years, a science education faculty member has had an appointment in aerospace engineering for the sole purpose of teaching this course. The engineering faculty regularly participate in the course, either as principal instructor or assistant to another faculty member.

Today, the course is institutionalized: It is a science elective in the university course listings for education and non-science majors, and it is currently a required science course for all elementary preservice teachers. The maximum number of students, 45, have generally enrolled for the course during the past 7 years, with students, particularly those outside of education, electing to place their names on the course waiting list. The course is offered during the fall and spring semesters each year.

Participants and Data Collection

To determine impact of the course, a section was selected for study during one semester. The researchers wanted to determine content knowledge gains and efficacy changes of participants as a result of taking the course. The participants in this course section were mostly freshmen and sophomore elementary education majors ($N = 32$) who were taking AEM 120 as 4 of their 12 credit hours of science electives. Only one participant was a secondary education major. The study involved collection of data from multiple sources. Data were collected using (a) pre- and post-instruction researcher developed content tests; (b) "draw-and-explain" pre- and posttests; (c) pre- and posttests using the Science Teaching Efficacy Belief Inventory (STEBI Form B) (Enochs & Riggs, 1990); (d) informal student journals collected at the end of the course; and (e) post-course general survey forms. The pre- and post-instruction multiple-choice test contained 28 questions about the solar system, phases of the moon, earth-space relationships, and principles of flight. The tests were created to correlate with the course content, and were used to determine content knowledge. A limitation of this test, however, is the more factual level of the questions. Samples of the questions include the following:

Example: If the average distance between the Earth and the Sun were increased, which would most likely result?

    (a) The length of the Earth's year would be shorter.
    (b) The observed (apparent) size of the Sun would increase.
    (c) The amount of solar energy received on the Earth would decrease.
    (d) The period of rotation of the Earth would increase.

The draw-and-explain pre- and posttests prompted students to make a drawing of their ideas about various concepts and then to write a description of what they drew. For this study, the researchers chose to look at two major concepts within the course: (a) the cause of the seasons and (b) the solar system.

The Enochs and Riggs (1990) Science Teaching Efficacy Belief Instrument (STEBI Form B) was also used for this study. Efficacy, as a measure of the extent to which teachers believe that they can affect pupil performance in the classroom (Piagge & Marso, 1994), can be an important tool in preparing preservice teachers to teach science. Efficacy may be related to teaching behaviors and pupil performance. Gibson and Dembo (1984, p. 570) suggest that:

> Teachers who believe student learning can be influenced by effective teaching (outcome expectancy beliefs) and who also have confidence in their own teaching abilities (self-efficacy beliefs) should persist longer, provide a greater academic focus in the classroom, and exhibit different types of feedback than teachers who have lower expectations concerning their ability to influence student learning.

These beliefs can have a profound impact on the professional future of preservice teachers. Teachers with low self-efficacy and low outcome expectancy beliefs are (a) more likely to shy away from teaching science, (b) more likely to have negative attitudes toward science as a discipline, and (c) less likely to foster a positive attitude toward science within their own students once they become teachers. Because teachers teach as they have been taught (Yager, 1991), it is important for college educators to use effective teaching methods to teach content science, thus improving the outcome expectancy efficacy and self-efficacy of the preservice teachers.

The STEBI Form B (Enochs & Riggs, 1990) was created specifically for use with preservice teachers to measure personal and teaching efficacy. There were 23 questions on the instrument. Participants rate each question on a Likert-type scale ranging from strong agreement to strong disagreement. Points are given for positive-type questions, as indicated in Table 28.1, with 1 point being given for strong agreement and 5 points for strong disagreement. Negative-type questions are scored in reverse. Because negatively worded statements are graded in reverse, it is best for scores to decrease, therefore showing an increase in either teaching or personal efficacy. The questions are included in Table 28.1.

Table 28.1.   Science Teaching Efficacy Belief Instrument Sample Statements with Scoring Guide (Enochs & Riggs, 1990)

Statement Type (Efficacy)	Statement	Positive or Negative
Personal	I will continually find better ways to teach science	Positive
	Even if I try very hard, I will not teach science as well as I will most subjects.	Negative
	I know the steps necessary to teach science concepts effectively.	Positive
	I will not be very effective in monitoring my students science experiments.	Negative
	I will generally teach science ineffectively.	Negative

Table 28.1. Science Teaching Efficacy Belief Instrument Sample Statements with Scoring Guide (Enochs & Riggs, 1990)

Statement Type (Efficacy)	Statement	Positive or Negative
	I understand science concepts well enough to be effective in teaching elementary science.	Positive
	I will find it difficult to explain to students why science experiments work.	Negative
	I will typically be able to answer students' science questions.	Positive
	I wonder if I will have the necessary skills to teach science.	Negative
	Given a choice, I will not invite the principal to evaluate my science teaching.	Negative
	When a student has difficulty understanding a science concept, I will usually be at a loss as to how to help the student understand it better.	Negative
	When teaching science, I will usually welcome student questions.	Positive
	I do not know what to do to turn students on to science.	Negative
Teaching Outcome Expectancy Scale	When a student does better than usual in science, it is often because the teacher exerted a little extra effort.	Positive
	When the science grades of students improve, it is often due to their teacher having found a more effective teaching approach.	Positive
	If students are underachieving in science, it is most likely due to ineffective science teaching.	Positive
	The inadequacy of a student's science background can be overcome by good teaching.	Positive
	The low science achievement of some students cannot generally be blamed on their teachers.	Negative
	When a low-achieving child progresses in science, it is usually due to extra attention given by the teacher.	Positive
	Increased effort in science teaching produces little change in some students' science achievement.	Negative
	The teacher is generally responsible for the achievement of students in science.	Positive
	Students' achievement in science is directly related to their teacher's effectiveness in science teaching.	Positive
	If parents comment that their child is showing more interest in science at school, it is probably due to the performance of the child's teacher.	Positive

For personal efficacy, the highest score possible is 65. The teaching efficacy score may be as high as 50. Results of a study by Enochs and Riggs (1990) suggested that the STEBI was a valid and reliable instrument. Reliability analysis produced an alpha coefficient of 0.90 on the Personal Science Teaching Efficacy Scale (self-efficacy). The alpha coefficient of the Science Teaching Outcome Expectancy Scale (teaching efficacy) was 0.76.

In a study by Ramsey-Gassert, Shroyer, and Staver (1996), 23 elementary teachers were given the STEBI Form A (for in-service teachers) to complete as a part of a professional development program. Findings suggest that in-service teachers with higher personal efficacy (PE) had more successful preservice teacher education training, were more involved in more professional development, and included more inquiry science-related experiments in their own classrooms (p. 304). These teachers were able to take negative early experiences in their own education and turn them into ideas about how not to teach science themselves. The authors also found that low personal efficacy was related to feelings of less competence to teach science. Lower personal efficacy was also linked to less independence and lower intrinsic motivation to teach science. In this same study, scores on the science teaching efficacy (TE) measure showed that teachers with the lowest teaching efficacy had taken fewer science courses, or the courses were more poorly taught. The negative experiences led these teachers to have a decreased interest in science, and even tended to foster negative attitudes toward science as a teaching discipline.

Findings

On the pre- and post-instruction course content test the mean of the pretest was 12.1, and the mean of the posttest was 14.5. The difference in the scores was found to be statistically significant ($p < .01$), but the small gains in number of correct questions is not practically significant. From the research, we can conclude that the students showed moderate improvement on the content-based measure.

On the STEBI, the teaching outcome expectancy efficacy measure showed a mean of 25.4 on the pretest, and a mean of 25.1 on the posttest. The two-tailed t test revealed no significant difference between the pre- and post-measures ($p = .61$). Of the 32 students participating in this study, however, 17 students had scores that decreased, which means that the students had higher outcome expectancy efficacy after the course than before. The AEM 120 students did not perceive that they had more effective teaching skills to help children to learn science following the class.

On the personal (self) efficacy measure, the pretest mean was 38.0. The posttest mean was 33.6. The two-tailed t test showed a highly significant difference between the two measures ($p < .01$). Of the 32 students, 25 had scores that decreased. This indicates that the students became more interested and comfortable with hands-on approaches, and were more interested in science than before taking the course. Teachers with a higher level of personal efficacy (lower scores), as reported in several other studies, tend to spend more time teaching science once they become teachers. They also become resourceful in acquiring science teaching materials to use with a wide variety of activities within their classrooms. The AEM 120 students did perceive that they were less anxious and felt more comfortable in understanding science following the class.

It was not anticipated that efficacy scores would be lower, indicating higher efficacy levels. This was a science course, and not an education course. The score decreased when the students understood the science more deeply and felt more comfortable with science but were not provided with knowledge or skills to create meaningful learning in this area with younger students. The course instruction modeled the learning cycle, but information about how to teach in this manner was not explicitly discussed as a part of this course.

Consistent with the results of the study by Ramsey-Gassert et al. (1996), low scores on personal efficacy did not necessarily correlate with low scores on teaching outcome expectancy efficacy on either pre- or post-course measures. Ramsey-Gassert et al. suggest that the lack of correlation between scores indicates little relationship between teachers' belief that they can teach science with the belief that they can affect change in the students.

The comments in journal reflections and the post-course survey forms indicated that many of the students echoed the Ramsey-Gassert et al. findings in their reflections. When asked on the survey form if they feel more confident in their knowledge of science concepts covered in the course, most students answered positively. Five students answered negatively with no elaboration and two responded with such comments as, "I needed more direct instruction and lecture," and "I totally didn't get the format of this course." These numbers reflect the personal efficacy scores reported earlier. Seven of the students had scores that either increased or remained the same (lower efficacy). These students did not feel comfortable with the format of the course, or did not feel more confident in knowledge or interest in science after taking this course. This was a common issue with many students, as the overwhelming majority of students wrote in their reflection journals that they were experiencing discomfort at the beginning of the course with respect to course format. Many students said that they were not used to "thinking this way" and having to figure things out for themselves. This discomfort is common with students not familiar with the learning cycle model (Sunal & Sunal, 2003). Interestingly, the complaints subsided as the students got used to the new teaching and learning format. The students had some difficulty getting used to working in groups, and most mentioned enjoying the interaction with peers and not having long lectures in the class. Several students, including the secondary science education student, said that working in cooperative groups was their favorite part of the course. Even one of the non-education majors mentioned that she had never learned as much in a science class before.

When asked if they would feel more confident teaching about these subjects as a result of this course, more than half of the students answered positively, but many did not elaborate. Those who did, responded with such comments as, "I really enjoyed working in groups and learning about these topics. I can't believe I didn't know any of this stuff. I would definitely feel more comfortable teaching about it now."

Results on the draw-and-explain pre- and posttests show significant change from the pretest in major elements displayed, and describe accurately for two of the courses key concepts, the cause of the seasons, and the solar system. While major misconceptions were displayed by all students on the pretest, few were recorded on the posttest.

IMPLICATIONS AND CONCLUSIONS

When developing and teaching courses that are radically different from those offered before on a university campus, barriers will have to be overcome. In the case of creating the AEM 120 course, the faculty were all concerned with the current state of science in K–12 schools, searching for ways to improve science content knowledge of preservice teachers, and they worked together to make it happen. The science educator supplied many of the course materials that he had previously published in educational journals (Sunal, 1984) and used in high school and college courses on other campuses. The engineering faculty supplied scientific and technical information and ideas for modules that would expand student knowledge on various course concepts. The members of the team met on a regular basis to plan and monitor the course, team taught the course, and knew the system through which they were working well enough to solve many of the barriers and limitations that occurred (Sunal et al., 2001).

The pedagogy used in this course had a positive effect on student achievement and efficacy in this course. While difficult for students to get used to, as reflected in students' journal reflections and post-course survey forms, the discomfort that the students experienced in the beginning of the course did not last and eventually disappeared for most students. This study demonstrated that by improving instructional techniques and making the course content more relevant and engaging, the students improved content knowledge and personal science efficacy. As a result, these preservice elementary teachers will be more likely to teach science effectively and engage their own students in hands-on science activities that stress content application. By using efficacy instruments with pre- and posttests on content knowledge acquisition, we can more effectively measure how well we are teaching content, while determining if our techniques are helping preservice teachers become more confident in their scientific knowledge and ability to teach.

REFERENCES

American Association for the Advancement of Science. (1989). *Project 2061: Science for all Americans.* Washington, DC: Author.

Appleton, K. (1997). Analysis and description of students' learning during science classes using a constructivist-based model. *Journal of Research in Science Teaching, 34,* 303–318.

Bazler, J., & Charles, M. (1993). Project 2061: A university model. *Journal of Science Teacher Education, 4*(4), 104–108.

Chiappetta, E.L. (1997). Inquiry-based science: Strategies and techniques for encouraging inquiry in the classroom. *The Science Teacher, 65*(2), 22–26.

Clough, M.P., & Kauffman, K.J. (1999). Improving engineering education: A research-based framework for teaching. *Journal of Engineering Education, 88,* 527–534.

Enochs, L.G., & Riggs, I.M. (1990). Further development of an elementary science teaching efficacy instrument: A preservice elementary scale. *School Science and Mathematics, 90,* 694–705.

Frey, J.T. (1997). Homestudy assignments: An experiment in promoting active learning in introductory chemistry. *Journal of College Science Teaching, 26,* 281–282.

Gibson, S., & Dembo, M.H. (1984). Teacher efficacy: A construct validation. *Journal of Educational Psychology, 76,* 569–582.

Gregory, E. (1992). Science literacy enhancement for nonscience majors—The Rollins College single-term experiment. *Journal of College Science Teaching, 21,* 223–225.

King, A. (1993). From sage on the stage to guide on the side. *College Teaching, 41*(1), 30–35.

McIntosh, W.J. (1996). The national science education standards as a referent to course revision. *Journal of College Science Teaching, 25,* 442–443.

National Center for History in the Schools. (1996). *National standards for United States history: Exploring the American experience.* Los Angeles: Author.

National Commission on Excellence in Education. (1983). *A nation at risk: The imperative for educational reform* (Stock No. 065-000-00177-2). Washington, DC: U.S. Government Printing Office.

National Council for the Social Studies. (1994). *How do we achieve excellence in social studies?* Retrieved August 7, 2003, from http://www.ncss.org/standards/1.2.html

National Council of Teachers of English and International Reading Association. (1996). *Standards for the English language arts.* Urbana, IL: National Council of Teachers of English.

National Council of Teachers of Mathematics. (1989). *Curriculum and evaluation standards for school mathematics.* Reston, VA: Author.

National Research Council. (1996). *National Science Education Standards.* Washington, DC: National Academy Press.

National Science Foundation (1996a). *Shaping the future: New expectations for undergraduate education in science, mathematics, engineering, and technology* (NSF Publication No. 96-139). Washington, DC: Author.

National Science Foundation. (1996b). *Shaping the future: Strategies for undergraduate education: Proceedings from the National Working Conference* (NSF Publication No. 98-73). Washington, DC: Author.

Olson, L. (1994, November 2). Learning their lessons. *Education Week,* pp. 43–46.

Piagge, F.L., & Marso, R.N. (1994). Outstanding teachers sense of teacher efficacy at four stages of career development. *The Teacher Educator, 20*(4), 35–42.

Ramsey-Gassert, L., Shroyer, G.M., & Staver, J.R. (1996). A qualitative study of factors influencing science teaching efficacy of elementary level teachers. *Science Education, 80,* 283–308.

Sunal, D.W. (1984). Star gazing—Basic activities for middle school astronomy, *Science Scope, 7*(3), 16–17.

Sunal, D.W. (1998). Innovative learning strategies. In D.W. Sunal (Ed.), *Project NOVA workshop manual.* Tuscaloosa, AL: Ferguson Center Press.

Sunal, D.W., & Sunal, C.S., (2003). *Teaching elementary and middle school science.* Columbus, OH: Merrill Prentice Hall.

Sunal, D.W., Hodges, J., Sunal, C.S., Whitaker, K., Freeman, L.M., Edwards, L., et al. (2001). Teaching science in higher education: Faculty professional development and barriers to change. *School Science and Mathematics, 101,* 246–257.

Weld, J.D., Rogers, C.M., & Heard, S.B. (1999). Semester-length field investigations in undergraduate animal behavior and ecology courses. *Journal of College Science Teaching, 28,* 340–344.

Yager, R. (1991). The constructivist learning model: Towards real reform in science education. *The Science Teacher, 58*(6), 52–57.

CHAPTER 29

A MODEL FOR REFORM IN TEACHING ENGINEERING AND TECHNOLOGY

With a Focus on Prospective Elementary Teachers

William Jordan, Bill Elmore, and C. W. Sundberg

ABSTRACT

We now live in a world where our interaction with high technology has become commonplace. However, many people have little understanding of the vast number of technology-based devices in everyday use. It is in the best interest of society in general and the engineering community in particular to improve the technology literacy of the general population. The future health of science and engineering fields critically depends upon improving students' mathematics and science preparation and problem-solving skills accompanied by a heightened understanding of the role technology plays in everyday life. In particular, the dilemma of "technology literacy" is evident in K–12 science education when noting the struggle many teachers face as they attempt to relate science concepts to real-world experiences. To address this

Reform in Undergraduate Science Teaching for the 21st Century, pages 511–524
Copyright © 2004 by Information Age Publishing
All rights of reproduction in any form reserved.

issue, we created a new 3-hour course in engineering problem solving specifically designed for education majors—Engineering Problem Solving for Future Teachers, co-taught by one science education professor and two engineering professors. In this course, preservice teachers are immersed in the integration of science and engineering principles applied to solving real-world problems. Techniques for improving pedagogy are included to strengthen their interest in science subject matter and to improve their ability to transfer this interest to their own future students.

INTRODUCTION

Using the theme "Our Material World," the authors integrate concepts involving the physical, mechanical, and chemical behavior of materials as a means to teach engineering problem-solving skills. Through the use of frequent laboratory exercises, our goal is to heighten the interest and skill level of preservice teachers to effectively communicate math and science principles to K–12 students within the context of everyday problem solving.

A key component of this course is direct outreach into the K–12 community. As part of the class, our students direct workshops for in-service teachers in our region, demonstrating the hands-on science skills that they have learned. Students also make presentations using simple experiments in local fourth-grade science classes. This contextualizes what they have learned and models new teaching strategies for in-service teachers. Elementary school children are impacted with an enlarged perspective toward materials science and engineering.

It was our hypothesis that a course designed to capitalize upon students' natural curiosity would enable them to make positive attitudinal gains toward science and engineering problem solving. Mean scores from a pre and post Survey of Attitudes About Problem Solving in Engineering for Teachers revealed a significant positive attitude shift on 17 of the 20 test items dealing with perceptions of the field of engineering, problem-solving ability, and ability to teach problem solving in the classroom within an engineering context. Students clearly gained an appreciation for the field of engineering along with a newfound confidence in their ability to solve real-life problems having to do with engineering. Qualitative gains in these areas were also observed from their daily journal writings.

RATIONALE FOR COURSE

One challenge in K–12 science education is relating real-world experiences to fundamental principles and problem-solving techniques. To address this issue, we adapted what we learned in developing our freshman engineering course sequence to create a new 3-hour course in engineering problem solving specifically designed for education majors. In this course we model innovative teaching techniques within the context of introducing fundamental principles and problem-solving techniques in mathematics, science, engineering, and technology.

LITERATURE REVIEW

Publications from the National Science Foundation (1993), National Science Teachers Association (NSTA, 1992, 1993), and the American Association for the Advancement of Science (1993), have called for certain curriculum strategies that go right to the heart of the teaching and learning experience. Anderson (1995) synthesized the perspective and recommendations of these publications by national science education organizations as (1) integrating themes in subject matter, (2) teaching for understanding, (3) making connections between subject matter and its applications, and (4) reaching all students—not just the elite—with rigorous content and attention to critical thinking. Thus, we believe the methodologies used in teaching this course are well founded.

Yager (1991) stipulated that most constructivist teachers promote group learning, where students in small groups discuss approaches to a given problem and work together to solve problems. He noted that students learn best in an environment that combines experimentation, dialogue with other students, and discussion with the teacher.

In each of our lessons we use small-group problem-solving activities, followed by discussions of the results, and the learning processes involved. Most lessons included the five essential elements to a constructivist lesson according to Zahorik (1995): (a) activating the prior knowledge of the learner, (b) having the learner acquire knowledge through direct interaction with materials and other learners, (c) conceptually developing the acquired knowledge, (d) applying the new knowledge, and (e) reflecting on the new knowledge. In terms of small group activities McIntosh (1995, p. 48) stated:

> We need to give students the opportunity to practice problem solving that is more realistic and requires them do to more of the work. I think this type of activity combined with meaningful post-lab discussion about what happened, what thinking processes were used, and what skills the students need to practice is a good way to give students good problem-solving experience.

While many of these reform concepts are common in education circles, they have only recently made a significant impact upon engineering education. Felder, one of the leading proponents of engineering education reform, has published a number of important articles that have influenced the engineering authors of this chapter. He has been writing on these subjects for many years, but many engineering faculties are only recently becoming aware of his work. Felder (1996) reported on the need for engineering faculty to adapt their teaching to the learning styles of their students, which he described in an earlier paper (Felder & Silverman, 1988). Felder, Rugarcia, Woods, and Stice (2000) noted that engineering faculty need to move from the lecture mode that has been common for more than a century; they described a number of teaching techniques that are more effective in communicating engineering, including strategies that many would consider active or collaborative learning (Felder, Woods, Stice, & Rugarcia, 2000).

IMPACT OF THE COURSE ENGINEERING PROBLEM SOLVING FOR FUTURE TEACHERS

The primary impetus for the research of the impact of our course was a response to the apparent need for future teachers to have a more extensive understanding about our technological world. Our response to this need was to create a new class whose title is Engineering Problem Solving for Future Teachers.

We initiated the process of "engineering problem solving" with laboratory-based activities by first forming teams. This was done to develop an early sense of collegiality among the preservice teachers and to provide a supportive framework for their entrance into potentially unfamiliar territory of problem solving from an engineering standpoint. Team formation was accompanied by a strong commitment to regular "teaming" activities providing ample opportunities for students to literally put their "hands to the task" of experimenting with the new concepts to be learned. Mixed with a lively interaction among the faculty members (and the students themselves) this quickly broke down many barriers to students' actively and cooperatively learning new concepts.

The course was taught in a cooperative learning environment, integrating numerous hands-on activities with brief lectures coordinated to provide "just-in-time" information for current team activities. By *doing* rather than merely observing, students engaged in constructivist instructional techniques. We designed the course with four specific goals in mind: (a) improving science and engineering problem-solving skills, (b) modeling effective teaching methods, (c) providing opportunities for students to create their own problem-solving strategies and modules and practice communicating them to others, and (d) extending outreach into the K–12 community. Outreach was obtained through (a) workshops for in-service teachers taught by our students, (b) presentations by our students in actual elementary classrooms, (c) a workshop on campus for engineering, science, and education faculty in our region, and (d) presentations at regional and national engineering or education conferences.

Problem Solving in Engineering Science for Teachers follows the guidelines set forth in the NSTA position statement on science teacher preparation standards (1993). While focusing on understanding and developing the major concepts and principles of properties of matter, it helps students conceptualize the inter-connectedness of the sciences, mathematics, engineering, and technology. Students relate the study of matter and materials to contemporary, historical, technological, and societal issues. Students are able to locate appropriate resources, design and conduct inquiry-based open-ended investigations in science, interpret findings, communicate results, and make judgments based on evidence.

Students learn about the atomic structure of metals, ceramics, polymers, and composite materials. They learn how internal structure affects the physical and chemical properties of the materials and how this internal structure affects the mechanical properties of the materials. Specific examples that are used include materials-related problems with space transportation systems and with the International Space Station.

Experiential learning was an important component of the course. The need for this is described in the National Science Education Standards (National Research Council, 1996): "Conducting scientific inquiry requires that students have easy, equitable, and frequent opportunities to use a wide range of equipment, materials, supplies, and other resources for experimentation and direct investigation of phenomena" (p. 220).

During the development of the course considerable care was used in the planning of instruction, selection of instructional materials, and evaluation of practices suitable for teaching elementary and secondary school students. Methods for teaching science, mathematics and engineering content to elementary and secondary students were evaluated for appropriateness and strengths, and limitations of a variety of teaching methods were considered. Methodologies included lecture, small-group activities, whole-group activities, individual participation, reflective writing, alternative assessments, cooperative learning, demonstrations, and technology-based assignments.

A significant aspect of the course is the extensive involvement of the preservice teachers with experimental work. Our goal is to introduce preservice teachers to principles, applications, and technologies that can readily be implemented in their future classrooms. Through these experiments, students practiced skills such as data measurement and analysis, graphing or tabulation, and fundamental statistics. Details of this research has previously been reported (Jordan & Elmore, 2002; Jordan, Elmore, & Silver, 2000b, 2001a; Jordan, Silver, & Elmore, 2001).

Experimental work is accomplished, in part, by the use of the Calculator Based Laboratory (CBL), an integrated system of measurement sensors or probes, a data acquisition unit, and a graphing calculator. The system was selected because of the modest cost, a fast "startup" for new users, and wide variety of options for measuring physical, chemical, and biological parameters. Experiments using the CBLs during the course included parameters such as temperature, pressure, light, voltage, pH, conductivity, and calorimetry.

The course content focused on strengthening the links between verbal descriptions, numerical data acquisition, and graphical representation as means of understanding physical phenomena. For example, one critical element in scientific inquiry is the fundamental understanding of rates of change in systems. The CBL system provided a clear approach to setting up an experiment for evaluating a process change with time (e.g., pH or temperature), collecting data at a desired frequency and range, and representing the process change graphically. Each of these aspects of experimentation and analysis was presented to the preservice teachers in a simple, hands-on approach to provide them with training and skills development.

Additionally, course activities included the use of video probes, computers, the Internet (including NASA's Strategic Enterprises resources), telecommunication technology, and software designed to enhance learning and presentation skills. Eventually, course designers plan to incorporate the compressed video program available on campus to offer this course and its outreach programs to classroom teachers and university faculties in other geographic locations.

Some of the laboratory experiences are described below:

1. Determination of rate of temperature change in aluminum, glass, and polymeric soft drink containers as they cool down when placed in an ice chest.

2. Modeling the crystal structure of metals by making models of unit cells using Styrofoam balls and wooden sticks.

3. Examining structural stability by making and testing structures made out of plastic straws and masking tape.

4. Assessing temperature effects on metal behavior by doing Charpy impact tests on aluminum and steel samples at room temperature and the temperature of liquid nitrogen (−320°F). Charpy impact tests are the most common test engineers use to measure the fracture toughness of metallic materials.

5. Evaluating the response of aluminum and steel samples to tensile tests.

6. Studying fatigue and elementary statistics by fatiguing several types of paper clips until they fail.
7. Observing viscoelastic behavior of materials by making a "silly putty" type material.
8. Measuring diffusion rates of different dyes onto filter paper.
9. Examining rates of change in pH, temperature, and humidity in specifically designed experiments.

The details of several of our laboratory exercises serve to further illustrate the experiences developed in the course.

Structural Stability Laboratory

Preservice teachers were introduced to the concept of stability through the design and building of a structure composed of plastic straws and masking tape. Their task was to build the tallest structure possible within a given time period (about 1 hour). The winning team was the one with the tallest structure that could hold up a soccer ball. When the structures failed at loads far below their actual strength levels, the students learned that strength is not the only issue involved in designing structures. The purpose of the experiment was to introduce the concept of stability, and how it differs from strength.

Fatigue Laboratory

The concept of fatigue was illustrated by having each student conduct repeated bending of two different sizes of paper clips. The experiment illustrated some of the problems involved in interpreting fatigue data. In particular, the wide range of results was used in two ways. We examined the results and discussed which lifetime we should use: the minimum lifetime or some statistical type of average. The minimum value would require too much material, for it greatly underestimates what the actual lifetime might be. On the other hand, using only a mean value of lifetime means that half of your parts will fail before expected.

The fatigue experiment was used to introduce the preservice teachers to several aspects of statistics. They determined mean, median, mode, and standard deviation of the lifetimes. One of the paper clip types had a smaller standard deviation on its lifetime and this was used to illustrate that its data was more uniform.

Since the first offering of the course (spring 2000), most of the class material has been placed on the Web (Jordan, 2003). This has enhanced student access to ideas and concepts. With further development, this can be used as the basis to provide the course to in-service teachers at remote locations.

Student-Led Workshops

One of the goals of the course was to create a workshop for area teachers that would be led by our students. All students participated in a half-day in-service workshop. Our students led small groups of in-service teachers through problem solving activities that they in turn could use in their classrooms. The preservice teachers modeled the same pedagogical techniques that they had experienced in their class.

Participating in-service teachers were delighted with the results. A few of the typical comments made on their evaluations of the workshop follows:

- This was a wonderful experience! The students have demonstrated their learning in very interesting ways. Many activities will be extremely useful and fun to use in my classroom. I know my students will love each activity. These are very useful both to demonstrate scientific principles and to use as activities when students want to do FUN activities.
- The students did a wonderful job, and their lessons were demonstrated in a very professional manner. Each one of these students will be a great teacher. PLEASE have more workshops. I would love more opportunities to explore new ideas and to get motivated to use these activities in my classroom.
- This was a wonderful opportunity for the Tech students to build their skills. The presentations were well prepared as were the lesson plans. I would love to see all of the presenters again. The material will be easy to use in the classroom. PLEASE do this again.
- I thought this was an excellent educational opportunity. The students were well prepared; there were a wide variety of activities presented. It gave the students experience and teachers were exposed to new content and methods.

In addition to the in-service teachers workshop described above, our students were required to create a hands-on/minds-on laboratory-teaching lesson that was presented to groups of fourth-grade students at a nearby elementary school. In this setting both the fourth-grade students and their teachers got to observe what we were doing. As part of this exercise, students were required to create complete lesson plans. It is a challenge to create experiments that demonstrate science and the scientific method to students of elementary age. Most of these engineering and science concepts were new to our students.

Examples of the student-created labs included the following:

1. Measuring the effect of small amounts of caffeine on pulse rate by having students drink different liquids and then measure their pulse. Students measured their pulse after drinking a small soft drink. Some drinks had no caffeine; others had differing amounts of caffeine. The effect of caffeine on increasing pulse rate could be clearly seen.

2. Illustrating viscoelasticity by making a Silly Putty-type material and then evaluating it. Our preservice teachers did not use the term viscoelasticity when this was presented to the fourth graders, but still introduced the concept of time-dependent material behavior.

At the outset of the presentation to the fourth-grade classes, many of our preservice teachers were very nervous. However, both the students and their teachers enjoyed the presentations. The following teacher comments were typical: "Students loved the activities and trying out the experiments." "Presenters were very enthusiastic. Students enjoyed the activities and the lesson."

The authors would like to see more engineering faculty incorporate active learning into their classrooms. We would also like to see more engineering faculty become involved with teacher education and outreach to the K–12 community. As a result of this goal, we have made presentations at regional and national conferences (Jordan & Elmore, 2002; Jordan, Elmore & Silver, 2000b, 2001a, 2001b; Jordan, Silver, & Elmore, 2001).

Another goal for the course was to promote reform-based teaching and assessment strategies among preservice teachers by immersing them in instructional techniques that modeled a constructivist approach. The focus was on *doing* science rather than merely acquiring isolated facts of content knowledge. The instructors encouraged the preservice teachers to connect new learning to previous experience, to ask questions, to explore a wide range of possible answers to their own questions, and to construct their own conclusions. Incorporated in this context were certain aspects common to other teaching models: cooperative learning, inductive thinking, nondirective teaching, teaching to multiple intelligences, and efficacy of instruction for all learners.

As was previously stated, our hypothesis was that students in this class would make gains in positive attitudes toward doing and teaching problem solving in engineering if we used a curriculum design that capitalized on the students' natural curiosity.

We have measured the student attitudes toward the content of this course by creating a survey. We have administered the survey at the beginning and the end of the course. There were 20 statements to which the students had to respond. For each statement they were to circle a number on a scale ranging from 1 (*strongly disagree*) to 6 (*strongly agree*). Some of the statements on the survey are shown in the list below:

- Basic engineering concepts are understandable to the average person.
- Engineering professors are a bit threatening to me.
- It takes specialized equipment to teach basic engineering concepts.
- I am strong in computational math.
- I plan to teach problem solving techniques and engineering concepts when I have my own classroom.

On a comparison of mean scores from a pre and post Survey of Attitudes About Problem Solving in Engineering for Teachers, the preservice teachers showed a significant positive attitude shift on 17 of the 20 test items dealing with their perceptions of the field of engineering, their ability to do engineering problem solving, and their ability to teach problem solving in the classroom within an engineering context. Our preservice teachers clearly gained an appreciation for the field of engineering along with a newfound confidence in their ability to solve real-life problems having to do with engineering.

The students changed their opinion on whether or not specialized equipment is needed to teach engineering concepts (decreasing about 0.9 points). This is important, for the students now realize they can teach some basic engineering concepts in the K–12 classroom even if they do not have fancy equipment. One interesting change in their attitudes that we did not expect was a significant decline (more than 1.2 points on the 6-point scale) in their perception of their math computational skills. It is not likely that there was any decrease in their skills, only that they realized they were not as skillful as they had thought once they were exposed to some actual engineering calculations.

Their gains were not only reflected quantitatively on the attitude survey, but also qualitatively in their daily journal writings. Typical entries from the preservice teachers were as follows:

- The things that I like most about this class were the experiments we did. From the silly putty to the sponge [creative thinking] activities.... I also found that I do have a little bit of science knowledge. From discovering that the scientific name for plastics is polymers to actually going to an engineering lab [I am now] able to relate to conversations that my friends are having. Before that I would just listen.

Now I can offer my two cents. After finishing this course I have a new found interest in science related areas.

- I thought it was so much fun learning and doing activities that we can do with our students when we become teachers. I also liked how we learned about chemical engineering and materials engineering and how the understanding of these made the activities not only fun, but also a learning experience!

- I enjoy many things in this class. Learning how to incorporate problem-solving techniques into the classroom was especially interesting. I enjoyed learning about cooperative learning. It provided me with valuable insight to how to incorporate cooperative learning and problem solving techniques.

- Everything we did was "hands on!" It was great. This is the best fun I've had in any class in my entire 4 years in college.

- [My favorite things about this class were] hands-on activities, interaction among students, interaction with teachers, and the comfort level in the classroom which made it so easy to ask questions. The combination of professors was just cool and greatly added to the learning.

- [What I liked about this class was] the cooperative group settings. The "laid back" environment approach in learning some of the different engineering concepts of this course material. I loved the sponge activities in the beginning of the class. They were fun and very relative [sic] for future use in our classrooms. Great motivators! I enjoyed the enthusiasm, the "down to earth" approach, and the "serious side." All three combined instructors made this course a SUCCESS! It was great, and I recommend this course to all education majors!

SUMMARY

We created an innovative course that introduced engineering concepts to future teachers. We do this using many active learning strategies. The preservice teachers learned how to do engineering science as well as how to teach it. The preservice teachers who were very apprehensive upon entering the course were able to learn engineering concepts in a non-threatening manner.

This type of project also benefits engineering faculty members who become involved in these activities. There are four significant benefits to engineering faculty members who become involved in activities such as those described in this chapter. First, activities similar to these help increase the technological literacy of our society, which will make it easier for engineers to do the things they really want to do. Secondly, K–12 students' exposure to engineering concepts may encourage some of them to

pursue engineering when they get to college age. This can only be a benefit to our engineering profession. Some faculty may object to the suggestions in this paper, because they have the attitude that this is not real
engineering. We would assert that our class, while not traditional engineering, certainly fits the category of pre-engineering. We are using this class to
present engineering in a way that can excite others to join our profession.

A third major, though indirect, benefit to engineering professors who
begin to do the sorts of things we have described in this paper, is in the
improvement of their teaching. The authors have been exposed to what
are new ways of teaching (at least new to us). The first author has incorporated many of these teaching techniques into several traditional engineering courses such as materials engineering and a combined statics/strength
course. Our engineering students are benefitting today from our more creative teaching styles.

A fourth benefit is the personal and professional growth of the engineering faculty member. We have found our involvement in these activities
to be one of the more rewarding and enjoyable things we have done since
we became professors. We have seen this benefit in other faculty members
who have also become more creative in their profession.

We have reported on what we believe are successful outreach efforts,
presenting engineering concepts to faculty, college students, in-service
teachers, and K–12 students who otherwise would not have learned such
concepts. This has been done at relatively low cost. These sorts of activities
are new to most engineering faculty members. We encourage other engineering faculty members to think outside the box and become involved in
similar activities.

FUTURE DIRECTIONS

One long-term goal of this project would be to follow the future performances of the preservice teachers as they begin their careers as educators,
to see if this course made a difference in the way they teach. While a
daunting task, the information gained from these teachers as they embark
on their careers would be invaluable in our efforts to continually improve
this effort.

AUTHOR NOTE

This work was in part supported by NASA Opportunities for Visionary
Academics (NOVA), a program funded by the National Aeronautics and
Space Administration, although the views expressed here are the authors'

only. We wish to acknowledge additional financial support for this project from the Louisiana Tech University's Center for Entrepreneurship and Information Technology (CEnIT) that has allowed us to continue the course and extend it into the area of engineering design, using our rapid prototyping system.

REFERENCES

American Association for the Advancement of Science. (1993). *Benchmarks for science literacy: Project 2061.* New York: Oxford University Press.

Anderson, R. (1995). Curriculum and reform: Dilemmas and promise. *Phi Delta Kappan, 1,* 33–36.

Felder, R. (1996). Matters of style. *ASEE Prism, 6*(4), 18–23.

Felder, R., & Silverman, L. (1988). Learning and teaching styles in engineering education. *Engineering Education, 78*(7), 674–681.

Felder, R., Rugarcia, A., Woods, D., & Stice, J. (2000). The future of engineering education I. A vision for a new century. *Chemical Engineering Education, 34*(1), 16–25.

Felder, R., Woods, D., Stice, J., & Rugarcia, A. (2000). The future of engineering education II. Teaching methods that work. *Chemical Engineering Education, 34*(1), 26–35.

Jordan, B. (2003, May 6). *Engineering 289c class page.* Retrieved August 6, 2003, from http://www2.latech.edu/~jordan/Nova/index.htm

Jordan, W., & Elmore, B. (2002, March). *Louisiana NOVA 2002 Phase II Report.* Paper presented at NASA Opportunities for Visionary Academics (NOVA) Leadership Development Conference (LDC), Greenbelt, MD.

Jordan, W., Elmore, B., & Silver, D. (2000a, January). *Problem solving in engineering science for teachers.* Paper presented at the NASA Opportunities for Visionary Academics (NOVA) Conference, Orlando, FL.

Jordan, W., Elmore, B., & Silver, D. (2000b, June). *Creating a course in engineering problem solving for future teachers.* Paper presented at the annual meeting of the American Society for Engineering Education, St. Louis, MO.

Jordan, W., Elmore, B., & Silver, D. (2001a, January). *Development of Web-based course material for a course in engineering problem solving for future teachers: NOVA Phase III Progress Report.* Paper presented at NASA Opportunities for Visionary Academics (NOVA) Leadership Development Conference (LDC), Arlington, VA.

Jordan, W., Elmore, B., & Silver, D. (2001b, January). *Report on the creation of a problem solving in engineering science course for future teachers.* Paper presented at NASA Opportunities for Visionary Academics (NOVA) Leadership Development Conference (LDC), Arlington, VA.

Jordan, W., Silver, D., & Elmore, B. (2001, June). *Using laboratories to teach engineering skills to future teachers.* Paper presented at the annual meeting of the American Society for Engineering Education, Albuquerque, NM.

McIntosh, T. (1995). Problem-solving practice: Challenging students to design experiments and organize data. *The Science Teacher, 62*(1), 48–50.

National Research Council. (1996). *National Science Education Standards.* Washington, DC: National Academy Press.

National Science Foundation, Directorate of Education and Human Resources, Division of Elementary, Secondary and Informal Education. (1993). *Science instructional materials: Preschool-high school.* Washington, DC: Author.

National Science Teachers Association. (1992). *Scope, sequence, and coordination of secondary school science: Vol. 2. Relevant research.* Arlington, VA: Author.

National Science Teachers Association. (1993). *Scope, sequence, and coordination of secondary school science: Vol. 1. The content core: A guide for curriculum designers* (Rev. ed.). Arlington, VA: Author.

Yager, R. (1991). The constructivist learning model: Towards real reform in science education. *The Science Teacher, 58*(1), 52–57.

Zahorik, J. (1995). *Constructivist teaching.* Bloomington, IN: Phi Delta Kappa Educational Foundation.

CHAPTER 30

A MODEL FOR REFORM IN TEACHING IN ENGINEERING AND TECHNOLOGY

Artificial Intelligence Systems in Science

Charles L. Karr and Cynthia Szymanski Sunal

ABSTRACT

Faculty from the University of Alabama's Colleges of Engineering and Education joined in a cooperative effort to develop a new interdisciplinary science
course, ESM 130: Artificial Intelligence (AI) Systems in Science. The course
introduces students to artificial intelligence systems in science through the
computer modeling of systems found in nature. The course presents a progression of natural systems, each of which has been modeled on a computer
to develop a mainstream artificial intelligence technique. The students are
involved in modeling the natural systems, using the resulting AI techniques
to solve scientific problems, and discussing published applications of the AI
techniques in various scientific disciplines. The AI techniques presented
include tree diagrams, expert systems, fuzzy logic, neural networks, and
genetic algorithms. Further, the course teaches the "process" of doing science, an approach that is fundamental to scientific literacy. To determine the

Reform in Undergraduate Science Teaching for the 21st Century, pages 525–539

impact of the course design and pedagogy, student feedback was collected in a pretest-posttest strategy. Overall the course achieved its goal to improve confidence, understanding, and transfer of learning in students taking the course. Weaknesses in the overall offering were determined and addressed in subsequent offerings of the course.

INTRODUCTION

As computers become more readily available and used both in schools and in homes, the comfort level and proficiency of most children with computers is expected to increase dramatically in the coming decade. Combine this fact with the exponential growth in research into artificial intelligence, discussions of AI in many home and personal products, computer and Internet games and simulations, and popular television shows and movies such as *Star Trek* and *The Matrix*, and the result will be a generation of inquisitive students looking for answers and guidance into a field that could potentially drive technological advances for decades to come. Unfortunately, few adults are prepared to spark and nurture students' desire to learn about this exciting field of science.

For example, consider a scenario in which a middle school student asks her science teacher, "Ms. Hollingsworth, Data on *Star Trek* is a neural network, what is that?" Based on the college science courses most teachers took or are taking, their response would probably be something like, "Celeste, that is science fiction; they really can't do that." Unfortunately, that response is not at all true; they can do that! A much more accurate (and better) response might be:

> Celeste, neural networks are rough models of the human brain. They attempt to model the characteristics of your brain that allow you to adapt, change, and improve with experience—to learn. Although neural networks have not been developed to the extent that we can build a Data on Star Trek, they have been used to solve a lot of problems from a wide variety of areas, including geology, aerospace engineering, chemistry, biology, and others.

LITERATURE REVIEW: ARTIFICIAL INTELLIGENCE SYSTEMS IN SCIENCE, STUDENTS CONCEPTIONS, AND INQUIRY PEDAGOGY IN HIGHER EDUCATION CURRICULA

A good many college students have encountered basic concepts used in modeling systems in science and engineering through artificial intelligence, although they may not actually "know" or understand these concepts (Thoresen, 1993). Such modeling can help students become aware of the types of scientific problems routinely solved using modeling with artificial intelligence techniques.

Watson, Prieto, and Dillon (1997) contend that conceptual change is marked by a state of transition. Accommodation does not occur across all contexts at the same time. It can be expected that accommodation may occur first with the most familiar events, those with which we have the most experience. Earlier, White (1991) described concepts as continuously developing over many dimensions. This process, in theory, is never ending. Others have noted that conceptual change includes elements such as developing precision in using relevant language, replacing aspects of the old ideas with aspects of the new, incorporation of the new concept, and sometimes retention of aspects of both the old and the new simultaneously (Garnett, Garnett, & Hackling, 1995; Watson et al., 1997). Bliss (1995) has described the effects of context on conceptual change. Since concepts develop over many dimensions and such development is marked by a state of transition, conceptual change has been explained as a dynamic process occurring over a period of time (Tyson, Venville, Harrison, & Treagust, 1997). This dynamic process takes into consideration the pre-instructional conceptions of the student, the science content, and the path between them as a student constructs learning.

In a constructivist view of learning students bring to every course experiences, ideas, and skills related to the content (Garnett et al., 1995). Because prior knowledge is brought by students into the course, teaching so that the session's content has meaning to students typically involves helping them to reconstruct their existing ideas rather than teaching totally new ideas. This is conceptual reconstruction. Inquiry-based teaching provides the environment for the students to confront their previous ideas, explore their ideas, reconstruct their ideas, and form new connections to their previous ideas (Chi, Slotta, & deLeeuw, 1994). Knowledge is viewed as found in the minds and bodies of thinking beings (Johnson, 1987). Learning is the construction of knowledge by individuals as sensory data are given meaning in terms of their prior knowledge. It is an interpretive process, involving constructions by individuals (Tobin, Briscoe, & Holman, 1990). Meaning can only be formed by students in their own minds (Saunders, 1992). Students cannot be passive during learning. Teachers facilitate meaningful learning by planning and using activities that engage students in working with ideas in their own minds (Yager, 1991). Meaningful learning is an active construction process. It is a process that involves and depends on the background knowledge the learner brings to a situation, the learners' attention focused on the ideas being presented, and the mental and physical actions of the learner as he or she works with objects and events in testing prior or new tentative versions of an idea.

To create learning in students, instruction must go far beyond traditional teaching strategies. Presenting information about an idea through
lecture or reading, followed by review of the information focuses primarily
on mental recall and produces science that is not meaningful to students.
In this course students are involved in activities that foster restructuring of
prior knowledge.

Specific pedagogical implications are related to this view of learning as
an active process. First, use prior knowledge in creating new, meaningful
learning. Second, motivate students to learn. Third, link new information
to prior knowledge. Fourth, integrate knowledge. Fifth, encourage growth
of cognitive and meta-cognitive processes. Knowledge structures include
ways of perceiving and processing knowledge and analytical skills and dispositions as well as concepts, generalizations, and theories. Strategies that
help form new thinking processes are reflection, restructuring of tasks, and
learning from others.

This constructivist model is typically incorporated into education
courses at universities, but is just now beginning to appear in courses
taught in different areas of instruction. This model is used exclusively in
the course described in this chapter. Currently, other than research by
Sunal, Karr, and Sunal (2003) no other investigations of an undergraduate,
non-majors course in artificial intelligence exist.

AN INVESTIGATION IN CREATING AN INFRASTRUCTURE
FOR TEACHING ARTIFICIAL INTELLIGENCE SYSTEMS
USING CONCEPTUAL CHANGE PEDAGOGY

Research Methodology for Development of the Course:
Artificial Intelligence Systems in Science

Artificial Intelligence (AI) Systems in Science was not designed to make
non-science majors experts in the development of AI systems. Rather, this
course was designed to prepare non-science majors who have solid information on the capabilities of, misconceptions about, potential uses for, and
documented uses of artificial intelligence techniques. Further, the course
was designed to demonstrate the relationship between natural systems and
AI systems and to demonstrate the role of modeling systems in modern science and engineering. The concepts investigated in this course are process
concepts used to explain how an event is structured (Chi et al., 1994). Conceptual change can be expected to be lengthier and more often inconsistently applied with process concepts because they are constraint-based
interactions. Specifically, there are four main goals for the course: (a) to
help develop students' understanding of how systems in nature can be mod-

eled; (b) to make students aware of the types of problems that are routinely solved in the natural sciences using modeling, and specifically, AI techniques; (c) to provide students with an introduction to modeling using AI techniques; and (d) to provide students with the opportunity to solve scientific problems in a laboratory setting through modeling that uses AI tools.

To investigate the effects of the course, students' conceptions of fuzzy logic, neural networks, and genetic algorithms as used in systems were elicited by asking them to consider everyday situations during which an individual comes into contact with one of these concepts through its incorporation into an object or event. These situations were presented as scenarios ending with a question to be considered, for example, using an automatic teller machine (ATM) to withdraw money from a bank.

A pilot study investigated the use of 20 scenarios with 18 students not enrolled in the course. Based on these students' responses, two scenarios on fuzzy logic systems, five scenarios on neural networks, and four scenarios on genetic algorithms were selected for this study. Criteria for selection were ability to discuss the scenario as an event even when unable to explain underlying concepts, affirmation of familiarity within the context of the scenario, description of the scenario as demonstrating scientific concepts in use, and ability to make suggestions regarding which concepts might be involved even if these were not appropriate to the situation.

Eleven students chosen randomly from two sections of the course taught by the same instructor were pre-interviewed at the beginning of the course and post-interviewed at its completion by a graduate student. The interviewees were 20 to 40 years of age, averaging 24 years, and included two males and nine females.

Each student was interviewed using the scenarios during an audiotaped, individually scheduled interview averaging 70 minutes in length. Pre-interviews occurred during the first week of classes in the semester prior to any content coverage in the course. Post-interviews occurred within 3 days following the course's completion. Probes asked for further explanation, requesting a real-life example similar to that in the scenario, and welcoming additional comments. Pre- and post-interview comments were matched. Evidence statements for each concept from all interviewees were grouped and trends identified. Statements indicating responses supportive of alternative conceptions were also identified and trends noted (Posner, Strike, Hewson & Gerzog, 1982.)

Course Model Investigated

The course focuses on the scientific concepts underlying various artificial intelligence techniques based on computer models of natural systems, and the relationship between these techniques and more traditional fields of scientific study including neuroscience, genetics, and life sciences. At its foundation is the concept of systems. The popular concept of systems is "a collection of things and processes (and often people) that interact to perform some function" (American Association for the Advancement of Science [AAAS], 1993, p. 262). Scientists use systems in a context implying "detailed attention to inputs and outputs and to interactions among the system components" (p. 262). Scientists and engineers seek to specify systems quantitatively, study their theoretical behavior through computer simulations and thereby define problems and investigate complex phenomena. Elementary children study many types of systems from ecosystems through political systems but do not begin to conceptualize interactions between systems until the middle school years. A secondary goal of this course is to help preservice education majors enhance their ability to attend to aspects of particular systems in attempting to understand and work with a whole system, and in so doing, to prepare them to assist students in constructing a network of interrelated concepts that will provide the foundations for understanding systems in science. The vehicle for accomplishing this goal is artificial intelligence. Specifically, the following five AI techniques are studied: (a) tree diagrams and tree searches, (b) expert systems, (c) fuzzy logic, (d) neural networks, and (e) genetic algorithms.

National Standards and Higher Education Curricula

This course addresses national standards developed by the American Association for the Advancement of Science and published in *Benchmarks for Science Literacy* (AAAS, 1993), in which systems and the use of technology are "really ways of thinking rather than theories or discoveries" (p. 261). This course promotes ways of thinking about a possible system, which will enhance college students' conceptualization of science as a way of knowing. Systems should be encountered through a variety of approaches, including designing and troubleshooting. "The main goal of having students learn about systems is not to have them talk about systems in abstract terms, but to enhance their ability (and inclination) to attend to various aspects of particular systems in attempting to understand or deal with the whole system" (p. 262). Additional connections utilize the recommendations for the use of technology from the *Curriculum and Evaluation Standards for School Mathematics* (National Council of Teachers of Mathematics, 1989). The content topics from these national standards include:

1. the scientific enterprise—science as a social system,
2. mathematics, science, and technology—modeling systems,
3. design and systems,

4. the living environment including interdependence of life—ecosystems and evolution of life—feedback,

5. the human organism including human development—mother-embryo feedback and basic functions—interacting organ systems,

6. human society including group behavior—emergent properties,

7. the designed world including information processing—feedback,

8. the mathematical world including numbers—number systems, symbolic relationships—symbol systems, and reasoning—logical systems,

9. earth systems—atmospheric, geologic, life—and how they are inter-related with other natural systems and with systems made by people.

These topics are woven throughout the course components represented by the AI systems covered.

Course Components Description

One of the major challenges of this course was to describe cutting-edge technological tools to general college students who often do not have the mathematical background to fully understand the details of techniques such as neural networks, which rely on calculus-based concepts. Thus, the goal is not to have the students program these techniques on the computer, rather the goals are to have the students (a) become familiar with the methods conceptually, (b) be familiar enough with the capabilities of the methods to direct their future students, (c) be able to use modern technology such as the Internet and the World Wide Web to locate more information on these and other scientific techniques, and (d) understand how these systems relate to larger systems in science. However, students who happen to have a mathematical or computer background are encouraged to investigate the computational underpinnings of the techniques around which the course has been built.

To ease the process of comprehending the material, the class draws on students' prior knowledge of natural systems. Virtually all of the students who take this course will have been exposed to biology courses of some type. Thus, the topics covered in the current course are all presented according to a common template. First, the students explore their prior knowledge of a particular system in nature (e.g., for neural networks, exploring their notions of how the brain works; for genetic algorithms, considering chromosomes and the genetic system). Second, their knowledge is explored as the instructor leads them into discussions of the particular aspects of the natural system to be modeled. Third, the students see how said natural system can be modeled on the computer yielding an AI technique. Finally, the computer models that result are used to solve problems related to the natural sciences. Thus, the fact that the AI systems considered are presented as expansions of natural systems with which the students are already familiar greatly eases the difficulty students might experience if the material were presented in an alternative manner. In this manner, the goals of the course are more readily achieved.

The result of virtually any decision-making process is a tree diagram originating with discrete event simulations. A prime example of this occurs in nature with the classification of plants and animals in the phyla structure. Students are introduced to classic tree-searching techniques including both depth-first and breadth-first search techniques. Tree-searching concepts are introduced using tree diagrams.

Expert systems are rule-based systems that model human decision making. Expert systems are perhaps the flagship of AI; they have been used successfully and have achieved notoriety in a wide variety of disciplines including medical diagnosis, aerospace engineering, game theory, and mineral exploration. During the course of their introduction to expert systems, students work with logic diagrams and flow-charting techniques.

Fuzzy logic represents an extension to expert systems in which the "rule-of-thumb" approach typically used in human decision making is better modeled. This mathematical technique provides computers with the ability to manipulate abstract concepts that humans so effectively manipulate. Fuzzy logic has been used to solve numerous decision-making and path-planning problems. In their exposure to fuzzy logic, students review traditional set theory. Further, they alter a functional computer-simulated fuzzy system applied to a real-world chemical-process control problem which allows them to investigate ideas and concepts about acids and bases from their chemistry courses.

Neural networks are computational paradigms of the human brain. They are computer algorithms that mimic the way humans learn to accomplish tasks. Thus, unlike traditional computer algorithms, neural networks are trained rather than programmed. Students have the opportunity to perform the simple calculations associated with neural networks using networking techniques. Then, they are exposed to the vast potential associated with these paradigms by accessing information about and simulations of successful neural network applications on the Internet and World Wide Web. The class exercise on neural networks allows students to refine their knowledge regarding the interaction of systems, and their knowledge obtained from biology on how the human brain works.

Genetic algorithms are search algorithms based on the mechanics of natural genetics. They effectively search through large search spaces in difficult problems. Genetic algorithms have been used to solve optimization and machine-learning problems in a variety of disciplines. Students investigate the mechanics of genetic algorithms and discover their relationship to both natural genetics and probability theory. As with neural networks, students locate working genetic algorithm simulations on the Internet and the World Wide Web. They are exposed to computer implementations of genetic algorithms on real-world problems. Their experience with genetic algorithms allows them to expand their understanding of the human genetic system.

Course Structure and Contents

The following is a description of the course's structure and contents by topics. Its initial offering was as a 5-week summer session course. The 4-semester credit hour course met Monday through Friday for 2 hours 15 minutes.

Topic 1.   The first course topic was an introduction to the course and to technological tools used in the course such as electronic mail and the Internet. Students were informed that all assignments and information dissemination in the course would be done electronically. Thus, a portion of the first week of class focused on providing students with information needed to use electronic mail to acquire and turn in assignments, and to use the Internet to obtain copies of course lectures which were made available on a course Web page and to locate and acquire computer software implementing the AI techniques to be introduced later.

Topic 2.   The second topic was an introduction to basic theory of systems. The concept of a system is such an important theme that it pervades all of the natural sciences. Students developed their own definitions of a system which they compared and contrasted with "standard" definitions. (Generally, a system is thought of as any collection of things that have some influence on one another.) Additionally, students participated in group-learning exercises in which they explored the relationship between various human systems. They developed a preliminary working model identifying which of these systems are necessary for intelligence.

Topic 3.   Topic 3 was an overview of systems in science. This second week of class was spent familiarizing students with some common instances of scientific systems that have been simulated by AI techniques, and a classic test for AI called the Turing Test. Here, the focus was on the wide range of scientific disciplines and systems to which AI has been applied, with special emphasis on the natural sciences, including geology, chemistry, physics, astronomy, and biology. The remainder of the course was devoted to the introduction and use of specific systems in science and the related AI technique heavily used to solve problems related to it. These were presented in units. Each unit consisted of the following four parts: (a) an introduction to a system in nature, (b) a presentation of a method for modeling said system, (c) a laboratory session in which the students were presented with a problem related to the system which they investigated and then solved by applying a computer model, and (d) a discussion of complex scientific problems which have been solved using the AI technique.

Topic 4.   Topic 4 worked with animal classification using tree diagrams during Week 2 of the course. The hierarchical structure by which animals are classified was studied using tree diagrams. The students constructed their own hierarchical diagrams representing the classification of animals. These hierarchical diagrams were converted into tree diagrams. Mechanisms for modeling and searching hierarchical diagrams were discussed. At this point, the students solved a classification problem using a computer model of their tree diagrams in conjunction with algorithms for searching both depth and breadth first. Aside from discussions of scientific problems requiring rapid searching of hierarchical structures as in medical diagnosis, game-playing computer programs were presented as a popular instance of systems in which computers can be used to search tree diagrams to make effective decisions rapidly.

Topic 5.   Topic 5 involved students in developing rules for modeling systems in science: expert systems during Week 3. Students drew on their experience in classifying animals to develop a set of rules for accomplishing this goal. Methods for modeling these rules were discussed. Students expanded this idea by working with another system, using computer software in which an expert system is able to successfully classify mineral samples. The students completed a laboratory experience in which they performed mineral classification, and then used the computer software to check their thought processes against that of the expert system. Finally, additional examples of modeling rules for systems from chemistry, medicine, geology, and engineering were presented. In the area of chemistry, for example, students used an expert system to categorize the strengths of various acids and bases, while in the area of medicine various diseases and their associated systems were studied.

Topic 6.   Topic 6 involved students in manipulating scientific concepts to solve complex scientific problems: fuzzy logic during Week 3. Humans effectively manipulate abstract concepts to solve complex problems through fuzzy logic. Computers also can use fuzzy logic to solve problems. Here, fuzzy logic was presented as an extension to expert systems. There are numerous classification tasks in which the evidence is not as concrete as in the above example of mineral classification. Fortunately, nature provides mechanisms for dealing with imperfect information, and for manipulating abstract concepts. For instance, in diagnosing illnesses physicians often must utilize subjective information such as "my head hurts." The students discovered the need for such fuzzy systems as they solved a simulated problem from chemistry in a laboratory session. Specifically, the students solved a titration problem in a computer-simulated environment. They worked with an imperfect pH sensor to determine the need for subjective descriptions as they developed rules for neutralizing a solution. Fuzzy systems that have been developed in chemistry, engineering, and physics were presented to provide the students with a feel for the wide-ranging applicability of fuzzy logic.

Topic 7. Topic 7 was the human brain as a system: modeling it using neural networks as was discussed during Week 4. Models of the human brain are found in the form of neural networks. This topic's study began with an exercise in which students described their impression of how the human brain works. Next, they diagramed a simple vision of how they thought the human brain makes a decision. Then, they developed an initial concept of what a neural network is and how it works. They received a description of the general way in which a neural network functions, and computer software for implementing a back-propagation neural network. They solved one of several problems from natural science, including determining the decay rate of nuclear material, tracking the trajectory of a planet, and determining the growth rate of a virus introduced into a population of organisms. For closure, examples of problem solving and simulation using neural network applications in chemistry, geology, astronomy, and engineering were investigated.

Topic 8. Topic 8 involved students in investigating the human genetic system: modeling them using genetic algorithms during Weeks 4 and 5. Human genetic systems were modeled using genetic algorithms. The students' understanding of genetics was explored and developed. The steps necessary for modeling the genetic system were discussed. Computer software for implementing a genetic algorithm was provided and used to solve one of several scientific problems including performing spectral analysis of chemical compounds. Finally, several examples were presented of systems where problems can be solved using genetic algorithms. Examples included minimizing the weight-to-thrust of an aircraft and the development of a mathematical relationship between various chemical bonds and the associated chemical properties of the resulting solutions.

Topic 9. Topic 9 focused on application of course concepts through projects during Week 5. The students completed a course project that required them to demonstrate a fundamental understanding of at least one of the topics presented in the class. Students developed a presentation consisting of exercises and discussion material designed to describe a system in science, a problem related to the system that has not been presented in class, and an AI technique that might be used to solve a problem related to the system.

Instructional Methodology

Artificial Intelligence (AI) Systems in Science has a constructivist focus and utilizes instructional methodology that assists students in constructing their own knowledge. The learning cycle is used in this course because it is a strategy involving experience, interpretation, and elaboration (Karplus, 1979). The learning cycle has three phases. In Phase 1, exploration, students are involved in laboratory-based activities that allow them to confront and make evident their own thinking and representation of the idea or skill to be taught. The activities focus students' attention, ascertain their prior knowledge, and relate previous learning to new learning. In Phase 2, invention, a more direct teaching format is used. Laboratory activities and lecture are integrated. Each is necessary to understand the other fully. Students are involved in experiences that develop the idea or skill more fully or to a higher order through providing explanations, examples, and closure. In Phase 3, expansion, activities allow students to practice the idea or skill just taught in the invention. In this course students participate in expansion experiences that cause them to extend the range, modality, and context of application of an idea or skill.

To foster students' reconstruction of their knowledge, cooperative grouping is a basic element in each meeting of the course. Students also work in small groups to develop and construct a final course project. The project involves them in the planning and construction of a presentation. They are asked to select a single concept presented in the course, and to describe this concept to an audience of their choice. They must develop presentation materials and group exercises to assist in their explanations.

RESULTS AND SUMMARY OF AN INVESTIGATION IN CREATING AN INFRASTRUCTURE FOR TEACHING ARTIFICIAL INTELLIGENCE SYSTEMS IN SCIENCE

As with the initial offering of virtually all courses, some weaknesses were identified in Artificial Intelligence (AI) Systems in Science. First, several of the students had never taken a course in which the learning cycle paradigm was used. This created a problem because these students seemed to take a good portion of the term adapting to the mechanism of material presentation. This weakness was addressed in later offerings by emphasizing the course format more strongly in the beginning of the term. Second, it became evident that the lengthier full semesters found in fall and spring terms were more appropriate for this course. Students need a longer period of time to reflect on the course's concepts and to reconstruct their ideas (Sunal et al., 2003). Third, a few students were "computer gun shy." This was evident in their initial reluctance to "surf the Internet" for information on the various AI techniques presented. However, as the course progressed, the students certainly became more comfortable using computers to do a variety of tasks. As with the first weakness identified above, this weakness is addressed by increasing the amount of time devoted in the beginning of the course introducing computer capabilities. Fourth, the students were a bit overwhelmed with the amount of material presented. This is, of course, a common problem, but an easily rectified one. In future offerings, the course presented only four AI techniques; the unit on expert systems was dropped since the popularity of these systems is waning.

Using a pretest-posttest strategy, students were interviewed with a set of 11 scenarios involving key course concepts. A course project was evaluated. These materials served as data points. The analysis indicted students moved from a pretest interview understanding of 5% of the key course concepts to a posttest interview understanding of more than 50%. A significant difference was found for all concepts tested. On posttesting, students appropriately used all of the concepts but their application was not consistent. The results support Watson et al.'s (1997) contention that conceptual change is marked by a state of transition. The concepts investigated in this study are process concepts used to explain how an event is structured (Chi et al., 1994). Since they are constraint-based interactions and since conceptual change can be lengthier with process concepts, the inconsistency evident in the results can be expected. Genetic algorithms are the most dynamic, causal, and constraint-based of the three concepts investigated in this study and therefore, the most difficult to understand. This point was characterized by student comments such as:

> I really liked the class. Most of the topics were very clear, and the presentation style helped a lot! However, I struggled with my understanding of the genetic algorithms. Maybe I do not understand the biology behind this technique as well as I thought I did.

The findings indicate that an investigative, inquiry-oriented course can create conceptual change in adult students. The posttesting found many more responses that indicated understanding and application of the concepts of fuzzy logic, neural networks, and genetic algorithms than were found in the pretesting. However, since half of the posttest interview responses were not appropriately demonstrating an understanding and application of these concepts, the conceptual change is not fully applied.

Among the three concepts, neural networks were most often applied in post-interviews. However, they also appeared most often in pre-interviews. This suggests an understanding of neural networks among some of the students prior to the course. A genetic algorithm is an abstract and complex concept incorporating fuzzy logic and neural networks. Yet, students appropriately used a genetic algorithm in their posttesting, albeit inconsistently.

AUTHOR NOTE

This work was in part supported by NASA Opportunities for Visionary Academics (NOVA), a program funded by the National Aeronautics and Space Administration, although the views expressed here are the authors' only.

REFERENCES

American Association for the Advancement of Science. (1993). *Benchmarks for scientific literacy*. New York: Oxford University Press.

Bliss, J. (1995). Piaget and after: The case of learning science. *Studies in Science Education, 25,* 139–172.

Chi, M.T.H., Slotta, J.D., & deLeeuw, N. (1994). From things to processes: A theory of conceptual change for learning science concepts. *Learning and Instruction,* 25 (special issue), 27–43.

Garnett, P., Garnett, J., & Hackling, M.W. (1995). Students' alternative conceptions in chemistry: A review of research and implications for teaching and learning. *Studies in Science Education, 25,* 69–95.

Johnson, M. (1987). *The body in mind: The bodily basis of meaning, imagination, and reason.* Chicago: University of Chicago Press.

Karplus, R. (1977). *Science teaching and the development of reasoning.* Berkeley: University of California.

National Council of Teachers of Mathematics. (1989). *Curriculum and evaluation standards for school mathematics.* Reston, VA: Author.

Posner, G., Strike, K., Hewson, P., & Gertzog, W. (1982). Accommodation of a scientific conception: Toward a theory of conceptual change. *Science Education, 66,* 211–217.

Saunders, W. (1992). The constructivist perspective: Implications for teaching strategies for science. *School Science and Mathematics, 92,* 136–141.

Sunal, C. S., & Karr, C.L., & Sunal, D. (2003). Fuzzy logic, neural networks, genetic algorithms: Views of three artificial intelligence concepts used in modeling scientific systems. *School Science and Mathematics Journal, 103*(2), 81–91.

Thoresen, K. (1993). Principles in practice: Two cases of situated participatory design. In D. Schuler & A. Manioka (Eds.), *Participatory design: Principles and practices* (pp. 271–298). Hillsdale, NJ: Erlbaum.

Tobin, K., Briscoe, C., & Holman, J. (1990). Overcoming constraints to effective elementary science teaching. *Science Education, 74,* 409–420.

Tyson, L.M., Venville, G.J., Harrison, A.G., & Treagust, D.F. (1997). A multidimensional framework for interpreting conceptual change events in the classroom. *Science Education, 81,* 387–404.

Watson, J.R., Prieto, T., & Dillon, J.S. (1997). Consistency of students' explanations about combustion. *Science Education, 81,* 425–443.

White, R.T. (1991). Episodes and the purpose and conduct of practical work. In B. Woolnough (Ed.), *Practical science* (pp. 78–86). Milton Keynes, England: Open University Press.

Yager, R. (1991). The constructivist learning model: Towards real reform in science education. *The Science Teacher, 58*(2), 16–27.

ABOUT THE AUTHORS

Ronald K. Atwood holds an Ed.D. in science education from Florida State University. He currently is a professor in science education at the University of Kentucky and Co-Principal Investigator of the Appalachian Mathematics and Science Partnership. Atwood's research interests are alternative conceptions and conceptual change. He has recent publications in the *Journal of Research in Science Teaching, School Science and Mathematics,* and the *Journal of Science Teacher Education.* He has recent paper presentations at the national meetings of the National Association for Research in Science Teaching and the Association for the Education of Teachers in Science. He can be contacted at rkatwo00@uky.edu.

Scott Badger holds a Ph.D. in counseling psychology from Oklahoma State University. He currently is an assistant professor of technology education at the University of Idaho. His research interests are in online learning, online course design, online research strategies, and student motivation. He is a co-investigator on several grants related to interactive online learning. He is the co-editor of the *Journal for Interactive Online Learning.* He can be contacted at sbadger@uiadho.edu.

Jeanelle Bland Day, Ph.D., serves as the Director of the Connecticut NASA Educator Resource Center, the coordinator of the NASA Faculty Fellowship Program at Marshall Space Flight Center, a board member of Northeast Corner Connecticut Audubon Society, and a board member of the Connecticut Science Teachers Association. She is also a co-program chair of the 2004 Tunza International Children's Conference sponsored by the

Reform in Undergraduate Science Teaching for the 21st Century, pages 541–556
Copyright © 2004 by Information Age Publishing
All rights of reproduction in any form reserved.

United Nations. Her research areas of interest include the use of the learning cycle, use of online learning cycle tutorials, development of learning cycle rubrics, and online teaching and learning. Jeanelle has made presentations at meetings of organizations including the National Science Teachers Association, the National Association for Research in Science Teaching, the Association for the Educators of Teachers of Science, and the Council for Elementary Science International. At Eastern Connecticut State University, Jeanelle teaches undergraduate and graduate science methods courses and graduate introductory educational research. She has developed and taught many of her courses through the Online Connecticut State University system. She can be contacted at BlandJ@easternct.edu.

John E. Christopher holds a Ph.D. in physics from the University of Virginia. He currently is an associate professor in physics at the University of Kentucky where he has served as Associate Dean of the College of Arts and Sciences and Director of Undergraduate Studies in Physics. Christopher's instructional innovations include the implementation of tutorials in large-group general physics courses. His research interests are alternative conceptions and conceptual change. He has a recent publication in the *Journal of Research in Science Teaching* and has recent paper presentations at meetings of the National Association for Research in Science Teaching and the Association for the Education of Teachers in Science. He can be contacted at jchris@uky.edu.

Monica Clement is an instructor in the Geology Department at Kansas State University. She holds a B.A. in Geology from Western Washington University, an M.S. in Geophysics from the University of Utah, and a B.S. in Science Education from the University of Minnesota. Her research interests include practical application of various aspects of non-science geophysics, geoscience undergraduate education, geoscience teacher education, and incorporating field experiences into undergraduate geology courses. Her various science teaching experiences include Geology of Planets, Earth in Action, Geology Laboratory, and Coordination of Geology lab program including teacher training for TAs. She can be contacted at mclement@ksu.edu.

George E. DeBoer holds a Ph.D. in science education from Northwestern University. He is deputy director for Project 2061 (http://www.project2061.org) of the American Association for the Advancement of Science. DeBoer joined AAAS from the Division of Elementary, Secondary, and Informal Science of the National Science Foundation, where he served as program director. He holds an appointment as Professor of Educational Studies at Colgate University where he has served as director of the Master of Arts in Teaching Program, chair of the Education Department, and acting director of the Division of Social Sciences. He is a member of AERA, NARST, and NSTA. His primary scholarly interests include clarifying the goals of the science curriculum, researching the history of science education, and analyzing the meanings of scientific literacy. Prior to becoming a university professor, DeBoer taught high school chemistry, biology, and earth science in Glenview, Illinois, and chemistry, biochemistry, and microbiology at the Evanston Hospital School of Nursing, Northwestern University Medical School. DeBoer is the author of *A History of Ideas in Science Education: Implications for Practice*, as well as numerous articles, book chapters, and reviews. He can be contacted at gdeboer@aaas.org.

R. Lynn Jones Eaton, Ph.D., is presently a K–12 senior program coordinator in the Division of Instructional Innovation & Assessment at the University of Texas at Austin. She was formerly a faculty member in the Science Education Center at the University of Texas at Austin and in the department of Curriculum & Instruction at Southwest Texas State University. She received her B.S. and M.Ed. degrees from Mississippi State University and her Ph.D. from the University of Alabama at Tuscaloosa. Her scholarly interests are in the areas of equity, diversity, and policy in K–16 education. Dr. Eaton has taught science at the high school and middle school levels. She can be contacted at LynnJ15@aol.com.

Bill Elmore, Ph.D., P.E., is associate professor and academic director for the Chemical Engineering, Civil Engineering, and Geosciences Programs, Louisiana Tech University. His teaching areas include the integrated freshman engineering, chemical engineering unit operations, reactor design, and the senior capstone design sequence. Engineering educational reform, enzyme-based catalytic reactions in micro-scale reactor systems, and biochemical engineering are his current research interests. He can be reached by e-mail at belmore@coes.latech.edu.

Timothy G. Ewers holds a Ph.D. in curriculum and instruction from the University of Idaho. He currently is an assistant professor of STEM education at the University of Idaho. His research interests are in mathematics and science assessment, pedagogy, and policy. He is a co-investigator of several grants related to interactive online learning. He can be contacted at tewers@uidaho.edu.

Kathleen M. Fisher is professor of biology at San Diego State University, Director of the Center for Research in Mathematics and Science Education, Co-Director of the Professional Development Collaborative, and SDSU's representative for the CSU/NASA collaborative. Dr. Fisher has a B.S. in science from Rutgers University and a Ph.D. in genetics from the University of

California–Davis. She completed a postdoctoral fellowship with the Atomic Energy Commission in 1970, had a Fulbright Scholarship to work with the Universiti Sains, Malaysia, in 1980, and served as a Program Officer in the Research in Science Education program at the National Science Foundation 1980–81. Kathleen has worked in biology education research since the early '70s when she produced a televised two-semester genetics course in a novel video-auto-tutorial format. More recently Kathleen collaborated with the Science Media Group at the Harvard-Smithsonian Center for Astrophysics in Cambridge. Kathleen has also developed a series of guided discovery biology lessons as a resource for practicing teachers and home schooling parents (www.BiologyLessons.sdsu.edu). Kathleen and the SemNetResearch Group designed and developed the SemNetTM knowledge construction tool (1983–87). The Macintosh-based SemNet software has been converted into a series of Java-based, cross-platform knowledge construction tools called the Semantica™ software series (http://www.semanticresearch.com). She can be contacted at kfisher@sciences.sdsu.edu.

April French is a graduate student at the University of Kansas. She graduated from the University of Tulsa cum laude with a B.S. in Chemistry and minor in Communication in 2002. At TU, she worked under the supervision of Dr. Gil Belofsky, isolating novel and neruopharmacologically active/opiate receptor binding compounds from the plant *Dalea purpurea*. At the University of Kansas, her thesis will describe the informal learning environment at Science City, a local hands-on science museum in Kansas City, Missouri, and examine the learning that occurs in this informal environment. She can be contacted at aprilf@ku.edu.

Dorothy L. Gabel holds a Ph.D. in science education from Purdue University. She is currently a professor of science education at Indiana University. Her research interests include chemistry education and teacher preparation. She has 14 funded proposals from organizations including NSF, ACS, NSTA, Indiana Higher Education Commission, Proffitt, NASA, and Indiana University. She is the director for the QUEST Program for preparing prospective elementary teachers, the Saturday Science Program for children, and the development of the CORE 40 Science Assessments for the State of Indiana. She has journal publications in the *Journal of Research in Science Teaching, School Science and Mathematics, Science and Children, Science Scope, The Science Teacher, Science Education*, and the *Journal of Chemical Education*. She has numerous research paper and teaching presentations at state and national meetings including the National Association for Research in Science Teaching, the Hoosier Association of Science Teachers, School Science and Mathematics, American Chemical Society, American Association for the Advancement of Science, and the National Science Teachers Association. Her books on science education include the *Handbook of Research on Science Teaching and Learning* and *Chemistry, the Study of Matter.* She can be contacted at gabel@indiana.edu.

Francis Gardner, Jr., is professor of biology and director of the Columbus State University Science Education Outreach Center. He currently teaches a course titled "Introduction to Life in Space," an elective science course in the Undergraduate Core Curriculum of the University System of Georgia. He has taught introductory biology and animal physiology courses for more than 28 years. Current interests include the exploration of space and the search for extraterrestrial signs of life. He is a member of the Challenger Center International Faculty, an active educator in the Solar System Educator Program (SSEP) of the Jet Propulsion Laboratory, and a NOVA Fellow. Research interests include the impact of the National Science Education Standards on college-level science instruction and the impact of on-line discussions in the development of higher order thinking skills. He is engaged in educational outreach programs associated with the Georgia Space Grant Consortium at local, state, and national conferences. He is a member of the Executive Board of the Society of College Science Teachers and has served as the President of the Georgia Science Teachers Association and member of its governing board. He is active in the incorporation of handheld technology into the K–16 classroom to enhance the practice of science. He can be contacted at gardner_francis@colstate.edu

Steven W. Gilbert is an associate professor at Virginia Polytechnic Institute and State University. His current research is on the impact of performance assessment on science teacher preparation programs, but he has also published on the use of models to develop understanding of the nature of science. He coordinates the national program review board for the National Science Teachers Association (NSTA) as part of the National Council for the Accreditation of Teacher Education (NCATE) accreditation process. He also chaired both the writing and revision of the current performance-base NSTA Standards for Science Teacher Preparation. With degrees in biological science and science education, he has taught numerous university biology, general education, and science teaching methods courses, written science and science education books and textbooks, and published in the *Journal of Research in Science Teaching, Journal of Science Teacher Education* and other related professional media. He consults on performance assessment systems at the university level. He can be reached at swgilbert@vt.edu.

M. Jenice "Dee" Goldston, Ph.D., is an associate professor of science education at the University of Alabama. Dr. Goldston's research interests focus on teacher beliefs and self-efficacy related to inquiry models of pedagogy and the nature of science, conceptual change related to teacher content acquisition and pedagogy adoption, preparation of high-quality science teachers and their ongoing professional development, reform in teacher education and in undergraduate science teaching in higher education, and sociopolitical influences affecting the status of science teaching in K–12 schools. She has authored articles in journals including *Science Teacher, Science and Children, The Physics Teacher, Journal of Science Teacher Education, Teacher Education and Practice*, and the *Journal of Research in Science Teaching*. She conducts conference presentations and workshops that range from exploring science inquiry to action research for K–12 teachers. Dr. Goldston is an associate editor for the *Journal of Science Teacher Education*. She is an active member of the National Association for Research in Science Teaching, American Educational Research Association, National Science Teachers' Association, Association for Educators of Teachers in Science, and Alabama Science Teachers Association. She is the President of the Council of Elementary Science International and is a Council member for the National Science Teachers Association. Dr. Goldston has been involved in numerous NSF grants for professional and curriculum development. She is also involved in research within NASA Opportunities for Visionary Academics (NOVA) to reform undergraduate science content courses (NASA). She can be contacted at dgoldsto@bamaed.ua.edu.

Scott Graves, Ph.D., is an assistant professor of education at the University of Idaho specializing in science and technology teacher professional development. Dr. Graves has initiated, a number of grant-funded science/technology education projects for the Institute for Mathematics, Interactive Technologies and Science Education (IMITS). In addition to managing projects and researching science and technology teaching, Dr. Graves teaches science methods, integrated science, Earth systems science, and educational technology for the University of Idaho. He can be contacted at sgraves@uidaho.edu.

William S. Harwood, Ph.D., is an associate professor at Indiana University. His research interests include scientific inquiry, inquiry teaching and learning in college science, and the nature and impact of undergraduate research experiences in science. He has functioned as a faculty member and as a campus-level administrator in two public research universities. In addition, he moved from a science department at one university to the science education department at another university. These different positions helped him to identify and look across some common elements that affect the ability of faculty to successfully engage in collaborative partnerships regarding science teaching and learning. He is currently actively engaged in two such collaborations involving science and science faculty (see http://www.reciprocalnet.org and http://hhmi.bio.indiana.edu) that focus on college science students. He also directs a collaborative project to improve teacher preparation (http://education.indiana.edu/dean/21stcenturyteachers.pdf). On the national level, he has co-authored a successful General Chemistry textbook and is program co-chair for the 2004 Biennial Conference on Chemical Education (http://www.chem.iastate .edu/bcce). His Web site is http://php.indiana .edu/~wharwood and he can be contacted at wharwood@indiana.edu.

Joseph A. Heppert, Ph.D., is a professor of chemistry at Kansas University. He is active in science teaching reform, teacher preparation reform, and educational research. He is currently the PI of the Kansas Collaborative for Excellence in Teacher Preparation (KCETP), in which he has managed teacher professional growth opportunities for middle school teachers. He is also PI on a NSF Teacher Enhancement project that is developing materials to help teachers apply GIS technologies in inquiry-based science curricula, and the PI on another project from the AT&T Foundation to develop and evaluate a hybrid teacher enhancement program involving both online and face-to-face learning environments. He directed a three-year project funded by the William and Flora Hewlett Foundation to reconceptualize the introductory undergraduate chemistry laboratory experience at the University of Kansas. Dr. Heppert is a member of the American Chemical Society's Committee on Education, and serves on the organizing committee for a Society-sponsored workshop on re-envisioning the undergraduate curriculum in the chemical sciences. He is the director of the University of Kansas Center for Science Education, an interdisciplinary center involving science and technology educators, scientists, and engineers in projects intended to improve science education for all students in Kansas and foster science education scholarship. He can be contacted at jheppert@ku.edu.

Mark C. James is an assistant professor in the Physics Department as well as a science educator in the Center for Science Teaching and Learning at Northern Arizona University. His research interests include analogy-making as cognition, physics education, and the nature of science. He has coauthored publications in nonlinear dynamics, physics education, and teaching the nature of science. Dr. James's undergraduate teaching responsibilities include teaching introductory physics courses, teaching a methods of teaching secondary science course, and the supervision of student teachers. He can be contacted at mark.james@nau.edu.

William Jordan, Ph.D., is a professor and the Mechanical Engineering Program Chair at Louisiana Tech University. His research interests include mechanical behavior of composite materials, failure analysis, engineering ethics, and engineering education reform. He has numerous publications and presentations dealing with traditional materials engineering research as well as educational research. His most recent educational presentation was "*Introducing Materials Science and Chemistry to the K–12 Community*," which was presented at the ASEE annual meeting in Nashville, June 2003. His professional offices include faculty advisor to student section of ASME for 8 years, former member of National Student Sections Committee of ASME, and session chair for the Materials Division of the ASEE national meeting in 2003 and 2004. His grants consist of 15 in the area traditional materials engineering research and four in the area of educational research. Dr. Jordan teaches undergraduate engineering courses as well as a class in Engineering Problem Solving for Future Teachers. His Web site is www2.latech .edu/~jordan/. He can be contacted at Jordan@coes.latech.edu.

Charles L. Karr, Ph.D., is a professor and head of the Aerospace Engineering and Mechanics Department at the University of Alabama. His research interests include genetic algorithms for search and optimization, control of engineering systems using fuzzy logic, machine learning and artificial intelligence, and engineering education. His recent notable publications include coauthoring the book *Practical Applications of Computational Intelligence for Adaptive Control* and the article entitled "Fuzzy Logic, Neural Networks, Genetic Algorithms: Views of Three Artificial Intelligence Concepts Used in Modeling Scientific Systems for the *School Science and Mathematics Journal.* He has worked on numerous funded grant projects including Perceived Gender Inequities in Freshman Engineering Design Projects, Systems in Science, Experimental and Numerical Investigation of Jets for Active Aerodynamic Control, and An Immunized Aircraft Maneuver Selection System. His Web site is http://ckarr.eng.ua.edu/ and he can be contacted at ckarr@coe.eng.ua.edu.

Kimberly L. Keller received her bachelor's degree from Wilmington College of Ohio, and her master's degree from Bowling Green State University (BGSU). Currently a doctoral student in biological sciences at BGSU, Kimberly's dissertation research involves studying DNA repair pathways. As a teaching assistant, she has taught various undergraduate, biology laboratory courses, including special sections for a learning community of middle childhood education majors. Kimberly worked on a National Science Foundation (NSF) grant awarded to BGSU to integrate inquiry-based education in introductory biology laboratory classes. As Master Teacher on this grant, she instructed teaching assistants in inquiry teaching methods and helped revise laboratory exercises. She has presented inquiry-based laboratory exercises at many conferences, including major and minor workshops for the Association for Biological Laboratory Education (ABLE). Kimberly is a NSF

Fellow assigned to the Partnership for Reform through Inquiry in Science & Math (PRISM) GK–12 Program at BGSU. PRISM places graduate students with local science teachers as a content/inquiry resource for school districts as they align their curriculum with National/Ohio Content Standards in Science/Math. She can be contacted at kkeller@bgnet.bgsu.edu.

Teresa Kennedy holds a Ph.D. in curriculum and instruction from the University of Idaho. She is currently an assistant professor of foreign language education at the University of Idaho. Her research interests are in online learning, online course design, pedagogy, and online tool design. She is the Deputy Chief Educator for the GLOBE program in Washington, D.C. She is also the director of the Center for Evaluation, Research and Public Service at the University of Idaho. She can be contacted at tkennedy@uidaho.edu.

Mitchell Klett holds a Ph.D. in curriculum and instruction from the University of Idaho. He currently is an assistant professor of science and technology education at the University of Idaho. His research interests are in online learning, earth science education, authentic assessment, and online courseware. He is a co-investigator of several grants related to interactive online learning. He can be contacted at mklett@uidaho.edu.

Christy MacKinnon, Ph.D., is a professor of biology at the University of the Incarnate Word, in San Antonio, Texas. She currently teaches introductory biology and genetics. Her scientific research includes collaborative studies in protein folding and molecular biology. Her recent research interests are the professional development of pre- and in-service teachers in sciences, and best practices for teaching undergraduate students in biology. Dr. MacKinnon developed and directs the M.A. program in Multidisciplinary Sciences, which is a graduate program designed to enhance content knowledge and improve instructional strategies for middle and high school teachers. She has obtained several grants from the Eisenhower Professional Development Program of the Texas Higher Education Coordinating Board to support teacher participation in this graduate program. She has also collaborated on projects funded by NASA (Project NOVA, MASTAP). Her other scholarly works include published laboratory activities and numerous presentations at national meetings. She can be contacted at mackinno @universe.uiwtx.edu.

Cheryl L. Mason is a professor of science education and biological sciences at San Diego State University. Mason received her Ph.D. in science education and educational computing from Purdue University, and her bachelor and master's degrees in biological sciences from Indiana University. Her teaching experiences include undergraduate and graduate courses in biological sciences, science teaching and learning, and interfacing technology in the science classroom. She has scholarly publications in numerous journals and books, and has made presentations to various audiences over the past 32 years. Mason is a member and leader of professional organizations such as AETS, NARST, NSTA, NABT, AERA, and PDK. Two of the highlights of her career are receiving the Association for the Education of Teachers of Science's Outstanding Science Teacher Educator Award and the Perham Indiana Women of Distinction Award. Overall, Mason's research focus is on the relationship of cognitive, including visual/spatial thinking skills, and attitudinal factors concerning successful science teaching and learning. She is concerned with helping girls and persons of color succeed in the science classroom and scientific community. She can be contacted at cmason@mail.sdsu.edu.

Deborah A. McAllister, Ph.D., is a UC Foundation Professor in the Teacher Preparation Academy at the University of Tennessee at Chattanooga. She holds a B.S. in zoology (University of Massachusetts, Amherst), and an M.S.Ed. in educational psychology and research and an Ed.D. in curriculum and instruction/educational technology (University of Kansas). Her research interests include the integration of educational technology in mathematics education and science education, and working with M.Ed. students on completing action research projects. Dr. McAllister has completed two books and eight articles, and has presented numerous workshops and conference sessions. Current grant activity includes Tennessee Higher Education Commission/No Child Left Behind, Tennessee Space Grant Consortium, NOVA Faculty Fellow, UTC Faculty Research, and UTC Instructional Excellence. She holds membership in AECT, NCTM, NSTA, and PDK. Currently, her main focus is on mathematics education and educational technology, but she has taught several science and science education courses: Teaching Strategies and Materials in Science, Teaching Science in Elementary and Middle School, Science Concepts and Perspectives, and Laboratory Procedures and Safety. She can be contacted at Deborah-McAllister@utc.edu, and her Web site is http://oneweb.utc.edu/~deborah-mcallister/.

Julia McArthur is an associate professor in the Division of Teaching and Learning at Bowling Green State University. Her research interests include undergraduate science education, technology in teacher education, and the use of portfolios in teacher education. She coauthored with Charlene Waggoner a laboratory manual for introductory biology course with an environmental focus. She and Dr. Waggoner have had several grants to improve instruction in biology laboratories. This work has been presented at conferences such as the National Association for Research in Science Teaching and Association for the Education of Teachers in Science. She currently teaches science methods for middle childhood teachers. She can be contacted at mjulie@bgnet.bgsu.edu.

Bonnie McCormick is an assistant professor and Chair of the Department of Biology at University of the Incarnate Word, San Antonio. She teaches both majors and non-majors in introductory and advanced biology courses. Her research interests are in the area of undergraduate teaching and learning in the biological sciences. She has made numerous presentations to professional organizations on integrating technology to engage undergraduate students in inquiry learning and on student learning in inquiry-based classrooms. Her grant awards have supported curriculum revisions in the introductory biology course, training of prospective science teachers and the acquisition of technological tools for improving undergraduate teaching through authentic research experiences. She can be reached at mccomic@univirse.uiwtx.edu. More information on course development can be found at www.uiwtx.edu/~mccormic/nova.

Wayne Morgan, Jr., holds a Ph.D. in curriculum and instruction with a science education emphasis from Kansas State University. He currently serves as assessment coordinator and chemistry instructor at Hutchinson Community College. His research interests are in effective development of critical thinking and problem solving skills by students in introductory science courses, assessment of student learning, and transfer of learning from the classroom to practice. He has been involved in several curriculum development projects for the American Chemical Society including *Science in a Technical World* and *ACS Chemistry*, a textbook for general chemistry. He has had an article on critical thinking published in the *Journal of College Science Teaching*. He has recent paper presentations at the national meetings of the American Association of Higher Education Assessment and the American Chemical Society. He can be contacted at morganwa@hutchcc.edu.

Michael Odell holds a Ph.D. in curriculum and instruction from Indiana University. He currently is an associate professor of science and technology education at the University of Idaho. His research interests are in online learning, online course design, pedagogy, and online courseware. He is the director of several grants related to interactive online learning. He is the director of the Idaho Virtual Campus and the Institute for Mathematics, Interactive Technologies and Science at the University of Idaho. He was selected as a NASA Summer Faculty Fellow at Kennedy Space Center in 2003 and was also selected as a National Space Grant Fellow at NASA Headquarters in 1995–96. He can be contacted at mirodell@uidaho.edu.

Carol Dianne Raubenheimer is assistant professor in educational studies at Brescia University, Kentucky. She teaches methods courses for prospective elementary and middle school teachers in science, mathematics, technology, and curriculum and classroom management. She also teaches an integrated physical science course and is director of the Master of Science in Curriculum and Instruction Program. She has two master's degrees, one in biological science and the other in curriculum development and evaluation, and is currently completing her doctorate through the University of Louisville. Dianne has over 20 years experience working in nongovernmental organisations and universities, particularly with historically disadvantaged communities in South Africa. She has been involved in a number of evaluations of educational programs in science and mathematics education, including the NASA Opportunities for Visionary Academics (NOVA); the Centre for Improvement in Mathematics, Science and Technology Education (CIMSTE) at the University of South Africa; the Science Foundation Progam (SFP) at the University of Natal, South Africa; the Mathematics, Science and Technology Centre (MASTEC) in South Africa; and the Toyota Teach Primary School Project, South Africa. She can be contacted at dianner@brescia.edu.

Monika Schaffner holds a Ph.D. in educational measurement and evaluation from the University of Iowa. She currently is a senior policy analyst at SRI, International, in Washington, DC. Her research interests are in program evaluation, research design and methodology, and policy analysis. She serves as principal evaluator on several grants related to research in online interactive learning, the teaching of science and mathematics, and contextual teaching and learning. She has published in the *American Journal of Evaluation* and in *Contemporary Psychology*. Most recently, she has been invited to present at the Oxford Round Table at the University of Oxford in England. She also has recent paper presentations at the annual meeting of the American Evaluation Association and the American Educational Research Association. She can be reached at m_schaffner@hotmail.com.

Lawrence C. Scharmann is a professor of education and chair of the Department of Secondary Education at Kansas State University. Scharmann teaches undergraduate and graduate courses in science education. His research interests include the promotion of active learning, alternative assessment, and approaches to teaching evolution and the nature of science/scientific theories. He can be contacted at lscharm@ksu.edu.

Jacqueline Spears, Ph.D., is an associate professor of secondary education at Kansas State University. Her research interests include gender and science, rural education, and higher education policy. She is currently working on two publications related to the involvement of undergraduate women in a project designed to involve middle school girls in STEM fields. She has published extensively in the field of rural education, including monographs that explored multicultural education, linking rural schools to community development, literacy practice in rural areas, and model programs for improving rural access to postsecondary education. She was also codirector and author of an Annenberg Telecourse on rural sociology. She has chaired the Rural Special Interest Group (SIG) in AERA. She has been awarded grants for work related to rural adult access to education, multicultural education in rural areas, rural literacy practice, and rural sociology. Recently she was awarded a planning grant and demonstration grant for a program designed to increase middle school girls' interest in education and a grant to develop an interactive CD-ROM on gender issues in mathematics and science teaching. She has taught elementary school science methods courses through the College of Education and content courses through several physics departments. She coauthored an introductory physics textbook entitled *The Fascination of Physics*. She can be contacted at jdspears@ksu.edu.

Ann Stalheim-Smith is an associate professor of biology and a university distinguished teaching scholar at Kansas State University. Stalheim-Smith teaches undergraduate courses in both beginning biology and human anatomy/physiology. Her research interests include the promotion of active learning and critical thinking in the large college classroom. She can be contacted at stalheim@ksu.edu.

Kimberly A. Staples, Ph.D., is an assistant professor of science education in the Department of Elementary Education at Kansas State University. Her research interests include science conceptual understanding, alternative conceptions, K–12 science teacher professional development, and undergraduate college science teaching. She is the author of research and conference presentations such as *Teaching with Scientific Discrepant Events, Nontraditional Undergraduate Science and Preservice Teacher Professional Development, The Effect of a Nontraditional Undergraduate Science Course on Teacher and Student Performance in Elementary Science Teaching, Effective Pedagogy for Online Learning Environments,* and *Students' Understanding of Science (SUE).* Dr. Staples has served as a member of the Board of Directors of the Alabama Science, Mathematics, and Technology Coalition. She is currently working on a grant with the Kansas Board of Regents entitled Supporting Elementary Teacher Professional Development and Student Success in Science. She can be contacted at kstaples@ksu.edu.

Cynthia Szymanski Sunal holds a Ph.D. in early childhood social studies education from the University of Maryland. She currently is department head and professor of Elementary Education Programs at the University of Alabama. Her research interests are in cognition, curriculum development, the effects of the online environment on conceptual change, and longitudinal investigation of the development of primary school education in developing nations. Dr. Sunal has authored 8 books, 19 chapters, and nearly 200 refereed articles in a wide range of journals, including *Theory and Research in Social Education, The International Social Studies Forum, Journal of Research in Childhood Education, African Studies Review,* and *School Science and Mathematics.* She has recently presented papers and organized symposia at the annual meetings of several professional organizations including the American Educational Research Association, the American Association of Colleges of Teacher Education, and the National Council for the Social Studies. Among her other professional activities is her role as executive editor of the *Journal of Interactive Online Learning.* She can be contacted at cvsunal@bama.ua.edu.

Dennis W. Sunal holds a Ph.D. in science education from the University of Michigan. He currently is a professor of science education at the University of Alabama. His teaching experiences include undergraduate and graduate courses in physics, astronomy, engineering, research in curriculum and instruction, teaching in higher education, and science teaching and learning. His research interests are in professional development, alternative conceptions and conceptual change in teachers and faculty, college student conceptions of the nature of science, and Web course design, pedagogy, and contextual factors in interactive online learning. He has been project director and co-director in several grants related to research in online interactive learning and in creating change in science, engineering, and mathematics courses in higher education. He has published in the *Journal of Research in Science Teaching, Journal of College Science Teaching, Journal of Interactive Online Learning, School Science and Mathematics, African Studies Review,* and *Science Education* among others. Recent research presentations have been at the annual meetings of the National Association for Research in Science Teaching, National Center for Online Learning Research, Sloan International Conference on Online Learning, and the American Educational Research Association. His books in higher education include *Teaching Elementary and Middle School Science* and *Integrating Academic Units in the Elementary School Curriculum.* He can be contacted at dwsunal@bama.ua.edu.

C. W. Sundberg holds a Ph.D. in science education from the University of Alabama. Currently she is an assistant professor of science education at

Louisiana Tech University in Ruston, Louisiana. She was a visiting professor (postdoctoral) at the University of Alabama in Tuscaloosa from January 2001–May 2002. She is author of the publication in online learning entitled "Utilization of Communication Technologies to Facilitate Follow-up to On-site Professional Development" in the *Journal of Interactive Online Learning*. She has been a coauthor and webmaster of innovative, interactive online learning modules on the learning cycle for the National Center for Online Learning Research (NCOLR). Currently, she is involved in research on the learning cycle, online interactive learning environments, and undergraduate science education. She has been involved in postdoctoral research in interactive online learning environments through NCOLR, acROLL (Alabama Center for Research on On-Line Learning), and ILE. She can be contacted at sundberg@latech.edu.

Charlene Waggoner holds a Ph.D. in biology from the University of Michigan. She is interested in how the structure of the laboratory exercises affects student attitudes toward doing science and how to prepare teaching assistants to facilitate learning. She has written two introductory laboratory manuals and a study guide to accompany an environmental science text. She has publications in *Tested Studies for Laboratory Teaching*. She is currently second vice president for the Association for Biology Laboratory Education. She serves on the Advisory Board of the Northwest Ohio Center of Excellence in Science and Math Education, is director of the BGSU Women in Science Program and serves on the Ohio Project to Enhance Undergraduate Science Education. She is a co-investigator of ESA-21: Environmental Science Activities for the 21st Century, funded by NSF. She is a lecturer in the Center for Environmental Programs at Bowling Green State University where she teaches environmental studies, environmental education and does workshops for faculty, graduate students, and in-service teachers. She can be reached at cwaggon@bgnet.bgsu.edu.

Emmett L. Wright, Ph.D., is a professor of science/environmental education and curriculum in the College of Education at Kansas State University. Dr. Wright's research interests include decision-making attitudes, problem solving, misconceptions and scientific discrepant events, and international education. He has over 130 publications including research articles, yearbooks, science curriculum guides and secondary and college-level textbooks, and other reports. Dr. Wright has served as president of NARST and on the board of directors of NSTA and SCST. He has received major curriculum-development, teacher-education, and research grants, from EPA, NSF, the U.S. Department of Education, U.S. Department of State and U.S. Department of Energy. Recently, he served as the Director of Research for the National Commission on Mathematics and Science Teaching for the 21st

Century. Dr. Wright currently teaches doctoral-level science education seminars, and staff development and curriculum theory courses. At the undergraduate level he has taught science methods for elementary and secondary teachers, general biology, general ecology, ethnobotany, conservation of natural resources, Earth systems science, and environmental decision making. He can be contacted at birdhunt@ksu.edu.

Dean Zollman, Ph.D., is the head of the Department of Physics at Kansas State University. He has over 25 years experience working with physics education projects. He has been honored by KSU by being named a University Distinguished Professor and a Distinguished University Teaching Scholar. He was named the 1996 Doctoral University Professor of the Year by the Carnegie Foundation for the Advancement of Teaching. In recent years, he has concentrated on the use of technology for teaching physics and on providing materials to physics teachers. He is coauthor of six videodiscs for physics teaching, was the director of the software development component of the *Physics InfoMall*, and has twice been a Fulbright Fellow in Germany. At Kansas State University he was the first recipient of the Burlington Northern Award for contributions to undergraduate education. In 1995 the American Association of Physics Teachers awarded Dr. Zollman its Robert A. Millikan Medal for outstanding contributions to physics teaching. At present, Professor Zollman is leading a project, called Visual Quantum Mechanics, to develop materials for teaching quantum physics to three different groups of students—non-science students, science and engineering students, and students interested in biology and medicine. Professor Zollman also leads the project which has resulted from KSU's Recognition Award for Integrating Research and Education from the National Science Foundation. His Web site is http://www.phys.ksu.edu/~dzollman/ and he can be contacted at dzollman@phys.ksu.edu.